Y0-DBV-332

Biology of Ourselves

Biology of Ourselves

A Study of Human Biology

Gordon S. Berry
Head of Science
Thomas A. Stewart High School
Peterborough, Ontario

Contributing Author
Harold S. Gopaul
Head of Science
Port Moody Senior Secondary School
Port Moody, British Columbia

Consultant
Margaret E. Hatton
Associate Professor of Genetics
Department of Zoology, Faculty of Arts & Science
Department of Orthodontics, Faculty of Dentistry
The University of Toronto

John Wiley & Sons
Toronto New York Chichester Brisbane Singapore

Care has been taken to trace ownership of copyright material contained in this text. The publishers will gladly receive any information that will enable them to rectify any reference or credit line in subsequent editions.

The metric usage in this text has been reviewed by the Metric Screening Office of the Canadian General Standards Board. Metric Commission Canada has granted use of the National Symbol for Metric Conversion.

Design and cover by Michael van Elsen Design Inc.

Illustrations by James Loates

Assembly by Margaret Kaufhold

Canadian Cataloguing in Publication Data

Berry, Gordon S., 1930-
Biology of Ourselves

For use in secondary schools and community colleges.
Includes index.
ISBN 0-471-79898-3

1. Physiology. 2. Anatomy, Human. I. Gopaul, Harold P. II. Title.

QP36.B47 612 C82-094277-4

Typesetting by Trigraph

Printed and bound in Canada by The Bryant Press Limited

10 9 8 7 6 5

Table of Contents

To the Student		6
Unit I	**What Are We Made Of?**	9
Chapter 1	Knowledge of the Human Body	10
Chapter 2	Tissues and Organs	39
Unit II	**Our Outer Protective Covering**	57
Chapter 3	The Skin	58
Unit III	**How the Body Is Supported and Moves**	83
Chapter 4	Bones and Joints	84
Chapter 5	Muscles	114
Unit IV	**Communication and Control of the Body**	133
Chapter 6	The Nervous System	134
Chapter 7	The Eye and Vision	175
Chapter 8	Other Senses - Hearing, Taste, and Smell	194
Unit V	**How the Body Transports Substances and Defends Itself**	210
Chapter 9	Composition of the Blood	211
Chapter 10	The Heart and Circulation of the Blood	226
Chapter 11	The Body's Defences against Disease	253
Unit VI	**How the Body Exchanges Oxygen and Carbon Dioxide**	273
Chapter 12	Breathing	274
Unit VII	**How the Body Obtains Energy and Material for Growth**	299
Chapter 13	The Digestive System	300
Chapter 14	Nutrition	327
Chapter 15	Diet - Knowing What and How Much to Eat	349
Unit VIII	**How the Body Removes Wastes from the Blood**	375
Chapter 16	The Excretory System	376
Unit IX	**Chemical Control of the Body**	394
Chapter 17	The Endocrine System	395
Unit X	**Reproduction and Heredity**	414
Chapter 18	Changes in the Reproductive System	415
Chapter 19	Pregnancy and Birth	444
Chapter 20	Human Genetics	465
Glossary		497
Index		508
Photo Credits		519
Text Credits		523

Acknowledgments

A very special thanks is given to Trudy Rising of John Wiley & Sons Canada Limited, who provided untiring support and encouragement throughout the writing and production of this text. Her endurance and capacity for giving both her time and her energy has been a constant inspiration. My thanks are also extended to Nancy Flood for her work on development of the text, to Frank English for his capable and thorough editing of the manuscript, to James Loates for his excellent and attractive drawings, and to Michael van Elsen for his imaginative layout and design.

I am greatly indebted to the contributing author of this text, Mr. Harold Gopaul, who not only read the initial manuscript and supplied numerous helpful suggestions, but prepared many of the interest boxes, experiments, the glossary, and half of the teacher's guide that accompanies the text.

For the valuable and thorough critical review by Professor Margaret E. Hatton of the University of Toronto, my very sincere thanks. Thanks are also due to the many persons who aided Harold and myself with their expert knowledge in specific areas: Dr. Eric Morton, Dr. David May, Mrs. Gail Hancock, the administrator and staff of St. Joseph's Hospital, Peterborough, Mr. J. E. Foote, U.B.C. Medical Genetics Department, Vancouver General Hospital, Dr. Margaret Thompson, Department of Genetics, Hospital for Sick Children, Toronto, Dr. J. Hoeniger, Department of Microbiology and Parasitology, the University of Toronto, and Dr. Barbara McGillivray of the U.B.C. Faculty of Medicine, Department of Clinical Genetics. To the teachers of human biology who read the initial drafts of the manuscript and gave such helpful advice, my grateful thanks.

My thanks are extended to my family and friends for their understanding during the writing of this book. Finally to my colleagues and my own classroom students, who have provided me with inspiration and delight over the years and from whom I have learned so much, my warmest thanks.

Gordon S. Berry
April 1982

To the Student

This book is an invitation to look at yourself, to get to know you. Here is an opportunity for you to understand what you are made of, how you are put together, and how your body works. The human body is, in most cases, a mystery to its owner. Few people understand why their bodies respond in a particular way to regular day-to-day activities.

You know that you sweat when you are hot; but do you know why? What mechanism turns on the sweating apparatus, what turns it off? Where does sweat come from, where does it go? Where is it made, what does it contain? Why do we sweat more in some parts of the body than in others? Why do some people sweat more than others? If you use an anti-perspirant that "prevents wetness" and odours, what effects does this have on the sweat-producing mechanisms of the body?

Control and Balance

Our bodies, and those of all living things, are self-regulating. This tendency for all of our body processes to keep the body in balance makes possible our survival. The tendency for the body of each living organism to maintain its internal make-up is called **homeostasis**.

(hom-mee-o-stae-sis)

If we lose blood cells through a wound or by giving blood at a blood donor clinic, these blood cells will be quickly replaced. If we use more oxygen, by running for instance, our respiratory system immediately responds with an increase in the rate and depth of breathing, thereby supplying more oxygen to our body cells.

The principle of homeostasis is important in almost every process in the body. Any small variation from the "normal" will trigger a reaction to correct the imbalance. If your temperature drops a degree or two, you start to shiver to generate heat. If your temperature increases, you start to sweat, thus, cooling yourself. As you study the systems of the human body, you will see that they all contribute toward the body's homeostasis.

Physical Differences

One of the most remarkable things you will discover as you use experiments to investigate the human body is that, while we all look very much alike, we are very different from one another. Everyone has the same number of eyes in the same general location in the head. Everyone has only one nose and a mouth, yet the subtle differences in shape, spacing, and proportion of these organs enable us to recognize each person as a distinct individual. If you take any single organ, the skin, for example, and study it experimentally, you will find that your skin is different in many ways from the skin of another person, although they have many features in common. Your skin may differ in its sensitivity to heat, or touch; it may vary in colour or texture. The responses of the sweat glands, or the presence or absence of hair, may be distinctly different from those of others around you. While each person has many features in common with every other human being, each person is unique.

Being unique does not mean being "abnormal". If everyone in the class takes their temperature and the results are recorded and then averaged, you probably would find that the average is very close to 37.4°C. The class results would show some students with temperatures below 37.4°C and some above. It is quite possible to have a temperature reading above or below this average figure and the result would be quite "normal" for you. The temperature 37.4°C is the average for a large sample of people. Do not be too concerned when you find that your experimental results vary from the results of others.

Changes in our Pattern of Activity

About 100 years ago, the majority of people in Canada and the United States were working on the land, clearing the bush, tilling the soil, and using enormous amounts of physical energy. Technical advances have provided machines to assist in every aspect of our working day. The average food energy intake of men and women at the turn of the century has been estimated at about twenty thousand kilojoules per day. They needed large amounts of food because of the energy expenditures their physically active lifestyle demanded. Today, with automobiles, machines, and the vast array of labour-saving devices, our energy expenditure is less than half that required in 1900. This change in lifestyle is a major reason for the overweight condition which is characteristic of so many North Americans today. Many people eat much more than is required to balance their much reduced output of energy.

Keeping Healthy

Health is not just the "absence of disease" as so many books define it. Good health provides a vital, positive quality to life. The efficient working of our physical, mental, and emotional selves is our health. Our state of health is most apparent to us when we are not feeling well. Unfortunately, many people spend their lives in a kind of "twilight zone" between health and sickness. They are not ill, not really in need of a physician's care, but neither are they healthy. Many people sit in a classroom or in an office tired, feeling low, not interested in very much. Poor nutrition can be a cause of this lethargy.

A very large number of the serious diseases that afflict Canadians are diseases of choice. That is, in many cases, we select our type of life and death by our choice of lifestyle. For example, choosing to smoke increases the likelihood of dying prematurely. Each year 30 000 Canadians die prematurely from the effects of smoking. Choosing a life of limited activity and a poorly balanced diet, with heavy emphasis on carbohydrates and fatty foods, also promotes the high potential for heart disease and circulatory problems.

About the Human Body

The study of the human body includes the study of **anatomy**, which deals with the structures of the body, and **physiology**, which is concerned with the functions of the body parts. Anatomy and physiology are interrelated – each structure has a special function or perhaps several functions to perform.

In this introduction you have been presented with some ideas about good health and the maintenance of the human body. The following chapters give specific, factual information about the human body that you will find especially useful if you apply it to your own body. We hope you find this a useful study.

Gordon Berry,
Harold Gopaul

Unit 1

What Are We Made Of?

Chapter 1

Knowledge of the Human Body

- The Chemical Composition of the Body
- The Cell and its Parts
- Cell Processes
- Cell Division

Knowledge of the Human Body

Early Egyptians (at least 2000 BC) learned a great deal about the structure of the human body from their practice and method of embalming (mummifying) the dead. The Greeks, especially in the fourth century BC, added to that knowledge. For more than a thousand years, however, the discoveries of these ancient civilizations were suppressed by the Christian church. Until the fifteenth century, little was known about the inside of the human body; understanding was limited largely to what people could see or feel on the surface of the body. Fortunately, from the late fifteenth century onward, the human quest for knowledge prevailed. Intensive studies of the human body began and have continued since that time.

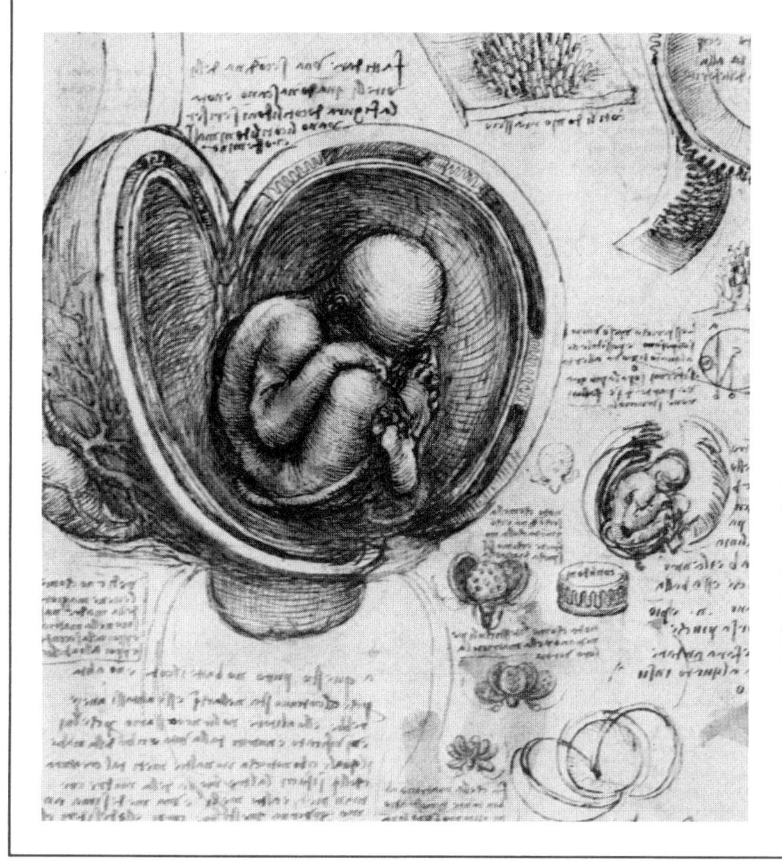

Leonardo Da Vinci, the brilliant artist, designer, and engineer of the 15th century, worked in a mortuary at night, dissecting and drawing detailed sketches of the human body. He worked secretly, by candlelight, since the Church forbade such dissection. Hundreds of Leonardo's drawings still exist. His fine detail and accurate observations are recognized as masterpieces both by artists and scientists.

The Chemical Composition of the Body

In modern times, specialized tools and equipment have allowed us to learn about the body in more detail. We now understand that all living things are composed of chemical elements. These elements are bonded together in complex arrangements to form a wide variety of chemical compounds. Although there are more than 100 known elements, only six of them are needed to make up 99% of the human body. A list of the entire range of elements found in the body is given in Table 1.1. As you can see, hydrogen and oxygen (which combine to form water) make up a very large percentage of the human body.

Table 1.1 The elements found in the human body.

ELEMENT		PERCENTAGE
Oxygen	O	65.00
Carbon	C	18.00
Hydrogen	H	10.00
Nitrogen	N	3.00
Calcium	Ca	2.00
Phosphorus	P	1.00
Potassium	K	0.35
Sulfur	S	0.25
Chlorine	Cl	0.15
Sodium	Na	0.15
Magnesium	Mg	0.05
Iron	Fe	0.004
Other elements		0.046

Expressed in another way, the average person contains the following elements and compounds:

- enough chlorine to disinfect 3 home-sized swimming pools
- enough fat to fill two 500-g tubs of margarine
- enough phosphorus for 1000 matches
- enough sulfur to rid a small dog of fleas
- enough glycerine to explode an artillery shell
- enough sugar to fill an average sugar bowl
- enough carbon to make two large bags of charcoal for a barbecue

- about 5 pinches of salt
- a small shovelful of calcium
- enough water to fill 25 wine bottles
- 20 kg of oxygen

Even at today's inflated prices the total of all the chemicals listed could probably be purchased for about $12.

What is Food Made Of?

If the body is to develop properly and work efficiently, it must be supplied with all these elements or with compounds that contain them. This supply is provided by two basic food sources: (1) plants, including grains, root crops, leafy vegetables, seeds, nuts, and fruits; and (2) animals, which supply meats of all kinds, fish, eggs, milk, butter, and cheese. All foods can be broken down into three major types: 1. **Carbohydrates**, which include starches and sugars, 2. **Fats**, including oils, and 3. **Proteins**. These three food substances provide us with the materials needed to manufacture new cells for growth and the replacement of old or damaged tissues. Food also supplies us with energy to carry out all the processes that go on within the body, as well as to walk, work, and perform our leisure activities.

(car-boe-hy-drate)

Each of the three major food categories is made up of the elements **carbon**, **hydrogen**, and **oxygen**, in various proportions. Proteins have the additional element **nitrogen** present. **Minerals**, **vitamins**, and **water** are taken into the body with the foods we eat. Provided that we have a varied and balanced diet, necessary supplies of all the other elements: calcium, iron, phosphorus, sodium, and others will be obtained from what we eat.

The proportions of these substances found in the human body are:

Water	70%
Protein	23%
Fats and Oils	5%
Inorganic Salts	1%
Other substances, including carbohydrates	1%

Note how little carbohydrate is found in the body. Carbohydrates form the bulk of the food in many diets; however, most of this is used up as energy and not retained in the body unless stored as fat.

Table 1.2. A summary of the major chemical substances in the human body.

SUBSTANCE	LOCATION*	FUNCTION
Water	In and around cells throughout the body.	Dissolve, transport, suspend, and regulate substances in the body.
Inorganic salts	Dissolved in fluids throughout the body.	Control water balance and the movement of many materials (e.g., sodium).
Minerals	Bones and teeth. Blood clotting.	Provides structural support, etc. (e.g., calcium).
Carbohydrates	Basis of other molecules. Non-living parts of cell.	Major energy source for body activities.
Lipids (Fats)	Membranes, Golgi apparatus.	Storage of energy. Provide insulation, protection. Give shape to the body.
Proteins	Membranes of cell and structures within the cell. Muscles.	Form enzymes and many important molecules. Provide strength and contractibility.
Nucleic acids. DNA RNA	In nucleus of cells. Nucleus and cytoplasm.	Direct all cell activities. Carry directions to other parts of the cell. Involved in protein production.
Hormones	Body fluids.	Numerous functions in partnership with enzymes. Activate or deactivate processes in many parts of the body.
Vitamins	Cytoplasm, mitochondria, body fluids.	Work with enzymes. Involved in many body activities.

*Some of these cell parts might be unfamiliar to you. They will be discussed later in this chapter.

Water

Water is the most abundant compound in every living organism. Without it, the average adult person can only live for a few days. (See Figure 1.1.) Water performs several vital functions:

1. *It acts as a solvent*. Food particles are dissolved in water for digestion. Various substances are dissolved in water for transport throughout the body in the blood plasma.
2. *Many chemical reactions take place in water*, especially those involving salts.
3. *Water plays an important role in controlling body temperature*, cooling the body by evaporating from the surface of the skin.
4. *Water is the basic component of all living tissues*. It forms the major portion of the fluid that fills cells and the spaces between them. It flows through organs and provides the means by which almost every substance crosses cell membranes.
5. *It is the major component of various body lubricants*. It is found in tears, in mucus, and in the fluids that lubricate the joints.

Water also functions to maintain the delicate balance between acids and bases.

All of the elements and compounds of the cell combine in a complex solution sometimes called **protoplasm**. This living substance, of which all plants and animals are made, is organized into the basic unit of life, the **cell**.

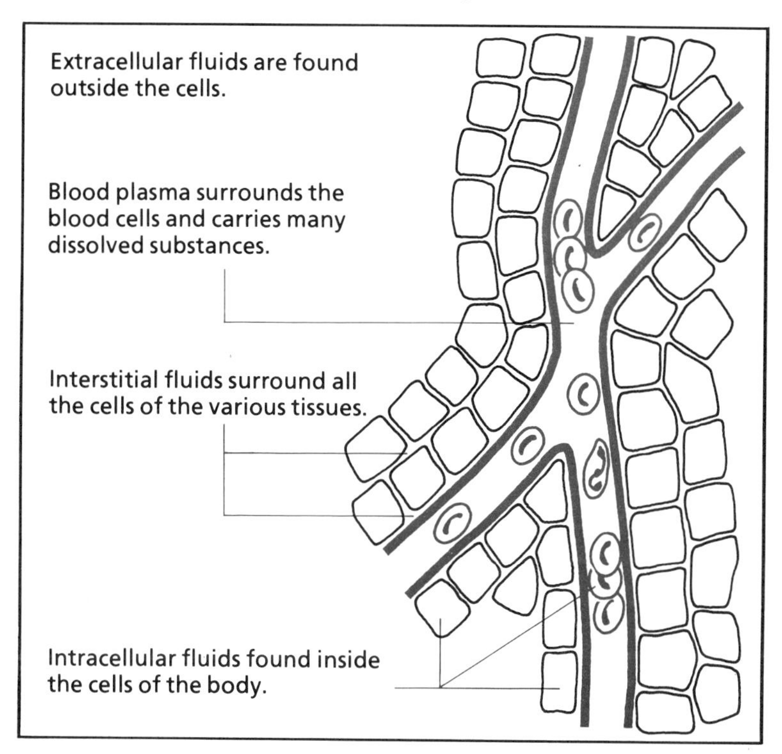

Figure 1.1.
This highly magnified area shows some body cells surrounding a small blood vessel and illustrates where water is found in the body. In a body with a mass of 70 kg approximately 50 kg would be made up of water. Of this 50 kg (50 L), about 37 L would be found inside the cells, about 11.5 L would be located in the interstitial spaces, and a further 3.5 L would form a major part of the blood plasma.

The Cell and its Parts

Under the light microscope, with a magnification of 600X (600 times), few of the details of the cell can be seen. Only the boundary membrane and large cell structures such as the nucleus, the nucleolus, some vacuoles, and perhaps the mitochondria are recognizable. The electron microscope magnifies objects 20 000 and 100 000 times or more, thereby enabling us to see many more details of a human cell. (See Figure 1.2.)

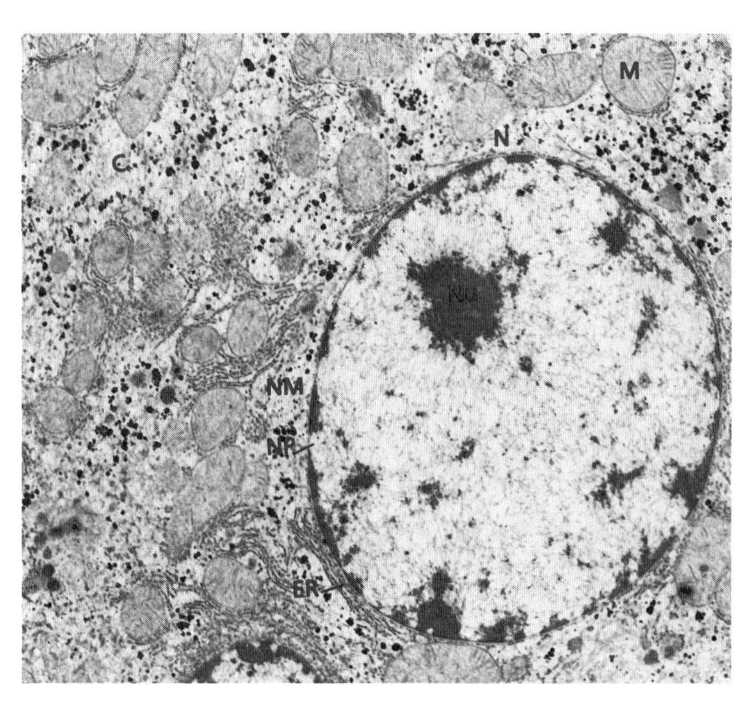

Figure 1.2.
Electron micrograph of a liver cell.

N nucleus
C cytoplasm
NM nuclear membrane
NP nuclear pores
ER endoplasmic reticulum
M mitochondria
Nu nucleolus

Electron microscope

The size and shape of a cell varies with its function. Most cells are so small it would take 100 000 of them to cover the head of a pin. It has been estimated that the average adult human body contains about 40-60 000 000 000 000 cells. An artist's impression of a typical cell appears in Figure 1.3. Cells are of many different types. Each type develops to a certain size and shape and has a special function. All cells, however, are made up of the same small parts called **organelles**. The different types of cells are able to carry out their specific functions because of the important specialized organelles that they contain.

(or-ga-nels)

Figure 1.3.
A typical cell.

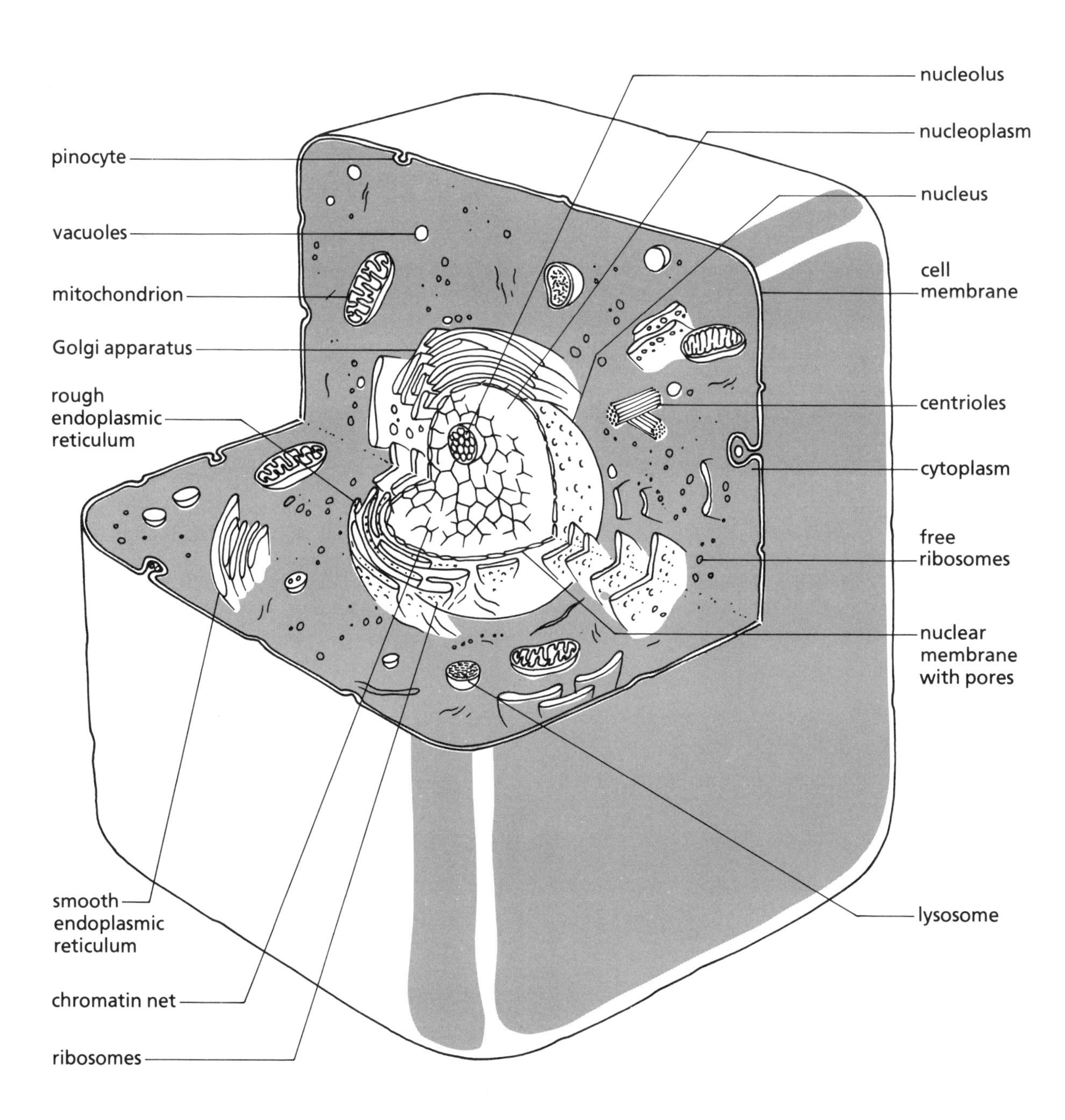

MICROSCOPY AND THE CONCEPT OF SIZE

The Dutch lensmaker, Anton van Leeuwenhoek, put together the first simple microscope in the seventeenth century. The Englishman, Robert Hooke, observed all kinds of living and non-living material by using a light microscope, and was the first to coin the word "cell" as he reported the make-up of cork tissue in 1664.

The light microscope uses light and the human eye to resolve objects with a maximum magnification of about 2000 times. But there are limits to the resolving power, that is, the ability to discriminate objects under the light mircoscope. Objects that are closer together than 0.0002 mm cannot be resolved even with a light microscope. With the light microscope, cell components such as ribosomes, Golgi bodies, or DNA cannot be seen.

The electron microscope uses electrons, not light as its reflective energy source. (See *A Remarkable Canadian Invention*.) Electron microscopes today can magnify 20 000 000 times with a resolution of 0.2 nm. In fact, the electron microscope was recently used to magnify material to such an extent that researchers were able to make a movie of uranium atoms in motion!

Under the light microscope, for example, the human hair at the best resolving power may appear to be 5–6 cm in width. An electron microscope, magnifying 10 000X (far below its maximum ability) would make that same hair appear to be 200 cm or 2 m thick. A red blood cell under the light microscope appears to be a fraction of a millimetre in diameter, but if magnified 30 000X with an electron microscope, it would be the size of a Frisbee. A grain of sand one millimetre in diameter would almost fill the field in a light microscope at 100X magnification. If magnified 30 000X it would have the proportions of a ten-storey building standing on a base of 4 m^2! The electron microscope, then, is used to resolve very small cell components. The cell membrane of blood cells has been resolved to be about 7 nm in width and the entire red blood cell is about 7000 nm in diameter. Viruses as small as 6 nm have been resolved. Biologists can even *see* normal cells changing into cancerous cells and identify genes that cause hereditary diseases. Fine structures of the cell, such as its DNA, have been resolved by a high-voltage electron microscope (see diagram).

The following is a scale starting with the centimetre. Each unit of measurement is one-tenth the size of the preceding one. This scale should put some of the tiny body parts into perspective for you.

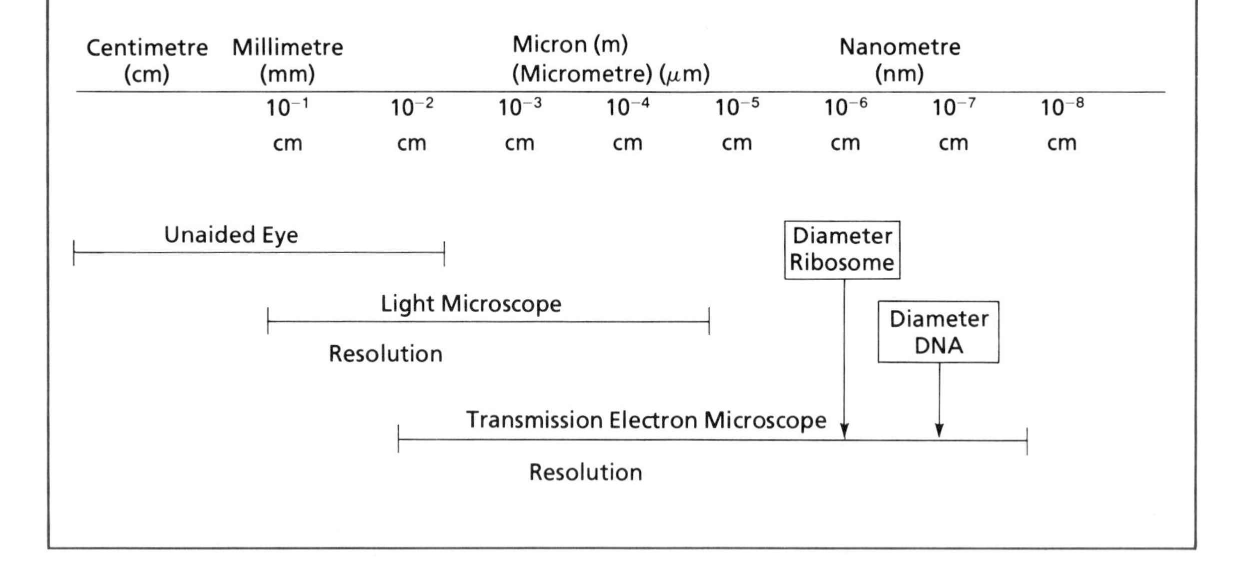

The Cell Membrane

The cell membrane, or unit membrane, as it is sometimes called, is a flexible, elastic envelope around the cell contents. It permits many molecules to pass through it in either direction. The membrane is composed of three layers. This "sandwich" has a thin layer of protein on each side with a double layer of lipid (fat) between. Large globular proteins are embedded right in the membrane and these proteins are believed to have channels through which water-soluble materials can be passed. The whole membrane is extremely thin. In spite of this, however, it is very strong and able to withstand wear, bumps, and abrasions. It also has the ability to seal tiny punctures made in its surface. Some cells have projections on their cell membranes. For example, cells along the lining of the small intestine have small fingerlike extensions called **microvilli**. These increase the surface area of the cell which, in turn, increases the cell's efficiency for absorption. Cells lining the breathing tubes have small hairlike structures called **cilia** which trap dust and move small particles back out of the air passages. (See Figure 1.5.)

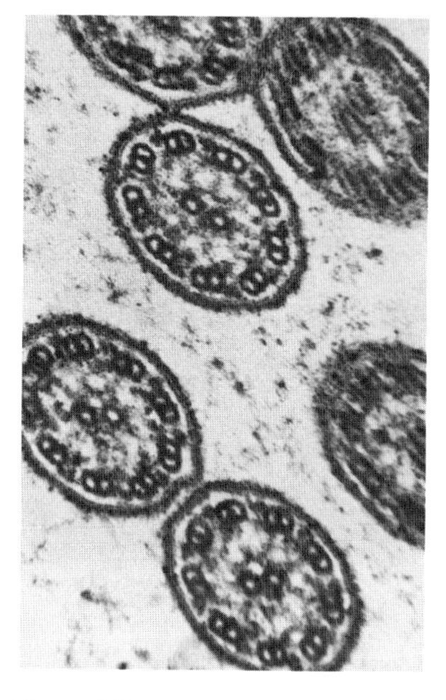

Figure 1.5.
Section of cilia. Note arrangement of fibrils.

Figure 1.4.
Summary of the structures and functions of cell organelles.

MITOCHONDRION	Round or oval, two-layered membrane. Inner membrane has shelflike folds. Power house. Produces "packets" of energy for cell activities.
ENDOPLASMIC RETICULUM (E. R.)	A double membrane which connects the plasma and nuclear membranes. Transports fluids and chemicals throughout the cytoplasm. Site of protein manufacture.
RIBOSOMES	Contain granules of nucleic acids active in protein synthesis.
GOLGI APPARATUS	Membrane-lined channels. Produces or accepts lipids and enzymes and packages these for export.
LYSOSOMES	Sacs of enzymes. Digest large molecules. Can destroy the cell if membrane is broken and contents escape.

VACUOLES

Bubblelike containers of water enclosed by a membrane. May also contain food or wastes.

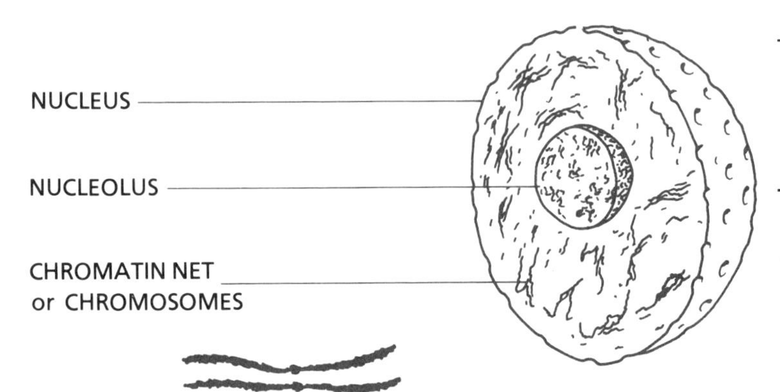

NUCLEUS — Control centre for the cell. Directs cell activities. Stores genetic information for the cell.

NUCLEOLUS — A collection of loosely bound granules of ribonucleic acid and protein. May be the site of the manufacture of ribonucleic acid. Has no membrane.

CHROMATIN NET or CHROMOSOMES — Nuclear protein bodies in the nucleus which bear genes along their length. The encoded plan for all cell activities and developments. The inherited genetic storehouse of information. Chromosomes show as dark bodies during cell division. They are normally seen as granular patches in the nucleus.

CENTRIOLES

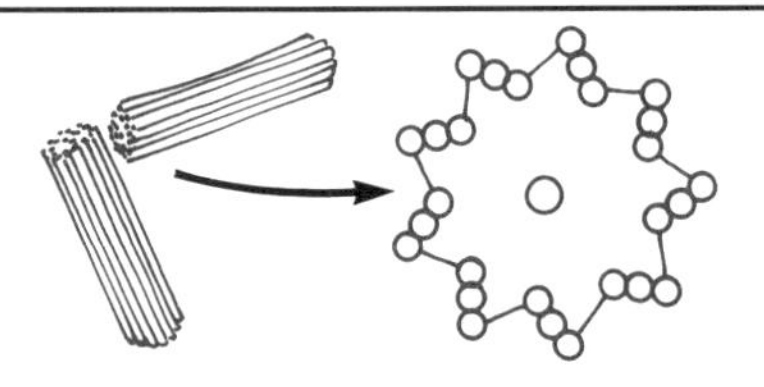

Two, found near the nucleus. Separate and migrate to the poles at cell division. Responsible for the organization of the spindles which may attach to the chromosomes during cell division.

CELL MEMBRANE

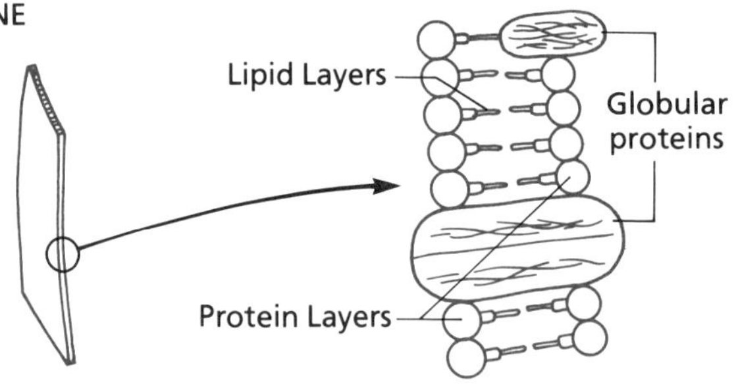

An elastic envelope composed of three layers. Protein, lipid, protein sandwich. It is semiporous and controls the passage of materials in and out of the cell. Numerous globular proteins are also present which aid water-soluble molecules in passing through the membrane.

CYTOPLASM

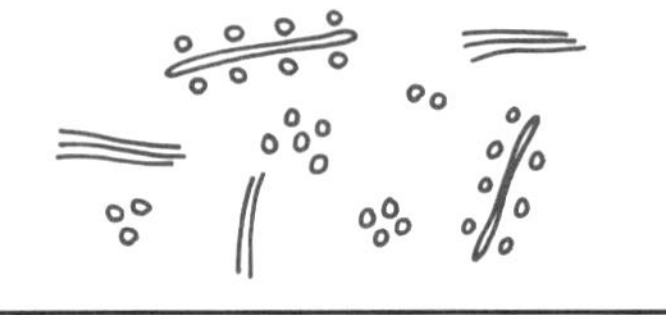

A complex mixture of materials,–water, gases, wastes, nutrients, raw materials, etc., for use in cell activities. Includes the cell organelles but not the nucleus.

MICROVILLI

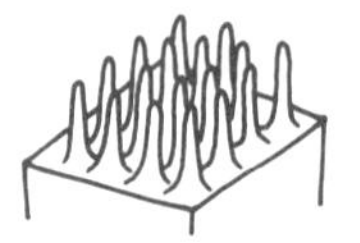

Fingerlike projections on the cell membrane which increase the surface area of the cell membrane for more efficient absorption.

CILIA	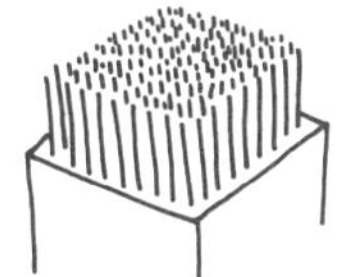	Fine hairlike structures which project from the surface of the cell and wave rhythmically, to move particles along a tube or across a surface.
PINOCYTE		Tiny sacs which bulge in or out of the cell. Part of a process by which particles are moved across the cell membrane.

A REMARKABLE CANADIAN INVENTION

In the late 1930's two Canadians at the University of Toronto took a primitive apparatus designed by D.E. Burton and his associates and began to remodel it to their own specifications. Thus, Dr. James Hillier and Dr. Albert Prebus produced the first working electron microscope. Many of the parts required were unobtainable or too expensive, so that both men quickly had to become expert machinists to produce their own specialized components.

The electron microscope uses electrons instead of light rays to magnify the object being viewed. The electrons are focussed by strong electromagnets and the image is then projected onto a fluorescent tube like a television screen. This image can then be photographed and permanent record of the image retained.

The original electron microscope produced by James Hillier and Albert Prebus can be seen on display in the Ontario Science Centre in Toronto.

The Cytoplasm

Inside the membrane is a transparant, granular fluid known as the **cytoplasm**. Under the electron microscope the granules are revealed to be the many organelles ("little organs") which will be described in detail. If the fluid part of the cytoplasm is analyzed chemically, it is found to contain a rich "soup" of sugars, proteins, minerals, and oils that supply the nutrients needed by the organelles for energy and building materials. It also contains cell wastes.

(sy-toe-plazum)

The Organelles

The Nucleus

(noo-klee-us)

The nucleus contains the chromatin net, nucleolus, and the fluid surrounding these structures (nucleoplasm). The nucleus is the most important part of the cell. Its major function is the control of all metabolic activities that take place within the cell. These metabolic functions occurring in other parts of the cell include the absorption and digestion of food, the conversion of sugars to a suitable form of energy, the production of new proteins and other substances, and the excretion of wastes – in fact, all processes that go on within the cell.

The nucleus is enclosed within a double-layered unit membrane, similar to the cell membrane, but with small openings which appear to connect with channels into the endoplasmic reticulum. The membrane provides a "compartment" separating the nuclear contents from the rest of the cytoplasm. It allows, however, a continuous exchange of substances across this boundary.

Chromatin

(kroe-ma-tin)

The chromatin lies within the nucleus. It derives its name from its ability to absorb stain (*chroma* means colour). When the cell is not dividing, but carrying on its normal functions, the chromatin appears as a mass of very fine threads called the chromatin net. During cell division, these threads contract and form thicker, rodlike structures called **chromosomes**, which can be easily seen and counted.

(kroe-mu-soem)

The chromosomes are made of complex molecules of **deoxyribonucleic acid** or DNA as it is usually known. These structures contain the blueprints for the architecture of the cell – the plans for its form and function, the inherited information from previous generations. These molecules are unique in that they possess the ability to replicate (reproduce themselves). They are responsible for the fundamental distinguishing characteristics of all living things. (See Figure 1.2.) (See Chapter 20, *Human Genetics*, for more on this topic.)

(de-ox-ee-ry-boe-noo-clay-ik)

The Nucleolus

(noo-klee-oe-lus)

The nucleolus lies within the nucleus and its function is not well understood. It appears in some cells and not in others. Sometimes there are two nucleoli present in one cell. It is considered by many researchers to be a "storehouse" for ribosomal nucleic acid. It may form a model for copies of the chromosomal plans which need to leave the nucleus and enter the cytoplasm. During the process of cell division it disappears entirely, then reappears when the two new cells are complete.

THE NATURE OF THE DNA MOLECULE DISCOVERED!

For many years scientists knew that chromosomes in the nucleus somehow duplicated themselves. Then scientists discovered that the chromosomes are made of a complex chemical, deoxyribonucleic acid (DNA). A great question remained: What is the special structure of DNA that allows it to make identical copies of itself? In 1953 Dr. Francis Crick and Dr. James Watson proposed a hypothesis on the structure of the DNA molecule. To establish their hypothesis they carefully pieced together many clues and discoveries made by other scientists.

Their work proposed that the DNA molecule is a double-stranded helix, made up of two sugar-phosphate strands linked together by pairs of nitrogenous bases. These bases are paired together in a special way due to their structure. *Adenine* and *thymine* formed pairs, as do *guanine* and *cytosine*. These base pairs then fit between the two chains of the helix like rungs on a ladder.

Doctors Crick and Watson advanced a theory that solved the mystery of how chromosomes (DNA molecules) replicate. They suggested that the double helix might separate into two strands by breaking the bonds between the pairs of bases. Then if other bases fit into the vacancies made by the separation of the strands, in the same sequence as was present before, two identical molecules would be formed.

Later research has confirmed their theory and added more knowledge to our understanding of this most important molecule. In 1962 Doctors Crick and Watson shared the Nobel Prize for Medicine with Dr. Maurice Wilkins for their imaginative explanation of the nature and replication of the DNA molecule.

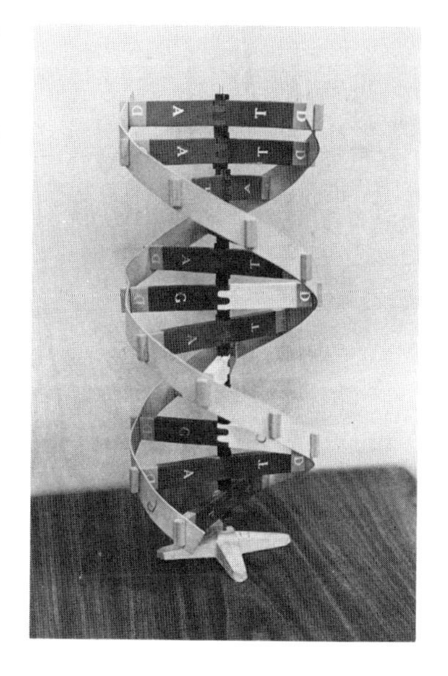

A good source for further information on this remarkable molecule is the book, *The Double Helix* by Watson.

The Mitochondria

These rod-shaped structures are the "power-houses" of the cell. By a complex series of reactions they use the energy trapped in food molecules to produce tiny energy packets of ATP (**adenosine triphosphate**). This ATP serves as fuel for the reactions and activities of the cell. (See Figure 1.6.) (See *Cell Respiration*, p. 30, for more on this topic.)

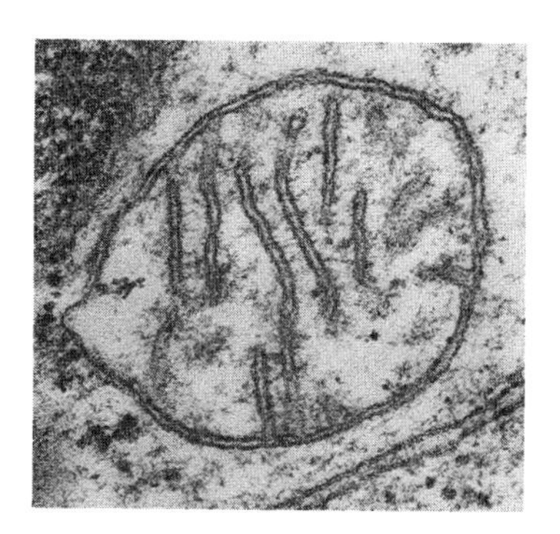

Figure 1.6.
Mitochondria showing cristae.

The Lysosomes

The lysosomes are tiny sacs containing powerful digestive enzymes. These enzymes are capable of breaking down many of the substances within the cell. They must therefore be kept safely enclosed within a unit membrane until they are needed. The enzymes are so concentrated that, if they are released by a rupture of the lysosome membrane, they can destroy the cell. For this reason they are sometimes called "suicide bags".

(ly-so-soems)

The Centrioles

(sen-tree-ohl)

The centriole is found close to the nucleus. It consists of two separate units placed at right angles to one another. They form a cylinder composed of nine groups of fibres, each group being made up of three smaller fibres. Prior to cell division, the centrioles migrate to opposite ends of the cell. Spindle fibres (microtubules) develop between them and connect with the chromosomes. (See Figure 1.7.)

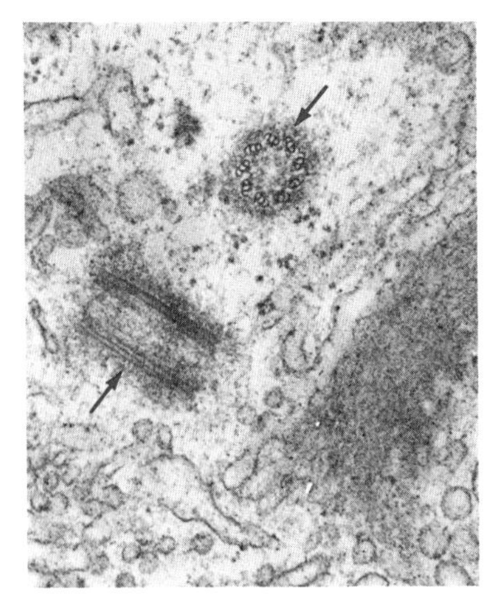

Figure 1.7.
Cross and longitudinal sections of the centrioles. Note the arrangement of microtubules in the centriole.

Vacuoles

Vacuoles appear as small bubbles containing liquids. The vacuole membrane helps to keep the liquids separated from other materials in the cell. They are sometimes involved in carrying fluids to the outer membrane for disposal or in maintaining water balance within the cell. It is worth noting that all the organelles within the cell are surrounded by unit membranes. The elastic, semipermeable nature of the unit membrane, as well as its ability to break down, or reseal when damaged, is remarkable. Membranes make possible the organization and separation of a great many activities within one microscopic cell unit.

The Endoplasmic Reticulum (E.R.) and Ribosomes

(en-doe-plaz-mik re-tik-yoo-lum)
(ry-boe-soemz)

The E.R. is a system of double membranes enclosing a narrow space. It is arranged in a network of interconnecting channels throughout much of the cytoplasm. The connecting canals appear to form a link, at some points, with both the membrane of the nucleus and the outer cell membrane. Other connections are made with the Golgi apparatus.

The outer surface of the most common kind of E.R. is covered with tiny spheres, the **ribosomes**, which contain small amounts of a special nucleic acid (ribonucleic acid or RNA). Ribosomes play an important role in the production of cell proteins. Those attached to the E.R. are believed to produce proteins for export from the cell. Ribosomes which are free in the cytoplasm produce proteins for use within the cell. (See Figure 1.8.)

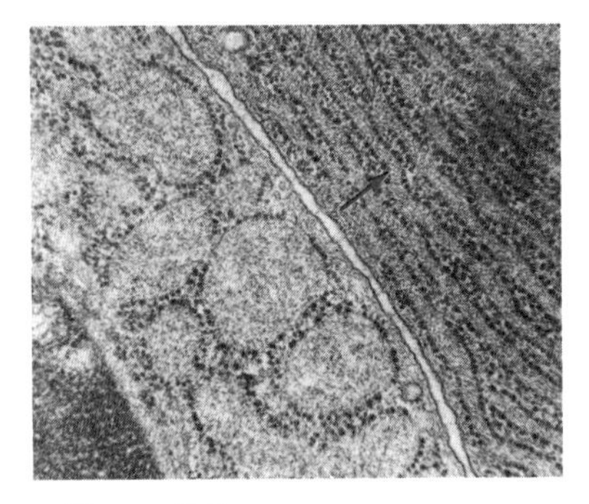

Figure 1.8.
Endoplasmic reticulum.

When ribosomes are present along the E.R. it is known as **rough endoplasmic reticulum**. When no ribosomes are seen adhering to the E.R. membrane it is called **smooth endoplasmic reticulum**.

The Golgi Apparatus

The Golgi apparatus is a collection of flattened, double-layered discs, often with small sacs loosely attached around the perimeter. These structures manufacture special products called glycoproteins, which are primarily a combination of carbohydrates and proteins. The apparatus then packages these substances for dispersal and use around the cell. (See Figure 1.9.)

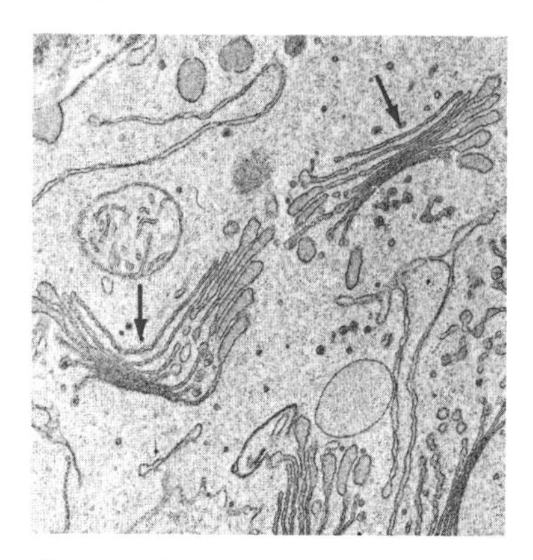

Figure 1.9.
Golgi apparatus.

Cell Processes

No cell can live in total isolation. It must exist within an environment which can supply the many nutrients that the cell requires. This environment must also be able to accept and disperse the waste products of the cell processes. Molecules that enter and exit from the cell must be small enough to pass through the tiny pores of the cell membrane. They must, for the most part, be carried in a water solution.

Most substances pass through the many membranes associated with cells by a simple process called diffusion.

Diffusion

If someone several tables away from you in the cafeteria opens a bag with a salami sandwich in it, you will know it within a short time. Even if you do not see the sandwich, you will be able to smell it because of diffusion.

Diffusion is the tendency of molecules to spread out from areas of high concentration to areas of low concentration. This spreading out of molecules is possible because of the kinetic energy that molecules possess. This energy drives molecules in liquids and gases to move randomly in straight lines until they either bump into other molecules or collide with the walls of their container. When such a collision occurs the molecule bounces off and continues in a straight line until another collision takes place. This is much the same as the behaviour of a ball rolling on a pool table when it

Figure 1.10.
Diffusion of a dye crystal.

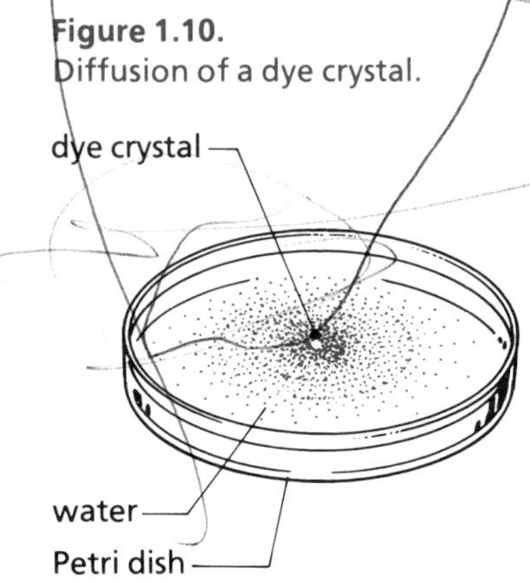

Dye spreading out by diffusion from an area of high concentration to areas of low concentration of dye molecules.

strikes another ball. After each collision, it takes a new direction and travels in a straight line until it strikes the cushion or another ball.

If the molecules of some substance are densely packed together (high concentration) the chance of striking other molecules is very great. The frequent collisions cause the molecule to spread out into areas where the molecules are more widely spaced. In such less concentrated areas the chance of a collision is less and the molecule will travel a greater distance before striking another molecule. As a result, molecules gradually spread themselves out until the mixture of molecules in a given solution is evenly distributed. This process of even mixing is called **diffusion**. (See Figure 1.10.)

The rate at which the molecules diffuse depends on several factors:

(a) Temperature. At higher temperatures, molecules possess greater energy, travel faster, bump into each other more often, and therefore diffuse more rapidly.
(b) The size of the molecules. Larger molecules have a greater chance of collision than smaller molecules. Smaller molecules move more rapidly and move further when they are bumped.
(c) The concentration gradient. This refers to the difference in concentration between two solutions. Diffusion occurs more rapidly when the difference in concentration between the solutions is high. As the concentrations approach similar proportions, the rate of diffusion decreases.

Diffusion accounts for many processes in the human body: for example, the movement of gases in the lungs and blood stream, as well as the passage of nutrients in the digestive tract, and the absorption of fluids in the excretory system.

Osmosis

(oz-moe-sis)

If you have ever seen a wilted plant, you have seen the result of another important process of all living things – osmosis. Osmosis is a special case of diffusion, occurring when water passes across a membrane. Membranes in the cells and tissues of the body are selectively permeable and semipermeable. This means that the membrane controls, to some degree, what will pass thorugh the pores in the membrane. Often the size of the pores in the membrane controls the passage of molecules. If the pore openings are small, small molecules such as those which make up water can pass thorugh easily whereas large molecules, such as those of starches or proteins, are too big to pass through. (See Figure 1.11.)

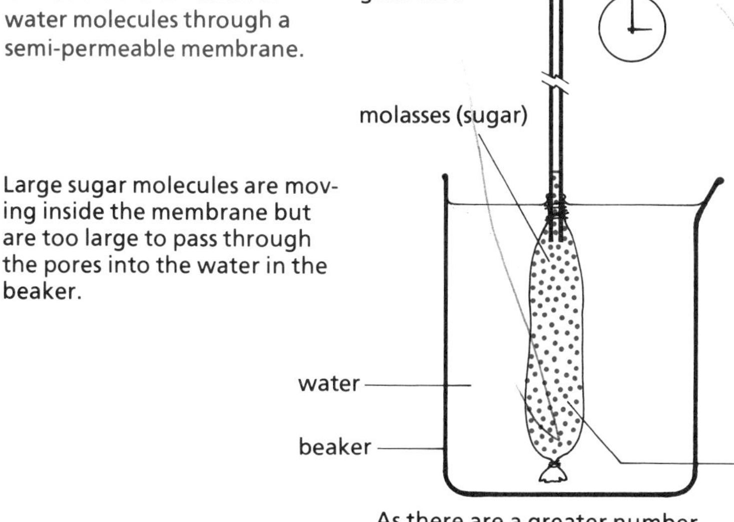

Figure 1.11.
Osmosis is the diffusion of water molecules through a semi-permeable membrane.

Most substances in the body are dissolved in water. (Some vitamins are soluble in oils.) Water is the **solvent**. The substances dissolved in the solvent are known as **solutes**. When sugar is dissolved in a soft drink, sugar is the solute and water is the solvent.

The Effect of Osmosis on Cells

When a red blood cell is placed in a solution in which the balance of solutes is equal on both sides of the membrane, the number of molecules entering the cell will equal approximately the number of molecules leaving the cell. The concentration of solute inside the cell equals that outside the cell. At the same time, the concentration of water inside the cell also equals the water concentration outside the cell. There is no concentration difference. This condition exists when red blood cells are present in the blood plasma.

If red blood cells are placed in water, the concentrations are not equal. There is a higher concentration of solutes inside the cell and a higher concentration of water outside the cell. Both solutes and water will try to move (in opposite directions, of course) from the higher concentration, through the membrane, into the areas of lower concentration. Water molecules are much smaller in size than the molecules of solute. They, therefore, will have a better chance of passing through the pores of the membrane – they will diffuse more rapidly. As a result, the red blood cells will begin and continue to swell with water until they burst (are *plasmolysed*).

When red blood cells are placed in a salt solution where the conditions are reversed, with relatively higher concentrations of solutes outside the membrane than inside the cell, the red blood cells will shrink as they lose water. The cells then become wrinkled (*crenated*). If a sailor is adrift in a small boat without a fresh water supply and he drinks salt water to satisfy his thirst, he will die sooner than if he did not drink at all. Can you explain why?

Dialysis

(dy-al-i-sis)

When solutes pass through a membrane and the membrane acts selectively to allow some solutes to pass but not others, we refer to the process as **dialysis**. This process occurs in the kidneys. It is the basis of the artificial kidney machine, which is called a dialysis machine.

Active Transport

Diffusion and osmosis are passive activities. The energy of motion of the molecules is sufficient to move them across the membrane. When molecules require added energy for the transfer across a membrane to take place, the process is known as **active transport**. For active transport to take place, an energy source is required in addition to certain enzymes and carrier molecules. The carrier molecules "ferry" the solutes across the membrane, while the enzymes help in the "loading" and "unloading" procedures which enable the solute molecules to become temporarily attached to the carrier substance. Energy is required because the molecules must travel *against* the concentration gradient. The molecules must move from areas of *low* concentration to areas of *high* concentration. Active transport is a vital process occurring in the kidneys where the final product (urine) is higher in concentration of waste solutes than the tissues around the tubules. Sodium, glucose, amino acids, potassium, and hydrogen ions are all actively transported across semipermeable membranes in the kidneys.

1 PINOCYTE FORMATION

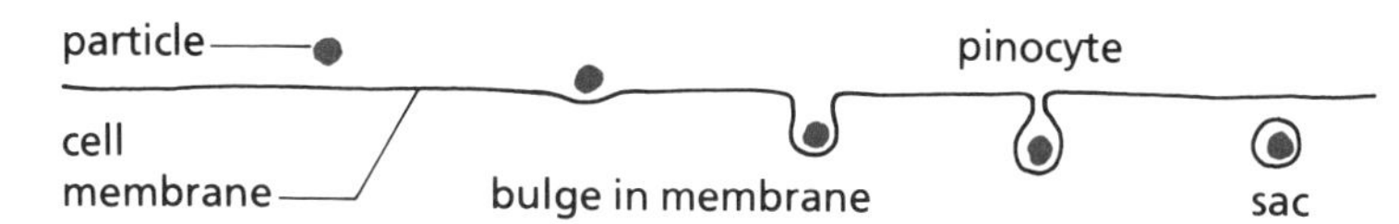

2 PHAGOCYTE CELL

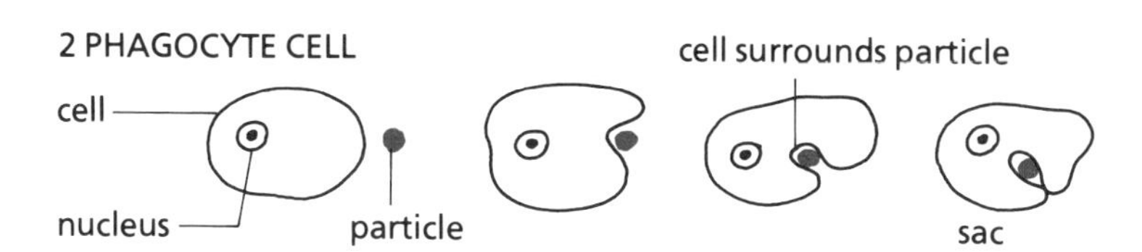

3 ACTIVE TRANSPORT

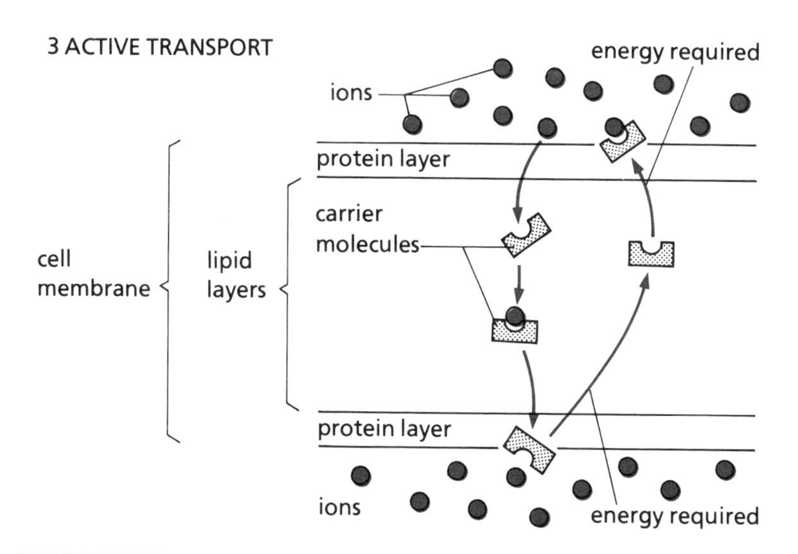

Figure 1.12.
Some ways that things get into and out of cells.

Pinocytosis and Phagocytosis

These processes help large molecules and particles pass across membrane barriers so that they get into and out of cells. Both these processes are discussed in more detail in other parts of the text.

(pin-oe-sy-toe-sis)
(fag-oe-sy-toe-sis)

Pinocytes are tiny bulges in the membrane which form around particles or large molecules. The membrane forms a tiny sac as it pushes in or out of the cell. This sac eventually pinches off, leaving the particle enclosed in a membrane bubble. The contents of the sac may then disperse in the cytoplasm or be broken down by enzymes within the cell.

Phagocytes are special cells that engulf large particles or even other cells. They surround a particle by extending part of the cell (pseudopodia) until the particle is completely enclosed within the cell. Certain white blood cells use this technique to capture and destroy bacteria. The *Amoeba*, a one-celled organism, takes food into its body by the same process. (See Figure 1.12.)

(a-mee-bu)

Cell Respiration

This section on cell processes has, until now, dealt with how things get into and out of cells. It is now important to explain how the glucose and oxygen that enter the cell are used by it to produce energy. This process is very complex. Here is a summary of the process that will give enough information for you to understand its importance.

Energy in the Cell

All cells constantly need energy to satisfy the demands of the hundreds of reactions that are happening in them. How do our cells obtain the energy necessary for maintenance and growth?

As you know, energy enters the body as food. The energy is trapped in the chemical bonds that hold the atoms of food molecules together. After food is digested, tiny molecules of glucose transport energy to the body cells by way of the blood.

For use in cell reactions, energy must be supplied in a controlled way and in small amounts. The cells have a special molecule to store energy and release it in small packets. This molecule is known as **adenosine triphosphate** or **ATP**.

ATP is made up of a chemical called adenosine linked to three phosphate groups.

The last two phosphate groups are easily attached and detached. The bonds that connect the last two phosphates to the adenosine contain much more energy than other bonds and this high energy value is represented by a wavy line instead of the usual straight line that indicates bonding. When ATP reaches a site where energy is needed, the phosphate group is detached, thus releasing energy and making it available for the reaction. The ATP molecule then has two phosphate groups instead of three and is called **adenosine diphosphate** or **ADP**.

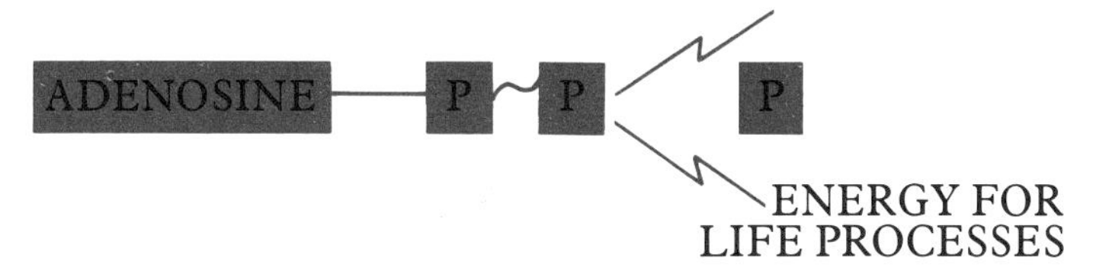

Figure 1.13.
Features of cell respiration.

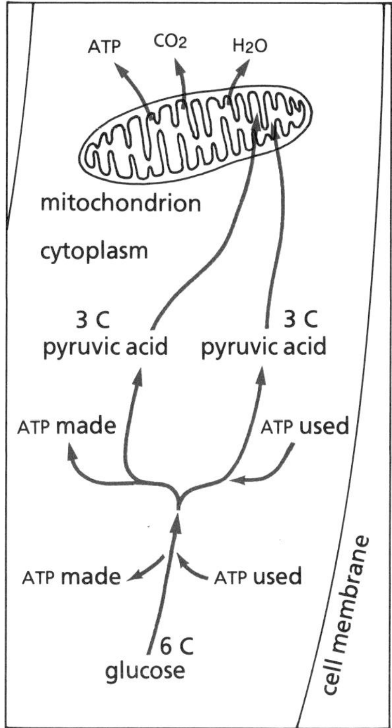

In the cytoplasm, each glucose molecule splits into two smaller molecules. This process releases energy (ATP).

The two smaller molecules move to the mitochondrion. Each of them breaks down, supplying an enormous amount of energy.

How Is ATP Produced?

The process of producing energy for use in the cells is called **cellular respiration**. It occurs in the cytoplasm and mitochondria of the cells.

First glucose is broken down in the cytoplasm of the cell by the process of **glycolysis**. By this series of reactions, the glucose molecule ($C_6H_{12}O_6$), a molecule with six carbon atoms, is split into two smaller three-carbon molecules of pyruvic acid ($C_3H_6O_3$). For this reaction to be possible, two molecules of ATP are required. By the end of glycolysis, four molecules of ATP have been produced. Thus, although some ATP is used to start the reaction, a net gain of two ATP molecules results from it.

Next, the molecules of pyruvic acid are further reduced to acetic acid ($C_2H_4O_2$). Acetic acid molecules contain only two carbon atoms; the third carbon atom of each pyruvic acid molecule unites with two oxygen atoms and is released from the cell as carbon dioxide. At this point, the acetic acid enters the mitochondria. In the mitochondria, the acetic acid molecules join with an enzyme (a special kind of protein), which carries them into the **citric acid cycle**, a complex series of reactions in which carbon dioxide is released as a product and large numbers of ATP molecules are produced. Finally, as water is formed and leaves the cell, many more ATP molecules are produced.

Biologists have found that, for every molecule of glucose that is broken down by the process described, 38 molecules of ATP are formed. The whole process can be summarized as follows:

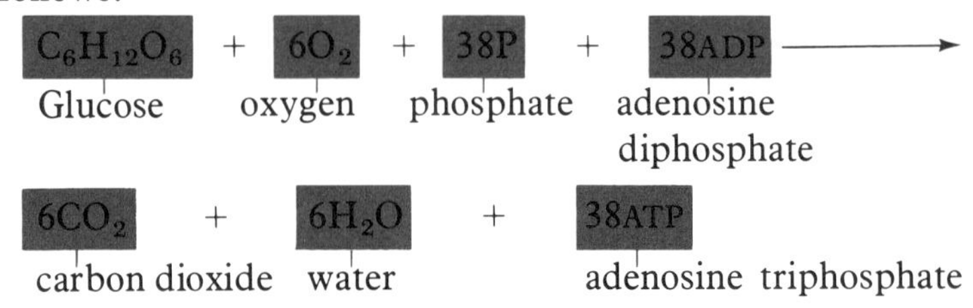

The ATP that is formed as a result of the reactions summarized above provides energy for all cell functions.

Enzymes

Enzymes are special proteins that act as catalysts. Catalysts cause chemical changes to occur and affect the rate of a change. However, they are not themselves changed by the reaction. Somehow, as yet unknown, enzymes lower the amount of energy required to cause substances to react. This means the reactions can occur more rapidly and with little heat required or released.

In order for two molecules to form a bond, the molecules involved must be oriented so that the parts of each molecule where the bond will form are brought close together. Chemicals in a cell are a little like a jigsaw puzzle; the right piece may be available but until it is turned around so that it is oriented correctly, it does not fit in. Enzymes are believed to help orient molecules so that they fit together more easily.

An enzyme attaches itself to one of the reacting molecules (called a **substrate**), and forms a temporary bond. Although there are many kinds of enzymes, each enzyme attaches to only one type of substrate. Once this association with the enzyme has taken place, a second substance, again a specific kind, associates with the enzyme complex and a bond is formed between the two substances. The enzyme, having completed its job of bringing the molecules together, is released and is immediately available to be used again. Enzymes act so rapidly that they are capable of interacting with about 1000 molecules each minute.

Not all enzymes bring other molecules together to form larger molecules. Many are involved in the opposite process, being needed to split larger molecules into smaller ones. Digestive enzymes are used in this way. The starch in the cracker you chew becomes sugar because a specific enzyme in your saliva helps break apart the starch molecule.

Figure 1.14.
Diagrammatic representation of enzyme action.

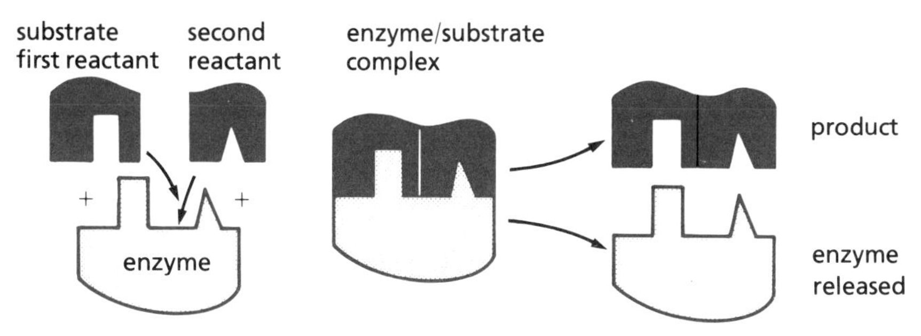

Enzymes are often likened to the key of a lock. A different key (enzyme) is required for each lock (reaction).

Most enzymes are named by adding the suffix -*ase* to the name of the substance with which they react. For example, sucr*ase* is the enzyme that acts on the sugar sucrose. Prote*ases* act on proteins helping to break them down into smaller molecules.

Cell Division

There is a constant need for new cells in the body, to replace worn or damaged tissues and to provide for growth. Most cells in the body are capable of dividing to form two new smaller cells where previously there was only one. This process of cell division is known as **mitosis**. When a cell divides, the genetic plans of the body, contained in the chromosomes, are duplicated and transferred to the new daughter cells. The cytoplasm and organelles are also divided relatively equally between the two new cells. At the completion of mitosis, the two daughter cells are identical to the single parent cell from which they originated.

(my-toe-sis)

Mitosis

Mitosis may be conveniently divided into 5 phases: (See Figures 1.15 and 1.16.)

1. **Interphase.** This is sometimes inappropriately referred to as the "resting phase". It is the period of time during which the cell is not actively dividing, but is carrying out the specific function for which it is adapted. All the

MITOSIS

cell membrane
centrioles
chromatin
nuclear membrane
nucleolus

interphase

aster

early prophase

spindle fibres
chromatids

mid-prophase

centromere

late prophase

equatorial plane

metaphase

anaphase

cleavage

telophase

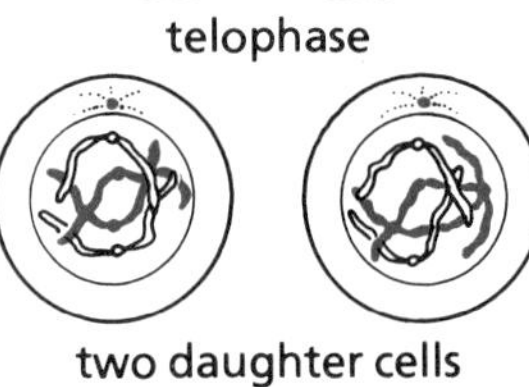

two daughter cells

Figure 1.15.
Mitosis. The division of one mother cell into two identical daughter cells.

INTERPHASE
Nuclear membrane and nucleolus visible. Chromatin present as granular mass. The cell is active, involved in its normal functional activities. Just prior to entering prophase the chromosomes duplicate.

PROPHASE
Aster rays appear. Centrioles move towards opposite poles of cell. Chromosomes appear as thin threads. Nucleolus no longer visible.

Centrioles form spindle fibres attached to chromosomes now seen as two chromatids. Centrioles reach the poles of the cell. Nuclear membrane no longer visible.

Chromatid pairs start to migrate towards the equator of the cell.

METAPHASE
Chromatid pairs line up across the middle of the cell. Separation of chromatid pairs occurs as the centromere splits. Each chromatid is joined by a spindle fibre from the centromere to the aster.

ANAPHASE
Two complete sets of chromosomes now are drawn towards the opposite poles of the cell. The contraction of the spindle fibres causes this movement.

TELOPHASE
Nuclear membrane reappears and surrounds chromosomes. Nucleoli reappear. Chromosomes become less distinct. Division of the cytoplasm occurs. Cell membrane starts to indent.

DAUGHTER CELLS ENTER INTERPHASE
Cells now resemble original mother cell, containing an identical set of genetic material. Cleavage and division of the cytoplasm is complete.

normal processes and activities are taking place in the cell; it is not, therefore, really at rest. One special activity that takes place toward the end of this phase is the duplication of chromosomes. This chemical process cannot be seen by using a microscope.

2. **Prophase.** The centrioles move to opposite poles in the cell. Thin spindles develop between the centrioles. The nuclear membrane breaks down and the chromatin net condenses into visible chromosomes. The chromosomes have duplicated and they can be seen side by side as two thin chromatids.

3. **Metaphase.** The chromosomes move to the centre of the cell and align themselves across the equatorial plane. They are attached by spindles to the asters formed from the centrioles at either end of the cell.

4. **Anaphase.** The two chromatids in each chromated pair are pulled apart as the spindles contract. One copy of each pair is drawn to each pole of the cell.

5. **Telophase.** The chromosomes become less distinct, returning to the dispersed chromatin net form. New nuclear membranes form around the chromosomes and the centrioles divide. The cell separates into approximately equal halves as the cell membrane pinches in to form the two new cells. The two cells then re-enter interphase and mature in size before further division takes place.

Figure 1.16.
Mitosis in whitefish showing several phases of cell division: a) interphase; b) anaphase; c) telophase.

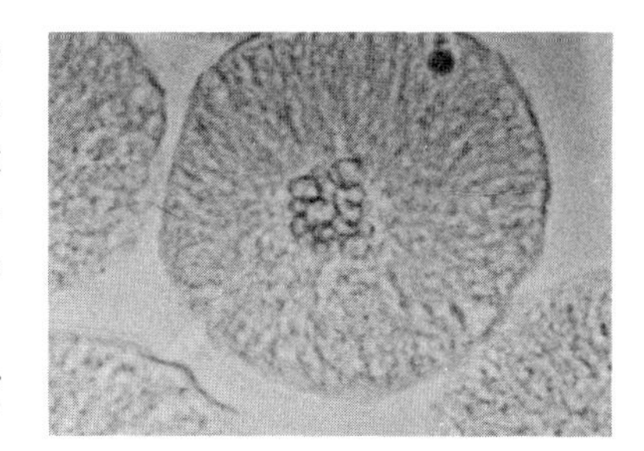

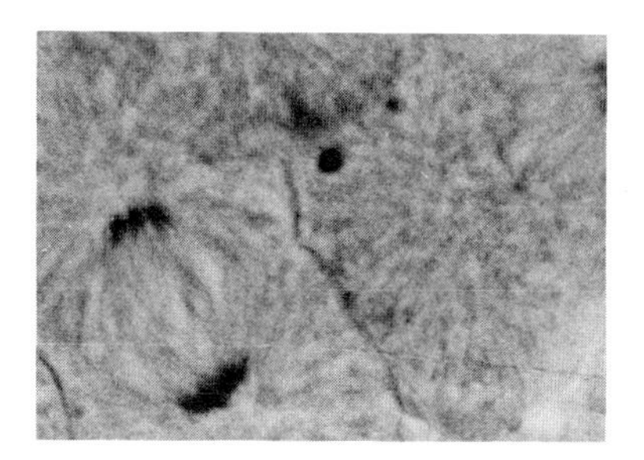

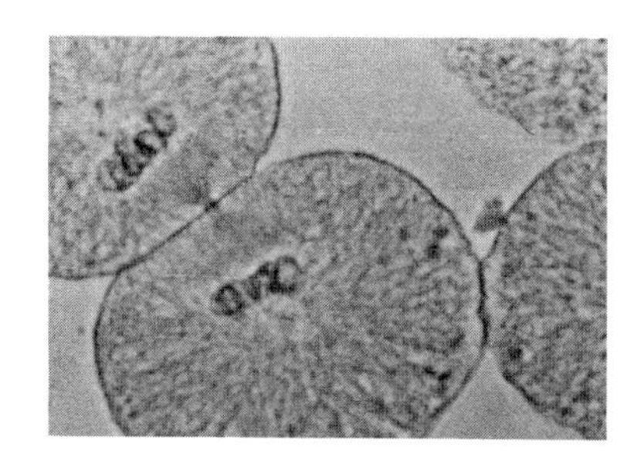

Meiosis

Mitosis occurs in all the cells of the body at some time during life. An additional kind of division, however, takes place in sex cells. The process of **meiosis** is confined to the cells of the ovaries and testes for formation of eggs and sperm. During this process, pairs of chromosomes are halved in number. The normal number of chromosomes in human cells is 46; in sex cells this is halved to 23 chromosomes. Meiosis is discussed in more detail in Chapter 20, *Human Genetics*.

(my-oe-sis)

QUESTIONS FOR REVIEW

SOME WORDS TO KNOW

Match each of the descriptions given in the left-hand column with a word shown in the right-hand column. DO NOT WRITE IN THIS BOOK.

1. "Bubbles" of water, etc., surrounded by a membrane.
2. Small sacs which help to pass large particles across the cell membrane.
3. Structures in which energy is converted for cell activities.
4. Membrane-lined channels which produce lipids and enzymes and package them for transport.
5. Double-layered membranes often found with ribosomes present.
6. Thin membranes made up of protein and lipid layers containing small pores.
7. Found in the nucleus, contains genetic material.
8. Mixture of fluids, structures, and materials found outside the nucleus, but inside the cell membrane.
9. Thin fingerlike extensions found on some cells which help to increase the surface area of the cell for more efficient absorption.
10. Two cylinderlike structures, found just outside the nucleus, which are active during cell division.

A. cell membrane
B. nucleus
C. mitochondrion
D. endoplasmic reticulum
E. Golgi bodies
F. cilia
G. vacuoles
H. nucleolus
I. ribosomes
J. centrioles
K. microvilli
L. pinocyte
M. chromosomes
N. cytoplasm
O. lysosomes

SOME FACTS TO KNOW

1. Draw a simple diagram of a typical cell and label as many parts as you can.
2. Make a chart listing all the organelles of the cell and their functions.
3. What reason can you give for the importance of the unit membrane which is found around the cell, the nucleus, and almost every cell organelle?
4. Why do cells vary so much in shape?
5. List four important functions that water performs for the body.
6. Each of the following terms refer to a method by which substances or other materials enter the cell. Briefly explain each term.
   a) phagocytosis,
   b) pinocytosis,
   c) osmosis.
7. List the five phases of mitosis and make a brief statement about what takes place in each phase of this process.
8. Where would you expect to find mitosis taking place in the human body?
9. What are cells doing while they are not actively engaged in mitosis?
10. What factors affect the rate of diffusion?

QUESTIONS FOR RESEARCH

1. Select one of the following topics and prepare a report on it:
   - microbiology
   - how cells move
   - Camille Golgi
   - histology
   - Hippocrates
   - forensic science
2. Research in more detail the structure and function of any one of the cell organelles.
3. There is a great deal more to learn about the functions of water in the body than can be given in this introductory chapter. Find out more about the importance of water in the body.
4. Find out more about either the transmission electron microscope or the scanning electron microscope. These are very important research tools in the field of medicine and biology.
5. Look up the list of elements found in the human body and then find some food sources for each of these elements.

Activity 1: INVESTIGATING THE STRUCTURE AND FUNCTION OF CELLS

The amoeba is a useful organism for study because it has many structures in common with human cells. Its method of feeding is similar to the process by which white blood cells destroy bacteria. With the exception of the nucleus only a very few cell organelles can be seen distinctly with the light microscope.

Part A

Materials
live amoeba (*Amoeba proteus* or *Chaos chaos*), live *Tetrahymenae pyriformis* (or a similar small protozoan for food), microscopes, slides, and coverslips, eye droppers, toothpicks, and 1% methylene blue dye

Method
1. Place a drop of the amoeba culture on a clean glass slide and cover with a glass coverslip. Do not press on the coverslip or you may crush the organisms.
2. Focus the microscope under low power and locate an amoeba. Observe the organism carefully using a low light intensity. Switch to a higher power if necessary.
3. *Draw an amoeba and label as many parts as you can. You should be able to identify the following structures: cell membrane, cytoplasm, nucleus and nuclear membrane, food and contractile vacuoles, pseudopods (false feet).*

4. *Observe the amoeba for a time and then describe the method by which it moves.* This process is called amoeboid movement and it is the method that some white blood cells use to flow out of the capillaries into the surrounding tissues.
5. Add a drop of culture water containing *Tetrahymenae pyriformis* and watch the amoeba feed. *Describe how the amoeba takes food into the cell.*

Questions

1. *Look up the word "phagocyte" and briefly describe what it means.*
2. *Give the functions of the structures that you have identified on your drawing.*

Activity 2: OBSERVING HUMAN CHEEK CELLS

Part B

Materials

toothpicks, glass slides, coverslips, methylene blue, iodine, compound microscopes

Method

1. Take a clean toothpick and gently scrape the inside of your cheek to remove a few dead cells.
2. Smear the sample onto a clean glass slide.
3. Cover the smear with 1 drop of methylene blue dye or iodine and put a coverslip on top.
4. Observe the slide under low power, then under high power. *Locate a few cells and carefully make a diagram. Label the parts that you can identify.*
5. Look for any tiny rod-shaped bacteria that may be present; include these in your drawing making sure that you maintain the correct proportions in size.

Questions

1. *What kind of cells line the surface of the body?*
2. *What general shape are the cells that you have drawn?*
3. *Account for the fact that cells you have observed differ in shape and that some of them are not complete.*
4. *If some cells are folded, what can you guess about the cell's thickness?*

Chapter 2

Tissues and Organs

- Tissues
- Organs and Systems
- The Cavities of the Body
- Directions in the Body

The Tissues of the Body

All livings things are made of cells, although not all cells are alike. Cells work together in groups called tissues. The cells of each kind of tissue are specially designed to do a particular job.

The different types of tissues are grouped and named according to their design and function. The tissues of the human body can be grouped into four major divisions.

(ep-i-thee-li-al)

Epithelial tissue acts as a *covering* or *lining* for other cells and tissues.

Connective tissue is often composed of cells that are soft and have a fluid base. They *bind and pack* cells together. Other connective tissues are formed of *supporting* cells. These are made up of firm or rigid materials such as bone and cartilage.

Photographs of tissues shown in this chapter have been magnified using a compound microscope.

Muscle tissues are composed of cells that *contract* to move parts of the body. **Nerve tissue** is responsible for *conducting impulses* to and from the brain and for co-ordinating the various activities of the body.

Epithelial Tissue

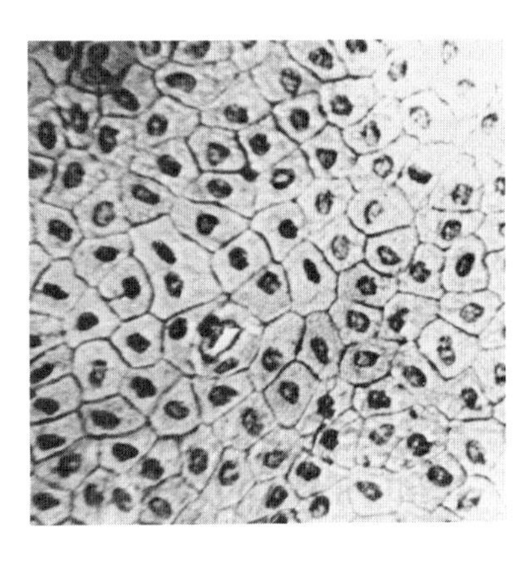

Figure 2.1.
Squamous epithelium. Surface view.

The **epithelial** tissues or **epithelium** cover the outer surfaces of the body or line the hollow internal organs and the cavities of the body. To form an effective covering, epithelial tissues must have cells that fit tightly together. The cells form flat sheets of tissue, sometimes single layers, sometimes with many layers one upon another. All epithelial tissues have a small amount of cementing substance between them, which helps to hold them together.

Epithelial cells are usually four- or five-sided (See Figure 2.1). The number of layers of cells in epithelial tissue is determined by the type of substance that will come in contact with the tissue. If a non-abrasive substance, such as air, is to pass over the surface, there will be little wear and tear on the tissue. A single layer of cells will then be adequate. A single-layered epithelium covers the surface of the air spaces in the lungs. Hands, which are often roughly used, need to be covered by many layers to provide toughness. Epithelial cells covering hands must be constantly replaced as they become damaged or worn. To handle all these different needs there are several kinds of epithelium.

Squamous Epithelium

Squamous epithelium is made up of flat, thin cells that fit together like tiles. These also cover surfaces that experience little wear and tear, such as the inside of tubes which carry urine to the bladder or the inside of capillaries which carry blood.

(skwae-mus)

Stratified Squamous Epithelium

Stratified, or layered, squamous epithelium is found on surfaces that are exposed to rough use. It is composed of many layers of squamous cells. When the top layer is worn away, there are then more cells underneath to take its place. The inside of the mouth and the palm of the hand are covered by stratified epithelium. Some areas of the skin, such as the sole of the foot, have very thick layers. On the inside of the arm, which is well protected, the number of layers is quite small. A simple scratch in this area will thus quickly draw blood. (See Figure 2.2)

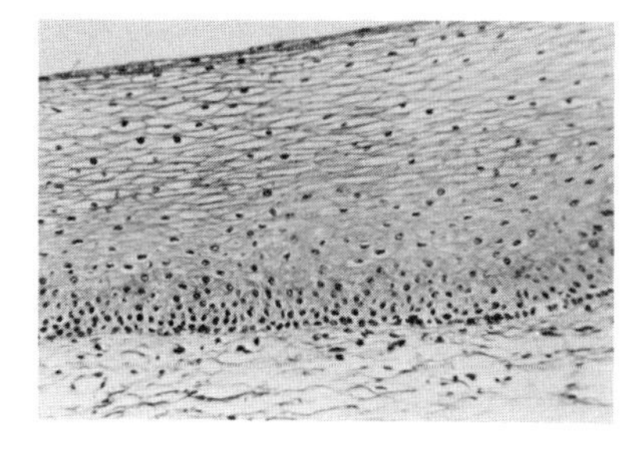

Figure 2.2.
Stratified squamous epithelium.

Cuboidal Epithelium

Cuboidal epithelium cells are thicker than squamous cells and rather like a cube in shape. (See Figure 2.3). Cuboidal cells often line glands. They frequently secrete fluids of one kind or another. Where larger quantities of a secretion are required, the epithelium may be formed into small pockets or indentations below the surface of the organ. These saclike chambers are called *glands* and are usually lined with cuboidal cells. The additional surface area that these glands provide below the surface enables larger quantities of the secretion to be formed, which can then be released through a small tube, onto a surface or into some organ. (See Figure 2.4).

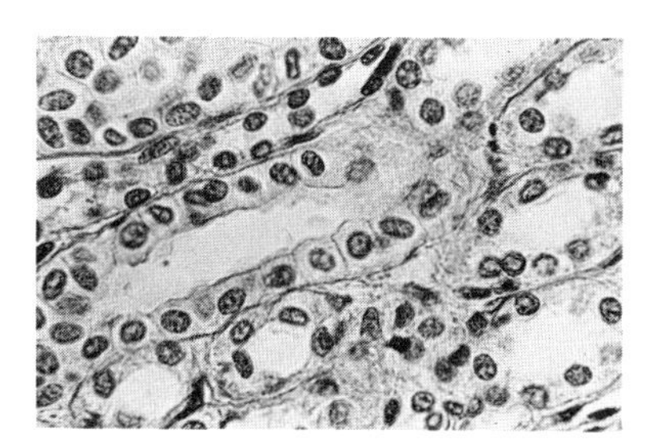

Figure 2.3.
Cuboidal epithelium. Gland ducts.

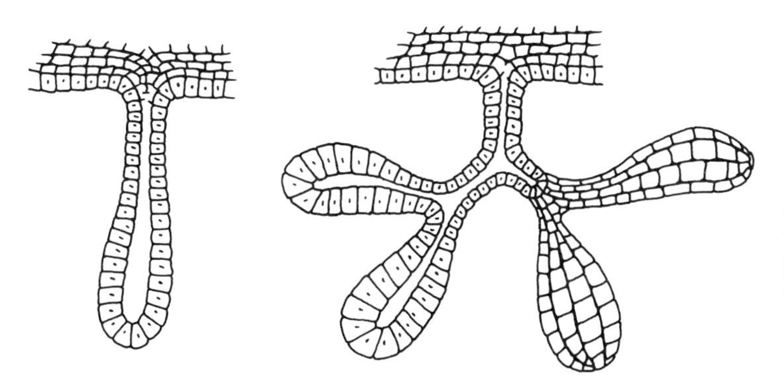

Figure 2.4.
Glands lined with cuboidal epithelial cells. These glands are small saclike pockets below the epithelial surface. They enable a large number of secretory cuboidal cells to be packed into a very small area of the surface tissue.

Columnar Epithelium

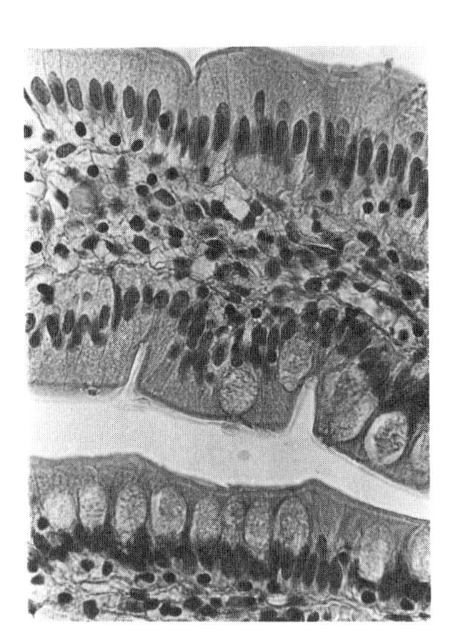

Figure 2.5.
Columnar epithelium. These cells are located along the villi in the small intestine.

Columnar epithelium cells are tall and shaped like elongated boxes. They frequently have some special feature on that side of the cell that is exposed along the surface of the tissue. (See Figure 2.5.) Some columnar cells have **cilia**, tiny hair-like projections. Cilia are found on the cells that line the trachea, for example. They beat back and forth trapping dust and dirt before it can enter the lungs where it would clog the respiratory surfaces. These particular epithelial cells are called **ciliated columnar epithelium**. Each word describes something about the cell: *epithelium*, that it is a lining cell; *columnar*, that it is tall; *ciliated*, that it has cilia.

Other columnar cells have very small (microscopic) extensions called **microvilli**. These structures project from the exposed surfaces of the cells, rather like the fingers of a glove. Microvilli serve to increase the surface area of the epithelial cells. They thereby make it easier to absorb needed materials (e.g., nutrients) across the surface membrane into the cell. Many of the cells lining the small intestine possess microvilli.

The small intestine epithelium also contains, along with the columnar epithelial cells, many **goblet cells**. These cup-like cells resemble a wine glass, with the base broken off. Goblet cells are designed to secrete mucus, which helps to protect the walls of the digestive tract.

In general the functions of the epithelial cells can be summarized under the following headings:

1) **Protection:**
from bacteria, dust, wear, etc. For example, the skin serves as a protection for the body within.

2) **Secretion:**
the linings of many tubes and surfaces secrete fluids. Glands are formed of epithelial cells.

3) **Absorption:**
specialized cells along the small intestine absorb the dissolved end products of digestion.

4) **Filtration:**
an important function of the capillary lining that allows materials to cross in both directions.

Muscle Tissue

Every movement made by the body results from the action of cells designed to contract. The muscles that enable us to walk or run are composed of cells that contract to pull two bones toward each other. Other muscles may contract in pairs to hold some part in place. For instance our head is kept from falling either backwards or forwards as a result of muscular control. Think of what happens when you "nod off" to sleep while sitting up! In smiling or frowning, many muscles combine their actions to move the tissues of the face. Other muscles pump blood or move food along the digestive tract.

There are three kinds of muscles, each of which can contract, relax, or extend. Each type is stimulated to contract by nerves and has the property of elasticity.

Skeletal Muscle (also called Voluntary Muscle or Striated Muscle)

Each of these names gives us a hint about the type of cells that make up skeletal muscle. The word "skeletal" reminds us that these muscles move the bones of the skeleton. They are also called "voluntary" because they are under conscious control. Our brain decides that we will move our arm or foot; therefore, the action is voluntary. The word "striated" describes the fine bands that can be seen in this type of muscle, when it is viewed under a microscope. (See Figure 2.6.)

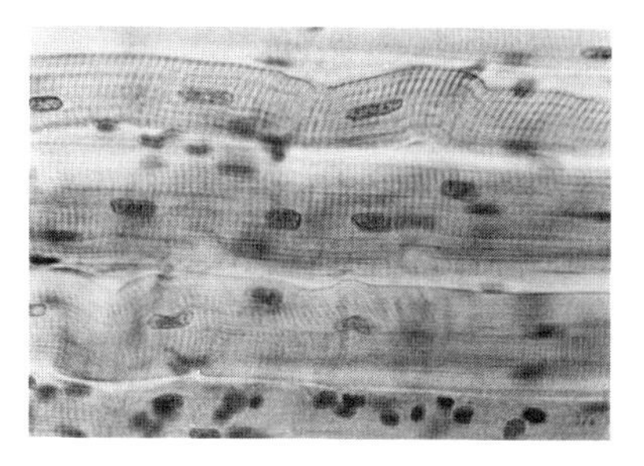

Figure 2.6.
Skeletal or voluntary muscle tissue.

Smooth Muscle

This tissue is also known as **involuntary muscle**. Again, these names describe some features of the muscle cells. Smooth muscle is found, for example, along the digestive tract in the walls of the intestine. There it contracts rhythmically to move food along the tube. Smooth muscle also plays an important role in the expansion and contraction of blood vessels throughout the body. Such action happens without conscious control, and it is therefore called **involuntary**. When viewed with a microscope, involuntary muscle cells appear smooth and lack striations. These cells, which often form a ring of muscle tissue, pull against each other when they contract. The ring then becomes smaller and controls the passage of substances along a tube.

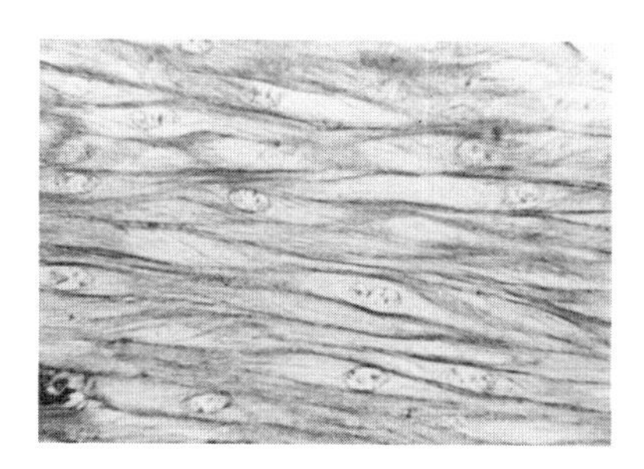

Figure 2.7.
Smooth or involuntary muscle tissue.

Cardiac Muscle

(car-dee-ack)

Cardiac or heart muscle is made up of cells which have enormous stamina. They start contracting about eighteen

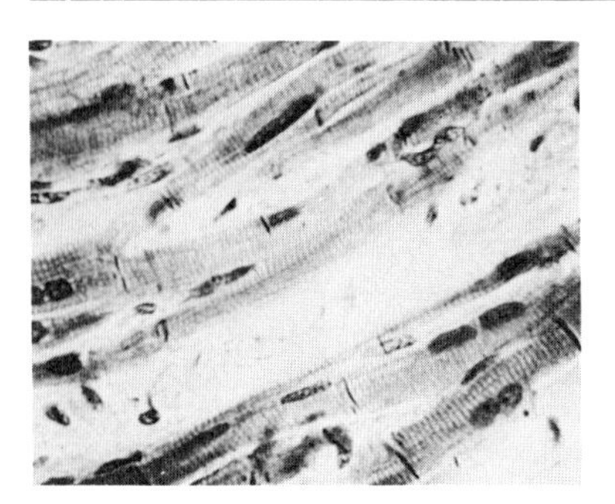

Figure 2.8.
Cardiac muscle.

days after we are conceived. They continue to contract and relax, until we die, about 70 times a minute or 50 400 times a day. This muscle cannot stop to rest when tired, nor can it be immobilized when damaged! Cardiac muscle cells have faint striations. They also branch and possess small discs that appear as lines occurring at right angles to the cells. (See Figure 2.8.) See Table 2-1 on page 47 for comparisons of these three types of muscle cells.

Connective Tissue

The connective tissues are among the most important and most widely distributed tissues of the body. They help to support the body, bind the joints together, and attach muscles. Connective tissues also act as packing around organs, join other tissues together, store nutrients, and (by white cell action) destroy bacteria.

Under the microscope, the most common types of connective tissue show two distinctive features: one, scattered and apparently disorganized, thread-like **fibres**; two, relatively wide intercellular spaces called the **matrix** which contain a fluid, a semi-fluid, or a rigid substance. There are scattered nuclei in the matrix but the cell membranes which must surround the nuclei are usually very difficult to distinguish.

There are several kinds of connective tissue which may be classified under the following headings.

Loose Connective Tissue (Areolar tissue)

Loose connective tissue is composed of fibres and a mixture of other types of cells embedded in a semifluid matrix. It provides strength and support for the body organs. If the body organs were not held in place by loose connective tissue, they would move about inside whenever we jumped or bounced around.

Fibrous Connective Tissue

Fibrous connective tissue is also called white fibrous tissue as it contains many closely packed white fibres. These fibres are very strong and flexible but have only limited elasticity. Fibrous connective tissue is found in **tendons**, which connect muscles to bones, and in **ligaments**, which hold the bones firmly together at the joints. Tendons are flexible but non-elastic. They allow the body a great deal of freedom of movement. For example, finger action is controlled by muscles that are found in the forearm. The finger bones are connected to these muscles by long tendons stretching back through the hand and wrist. The tendons allow the fingers to move freely without being surrounded by thick muscles. If you bend your hand back, you can see and feel these tendons on the back of the hand.

Ligaments like tendons, contain closely packed bundles of parallel fibres. Ligaments are very flexible, but are not quite as inelastic as tendons. Ligaments bind bones together at the joints so that they do not separate.

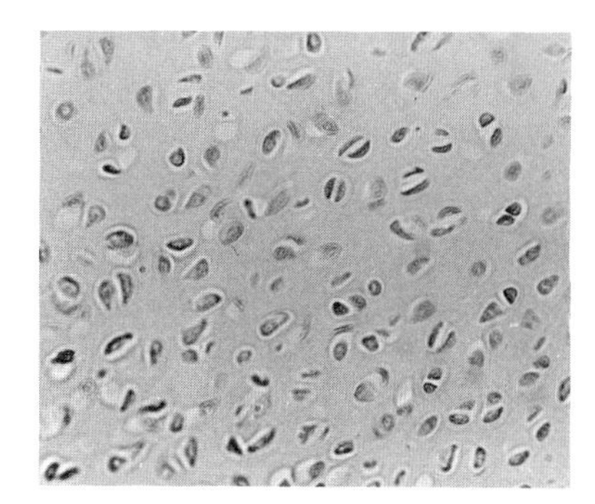

Figure 2.9.
Cartilage.

Cartilage

Cartilage, which is often called gristle, is a flexible, but firm, supporting tissue. It forms the base on which bone is built and it provides the rubbery support found in the ear and at the end of the nose. The end of the nose would be a very droopy affair without the cartilage that reinforces it! If you twist your ear about, you can easily feel the flexible nature of this material. It is still rigid enough, however, to maintain the shape of the ear. (See Fig. 2.9.)

There are three main types of cartilage.

Hyaline cartilage contains no fibres. It is a white, glossy substance that occurs around the ends of bones or in the trachea. **Fibrocartilage** contains scattered bundles of collagen fibre and is found in the vertebral discs. **Elastic cartilage** contains springy elastic fibres. It is present in the external ear and the epiglottis.

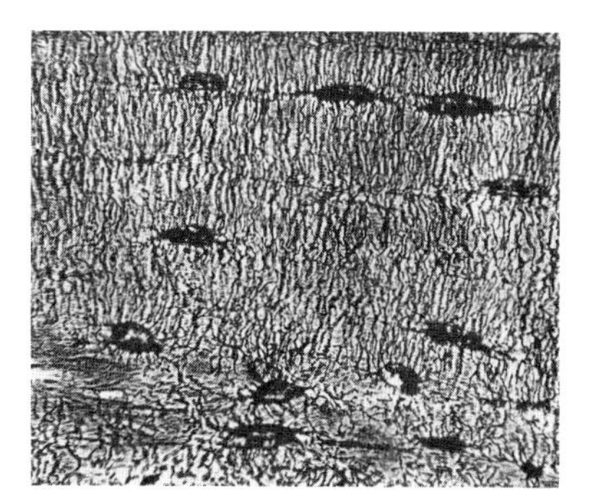

Figure 2.10.
Longitudinal section of bone.

Bone

The bone provides the framework of the body. It is initially formed on a pattern made up of cartilage cells. Once complete, it is not permanent but is constantly broken down and renewed. The tissue is formed by bone cells, called **osteocytes**. These cells are surrounded by tiny pools of blood, supplied by small canals running throughout the bone tissue (See Figures 2.10 and 2.11.) Between the cells is a matrix of calcium and phosphorus which forms the hard, rigid portion of the bone. A more detailed explanation of bone tissue is found in Chapter 5.

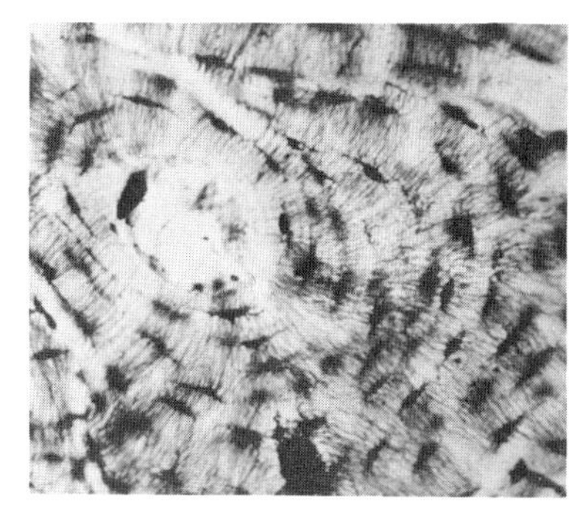

Figure 2.11.
Transverse section of bone. Note the Haversian canal and the tiny blood-filled canaliculi.

Adipose Tissue

This tissue is made up of large numbers of cells which act as storage containers. Adipose cells are often called fat cells.

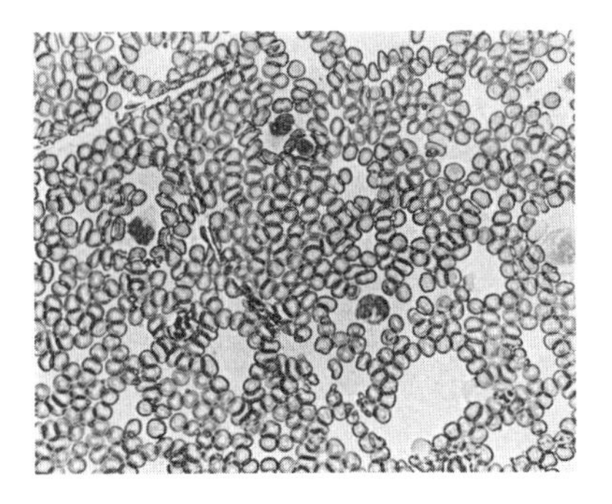

Figure 2.12.
Blood cells. The white cells are larger than the red blood cells and have stained nuclei.

Adipose tissue is found beneath the skin where it acts as an excellent insulator. It also stores food energy and acts as packing around organs helping to protect them from injury.

Blood

Blood is also generally grouped under the heading of connective tissue. Its major functions relate to the transport of oxygen and carbon dioxide by the red blood cells. Different types of blood cells perform a variety of tasks. Some are responsible for fighting infections, while others transport nutrients, wastes, and body chemicals. (See Figure 2.12.) The blood is dealt with in greater detail in Chapter 9.

Nervous Tissue

Nerve cells, called **neurons**, have the task of transmitting impulses between the brain and all other parts of the body. Some of these cells must therefore be quite long. Each nerve cell is made up of three parts: the *cell body* and two types of extensions or processes. The nerve cell processes extend out from the cell body and transmit nervous impulses. Sometimes these extensions, called *dendrites* and *axons*, can reach a metre in length. The cell body of such long neurons, however, may be no larger than that found in other types of nerve cells.

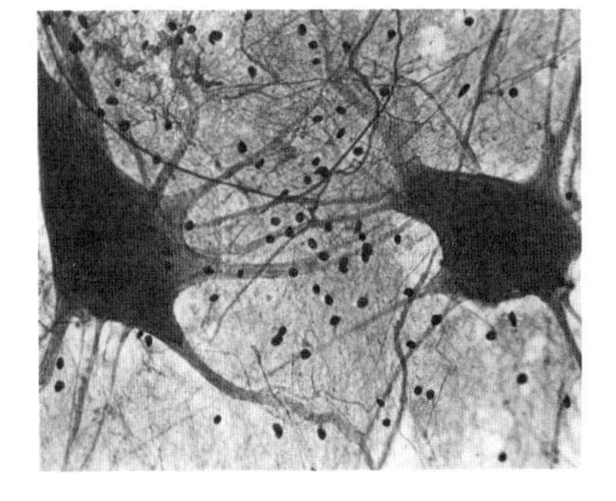

Figure 2.13.
Nerve cells showing dendrites and cell bodies.

Dendrites are one type of nerve cell process. They are short, fine, branching extensions that collect impulses and bring them *to* the cell body. (See Figure 2.13.) Each neuron has one or more dendrites. The single long extension which carries impulses *away* from the cell body is called the **axon**. Each axon terminates with short branches called nerve endings. Both types of nerve cell processes contain a living, jellylike substance. This may be surrounded by an insulating layer, called the **myelin sheath**, which is composed of fatty substances and proteins. This sheath permits nervous impulses to be transmitted more rapidly.

(my-e-lin)

When we speak of nerves we are actually referring to a bundle containing many nerve fibres (axons, dendrites, or both). The very large nerves of the body may contain many of these bundles bound together. Some may thus be quite large units, visible to the naked eye whereas the individual nerve fibres are microscopic in size.

Table 2.1. The tissues.

TISSUE	APPEARANCE	SHAPE & STRUCTURE	FUNCTION	WHERE FOUND
EPITHELIUM				
Squamous epithelium		Single sheets of thin flat cells. Scalelike with central nucleus.	Lines surfaces where rapid diffusion is needed and where there is little wear and tear.	Lining of capillaries and lymph vessels. Alveoli of the lungs. In locations where filtration or absorption occur.
Stratified squamous epithelium		Many layers of thin flat cells.	Protects against wear and tear.	The skin. In the mouth, esophagus, and vagina.
Cuboidal epithelium		In the shape of a cube with central nucleus.	Secretory cells and for absorption.	The walls of glands. Kidney tubules.
Columnar epithelium		Single layer of tall rectangular cells. Nuclei at base of cells. May have microvilli, and goblet cells may be present.	Secretion and absorption.	Lining of the stomach, small intestine, and digestive glands.
Ciliated columnar epithelium		Columnar cells with a fringe of fine hairs or cilia on the surface.	Move dust particles. Move ova or sperm.	Upper respiratory tract. Fallopian tubes and ducts of the testes.
MUSCLE				
Skeletal muscle (striated, voluntary)		Long fibres with striations. Nuclei around the edges of cells.	Contraction. Move bones and body parts under voluntary control.	Skeletal muscles, e.g., biceps, triceps, abdomen, face.

(continued on page 48)

TISSUE	APPEARANCE	SHAPE & STRUCTURE	FUNCTION	WHERE FOUND
Smooth muscle (involuntary)		Long cigar-shaped cells with central nucleus.	Contract to move food, wastes, etc., along tubes. Peristalsis.	Stomach, intestines, uterus, blood vessels.
Cardiac muscle		Branched, striated cells, scattered nuclei and discs.	Contract, pump blood through blood vessels.	Heart
NERVOUS TISSUE				
Nervous tissue		Neurons contain cell body, axon, and dendrites. Many different specialized cells.	Receive stimuli, conduct impulses, co-ordinate body activities, etc.	Brain, spinal cord. Specialized organs – eye, ear, skin (sense of touch).
CONNECTIVE				
Blood		**Red blood cells** Round, with central depression, no nucleus.	Carry carbon dioxide and oxygen.	Heart, arteries, veins, etc. W.B.C. and plasma in interstitial spaces.
		White blood cells Round or irregular, one or more nuclei.	Combat infections.	
		Plasma: Fluid medium.	Fluid medium carries nutrients, wastes, etc.	
		Platelets: Small irregular structures.	Platelets – initiate clotting.	
Supporting tissue Bone		Osteocyte cells embedded in matrix of solid inorganic material with many blood vessels.	Support. Movement of limbs. Protection, production of blood cells.	Bones, skull, ribs, vertebrae, etc.

Hyaline cartilage		Bluish white, glossy matrix, with cells (chondrocytes).	Allows flexibility in joints. Gives non-rigid support.	Ends of long bones, ends of ribs, nose, trachea, and embryonic skeleton.
Fibrocartilage		Cells (chondrocytes) scattered among bundles of collagen fibres.	Support and fusion between bones.	Pubic symphosis, discs between vertebrae.
Elastic cartilage		Chondrocyte cells in a network of elastic fibres.	Gives shape and support where flexibility is needed.	External ear, larynx, epiglottis.
Binding tissue Collagenous connective tissue		Many fibres in bundles, with fibroblast cells in rows between the bundles.	Attachment between structures: muscle to bone and bone to bone.	Tendons, ligaments. Membranes around organs.
Fibrous tissue Loose or Areolar tissue		Mixture of fibres and cells (fat, blood cells, mast cells) embedded in a semi-fluid matrix.	Strength, support, produces anti-coagulant.	Subcutaneous skin layers, blood vessels, body organs.
Adipose tissue (Fat)		Special storage cells, and fibres.	Insulation, protection, storage, support.	Subcutaneous layers of the skin. Around kidney, heart, etc., breast, padding in joints.

Organs and Systems

The several kinds of tissues already discussed may be found working together to perform some particular job. The heart, for instance, contains muscle tissue which contracts to provide the pumping action. It also has an inner and outer lining of epithelial cells. Nerve tissue stimulates the muscles and controls the rate at which the heart beats. Connective tissue gives the heart its structure and fills it with blood. All of the tissues forming the heart have a common purpose: pumping blood. When several tissues work together in this way they form an **organ**. Organs are then grouped into specialized **systems**, each with its own task to perform. The circulation system, for example, has the job of providing a transport and delivery system. The excretory system rids the body of wastes. Each system is made up of several organs helping to fulfil the overall purpose of the system. (See Table 2.2.)

Table 2.2. Body systems, their organs and functions.

SYSTEMS	MAJOR ORGANS AND TISSUES	FUNCTIONS OF THE SYSTEM
Integumentary	Skin, hair, nails, skin glands, and receptors.	Protection, temperature control, environmental stimuli.
Skeletal	Bones, joints, cartilage.	Support, protection, stores minerals, produces blood cells.
Muscular	Skeletal muscles.	Body movement, heat production.
Circulatory	Blood, heart, blood vessels.	Transports nutrients, wastes, gases, circulates the blood.

Lymphatic	Lymph, lymph nodes, tonsils, spleen, thymus, lymph vessels.	Return tissue fluids to blood, aid immunity, form white blood cells.
Respiratory	Nose, throat, larynx, trachea, lungs.	Exchange of oxygen and carbon dioxide.
Digestive	Mouth, esophagus, stomach, liver, pancreas, intestines.	Breakdown and absorption of nutrients, excretion of wastes.
Urinary or excretory	Kidney, ureters, bladder, urethra.	Excrete wastes and regulate blood composition.
Nervous	Brain, spinal cord, nerves, and specialized sense organs.	Control of body activities, monitor internal and external changes.
Endocrine	Hormone secreting glands, pituitary, thyroid, adrenals, pancreas, ovaries, testes.	Controls metabolism, growth, co-ordination with nervous system.
Reproductive	Testes, ducts, and glands. Ovaries, uterine tubes, uterus, vagina, breasts.	Produces sperm, semen, and hormones. Produces ova, hormones, and nourishes offspring.

So far in this chapter, we have discussed the way the body is organized. The basic unit of all living things is the cell. **Cells** are organized according to their structure and function into **tissues**. Several tissues working together at a common function are called **organs**. Organs that work together to achieve a common purpose are referred to as **systems**.

Figure 2.16 summarizes how the body is organized.

The Cavities of the Body

The human body can be divided into several major internal cavities, rather like a house with several rooms. Fig. 2.14 shows the names and locations of these cavities.

The **thoracic cavity** lies in the chest and is surrounded by the rib cage. It contains only two organs, the heart and the lungs. The thoracic cavity is separated from the cavity which lies below it by a very important sheet of muscle, the **diaphragm**. Below the diaphragm is the **abdominal cavity**. It is usually divided into two parts, although there is no partition between them. The upper part of the abdominal cavity contains the organs of digestion: the stomach, small and large intestines, liver, and pancreas. The spleen, which is not a digestive organ, is also present in this cavity.

(thor-ass-ic)

(dy-a-fram)

Figure 2.14.
The major cavities of the body.

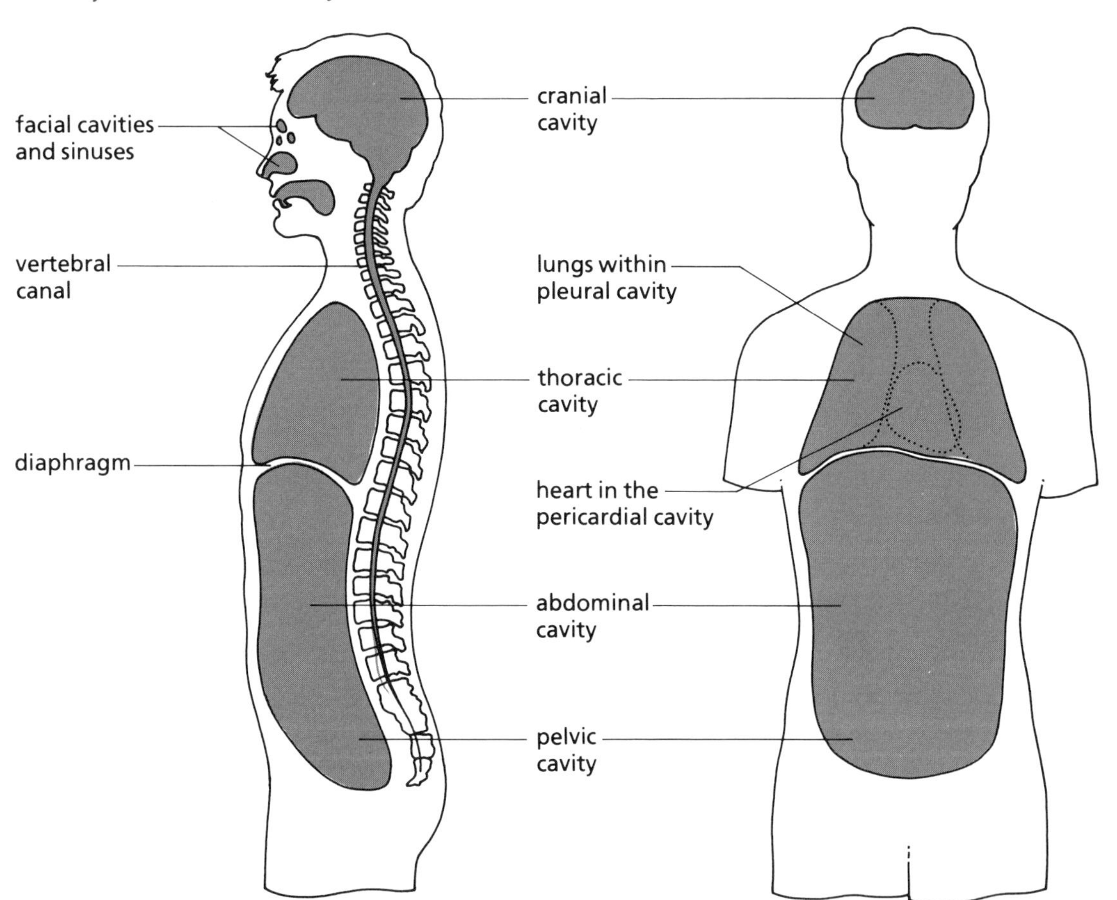

The lower part of the abdominal cavity is called the **pelvic cavity.** It is enclosed by the pelvis and contains the reproductive organs and the urinary bladder, as well as the lower part of the digestive tract and the rectum. The kidneys are found very close to the midpoint of these two abdominal cavities.

The skull contains one large cavity which surrounds and protects the brain. It also contains many small spaces known as **sinuses**. The many skull cavities, which vary greatly in size, are part of the **cranio-facial** complex. Some of the spaces, such as the sinuses, appear empty, whereas the cranium proper which surrounds the brain and its glands is obviously quite full. The cranio-facial complex also includes the oral cavity (with the tongue and teeth), the nasal cavity, the orbits, which contain the eyes, and the auditory cavities, which contain the organs of hearing and balance.

Directions in the Body

Before using a map, we must first be sure that North is located at the top of the map. We can then easily use the points of a compass to describe or give directions. Similarly, when describing the body, we need to position the body in a way that everybody will recognize. This is known as the **anatomical position**. The anatomical position implies that the body is upright and facing us with palms turned frontward. Instead of compass directions the following terms are used.

Superior means above or higher (near the head of a standing person). **Inferior** means below or lower in the body.

Anterior or **ventral** are terms used to describe the front of the body, from the face down to the feet. It may also be used to describe an organ that is in front of another. **Posterior** or **dorsal** refers to locations behind some organ or at the back of the body.

Medial is used to describe points near the centre or midline of the body. Such an imaginary line divides the body into two halves, right and left. **Lateral** is a term used to describe positions away from the midline towards the sides of the body.

Proximal implies toward, or near the origin of some structure. **Distal** is the opposite, meaning away, or further from the point being discussed. (See Figure 2.15.)

Figure 2.15.
The anatomical position and some basic terms used to describe locations in or on the body.

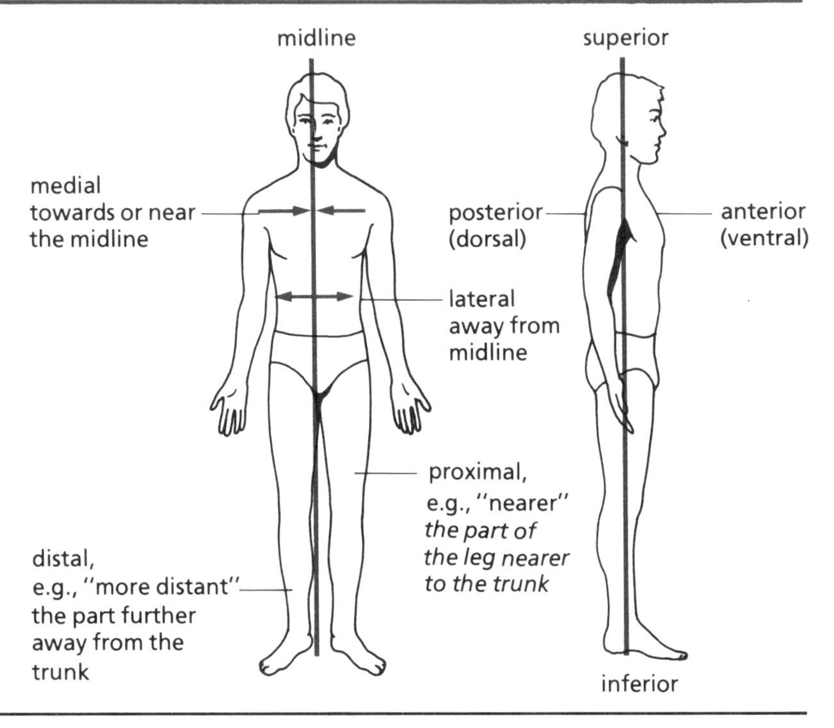

Figure 2.16.
How the body is organized.

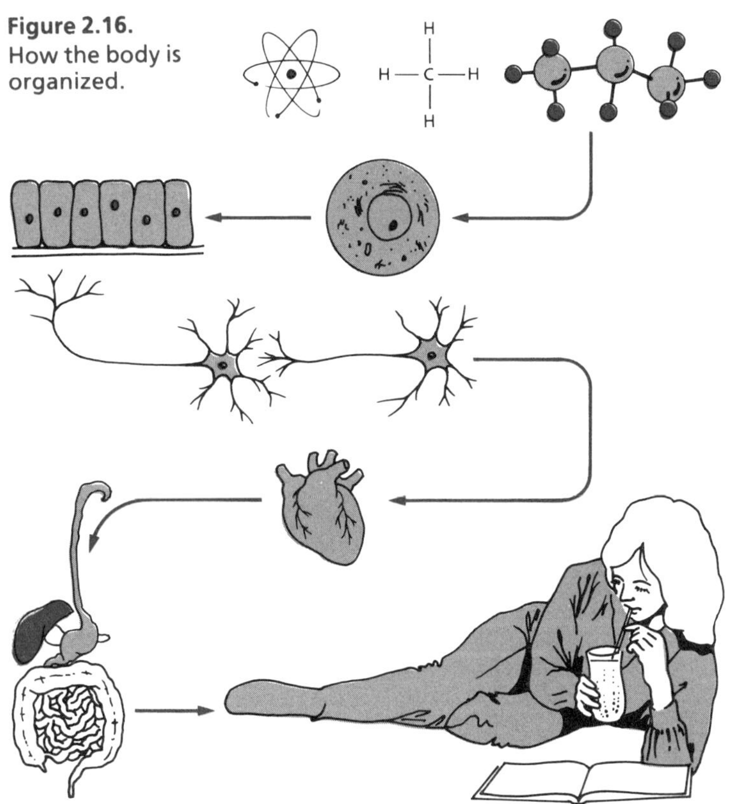

The smallest units of living things are *molecules*, chemical elements bonded together into compounds, e.g., oxygen, carbon dioxide, glucose, proteins.

Molecules are organized into functioning, living units called *cells*, e.g., muscle cells, nerve cells, etc.

The cells are grouped together by their specialized structures and functions into *tissues*, e.g., epidermal tissues, nervous tissues, etc.

Tissues are assembled into *organs*, each with a special function to perform, e.g., heart, liver, kidney.

Organs are organized into *systems*. Each system has a major function, e.g., respiration, circulation, digestion.

The systems of the body are coordinated and integrated into the total living body. Each is important but dependent upon the others.

QUESTIONS FOR REVIEW

SOME WORDS TO KNOW

Match each of the descriptions given in the left-hand column with a word shown on the right-hand column. DO NOT WRITE IN THIS BOOK.

1. Tissues which line the surfaces of the body or organs.
2. Tissues which move the bones of the body.
3. Main tissue found in the heart.
4. Layers of thin, flat cells found on the palm of the hand.
5. Storage cells.
6. A flexible support tissue.
7. Receives and transmits impulses.
8. Rigid support tissue.
9. Helps move food along the digestive tract.
10. Found along the lining of the trachea and other respiratory tubes.

A. bone
B. nerve tissue
C. epithelium
D. stratified squamous epithelium
E. ciliated columnar epithelium
F. skeletal muscle
G. tendon
H. smooth muscle
I. cardiac muscle
J. adipose tissue
K. cartilage
L. blood
M. cuboidal tissue

SOME FACTS TO KNOW

1. What general function does epithelial tissue perform? Give examples of epithelial tissues and how they are specialized to do a particular job.
2. How can you distinguish between voluntary and involuntary muscle?
3. Element - E, Tissue - T, Organ - O, Compound - C, System - S, Cells - CE. Select one of the letters from the list above which you think identifies each of the following terms.
   a) columnar epithelium,
   b) neuron,
   c) cardiac muscle,
   d) fat,
   e) liver,
   f) circulation,
   g) oxygen,
   h) digestion,
   i) protein.
4. Distinguish between the terms *cell, tissue, organ, system,* and *organism*.
5. Select one organ in the body and state its general function. List the tissues that it contains.
6. Select any body system and state its general function. List the major organs or structures that make up the system you have chosen.
7. In which cavity of the body are the following organs found?
   a) pancreas,
   b) lungs,
   c) stomach,
   d) liver,
   e) heart,
   f) brain,
   g) ovaries,
   h) intestines.
8. Describe the anatomical position.
9. What are the sinuses? Where are they located?
10. Write sentences containing each of the following terms to show that you understand the meaning of the word: lateral, medial, distal, anterior.

QUESTIONS FOR RESEARCH

1. Select one of the following topics and prepare a report on it:
   - tissue cultures
   - biopsy
   - radiation and tissue damage
   - cloning cells
   - the work of a cytologist
   - radiation and cancer treatment
   - adipose tissue and the storage of fat.
2. Examine some of the prepared slides that are in your school laboratory. Sketch several different types of cells and determine what types of cells they are. Try to find out how the shape and structure of the cells that you draw are adapted to the function they perform.
3. Aging starts at adolescence! It is a very slow process in which tissues and organs gradually atrophy (dry up and stop functioning). Find information on the different theories of aging.

Activity 1: EXAMINATION OF PREPARED SLIDES OF BODY TISSUES

Materials

Your teacher will provide a number of tissue slides for you to examine: a blood smear, epithelial tissues (skin and small intestine), three kinds of muscle, bone sections, adipose tissue, or others.

Method

Observe each slide carefully under low power magnification. Examine the cells present and note how their structure is related to their function.

Compare the slides with the micrographs and the tissue chart in the text and make simple sketches that illustrate the major features present in each type of tissue.

Make large diagrams, drawing a few of the cells in as much detail as possible. Do not attempt to draw all the cells in this much detail.

Make a chart naming each of the tissues that you have studied. State the major characteristics of each type of tissue and give at least one example of where each type of tissue may be found.

Unit II

Our Outer Protective Covering

Chapter 3

The Skin

- The Characteristics and Functions of Skin
- The Structure of the Skin
- Regulation of Body Temperature
- Some Common Skin Features
- Some Skin Problems

The Characteristics and Functions of Skin

The skin is an elastic organ, stretching to accommodate each bulging muscle, then retracting as the muscle relaxes. It can expand to adjust to extra size during pregnancy or to cover a bulging bump on the head. The skin is semi-translucent, depending upon the amount of pigmentation present. The skin contains vast numbers of tiny blood vessels, the capillaries. The blood flowing through these vessels gives the skin a rosy colour. If these capillaries dilate (expand) the colour deepens. This is apparent when we blush or are very hot. Fear or cold causes the capillaries to contract. The amount of blood in the skin then decreases, giving the skin a pale or 'blue' appearance.

The outer surface of the skin is constantly being worn away. After a shower or bath, if you rub your skin roughly with a towel, you can see the dead skin cells rub off. This is more obvious in summer when you have been exposed to the sun and may be sunburned. The speed with which cuts and wounds generally heal is evidence of the ability of skin cells to regenerate (replace themselves) after injury.

Table 3.1. *The Functions of the Skin.*

- Protects soft inner cells from wear and tear.
- Shields the body from the harmful rays of the sun.
- Excretes small amounts of body wastes.
- Acts as a defense against bacteria.
- Indicates what is happening around us: e.g., temperature or pressure change, touch.
- Heals wounds, replaces damaged cells around cuts.
- Acts as an insulator when cold.
- Releases heat when warm.
- Produces hair and nails.
- Indicates health: e.g., the body is paler when one is ill.
- Produces vitamin D in sunlight.
- Produces oils.
- Prevents fluid loss.
- Indicates our feelings: e.g., pale with fright, darker with rage.
- Keeps out the rain: - sheds water.

The texture and thickness of the skin varies according to where it is found on the body. Skin that is normally covered by clothing will be thinner than that exposed to air. Areas of skin that receive considerable wear, such as on the hands and feet, are much thicker than those found on the arms or back. We are all aware of the tender texture of the skin of a healthy baby compared to that of adults.

The functions of skin are summarized in Table 3.1.

The Structure of the Skin

(ep-i-der-mis)

The skin is composed of two layers. The external sheath of cells, called the **epidermis**, is made up of layers of squamous epithelium. Immediately below the epidermis is a thicker layer of connective tissues called the **dermis**. This layer contains a mixture of fibres, fat cells, blood vessels, hair follicles, and specialized nerve endings. (See Figure 3.1.)

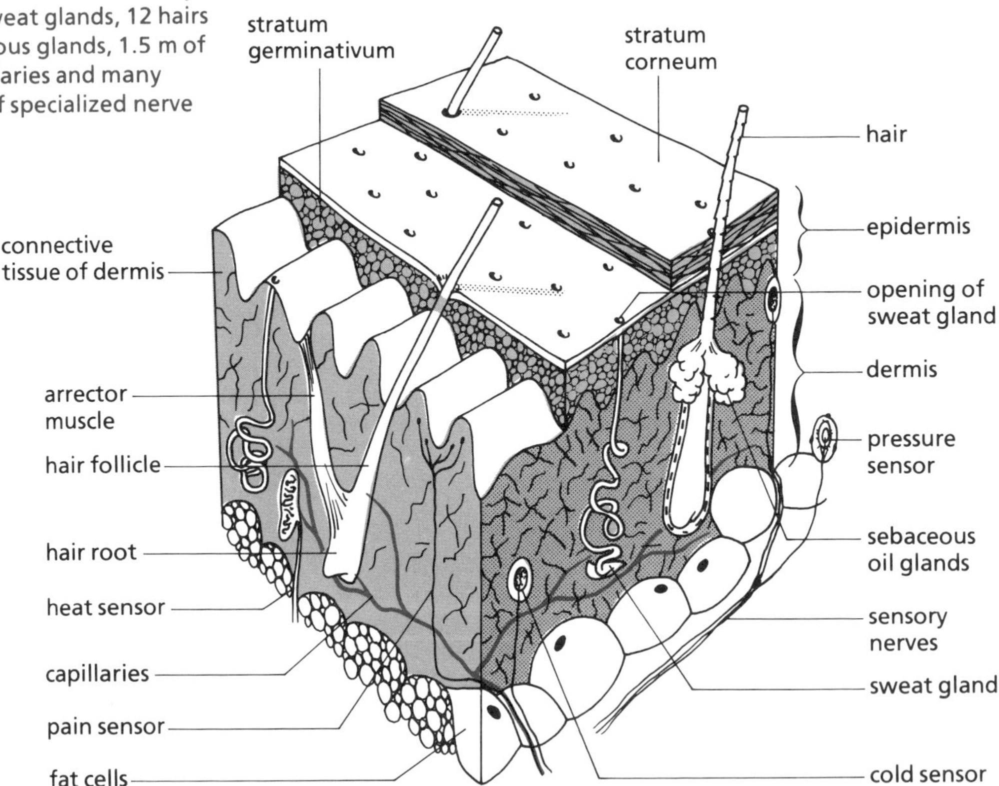

Figure 3.1.
The structure of the skin. One square centimetre of skin may have 100 sweat glands, 12 hairs and sebaceous glands, 1.5 m of blood capillaries and many hundreds of specialized nerve endings.

The two layers are quite firmly cemented together. Excessive rubbing of the skin, however, such as occurs when a shoe fits improperly and chafes the skin, may cause the layers to separate. When the dermis and epidermis are forcibly separated in this way, the resulting space fills with interstitial fluid and a blister forms.

The Epidermis

The outer layers of the epidermis called the **stratum corneum**, which we touch, caress, and care for, are only dead cells! These cells are, however, able to transmit sensations such as pressure to nerve endings in the layers below the epidermis. The layer of dead cells contains no blood vessels to supply nutrients needed by epithelial cells. (See Figure 3.2.) The cells then die and undergo a chemical process that changes them from soft, easily damaged cells into hard, tough, cornified cells. A substance called **keratin** hardens the cells and makes them waterproof, thereby helping to prevent water loss from the body.

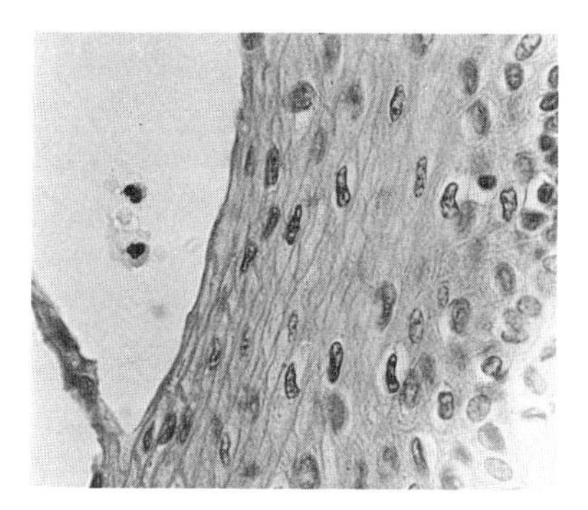

Figure 3.2.
Section of the skin. Note the dry outer layers peeling away.

The bottom layer of cells in the epidermis is composed of cells which divide to produce new cells. This deep layer of cells, the **stratum germinativum**, constantly replaces the cells that are worn away from the surface of the body. It also produces **melanin**, the pigment that colours the skin.

(jer-min-a-tiv-um)

Melanin

Production of melanin is stimulated by exposure to ultraviolet light. Large doses of sunlight thus result in production of this pigment causing the skin to tan. The tan then protects the skin during further exposure because the hot rays of the sun may destroy several layers of epidermal cells and cause painful sunburn and blisters. Melanin also accounts for the dark colour of the tissue around the nipples and for freckles, which are small irregular patches of melanin.

The Dermis

The **dermis** is composed of living tissues which perform a variety of functions. The structures in this layer are specialized to monitor the changes that occur in the environment immediately around the body. Blood vessels regulate body temperature in response to impulses from special heat and cold receptors. Touch, pressure, and pain receptors are located at different levels within the dermis to protect the body from potential danger. Glands, hairs, and fat cells are held in place by a matrix of collagenous and elastic fibres that give structure and elasticitity to the skin.

Where the epidermis and the dermis meet, there is a wavelike layer formed of many tiny cones and ridges. These patterns show through to the surface of the skin on the hands and feet, some of them becoming fingerprints. The patterns emerge while the baby is developing in the uterus and never change, except with respect to size.

The Patterns of Skin

All humans share the same major features. The eyes, nose, and mouth are all located in the same approximate positions on the face. Yet the diversity which exists in even these few features is striking. With the exception of identical twins, close observation will usually reveal sufficient variation in facial features to identify every individual. Also, we differ from others around us in more ways than facial appearance.

As everyone knows from watching police dramas on television or reading detective novels, fingerprints (or dermal prints) are different for every individual. Patterns on the skin of fingers can be classified into ten basic arrangements of loops and whorls. (See Fig. 3.3.) Police officers use several additional, more advanced characteristics to identify a particular set of prints. The chance that someone will have eight of these special characteristics in common with you is more than 39 000 000 000 000 to 1!

Figure 3.3.
Part of the official police identification form used to show the fingerprints. The form would normally include a print of the complete palm of the hand.

The patterns of whorls and loops may differ from finger to finger on each hand. Like the patterns on fingers and toes, the arrangement of ridges on the palm of each hand and on the sole of each foot differs among individuals. The print made by the lips and the pattern of hair follicles on the head are also unique.

Appendages of the Skin

A number of structures are classified as appendages of the skin. These include skin glands as well as the hair and nails.

Skin Glands

Skin glands are of two main types.

Sebaceous glands occur over most of the body surface, with the exception of the palms and soles. The openings of these glands are found around hair follicles. (See Fig. 3.2.) Sebaceous glands produce an oily secretion, which helps to keep the hair from becoming dry and brittle. The oil also keeps the skin soft and helps to waterproof its surface. Underactive sebaceous glands cause the skin and hair to be dry, while active glands result in oily hair and skin. When the secretions of these glands come in contact with oxygen in the air they become black and are known as **blackheads**. If the glands become infected, they cause pimples or **acne**.

(seb-ae-shus)

(ack-nee)

The **sweat** glands are found over the entire body surface but are especially numerous under the arms, on the palms, soles, and forehead. These glands are located in the dermis and have tubes leading up to the surface through the epidermis. The sweat glands are stimulated by nerves that cause the glands to operate when the body temperature rises or when a person is under nervous tension.

Hair

Humans do not appear to be particularly hairy, but, in fact, we have approximately the same number of hair follicles as chimpanzees. (See Figure 3.4.) Human skin, however, produces quite small, transparent hairs that are not easily seen. Men and women have about the same amount of hair, but it is usually finer on women than on men. Hair is a stained (pigmented) shaft of keratinized cells. It is non-living. (See Figures 3.5 and 3.6.) Human hair grows at a rate of about 0.3 mm each day. Each hair follicle seems to have a growing

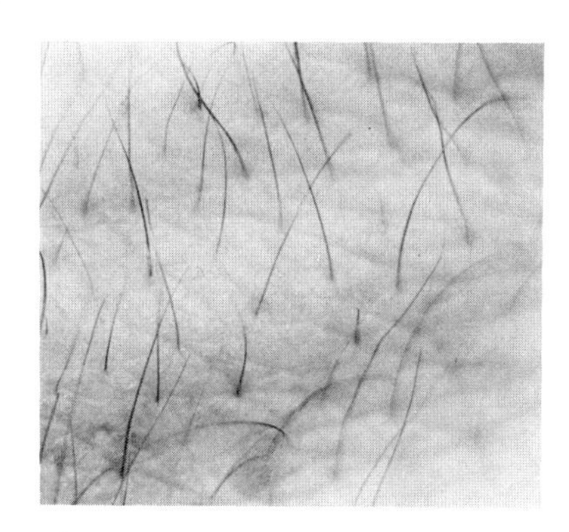

Figure 3.4.
The hairs and skin texture pattern on the forearm of a male.

Figure 3.5.
A human hair showing the hair follicle, nerves, blood vessels, glands, and muscles associated with the hair.

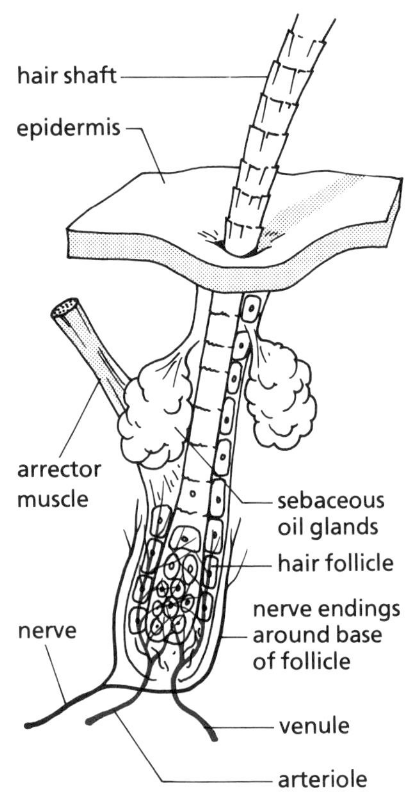

period followed by a "rest period". Hairs with a bulbous base are in the resting phase.

Hair is continually being shed and replaced. It is estimated that adults lose about 40 hairs each day. Baldness is often a hereditary condition, which commonly progresses with age. A tendency to become bald is inherited by men in particular. Loss of hair may be affected by the male hormone testosterone or be caused by illness, poor circulation to the hair follicles, or infections of the scalp.

Hairs are sensory organs with nerve endings located around their roots. Hairs act as tiny triggers to indicate that some light object is in close contact with the body, see Activity 8, Point 9. Each hair emerges at a slight angle to the skin surface.

A tiny arrector muscle is attached to each hair. When these muscles contract they cause the hairs to stand erect. Human hair has largely lost the ability to stand on end, although the hair muscles are still present. Contraction of these tiny mus-

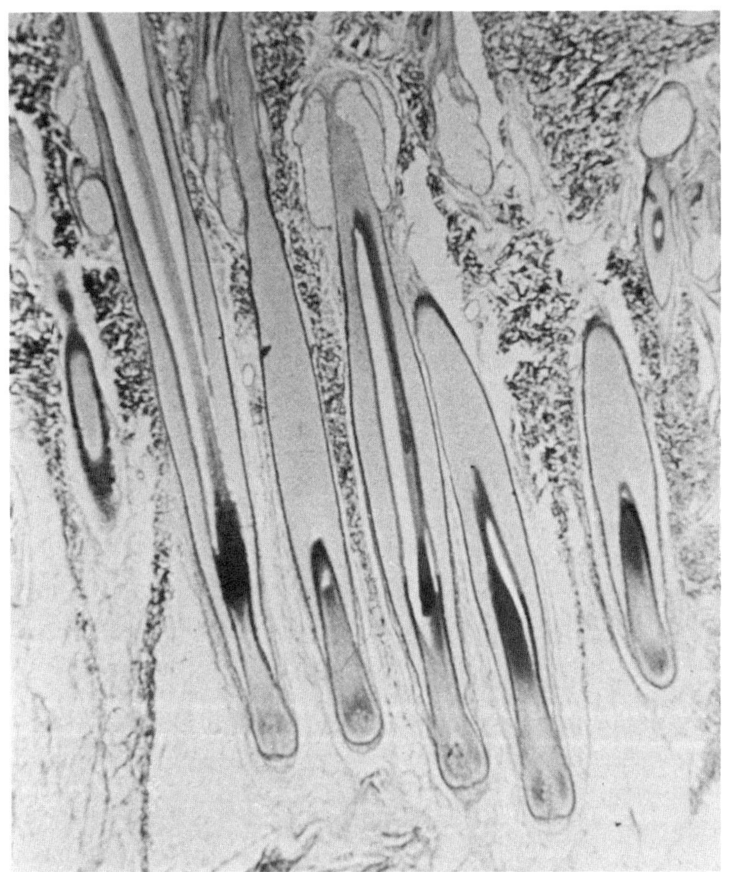

Figure 3.6.
A section through the skin showing several hair follicles.

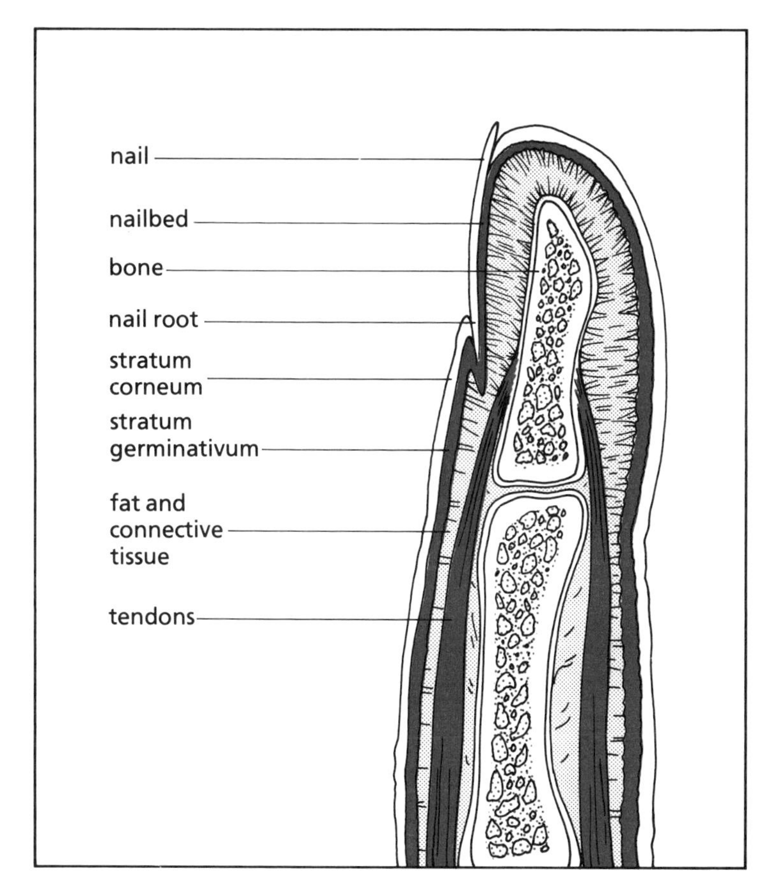

Figure 3.7.
A longitudinal section through the finger and nail.

cles sometimes causes the skin to pucker up into little bumps ("goose bumps"). In animals, such as dogs or cats, the hair can still be made to stand erect. You may have noticed that the fur along the animal's back will stand up if the animal is frightened or about to fight. The upright hairs make the animal look bigger and more ferocious. This process also helps to insulate the animal's body during cold weather.

Nails

Nails are hard structures, which are slightly convex on their upper surface. They are produced by the epidermis, originating as elongated cells which later fuse into flat plates. The living cytoplasm of these cells is replaced with keratin which makes them hard and more durable. The nailbed constantly produces more cells which results in the elongation of the nail. Fingernails grow more rapidly than toenails. (See Figure 3.7.)

Sense Receptors in the Skin

The skin provides a wide range of information about the immediate environment in which we live. Specialized nerve endings monitor conditions around the body and send impulses to the brain for interpretation.

Some nerve endings are sensitive to heat and cold, while others respond to touch and pressure. (See Figure 3.8.) Pressure-sensitive nerve endings occur at a greater depth in the skin than do the touch sensors. These nerve cells are capable not only of indicating contact with some object, but also register the texture of the material. We can, for example, distinguish between the textures of wool and silk, glass and wood, or skin and fur. Sometimes temperature, texture, and pressure combine to identify a substance. A heavy, cold, smooth piece of metal is easily recognized as being different from a piece of wood, which is lighter, rougher, and warmer to the touch.

The distance between **touch** receptors can be easily demonstrated with a small pair of dividers (see Activity 4). The lips and fingertips have receptors which are so close together that the divider tips must be almost touching. On the back, however, the distance for two-point discrimination increases to several centimetres.

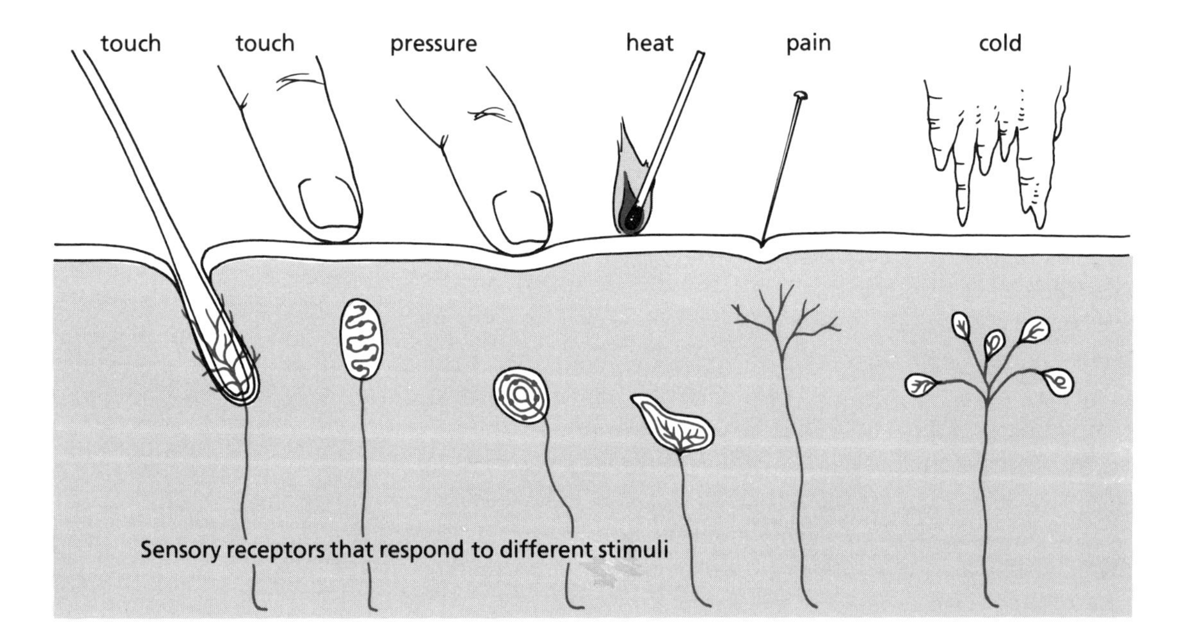

Figure 3.8.
Special sensory nerve endings in the skin.

There are more **cold** receptors than **heat** receptors, because cold is a greater threat to body temperature stability. Heat receptors respond best between 37° and 40° C. Cold receptors have two peaks, working well at 15 - 20°C and again at 46 - 50°C. The latter range may indicate a drop in temperature. For instance, if you step out of a hot shower you may suddenly feel cold, although the room temperature is quite normal. This is because the receptors are triggered to indicate a drop in temperature, not just "cold" as in a point on the thermometer.

There are also many free nerve endings which are sensitive to pain. These are found just under the epidermis and around the hair follicles. Not every part of the skin has the same sensitivity. The forehead has about 200 pain receptors per square centimetre of skin. The breasts and lower arms are also well supplied with nerve endings. The nose, however, has only about 50 receptors per square centimetre and the lobe of the ear contains even less.

Pain is a protective sensation (as are the other sensations in many instances); it acts as a warning device or an indicator of potential danger. Three types of pain are recognized. "Bright" pain is an intense pricking sensation, usually of short duration and localized, such as occurs at a cut or wound site. "Burning" pain, which results from burns, develops less rapidly but lasts much longer and is more widespread. "Aching" pain is often hard to describe and localize. It is persistent and may be nauseating.

Aching pain can be "referred" to another area. That is, the cause of the pain may originate in one area but be felt or experienced in some other location. Impulses from the painful area enter the spinal cord and connect with other nerve cells. These other cells are then stimulated to react as if they themselves were sensing pain. Such referred pain is of considerable concern to a physician as it makes diagnosis more difficult. Since it indicates overstimulation of the nerves, it may signal a more serious condition, perhaps of some internal organ.

Sensors in Other Parts of the Body

The types of receptors described so far are in the skin. Many of them, however, are also found internally, around the visceral organs and the joints. Pressure receptors in the muscles, tendons, and joints give us a sense of the body's position and its movements. Because of these receptors we can determine the position of a light switch even in the dark. Thus pressure receptors indicate to the brain how much the joints are bent and how high the arm is raised, so that we can judge the hand's position in relation to our "memory" of the switch's position.

An **itch** or **tickle** is not recognized by specialized nerve endings. These sensations are thought to result instead from a pattern of stimuli in several receptors. An itch seems to occur when chemical stimulation disturbs pain nerve endings in the skin. A light, moving stimulus, such as a moving feather just touching the skin, produces a pattern of responses in many touch receptors, resulting in a tickling sensation.

Regulation of Body Temperature

It is vital that the body temperature remains within very narrow limits. Variations of even one-half degree Celsius can affect our feeling of well-being. The body has automatic "thermostats" that register changes in body temperature and respond by producing other changes to regain normality as soon as possible. The maintenance of constant temperature is known as **homeostasis**.

(hy-poe-thal-ah-mus)

The major heat-regulating centre of the body is located just above the pituitary gland in the **hypothalamus** of the brain. One group of cells in this organ controls heat production; another group of cells is responsible for controlling heat loss. Nerve impulses transmitted from sensors throughout the body and the temperature of the blood reaching the brain are used by the hypothalamus to determine body temperature. If these indicators show that too much heat is being lost, some of the skin capillaries will close up (constrict) and some muscles may be activated to shiver, thus generating heat. Other stimuli may cause the skin to pucker into bumps of "goose flesh". These activities help to bring a lowered body temperature back to normal.

If the body gets too hot, blood vessels may expand (dilate), sweating may increase, and muscles will relax, thereby producing less heat. Unless we are ill and have a severe fever, the body is usually able to maintain a constant temperature with only minor variation. About 87.5% of body heat is lost through the skin surface and another 10.7% through the lungs. The remaining heat is lost by excretion of waste products from the body.

Deposits of fat beneath the skin help to maintain a normal body temperature by forming an insulating barrier to reduce heat loss. Women generally have a higher percentage of fat cells than do men. This fat serves to make female bodies more resistant to cold.

Under normal conditions, young people rarely suffer from **hypothermia**, which involves the lowering of the body core temperatures. However, extreme cold can be encountered during accidents and is a potentially fatal threat to anyone. Heat energy can be lost or gained by three different methods: **conduction**, **convection**, and **radiation**. For heat to be transferred by conduction there must be direct contact with another substance. If the body is in direct contact with the cold ground, without any insulating material present, for example, heat will be lost by conduction.

Convection requires the presence of a current of air or water. These substances absorb heat energy, become less dense, and as a result carry the heat away. Sweat evaporating from the skin surface is an example of heat loss by convection. Heat causes sweat (water) to turn to vapour. The heat required for this change is drawn from the skin and its loss rapidly reduces the body temperature.

The body absorbs heat by radiation when warm sunshine strikes the body. If you place your hand near the skin's surface when you have sunburn, or have been exercising heavily and are very hot, you can feel the body radiating heat.

If you are exposed to severe cold, there are certain basic principles which you should follow. These principles arise from the ideas discussed above. You may not always have the things you need to cope properly with an emergency situation. If you understand the problems involved, however, you may be able to improvise and avoid a potentially dangerous situation.

How to Conserve Body Heat in an Emergency

1. Find or make a shelter. Sit out of the wind. Cold, moving air quickly drains away the body heat.
2. Remove wet clothing. Body heat will be used up trying to dry wet clothing.

3. Put on *layers* of dry clothing. The extra clothes trap insulating layers of air and help to prevent heat loss. If the body is exposed to extreme cold and the body core temperature drops (hypothermia) it is necessary to raise the temperature of the skin before insulating it with more layers of clothing. This is similar in principle to placing frozen foods in an insulating bag. The layers, in this case, are keeping the cold in and preventing external heat from warming the frozen foods.
4. Keep moving. Heat is generated when muscles contract. Keep this within reason, you need to conserve energy, so sufficient movement to maintain good circulation is necessary without needless waste of energy reserves.
5. Stay awake. We generate less heat when we are asleep.

CASE STUDY. Helen, 18 years of age.

It was a miserably cold day with temperatures dipping below −30°C. A "bone chilling" wind was blasting across the frozen lake, whipping up trails of drifting snow. Several hundred people were impatiently awaiting the start of a 25-km cross-country ski tour. Helen, a tall, thin, athletic high school student, was determined to make a good showing in the race.

What to wear had been a source of worry for Helen. The weather dictated warm, wind-resistant clothing, but she knew she would quickly work up a sweat when the tour got underway. Her windbreaker had no vents and would not allow perspiration to escape and her upper body would quickly become wet with sweat. She decided to wear two layers of light clothing and hoped that she would not get too cold before the race started.

There were a few delays, and time was spent giving directions to the participants, but at last they were off. Helen quickly got into stride and felt she was making good time. After half an hour, she was perspiring heavily and shed her outer sweater and gloves, tucking them quickly into the small knapsack she carried. Later, she also took off her woollen cap. Along the wooded part of the trail, she had some shelter from the wind. By the time she had covered 15 km she began to wonder about the wax on her skis. She did not seem to be able to keep the rhythm she was accustomed to and she fell repeatedly.

Another skier came up behind and yelled "Track", then, as she passed, asked Helen if she was "okay", because she had noticed that Helen was stumbling and that she seemed poorly co-ordinated. Helen wasn't too pleased at what she felt was criticism of her skiing style and shrugged a reply which came out as a mumble.

She couldn't understand why her temperature seemed to vary. After working up the initial sweat, she had felt cold again. Now she felt warm and drowsy. Before long, Helen was having difficulty knowing how far she had come and how far she had to go. Soon her mind was a confused muddle. The last thing that Helen remembered about the ski tour was being held up and helped along by some other skiers. She eventually collapsed in the snow, unable to continue.

Helen awoke in hospital, her feet resting on a flannel-wrapped hot water bottle and there were other heating pads around her body. Slowly her mind began to function normally and she remembered what had happened.

Helen was a fit and experienced cross-country skier and she had difficulty accepting the fact that somehow she had got herself into serious trouble on the trail.

The extreme temperatures had drawn more heat from Helen's body than she realized. Without hat or gloves and with profuse sweating that produced rapid evaporation, her body had lost more heat than it could replace. Her skin became dull gray from lack of circulation and eventually the core (or inner body) temperature had dropped dramatically. The lack of blood to the brain had caused her confusion and disorientation. This also contributed to her loss of co-ordination, causing her to stumble and fall. Eventually, the muscles lost their strength, and drowsiness and eventual unconsciousness had occurred. If Helen had not received help, her heart might have stopped and caused her death.

Helen experienced a condition known as **hypothermia** in which the core temperature of the body drops to dangerous levels. To counter the effect of hypothermia, first of all, the core temperature must be raised. Warmth must be added to the skin surfaces, but not so rapidly that the blood vessels dilate excessively and cause even greater heat loss. Warm, non-alcoholic drinks can be used if the victim is awake and able to swallow. This will help to raise internal body temperatures.

Questions for research and discussion:

1. Find out the areas of the body that lose heat most rapidly.
2. Find out what types of clothing are best for withstanding cold, both during exercise and when less active.
3. Does cold have an effect on the lungs and respiratory passages?
4. In an emergency, what can you do to keep warm if exposed to severe cold?

Some Common Skin Features

Moles are small spots of pigmented skin. They are sometimes like small bumps and may have a hair or hairs growing from them. Moles are best left alone; constant irritation of the mole may affect the tissue and lead to a cancerous condition. Colour changes, bleeding, or rapid growth of a mole should be reported to a physician. If necessary, the mole can be removed.

Birthmarks are made up of unusual patterns of capillaries, heavily pigmented skin, or raised, bumpy layers of skin. Some disappear during childhood. Other types can be removed by cosmetic surgery.

Freckles are small harmless spots or patches of pigmented skin. The colour is a result of melanin formed by the skin cells as a protection from the harmful effects of ultraviolet light present in the sun's rays. People with red or blonde hair are more prone to develop freckles than people with other hair colours. Freckles usually begin to appear at about seven or eight years of age.

Some Skin Problems

Acne, which troubles many teenagers, is not necessarily caused by lack of cleanliness. Acne is caused by oily secretions of the glands becoming trapped below the skin. The area becomes swollen and red in fair-skinned people and may appear more bluish-purple in others. Often the site becomes infected and may result in permanent scarring of the skin.

Various types of food have been blamed for many of the acne problems suffered by adolescents. Recent research, however, has found no evidence to support the suggestions that chocolate or fatty foods, for example, are responsible for the condition.

The main culprit appears to be **androgens** – sex hormones. These hormones are first secreted at the age of puberty (the time at which both sexes become functionally capable of reproduction) and not long after that acne makes its appearance in some young adolescents. Many studies have shown that the quantity of androgens secreted by the sex glands influences the amount of sebum produced. It seems clear that the testes, ovaries, adrenal and pituitary glands all play a role in determining the activity of the sebaceous glands and the incidence of acne.

As yet, however, no reliable control for most types of acne has been discovered. Non-prescription acne "cures" have a poor record of success. If you have an oily skin, frequent washing of the face, neck, chest, and back, first with warm water and soap, then rinsing with cold water, is still the best method of keeping acne from spreading. (Never share a washcloth with others.)

Boils are round, tender, reddened areas of skin containing a central core of pus and bacteria (staphylococci). Some boils erupt and exude the central core before disappearing. Others simply regress and disappear. Boils should not be squeezed, as this may force some of the bacteria from the core into the surrounding tissue thus initiating other sites of infection. Boils should be covered to prevent the spread of the infection. A small boil will usually heal without treatment. Severe cases, involving several boils, should be treated by a doctor who may prescribe an antibiotic salve.

Athlete's foot is a fungus infection that begins especially in the warm, moist areas between toes. It produces a painful itching or burning sensation. The infection causes the skin to peel, leaving a red, shiny layer exposed. It may spread to cover the whole sole of the foot. Treatment requires that the foot be kept dry with pads of cotton and powder between the toes and frequent changes of fresh socks. Careful drying between the toes after a bath or shower is important. The application of non-prescription remedies is usually sufficient to cure mild cases. Serious cases of athelete's foot require medical attention. The fungus can be picked up in public showers, in gyms, or in swimming pools. These centres of activity should receive regular cleaning with disinfectants. The use of other people's infected footwear or socks can also result in infection.

Dandruff is caused by shedding the outer layers of the scalp, which produce rather obvious white flakes of skin. Poor circulation can increase the rate at which this shedding takes place. Infections, poor diet, lack of regular washing, strong shampoos, or insufficient rinsing of the hair after washing, may all promote this scalp condition.

Care of the Skin

The best care for skin involves regular bathing to remove the secretions of the sebaceous and sweat glands. Because these secretions are of an oily nature, dirt clings to the skin surface. The use of soap helps to free the skin of these oily substances.

Skin care is mostly a matter of common sense. If the skin is dry, it should be lubricated. Simple oils, such as mineral oil or baby oil may be used. Excessive exposure to the sun should be avoided. It can increase the risk of skin cancer and also ages the skin quickly. Smoking, too, can act to age the skin of the face.

Skin thrives on stimulation. Showers, massage, and brisk rubbing with a towel are excellent skin treatments.

The skin works hard! It renews itself about every 3 or 4 weeks! This constant production of new cells to replace dead and dry layers of epidermal tissue requires a rich supply of nutrients. Proteins, leafy green vegetables, whole wheat grains, dairy products, such as milk, fresh fruits, and other vegetables are excellent foods for the skin, as they are for the rest of the body.

QUESTIONS FOR REVIEW

SOME WORDS TO KNOW

Match each description given in the left-hand column with a word shown in the right-hand column. DO NOT WRITE IN THIS BOOK.

1. Thick underlying layer of the skin containing nerve endings, glands, fat cells, etc.
2. Colour pigment found in the skin.
3. Outer layer of dead epidermal cells.
4. Tubular structures that secrete sebum.
5. Layer of cells that constantly replaces the outer layers of dead epithelium.
6. Protein that hardens the cells of the skin, nails, and hair.
7. Tissues that line the surfaces of the body and organs.
8. Glands that secrete oil.

A. hair follicle
B. keratin
C. dermis
D. epidermis
E. stratum germinativum
F. stratum corneum
G. melanin
H. hypothermia
I. sebaceous gland
J. sweat gland
K. acne

SOME FACTS TO KNOW

1. List seven functions of the skin.
2. Explain how and why skin cells are replaced.
3. In what ways is your skin unique and different from that of other people?
4. What causes body odour?
5. Make a chart showing the different sensory nerve endings found in the skin and give their functions.
6. What changes take place in the skin
   a) when you are angry?
   b) when you are very hot?
   c) when you are very frightened?
   d) when you have been out in the hot sun for several hours?
7. Why do the numbers of sweat glands and the numbers of sensory nerve endings vary from one place in the body to another?
8. Why is the elbow rather than the hand a better part of the body to use when testing the temperature of a baby's bath water?
9. List four types of cells found in the skin and explain how the structure of each cell is related to its function.
10. Explain why the skin is classified as an organ.
11. Explain the difference between sebaceous and sweat glands.

QUESTIONS FOR RESEARCH

1. Select one of the following topics and prepare a report on it:
   - dermatologist
   - beauty treatments
   - allergies and problems with cosmetics
   - face lifts
   - birth marks and their removal
   - acne
   - psoriasis
   - cosmetologist
   - skin care
   - skin disorders
   - warts
   - smoking and premature skin aging
2. Select one of the skin sensation experiments from the following Activities section and read about the topic in depth. Design an experiment to show how a particular sensation may vary from one person to another. Graph the range of the senstion using a reasonable sample of people, and determine the upper and lower thresholds where appropriate.
3. Research the role of the skin in controlling body temperature. Select a person who smokes and another (a control) who does not. Then determine how smoking affects skin temperature. Graph your results. (Take the temperature of the hand at regular intervals for about half an hour after the cigarette has been smoked by the person who smokes. Take similar readings with the non-smoker control.
4. Determine what causes "goose flesh". Find out what will cause this reaction in people. You might also investigate how it is caused and used in other animals.
5. The most effective sun screen is reported to be PABA (para-amino benzoic acid). It is found in varying amounts in many sun creams. Do a survey of the creams available and of the data available on their effectiveness. Can you think of a means of testing their effectiveness by using several creams on your own arm?

Activities: THE SENSE OF TOUCH

We use our skin sensations continuously to judge hardness, softness, temperature, texture, and the shape of objects. We depend upon them to help us with our daily routines and to warn us of potential dangers. In these experiments you will investigate the sensory receptors in the skin.

Activity 1: TEMPERATURE

Materials (for Activities 1, 2, 3, and 4)
Straight pin, small finishing nail, beaker of hot water (60°C), beaker with an ice cube, a pair of dividers (mathematical), ruler, cotton thread, soap and warm water, rubber stamp (2.5 cm^2), divided into 100 equal squares.

Method

1. *Stamp two grids in your lab notebook. Label one "hot" and the other "cold".* Stamp a grid on the hairless inner surface of your wrist.

2. Heat the head of a small finishing nail in a beaker of hot water (about 60°C). Remember to reheat the nail frequently. First try touching the hot nail to several parts of your arm and you will be able to find a "hot" reaction in some spots, but little more than an awareness of pressure in others. Now methodically touch the centre of each of the squares in the grid on your wrist. *Mark an X in the appropriate square in your lab notebook, every time you feel a "hot" sensation.*
3. Using a small beaker with an ice cube, cool down the nail as much as possible. Then proceed as before, testing each square of the grid on your arm and *recording the "cold" spots in the grid in your notebook.*
4. Select another area of your body, perhaps the calf or neck, and test again.
5. *Count and record the number of X's in each grid.*
6. *Examine the grids and decide if the sensory endings form a regular pattern or are quite randomly arranged.*
7. *Can you discover an approximate mean distance between the sensitive points? (You may find that they are in small groups.)*
8. *Did you get the same results when you tested other parts of your body? Were your results similar to your partner's?*
9. *Why should the temperature of the bath water of infants be tested with the elbow rather than the hand?*

Activity 2: PAIN AND PRESSURE

Method

1. Wash the area to be tested with soap and warm water. Keep the area moist during the experiment.
2. Stamp a grid on your arm and *two grids in your notebook.*

3. Use an ordinary straight pin and, holding it upright, press it lightly against the skin. The pin should make a small depression but not pierce the skin! You will feel the pressure easily but in some squares there should be a much sharper "pain" response. *Mark an X in each square of the grid in your notebook to correspond to the "sharp" responses.*
4. Repeat using the head of the pin. Try to distinguish between dull responses and distinct sensations.

Compare the number of X's in these grids with those for heat and cold. What differences can you discover?

Activity 3: TACTILE DISCRIMINATION

Method

1. Close your eyes. Draw a cotton thread across the palm of your hand, first with your eyes closed, then with your eyes open.
   Record your impressions.
   Now draw the thread across the back of your hand, or the back of your wrist.
   Record the differences and explain why these differences occur.

Activity 4: TWO-POINT SENSIBILITY TEST

This test will enable you to discover whether some areas of your body are more sensitive than others.

Method

1. Extend the legs of the dividers until they are about 5 cm apart. (Dividers found in mathematical sets are quite suitable. Sterilize the tips in alcohol.)
2. Work in pairs for this experiment and have your partner close his or her eyes. You will be touching your partner with the dividers, and he or she will try to tell you if you are touching the surface of their skin with one point or both points at the same time. Try to trick your partner occasionally by using only one point.
3. Gradually decrease the distance between the points until they can no longer accurately distinguish whether you are using one point or two. *Measure the distance between the points in centimetres and record this distance in your notebooks.*
4. Test the following areas of the body: arm, back of the hand, fingertip, cheek, and lips. Steady the hand as much as possible so that the points make contact at the same time, rather than one after the other. If you want to pull your shirt out and try the middle of your back, you may have to make the points of the dividers even wider than 5 cm!

Examine the results that you have recorded and try to explain why the areas differ in sensitivity.

Activity 5: WHAT STIMULI ARE REQUIRED TO RECOGNIZE DIFFERENCES IN SIMILAR OBJECTS?

Materials

About 20 pairs of similar objects are prepared; each pair will differ in at least one characteristic. In order to determine what the differences are, the senses must be used.

Examples

a. Two pieces of paper, both white, but varying in texture.
b. Two pieces of cloth, the same colour and material, but one soaked in vinegar (and dried), the other in ammonium carbonate.
c. Two sealed boxes containing different packed weights.
d. Two clear, colourless liquids differing in taste.
e. Two sealed flasks, one containing hot water, the other cold water.

The objects should only vary to a small degree; some may require more than one of the senses to identify the difference. Do not forget sound, mass, and size variations, and remember that the sense of touch is made up of several kinds of sensations.

The class may be divided into two groups for this experiment. Group A will be permitted to use any or all of the senses they wish to identify the objects. Group B will be limited by blindfolds, or by not being able to use their hands. Decide exactly what senses you will restrict before you start.

Method

Group A

1. *Examine each of the pairs of objects provided and record the information suggested in a suitable table in your notebook.*

Items	*Differences A*	*Sense receptor employed*	*Part of body used*

Group B

Exclude one of the senses, such as vision, by using a blindfold, or allow your partner to touch the objects behind his or her back. *Record the differences recognized in the same type of table as used by Group A.*

2. *Compare the accuracy of the results when all of the senses were used, with those obtained when only some of the senses were used. Express the comparison as a percentage.*
3. *Which of the senses were used the most? Grade the senses in order of their usefulness in this test.*
4. *How many of the items required more than one of the senses to accurately identify the difference?*
5. *Make a list of some stimuli which are beyond the normal threshold of our senses. For example a dog whistle or certain wavelengths of light.*

6. *List all the senses, then state what variations can be recognized by that sense. For example: Sound: we recognize variations in pitch, loudness, and quality of the sound.*

Activity 6: AN EXERCISE IN OBSERVATION

The surface structure of the skin.

Materials
dissection microscopes, centimetre ruler (transparent).

Methods
1. Examine the skin surface of the palm of the hand with a binocular microscope. *Describe the magnified appearance of the ridges and furrows present.* Place a ruler across the ridges and count the number of ridges present in one centimetre. *Compare your result with those of other students. What useful purpose do the ridges perform?*
2. Examine the back of the hand. Can you discover any regular patterns there? *Describe what you see.*
3. *Is the skin on the back of the hand firmly attached to the underlying tissues, or can it be lifted away? How does this compare with the attachment of the skin on the palm of the hand? Try to explain the differences that you find.*
4. Look closely at the ridges on the finger tips and on the palm of the hand under the microscope. Rock the finger back and forth, or vary the position of the light, while you look for beads of perspiration. *Can you see the openings of the sweat glands? Describe their appearance and exact location.*
5. *Where does the skin appear to be thickest? Where is it least thick? Why does the skin vary in thickness?*
6. *Is it possible to cut a thin slice from the surface of the thick cornefied pads on the palms of the hands without drawing blood or causing pain? (No need to operate, use your experience!) Explain your answer.*
7. Examine any cuts or scars that may be present on your skin. *Describe the appearance of these features.*
8. If your skin is dry, you may find scaly patches on the surface of the skin. *Examine this tissue and explain how the condition is caused. Is this a natural feature of the skin?*
9. On the back of your hand you will find fine hairs. Examine the point at which the hair protrudes from the skin. *Describe the area immediately around the hair and the angle at which the hair leaves the skin.*

Some other features you may wish to examine are: fingernails and the cuticle, a hair shaft (use a compound microscope – can you tell how blondes differ from brunettes?) a hair root, split ends, an eyebrow hair, a bruise, a freckle, or a mole.

Activity 7: TEMPERATURE STIMULATION

Materials required
beakers, warm and cold water.

Method

1. Immerse the index finger of one hand in some warm water for two minutes. Now dip the other index finger into the same container of water.
   What differences in sensations can you distinguish between the two fingers? Record your observations.
2. Immerse one index finger in a beaker of warm water and the other index finger into a beaker of very cold water. After two minutes, change the beakers over and dip your fingers into the water of the opposite temperature.
   Record the sensations received.
3. Again place the finger of one hand in the warm water and the finger of the other hand in cold water. After two minutes immerse both fingers in another beaker containing cool water. *Explain the sensations experienced.*
   You may also wish to try placing your elbow in a tray of ice water. Keep it there until you feel sensations in the fingers.
   Describe the progression of sensations and exactly where these sensations occur.

Activity 8: DISTRIBUTION OF SWEAT GLANDS ON THE SURFACE OF THE SKIN

Materials
Iodine solution: 0.01 mol/L, corn starch, small paint brushes, dissection microscopes or hand lens, small pieces of bond or glazed paper.

Method

1. If your hands feel sticky or sweaty, wash them and dry them well.
2. Paint a small square with iodine on the palm of the hand and another on the wrist, neck, or forehead. The area should be about 1.5 cm^2. Allow the iodine to dry.
3. If your hands are cold, shake them vigorously or clap them together to warm them. Dust the painted areas with corn

starch and leave this on for several minutes. Blow off the loose starch.

4. Press the painted area of your hand firmly against a page taken from your notebook. Hold it there, without moving, for about thirty seconds. Use a small piece of paper for the neck or forehead prints. *Label the prints you have made to show from what part of the body they were taken. Rows of small dots should appear on the paper. Select a part of the print in which the dots show up clearly and draw a* 1-cm *square around them. Count the number of dots in each square and record your results. Which areas have the most sweat pores?*
5. *Examine the prints and the painted areas of your skin carefully. Are the dots found at regular intervals? Are they present on the ridges or in the valleys of the lines on the hand? Try a fingerprint if you are not sure. Describe how the patterns on the hand differ from those on the wrist or forehead.*
6. Try the same test after light exercise (i.e., running in place for two minutes). *Examine and compare the results with the earlier tests. Is there any change in the number and size of the dots? Explain your observations.*

   Why should bond paper be used for the tests. If you are not sure try to do the test with a piece of paper towel.

   Although sweat glands are not found in equal numbers on all parts of the body, a rough estimate of the total number of glands can be made as follows. You need two pieces of data. 1. The number of sweat glands in one square centimetre (use your lowest score from the tests above). 2. The total surface area of the body. A person with a body mass of 55 kg *and* 155 cm *in height has a surface area of about* 1.5 m^2. *Find the approximate number of sweat glands in a person of this size.*

Activity 9: THE EFFECT OF EXERCISE ON BODY TEMPERATURE

Method

1. Take your temperature after sitting quietly for at least 5 min. *Record your temperature.*
2. Run in place or use a step test box and do 2 min light exercise. *Record your temperature.*
3. Run around the building or arrange to carry out this exercise during a gym class. Take your temperature after heavy exercise. *Record your results.*

   (Note: If your doctor has advised against your participation in physical education classes, do not take part in this experiment.)

4. Allow at least 10 min to elapes while sitting quietly at your desk and again take your temperature and *record the result. Compare your results with others in your class and record the average changes and the range of results obtained.*

TEMPERATURE CHART

	At rest	*After light exercise*	*After heavy exercise*	*After* 10-min *rest*
Student				
Class average				

What caused the changes in temperature?
Account for the results that you obtained after 10-min *rest. What correlation is there between the type of exercise and the change in temperature. Explain the value of warm up exercises and cooling down exercises; remember they concern a great deal more than just temperature.*

Unit III
How the Body Is Supported and Moves

Chapter 4

Bones and Joints

- The Skeletal System
- The Skeleton
- The Joints
- Skeletal Injuries and Disease

The Skeletal System

A casual examination of bone gives the impression that it is dead material. Actually, bone is far from dead and inactive. In spite of its solid and rigid appearance, it is constantly changing, growing, and being reconstructed. It is supplied with blood vessels and nerves just like any other body tissue. The bones also produce more than a million red blood cells per second.

All organisms require structural support, although in some organisms that live in water the need is quite limited. The arthropods (insects, crayfish, etc.) possess an exoskeleton for support. This is a rigid outer case that provides excellent protection as well as support. However, the exoskeleton severely restricts growth. At times during growth, the exoskeleton must be shed; at these times the organism becomes extremely vulnerable to predators. (Soft-shelled crabs, a restaurant delicacy, are crabs that have recently shed their exoskeletons.) The vertebrates (fish, birds, mammals, etc.), with an internal skeleton, have less protection, but much greater potential for growth.

The skeletal design and muscle systems of the human body permit the supple grace of a ballet dancer, the speed and endurance of an olympic swimmer, and the power and strength of a hockey player or a gymnast.

Functions of the Skeletal System

The skeletal system is a remarkable and versatile bony structure which performs a number of important functions. It provides a framework for the attachment of muscles which support the body and make movement possible. Bone protects some delicate internal organs, such as the brain. It produces blood cells and also acts as a reservoir for the storage of minerals such as calcium and phosphorus. The entire skeleton acts as a built-in shock absorber taking the brunt of impact when a person is jumping or receiving hard knocks.

The Structure of Bone

Bones vary greatly in size and shape. Some, such as the bones of the middle ear, are extremely small. They are less than a centimetre in length. Others, such as the femur in the upper leg, are long and comparatively heavy.

There are two types of bone. **Compact bone** is hard and very strong. **Spongy bone** is lighter and full of tiny spaces; in fact, it looks like a sponge. Spongy bone is also very strong but contains many more blood vessels than compact bone.

Neither compact nor spongy bones are completely solid, however. If they were, bones would be so heavy that movement would become difficult. In the centre of most bones is a hollow canal containing the bone **marrow**. The bone marrow is made up of many blood vessels, fat cells, and some blood forming tissues. Covering the bone is the **periosteum**, a thin, double membrane that contains blood vessels, nerves, and cells which form bone. It is this membrane that controls the development of bone. (See Figure 4.1.)

(per-ee-os-tee-um)

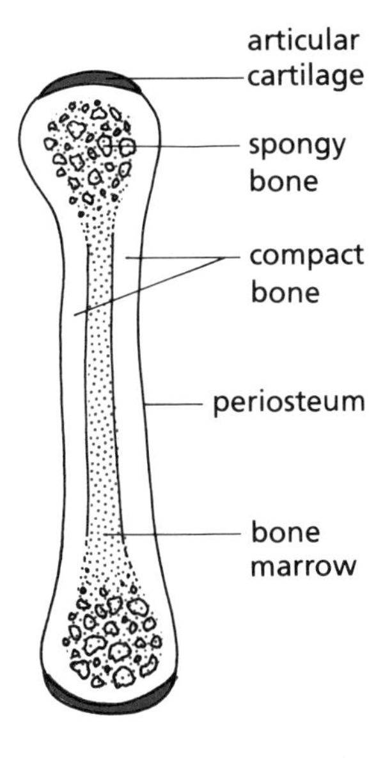

articular cartilage
- protection in joints

spongy bone
- porous, light bone

compact bone
- hard and strong, major support feature

periosteum
- a tough membrane containing blood vessels, nerves, and bone-forming cells

bone marrow
- soft centre, contains blood vessels. Red-blood-cell production site in flat bones. Some yellow marrow, which stores fats, is also present.

Figure 4.1.
A longitudinal section through a bone to show the internal structure.

Bone Formation

In the embryo, a model of the future skeleton is produced and is made up of cartilage and fibrous tissues. This model acts as a pattern for each bone as it is produced. Even at two months of age, the embryo may start to develop bone cells or **osteocytes** in this mold (Figure 4-2). Once started, the process of ossification (bone formation) continues throughout life, although the major part of it occurs during our first twenty years or so.

(os-tee-oe-site)

As they develop, bone cells become buried in a mixture of minerals, mainly calcium and phosphorus. The cells are

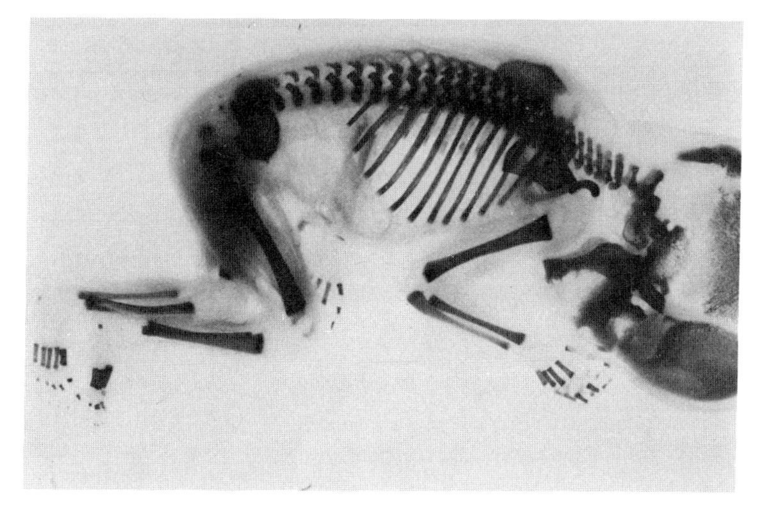

Figure 4.2.
Bone formation in a fetus of about 16 weeks. Note that the bones do not meet, but are extending from the centre of the bone as they grow.

separated within this matrix of mineral deposits, but are connected to each other by tiny canals called **canaliculi**. The canaliculi deliver blood supplies to the bone cells and are connected to large tubelike vessels, the **Haversian canals**. In compact bone, tissue is laid down in a circular fashion resembling the cut section of a tree with its annual rings. (See Figure 4.2.)

The *Haversian system* (see Figure 4.3) connects with large blood vessels that bring new supplies of oxygen, as well as

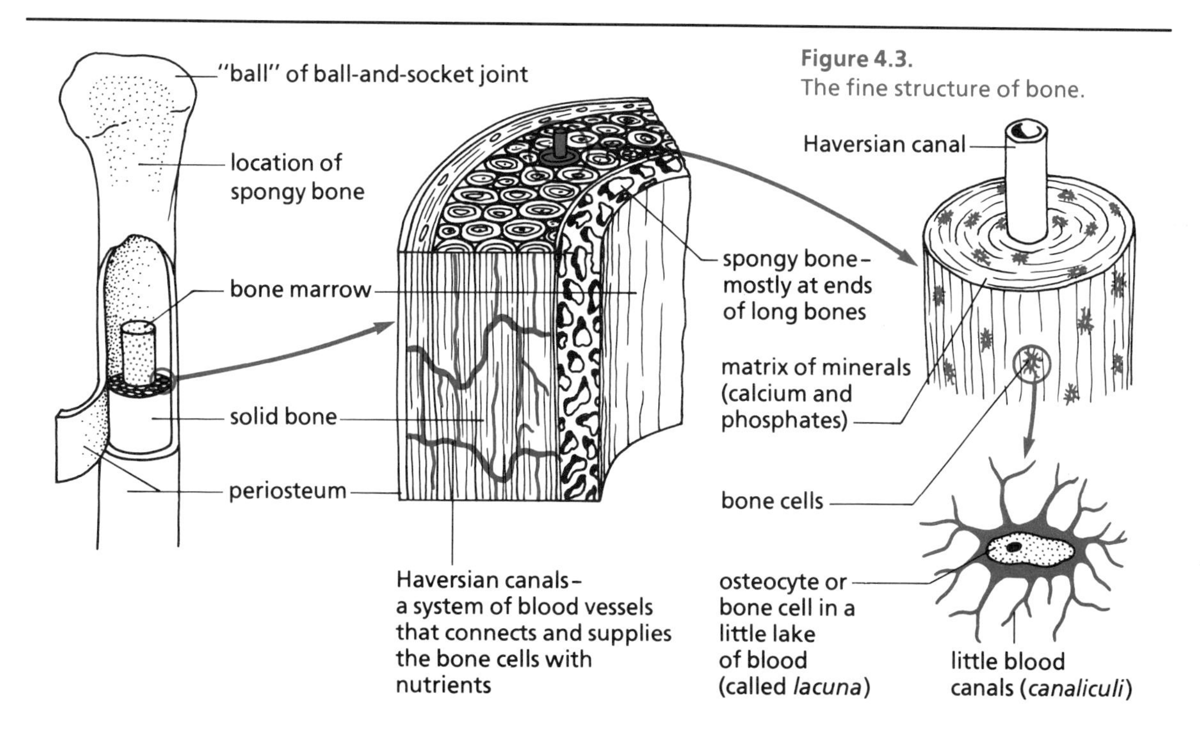

Figure 4.3.
The fine structure of bone.

calcium, phosphorus, and other essential nutrients to the bone tissue. Blood vessels also carry away carbon dioxide or cells produced in the bone marrow. The minerals, too, may be taken away if they are urgently needed elsewhere. For example, if pregnant women eat an inadequate diet (low in quantities of milk for instance), the bone tissue may break down to supply needed minerals for the baby. A mother with a young baby may find that her teeth need special attention after the baby is born. The demands of the growing fetus for calcium may have had priority over the needs of her own body.

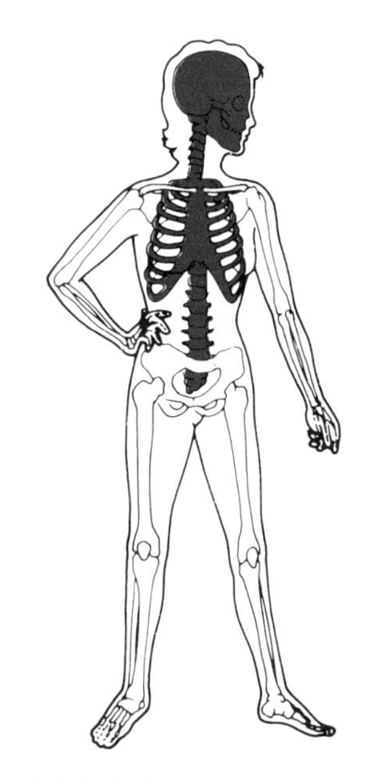

axial skeleton

The Skeleton

The skeleton is made up of two parts.

The axial skeleton contains the following bones: the skull, including tiny bones of the ear (29 bones), the vertebral column, sacrum, and coccyx (26 bones), the rib cage (12 pairs of vertebrae plus the sternum).

The appendicular skeleton includes: the arm, hand, and pectoral girdle, including the scapula (shoulder blade) and clavicle (collar bone) (32 x 2 bones), the leg, foot, and pelvis (31 x 2 bones).

For most of us, it is more important to understand the function of bones, than to memorize their names. However, the names are useful for identifying bones; knowing these names therefore reduces the amount of description needed to identify each bone. (See Figure 4.4.)

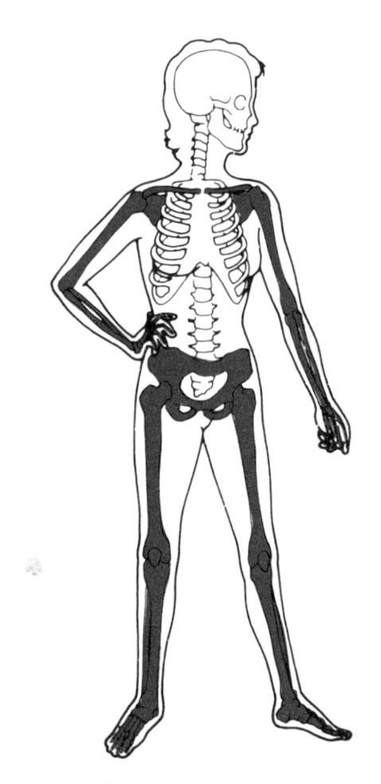

appendicular skeleton

The Bones of the Skull

At first, the skull appears to be made up of just two bones, a large irregular ball-shaped part and a loose fitting movable jaw. In fact the skull is made up of many bony plates joined permanently together by **sutures**. If you look at the diagram of the skull, you can see some of the wavy, zig-zag lines. (See

Figure 4.4.
The major bones of the skeleton.

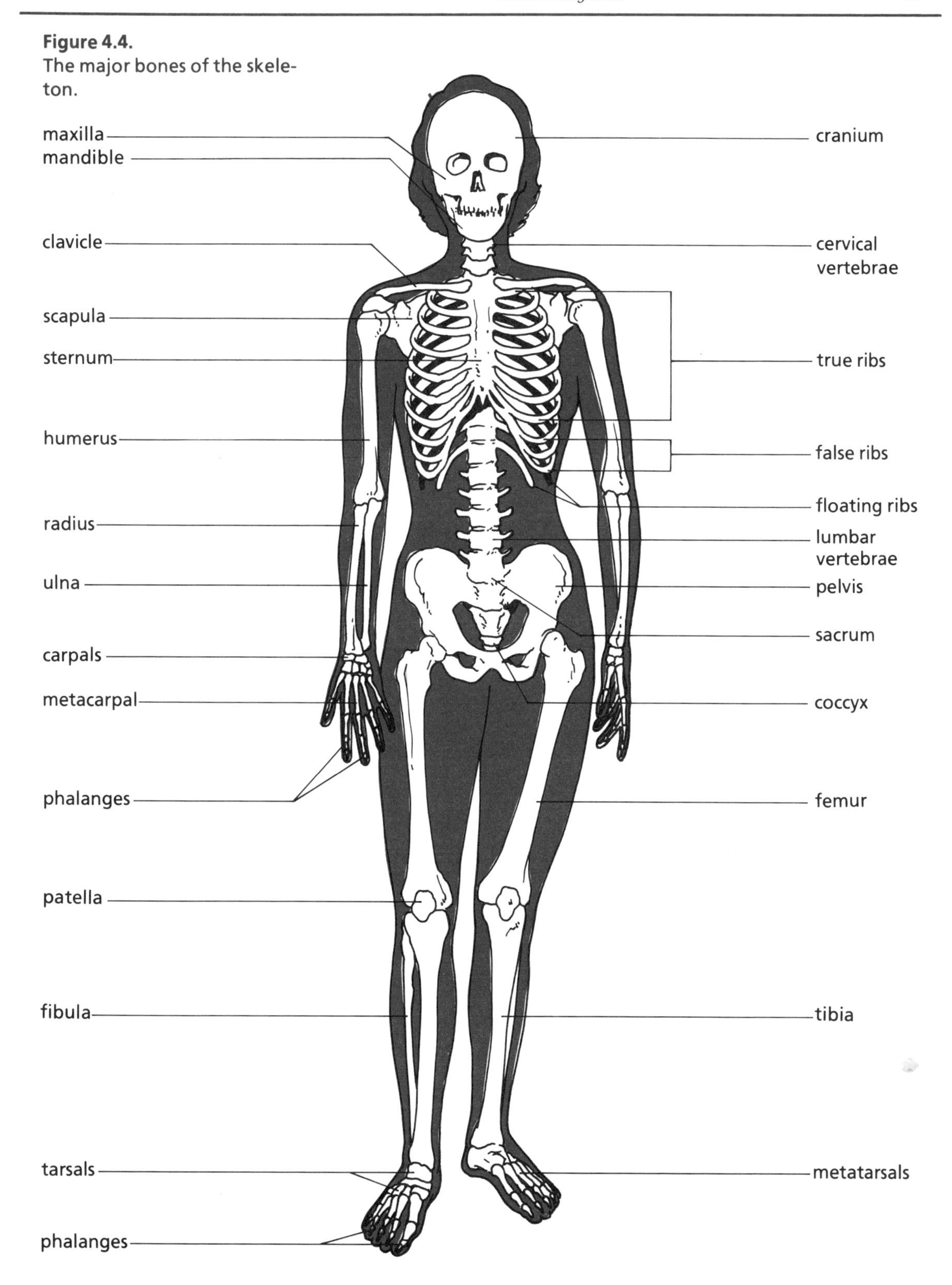

Figure 4.5.) These are suture lines, which mark the points where the bony plates have fused together. In some places, where three plates come together, there are small triangular depressions. In a new-born baby these indicate areas in which the separate bony plates have not yet completely joined. It is important the the skull bones do not become fused too early in life. During birth the separate plates can glide over each other slightly without injuring the brain. The skull must increase its size after birth to make room for the growing brain. When brain growth is complete the bones

Figure 4.5.
The major bones of the skull.

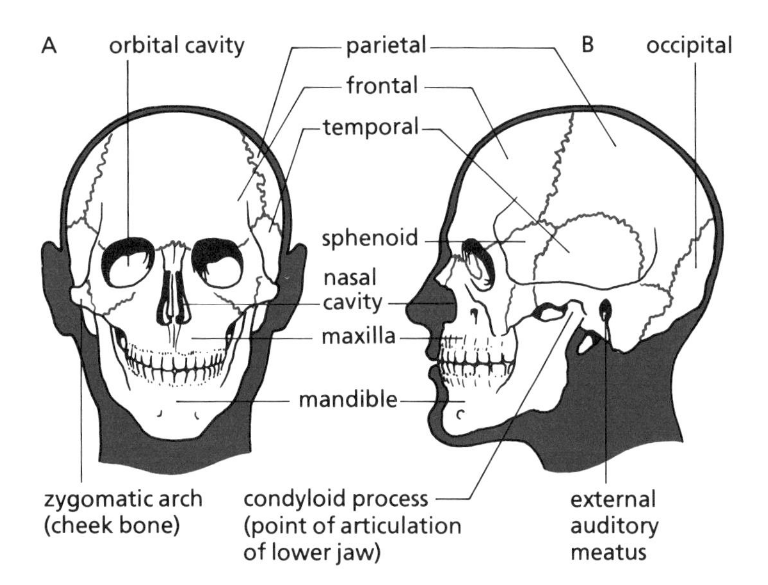

Front and side views of the skull showing the major bones and suture lines.

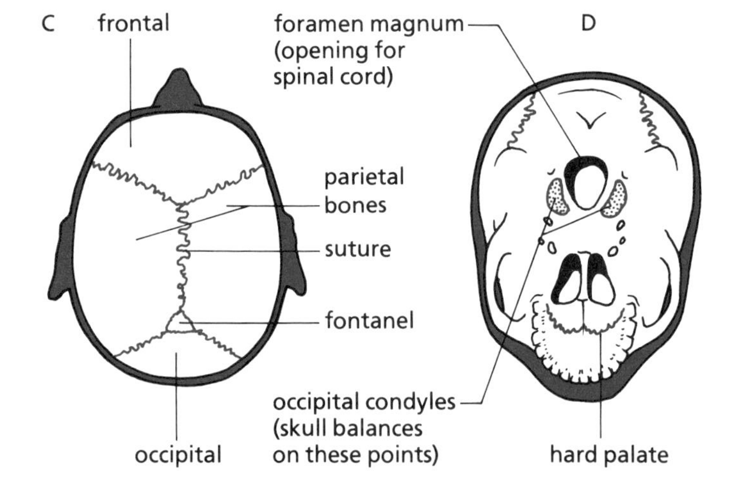

A top view of the skull showing the principal sutures joining the bones of the cranium. The right-hand diagram shows the two occipital condyles which rest on the top bone of the spinal column. It also shows the large opening through which the spinal cord passes in to join the brain.

fuse. If you feel along the centre of your skull, moving your fingers back towards the crown, you can probably still find a small depression called the **fontanel**. (See Figure 4.6.)

The rounded surfaces of the skull are quite thin, but surprisingly effective in protecting the brain. A curved surface resists impact more effectively than a flat surface since curved surfaces tend to cause blows to glance off.

The skull can be divided into two regions: the **cranium**, which contains the brain, and the **facial bones**, which provide a form for the eyes, ears, mouth, and nose.

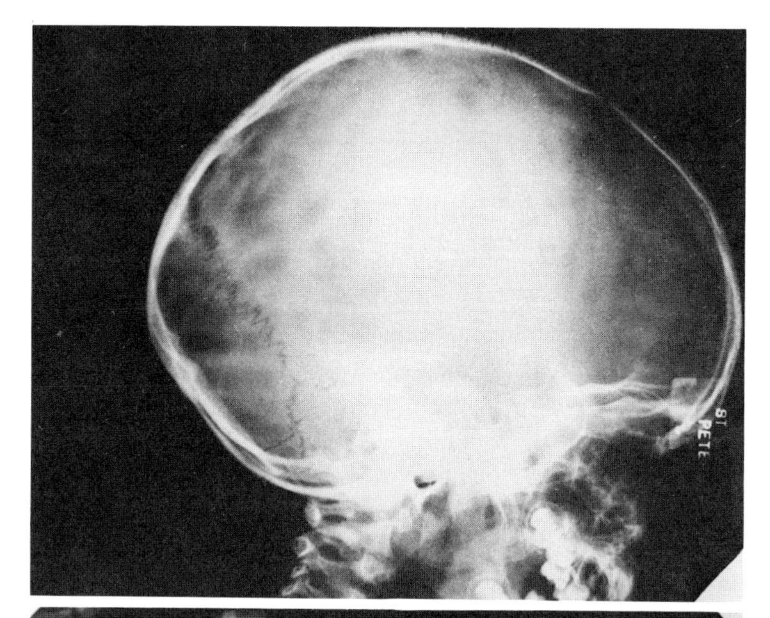

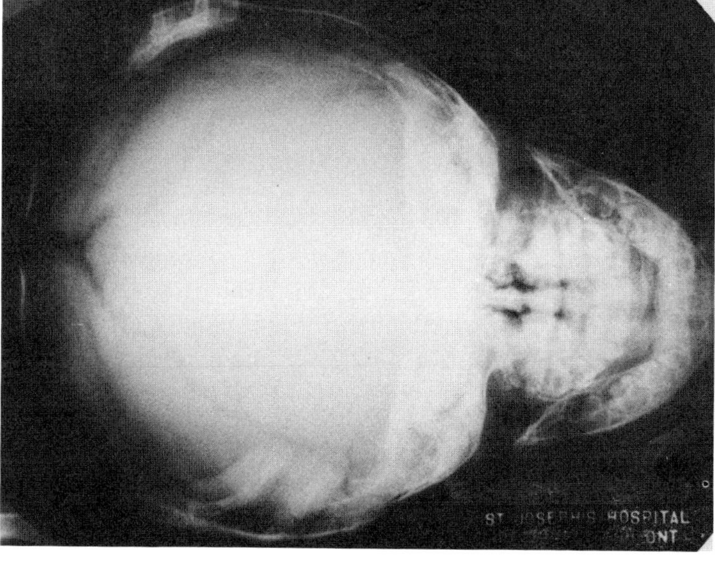

Figure 4.6.
X-rays of the skull. Note the fine suture lines that indicate the fusion of the cranial bones.

THE DISCOVERY OF X-RAYS

In 1895, Wilhelm Roentgen was performing experiments using a vacuum tube, passing an electric current through it. He noticed that a specially coated fluorescent paper nearby began to glow. He found that the paper glowed brilliantly when he held it near the tube (called a cathode ray tube) even when the cathode tube was covered with black cardboard. He realized that the paper was being illuminated by invisible rays that could pass through substances such as the cardboard that light could not penetrate. Roentgen found that the rays would pass through light metals but not heavy ones.

Dr. Roentgen called the unknown rays X-rays and began further experiments. He found that if he placed his hand between the cathode tube and the fluorescent paper that he could see the bones of his hand. Although he did not know it at the time, this experiment was extremely dangerous as the rays could cause serious burns, ulcerations, and cancer. Some early experimenters had to have their fingers or a hand amputated because of the damage caused by exposure to excessive radiation.

Modern X-ray tubes use a tungsten filament heated to a high temperature by high voltages (between 25 000 and 2 000 000 V) passed through it.

There are two major types of X-ray techniques used in medical practice today. **Fluoroscopy** provides a means of direct observation of an organ as it is functioning or some substance as it moves inside the body. A dense dye is used to fill a cavity or tube such as the digestive tract. The patient is placed between the X-ray source and the fluorescent viewing screen. As the liquid passes through the digestive tract any irregularities can be observed by the doctor watching the screen. The dye blocks the path of the X-rays making a contrast with the less dense tissues around it. The tissues without the dye allow the X-rays to pass through the body and strike the screen.

In **radiography**, a photographic film is used, which provides a permanent record for the doctors to examine. The film is exposed by X-rays. Tissues such as bone are much more dense than softer tissues, such as muscle. The greater the density the fewer X-rays can pass through the tissue and cast a "shadow" on the film.

The use of X-rays allows doctors to see inside the body and diagnose many types of internal problems. Fluoroscopy can be used to diagnose disorders of the heart and vascular system, the lungs, and the digestive tract. Radiography is more commonly used to determine bone injuries, dental decay, and digestive tract disorders.

The Cranium

(oc-sip-i-tal)
(pah-rye-e-tal)

In Figure 4.5c the **frontal**, **occipital**, and two **parietal** bones meet to form the top of the cranium. The frontal bone comes down and forms the heavy ridge above each eye. These ridges protect the eyes from large objects. The occipital bone curves down at the back and continues underneath the brain where there is a large opening, the **foramen magnum**. This opening provides a passageway for the spinal cord to connect with the brain. (See Figure 4.5d.) There are also two bony bumps on either side of this opening, the **occipital condyles** which help to balance the skull upon the atlas bone of the spinal column. The **temporal** bones contain the ear canal (auditory meatus) and articulate with the lower jaw. The **sphenoid bone** helps to bind the other cranial bones together

(sfee-noid)

and forms part of the cranial floor. The **ethmoid** bone (not shown) forms part of the cranial floor and surrounds most of the nasal cavity.

(eth-moid)

The Facial Bones

The facial bones support and protect the organs of sight, hearing, and smell. The prominent ridge of the **zygomatic arch** gives shape to the cheeks and helps protect the eyes. The **maxilla** forms the upper jaw, which is part of the skull and does not move even while you are chewing or talking. The lower jaw, the **mandible**, is hinged at the back (at the condyloid process) and performs all the active work when we eat or speak. (See Figure 4.7.)

(zy-goe-ma-tik)

(mak-sill-uh)

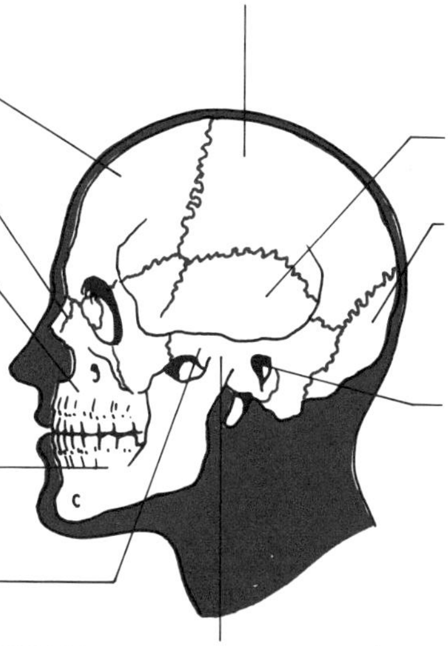

FRONTAL BONE forms the arches over the eye socket and determines the shape of the forehead.

PARIETAL BONES, one on each side, meet at the top of the skull, form most of the sides and top of the head.

NASAL BONES form the bridge of the nose and vary in shape in different people.

MAXILLA is a fixed bone, and does not move for speech or for chewing.

MANDIBLE is the only major, freely moving bone in the skull. (There are small moving bones inside the inner ear.)

ZYGOMATIC BONES give shape to the cheeks and form lower part of the eye socket.

CONDYLOID PROCESS, one on each side, point at which the lower jaw pivots with the temporal bone behind the zygomatic arch.

TEMPORAL BONES form the lower sides of the skull.

OCCIPITAL BONE forms the back of the skull and curves under where it has a hole through which the spinal cord passes to connect with the brain.

AUDITORY MEATUS is the opening for sound vibrations to reach the inner ear.

Figure 4.7.
The functions of the main bones of the skull.

The Vertebral Column

The vertebral column, also called the spine, is made up of 26 small, irregularly shaped bones. Each bone, called a **vertebra**, has a round drum-shaped body with three winglike projections, two lateral and one projecting at the back. The vertebra are tied together by bands of ligaments.

(ver-te-brah)

Figure 4.8.
The bones of the vertebral column.

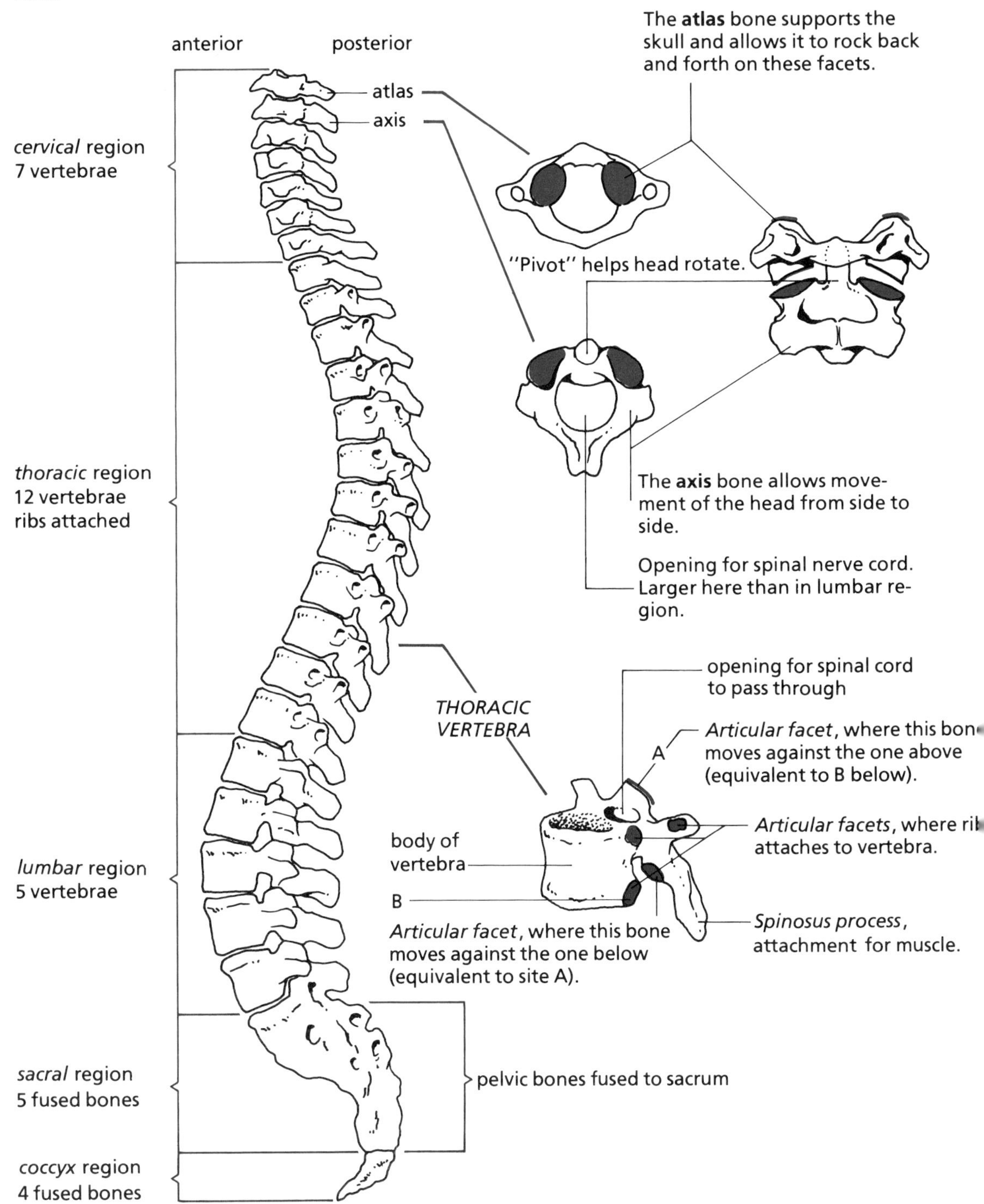

In the middle of each vertebra is a hole, through which the spinal nerve cord passes (rather like a bony necklace on a string). The vertebrae protect this vital nerve cord from knocks, but, because each vertebra can move a little and slide sideways, a severe blow can cause a shearing action and damage the cord. If the cord is severed, paralysis of the body below this point would result. For this reason it is most important not to move an accident victim if any back injury is suspected.

The Cervical Vertebrae

(ser-vi-kal)

The top two vertebrae below the skull have special functions. (See Figure 4.8.) The uppermost one, called the **atlas**, supports the skull on two flat discs and allows the head to move up and down, making "yes" movements. The head is held up by muscles when we are awake, but if we "nod" off to sleep these muscles gradually relax and the weight of the jaw and face causes the head to gradually drop down until the chin rests against the chest. The second bone is the **axis** which allows side-to-side movements of the head, such as the "no" movements. The atlas and axis plus five other cervical vertebrae make up the bones of the neck. These last five vertebrae provide only very limited movement in comparison to the atlas and axis. (See Figure 4.8.)

Thoracic Vertebrae

(thor-ass-ick)

There are 12 thoracic vertebrae, each of which has 2 processes. These processes extend from either side of the vertebra and connect with the 12 pairs of ribs to form the thoracic cage.

The ribs are flat curved bones. The first seven pairs are attached by cartilage to the **sternum** at the front of the body. (See Figures 4.9 and 4.10.) The next three pairs lie below the sternum; each pair is attached by cartilage to the pair of ribs above. The last two pairs are called floating ribs, because they do not complete the circle and are not attached in front. For this reason they are more easily broken; even an exceptionally hard hug can crack one of these springy, floating ribs.

Lumbar Vertebrae

The five lumbar vertebrae are easily recognized by their large, thick bodies. These vertebrae are located between the last vertebra, which is attached to ribs, and the top of the

Figure 4.9a.
The ribs, sternum, and thoracic vertebrae from above. Note how far the vertebrae extend into the cavity formed by the circle of ribs.

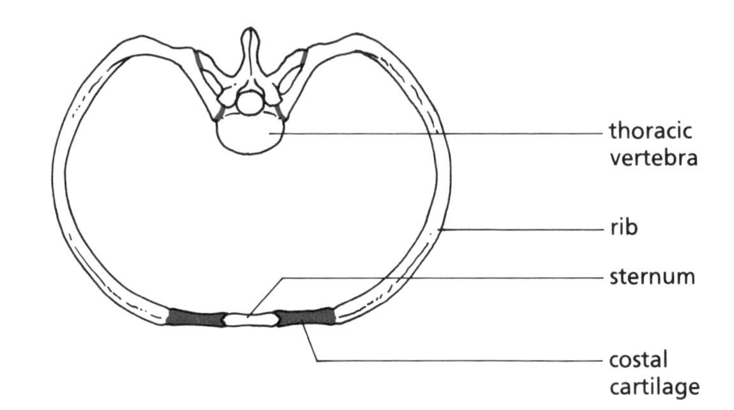

Figure 4.9b.
The sternum and ribs. The costal cartilages connect the ribs to the sternum and offer some flexibility to the movement of the rib cage during breathing.

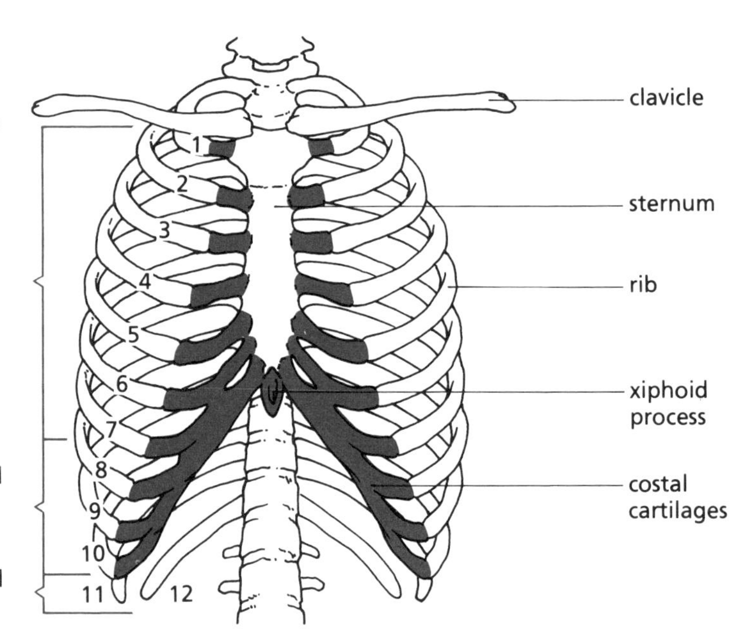

pelvis. They form the "small" of the back, and bear most of the weight of the upper body.

The Sacrum and Coccyx

Below the lumbar vertebrae at the base of the vertebral column are two more bones. The **sacrum** is really five bones fused together into a long triangle; it is also fused to the pelvic girdle. The **coccyx** is made up of four small bones fused together to form a small, unseen tail.

(sa-krum)
(cock-siks)

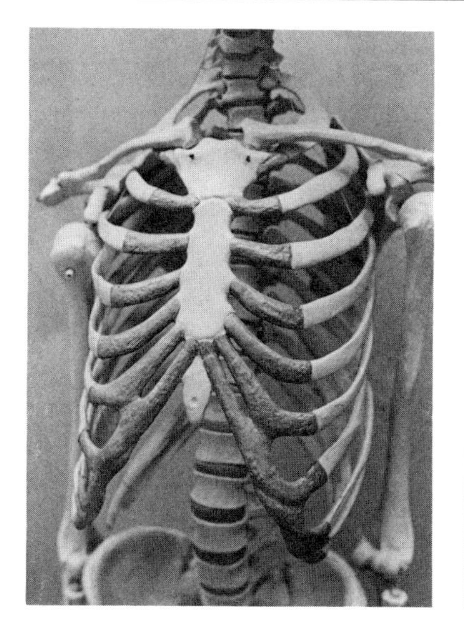

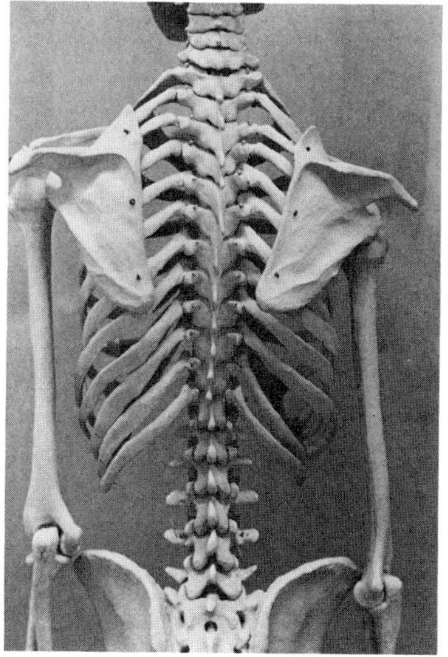

Figure 4.10.
Front and rear views of the thoracic bones.
a) Front view - Note the attachment of the ribs to the sternum by cartilage.
b) Rear view - Note how the scapula is loosely attached over the ribs, which allows more arm flexibility.

Intervertebral Discs

The intervertebral discs are cushions of compressible fibrous cartilage between the vertebrae. These discs act as shock absorbers and separate the bones of the vertebral column. At times they are subjected to considerable stress, especially between the fourth and fifth lumbar vertebrae. Sometimes this pressure may cause the disc to be displaced. Usually the disc slips backwards causing some nerves in the spinal cord to be pinched between the vertebrae. This results in very acute pain. (See Figure 4.11.)

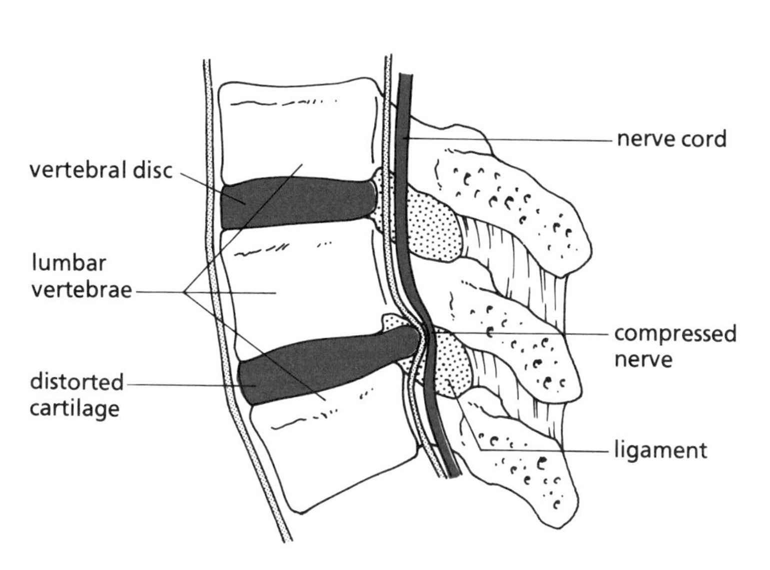

Figure 4.11.
Compressed spinal nerve due to an injury or poor posture habits. The amount of pain and discomfort experienced will vary with the degree of distortion and pressure created by the displaced cartilage of the vertebral disc.

The bones of the vertebral column have limited sliding action; only enough to permit the whole spine to bend in a shallow curve to the left or right. The column has excellent forward bending ability but very limited capacity for twisting and bending backward. This is explained in Figure 4.12.

The curves of the spine are very important. It might seem that a perfectly straight spine would be more efficient. It is the curves, however, which provide much of the body's springlike resilience and its resistance to shock when running or jumping.

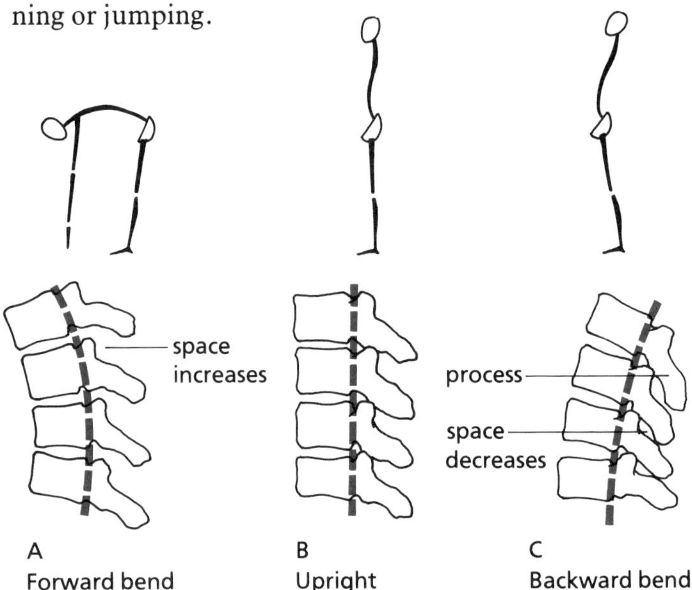

Figure 4.12.
This diagram illustrates why we can bend forward but scarcely backward. When we bend over to touch our toes, the spine curves easily as the processes at the back of the vertebrae open and become more widely separated. When we try to bend backward the processes at the back of the vertebrae may touch and prevent further curvature of the spine in this direction.

If the human body were supported by four legs, rather than being upright on two legs, the curve of the vertebral column would arch upward like a bridge. (See Figure 4.13.) The weight of the abdomen would then be slung naturally beneath the spine where it could be borne easily. The added weight carried during pregnancy could also be handled efficiently. Our upright posture, however, forces the vertebral column to arch backward in order to bear such weight. The pelvis is also tipped in this direction. The weight of the abdomen thus falls on the lower back in the lumbar region. During pregnancy, women frequently suffer strain and feel discomfort in the small of the back. Lower-back pain is, in fact, a very common ailment in both men and women. Our sedentary way of life has contributed greatly to the weakness of the support muscles and the ligaments of the spinal vertebrae. Figure 4.14 shows some of the disorders that distort the normal curvature of the spine.

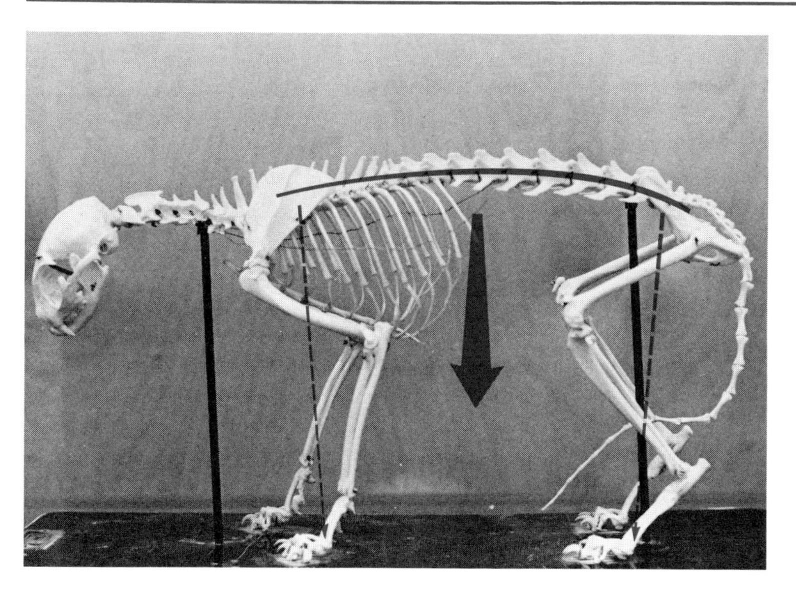

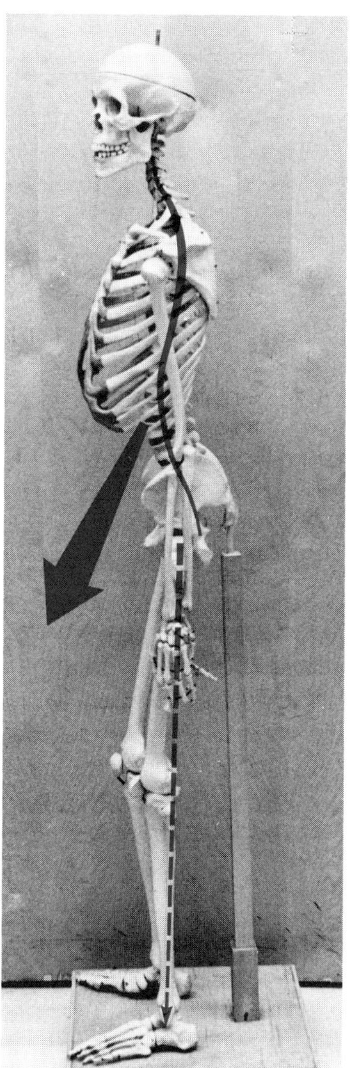

Figure 4.13.
The differences experienced when body mass is supported by a four-legged animal or by a human being in an upright position.

Figure 4.14.
Some disorders of the spine.

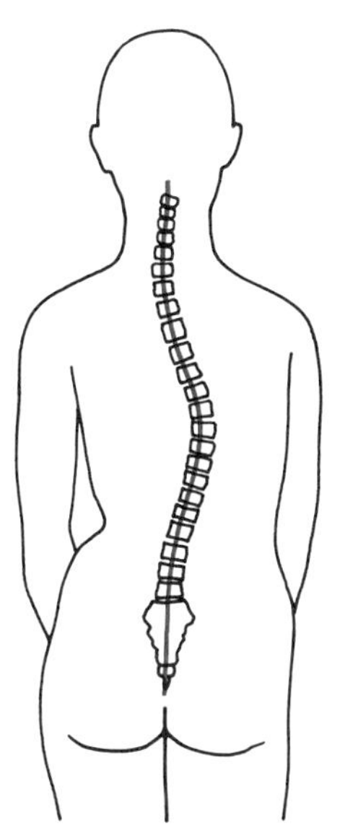

scoliosis
(lateral deformity)

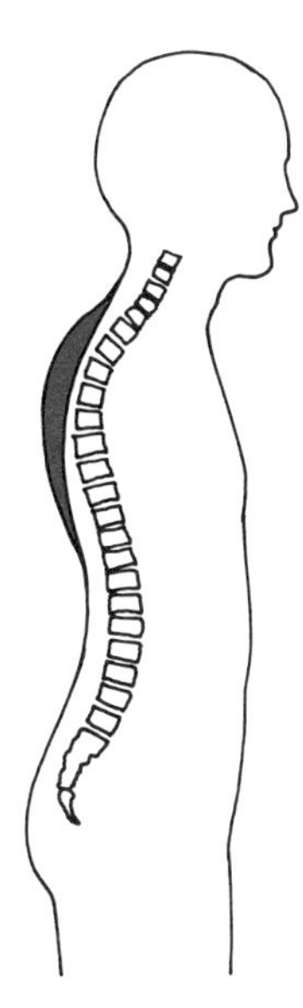

kyphosis
(hunchback)

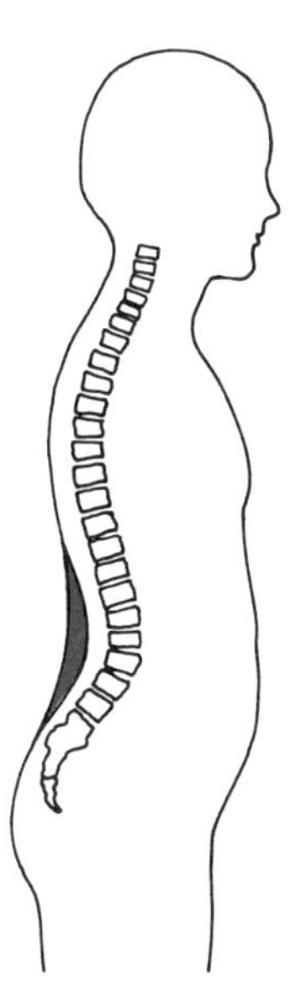

lordosis
(swayback)

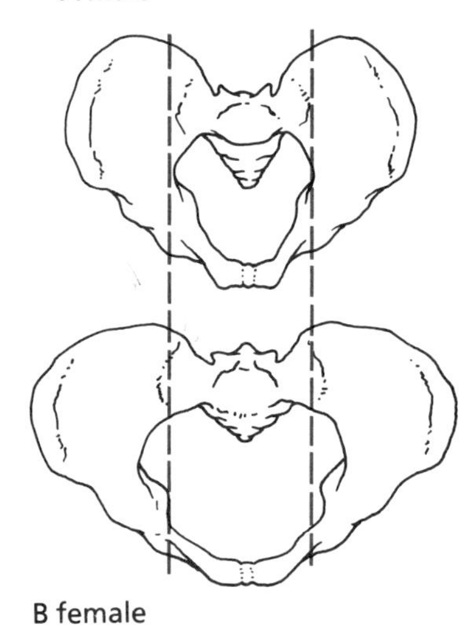

▲
Figure 4.15.
The human female pelvis is broader and lighter than the male. It has a wider angle to the pelvic arch and the opening in the pelvis is appreciably larger than in the male.

The Pelvis

The pelvis is made up of the **ilium**, **ischium**, and **pubis** which in the adult are fused together to form a bowl-shaped dish without a bottom. The pelvis provides attachment for the bones of the lower limbs. It is extremely strong and bears most of the weight of the body, as well as the thrust of all leg movements (running, jumping, walking, etc).

One of the most obvious differences between the skeletons of males and females is found in the pelvis. In the male, the pelvis is deeper than that of the female, but it is not as broad. In the male, the angle formed by the 2 bones at the front (the pubis) is sharp, about 90°. In the female, the pubic angle is about 120°, which creates a larger opening in the bottom of the pelvis to allow for childbearing. (See Figures 4.15 and 4.16.)

Figure 4.16. ▶
The pelvic girdle.

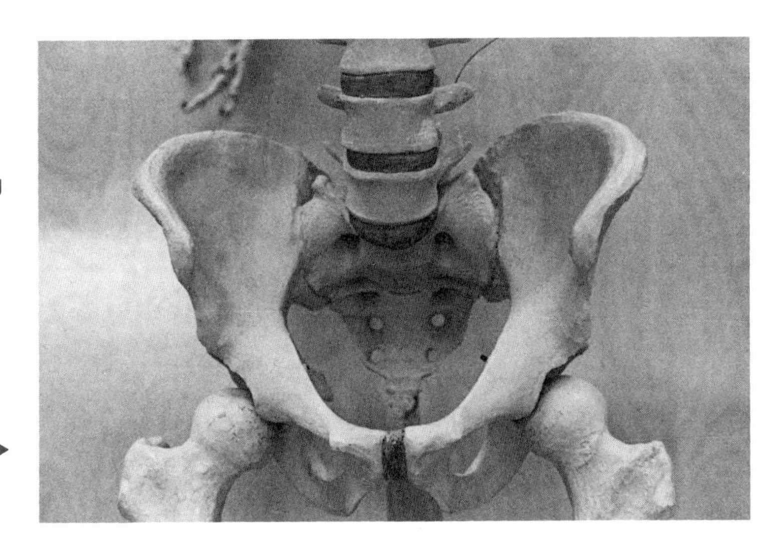

The Lower Limbs

The **femur**, or thigh bone, is the longest and strongest bone in the body. Its round, smooth head fits into a socket formed in the ilium of the pelvic girdle. The lower end of the femur connects with the larger of the two bones of the lower leg, the **tibia**. A flat, disc-shaped bone, the **patella** or kneecap, is found just in front of this joint. It is loosely attached (you can move it a little with your fingers) and offers a protective pad for the joint. The patella develops later than most bones, at about three years of age in girls and sometimes not until six years of age in boys. The **fibula** is the other bone found in the lower leg. It improves stability and locomotive capacity.

The ankle is made up of seven **tarsal** bones, which are very similar to the carpals of the wrist. These bones provide a sliding joint which enables the foot to be extended and flexed with every step we take. The foot consists of **metatarsals**, which are the larger bones of the foot, and the **phalanges**, the small bones at the ends of the toes. (See Figure 4.17.)

(fa-lan-jeez)

The bones of the foot should curve in two directions forming natural arches. One arch spans the ball of the foot and the heel; the other is at right angles to this, across the width of the foot. (See Figure 4.18.) These arches provide the "spring" in our step.

If these arches break down and lose their muscle tone they are sometimes called fallen arches: the result is a painful ache in the foot. This disorder may be caused by a number of things, such as poor prenatal nutrition, poor posture, extra body mass, improperly fitted shoes, or other factors. Well-fitted shoes are very important, especially for young children. Extreme shoe styles can also result in permanent distortion of the foot and thus the spine. Excessively high heels may cause a shortening in the ligaments of the calf muscles.

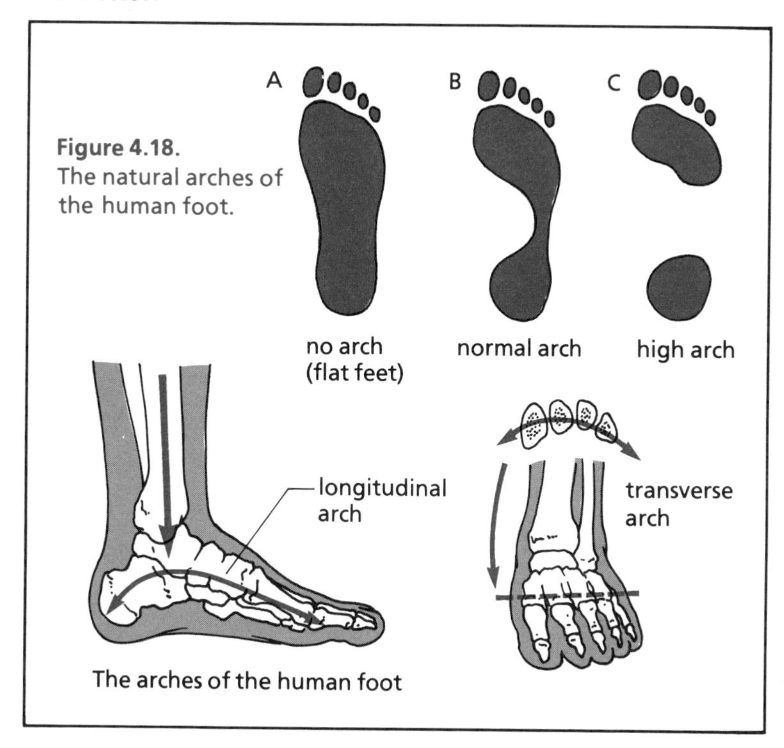

Figure 4.18.
The natural arches of the human foot.

Figure 4.17.
Bones of the lower limb.

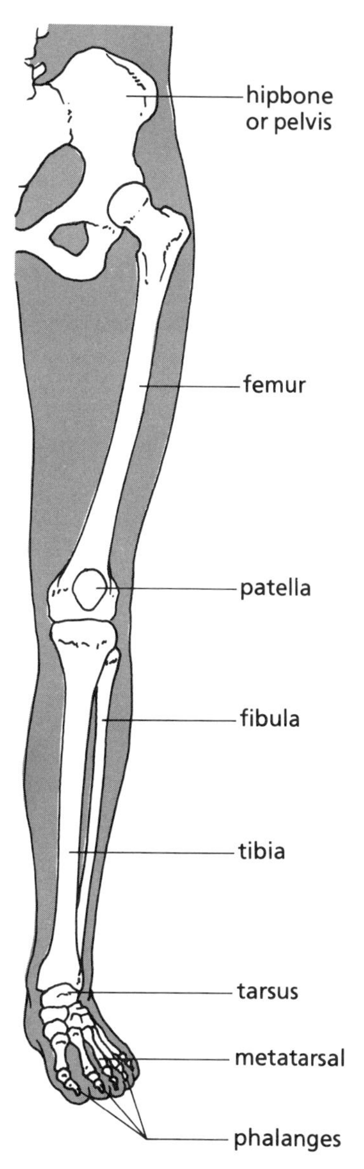

MARATHON MAN – TERRY FOX (1958 – 1981)

Recently, a young Canadian became a great source of inspiration for people across the country. It has been said of him that "once in a while there appears an exceptional human being whose words and deeds restore faith in the human race." Terry Fox was this exceptional human being.

Terry Fox was born in Winnipeg, Manitoba in 1958, and moved to Port Coquitlam, BC, where he grew up. Terry was a determined athlete who worked hard at this interest.

While at university in 1977, in his first year, Terry made it into the Junior Varsity Basketball team. In February, his right knee began to hurt and Terry thought that it might be cartilage damage as a result of his basketball and running. A pain-killing drug was prescribed but the pain became more severe; the pain was soon diagnosed as bone cancer.

After an operation to amputate his cancerous right leg above the knee, Terry was fitted with a temporary artificial leg and some weeks later he had received his permanent prosthesis. When he got used to his new leg, he learned to walk, jog, and play golf.

By September, Terry had returned to Simon Fraser University where he continued to achieve good grades. He played wheelchair volleyball and basketball, and drove a car with a left-foot accelerator pedal. Terry went jogging and began to feel physically and emotionally stronger. In 1979, he told his parents that he was going to run across Canada to raise money for cancer research. He spoke with Blair MacKenzie of the BC and Yukon Division of the Canadian Cancer Society to obtain the necessary funds to carry out his mission. Funds were granted and he was given full support by the Canadian Cancer Society.

On April 12, 1980, Terry set out from St. John's, Newfoundland, to run 8000 km aross Canada accompanied by a friend who followed him in a camper van. On May 6, twenty-five days and 800 km later, Terry and his friend boarded the CN ferry for North Sydney, Nova Scotia. Already, people were sending donations in response to Terry's *Marathon of Hope*. Terry ran through PEI and New Brunswick, and by mid-June he was in Quebec. Terry ran about 40 km every day, passing the halfway point at French River, Ontario.

On August 29, running southwest toward Thunder Bay, Terry began to feel a tightness in his chest and had developed an irritating cough; he thought that he had a cold. He felt better in a short time and ran another 35 km. The next day he set out again; there were people along the road to watch him and he kept going even though he was exhausted. In the afternoon he ran another 12 km, rested for fifteen minutes, then started off again; but this was to be Terry's final run of the trip. He was running even though he was experiencing tremendous chest pains. At last, totally exhausted, he climbed inside his van and asked to be taken to a doctor.

Terry was taken to the Port Arthur General Hospital where he was examined, and the X-rays showed that the cancer had spread to his lungs. On September 2, after 144 d and 5340 km on his *Marathon of Hope*, Terry was on his way back to the New Westminster Royal Columbian Hospital to start a new session of chemotherapy.

Terry's courage inspired thousands of others across the country to embark on walk-a-thons, swim-a-thons, and other money-making ventures in his honour, all donating funds raised to the Canadian Cancer Society. The Terry Fox Fund raised over 22 million dollars for the Canadian Cancer Society in 1980 and 1981.

In his last few months Terry Fox received the highest awards the country has to offer, including the Order of Canada and many others. His emotional and physical strength and endurance will not be forgotten.

The Upper Limbs

The bones of the arms are attached to the **pectoral girdle**. The **clavicle**, or collar bone, connects at one end with the top of the **sternum** and at the other end with the **scapula** or shoulder blade. The clavicle is just below the skin surface and you can feel its outline quite clearly. The scapula, which is also close to the surface, lies on top of the ribs. This bone is only very loosely attached and if you ask a partner to move an arm in a circular motion, you can feel the movement of the shoulder blade on the upper back. Because the scapula is so loosely attached it is held in place by ligaments and muscles; thus the arm has great freedom of movement in almost any direction.

(clav-i-kul)
(skap-yoo-lah)

The ends of the clavicle and the scapula together form part of a socket for the bone of the upper arm. The rest of the socket is composed of a capsule of cartilage. The bone of the upper arm is called the **humerus**. It connects with the scapula at the shoulder and with the two bones of the lower arm at the elbow. (See Figure 4.19.)

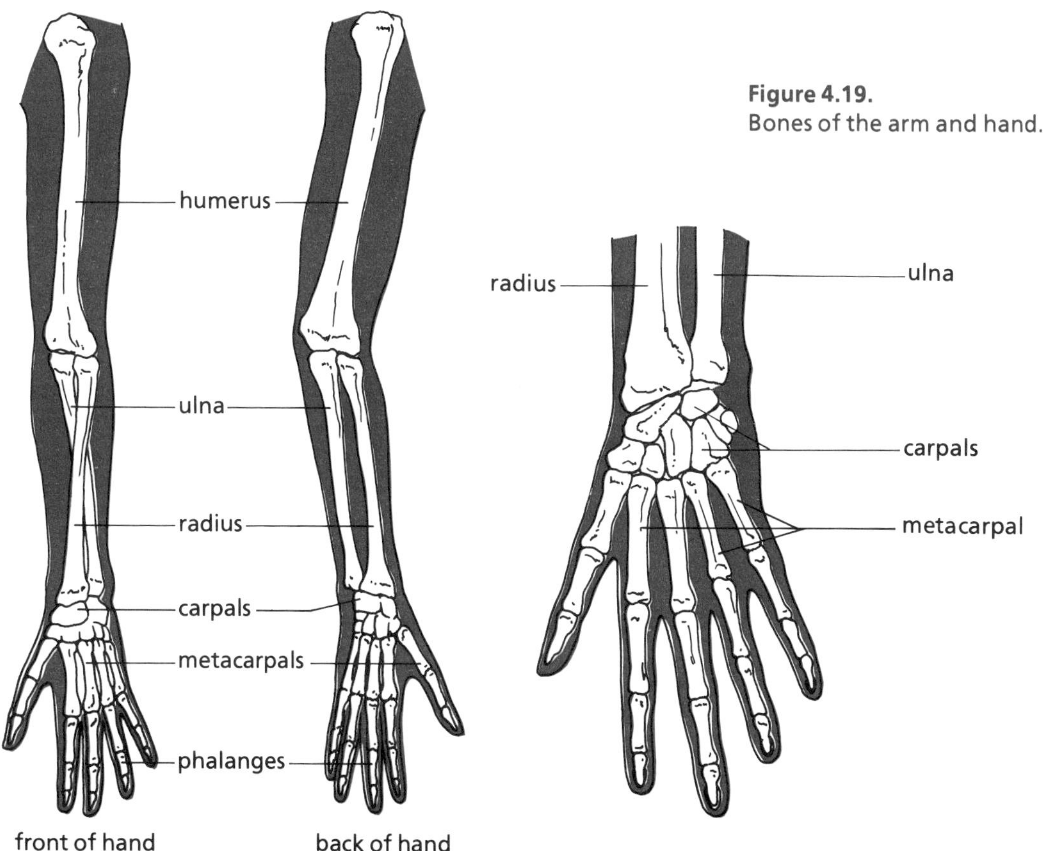

Figure 4.19.
Bones of the arm and hand.

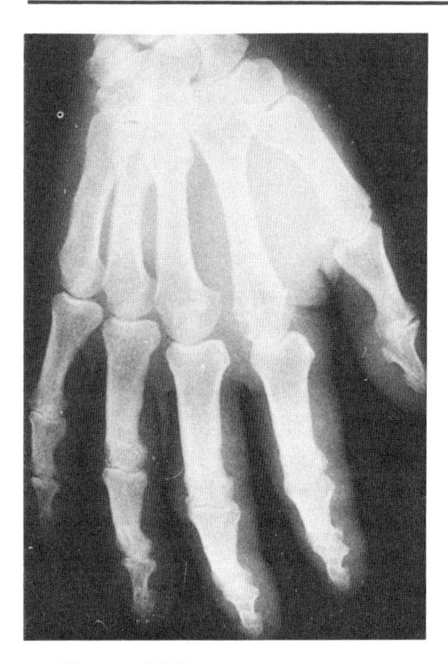

Figure 4.20.
An X-ray of the bones of the hand.

The **ulna** is the main supporting bone of the forearm. It is attached to the humerus at one end and the bones of the wrist at the other. Beside it is a slightly shorter bone called the **radius**, which rotates around the ulna so that the hand can be turned. If you stretch out your forearm with the palm of your hand facing up, then turn your hand over, you can get an idea of how the radius moves by watching the skin of your arm.

The wrist is made up of eight small bones or **carpals**, which are joined to the five **metacarpals** that form the hand. The bones forming the fingers and thumb are the **phalanges**. There are three phalanges in each finger and two in the thumb. (See Figure 4.20.)

The Joints

Where two bones come together they form an **articulation**. Some bone articulations are fixed and allow no movement. Others either permit slight movement or are freely moveable.

Immovable Joints

These joints are fused so that no movement occurs between the bones thus joined. Immovable joints are found, for example, between the sacrum and the pelvis and between the fused bones of the skull.

Slightly Movable Joints

In these joints a small amount of movement is possible and the articulating surfaces are protected by a pad of cartilage. The lumbar vertebrae, for example, are connected by this type of joint.

Freely Movable Joints

These joints provide great freedom of movement such as is found in the joints of the shoulder, hip, knee, and elbow. The joint is contained within a special capsule, the **capsular ligament**, which helps to hold the bones in place. The chief function of these moveable joints is to permit the body the mobility needed for running, jumping, reaching, bending, etc.

Figure 4.21.
The functions of bone, cartilage, ligaments, and tendon.

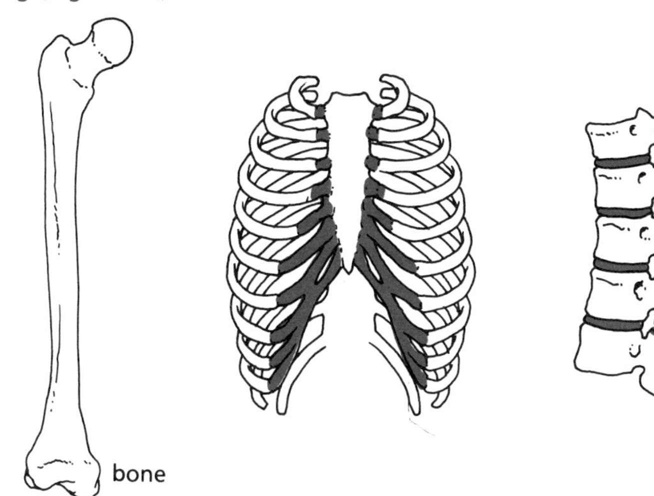

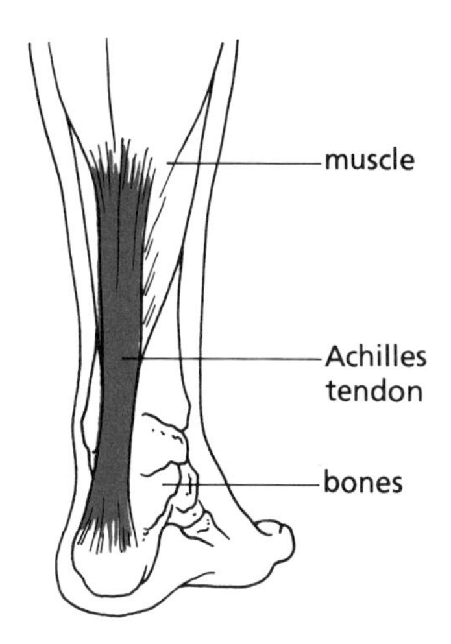

BONE is a hard, rigid, supporting, and protective tissue, composed mainly of calcium and phosphorus.

CARTILAGE (gristle) is strong, supporting, and attaching tissue, more flexible than bone and much less hard. The ear and the end of the nose are also composed of cartilage.

TENDONS are tough and flexible. They attach muscle to bone.

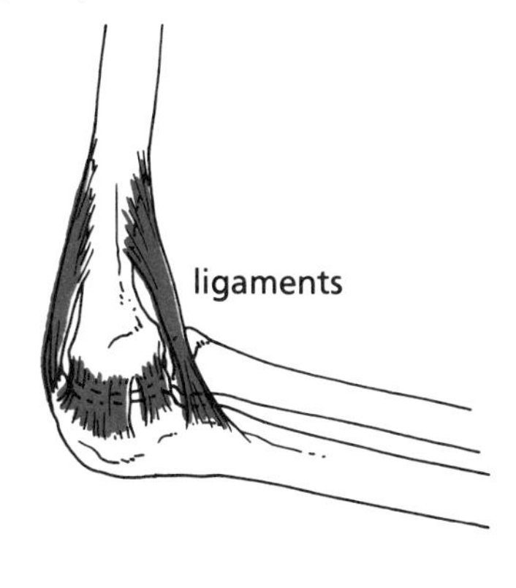

LIGAMENTS are strong bands of tissue which hold bone to bone. They are flexible and allow small amounts of movement between the bones.

The joints are held in place by straplike ligaments. These connective tissue structures are flexible and have a small amount of elasticity. They bind the bones together and help to prevent dislocations while still allowing the bones to move. (See Figure 4.21.)

If two bones could move freely the ends might be expected to rub together causing friction and wear. This is prevented in freely moveable joints by covering the ends of the bones (the articulating surfaces) with **articular cartilage**. Inside the capsule is a smooth **synovial membrane** that lines the capsule and secretes **synovial fluid**. This fluid lubricates the joint, bathing the tissues, and helping to reduce the friction

(sy-noe-vee-al)

produced by the moving bones. The capsule forms a tight seal around the joint to prevent the fluid from escaping. (See Figure 4.22.)

Joints are also often grouped according to some special characteristic that they possess. (See Figure 4.23.)

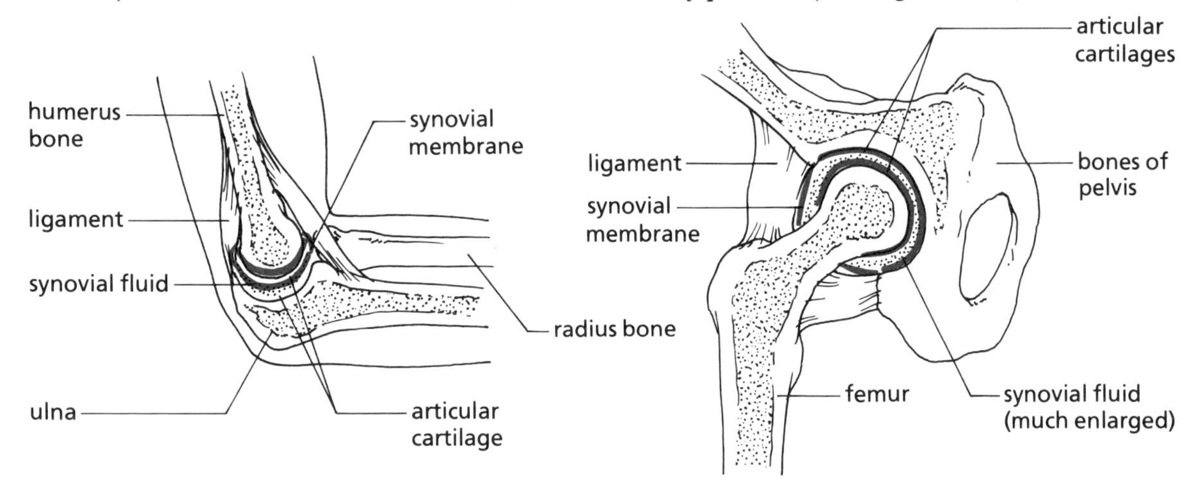

Figure 4.22.
The hinge joint of the elbow and the ball-and-socket joint of the hip.

In the synovial capsule, the fluid lubricates the joint. The membrane keeps the fluid from escaping and the articular cartilage makes a smooth bearing surface and prevents the bones from rubbing together.

Ball-and-Socket Joints

This type of joint allows movement in almost any direction. The rounded, ball-shaped head of one bone fits into the hollow depression found in another. Sometimes more than one bone is involved in forming the socket. Ball-and-socket joints are found in the shoulder and the hip.

Hinge Joints

These joints allow movement in one plane only. The knee and elbow joints are examples of hinge joints.

Gliding Joints

These joints are more limited, allowing only small sliding movements of one bone over another. Gliding joints are found in the wrist and ankle for example.

Pivot Joints

Pivot joints permit radial or circular motion. An example of a pivot joint is in the forearm where the radius articulates with the humerus. (See Figure 4.23.)

Figure 4.23.
Types of joint found in the human body.

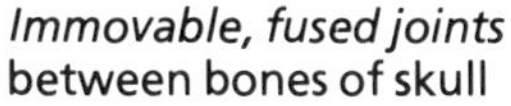
Immovable, fused joints between bones of skull

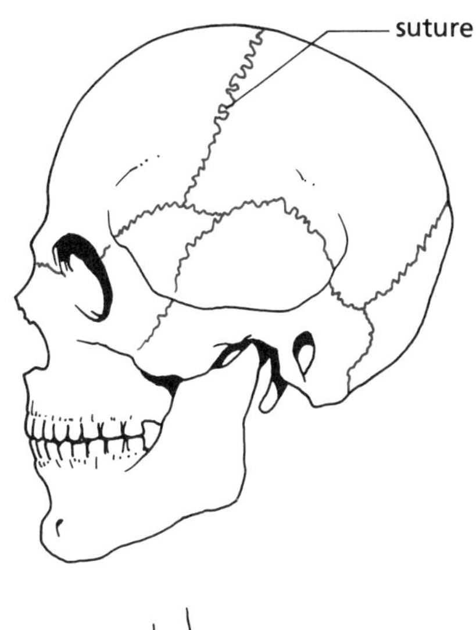

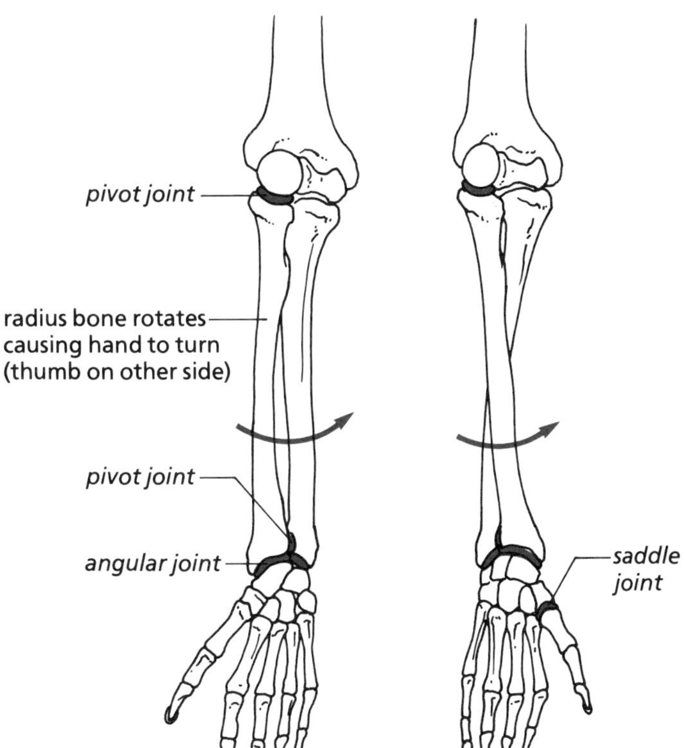

hinge joint of the elbow
Allows only two-way movement

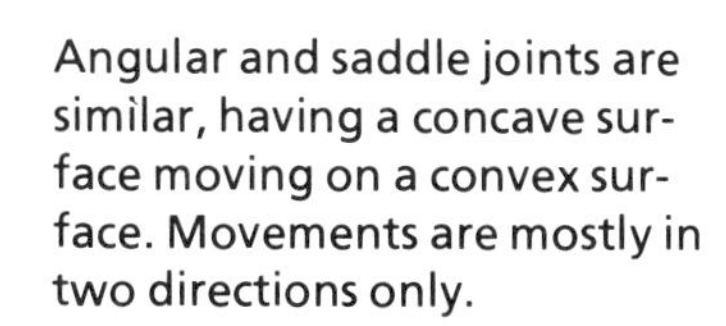
Angular and saddle joints are similar, having a concave surface moving on a convex surface. Movements are mostly in two directions only.

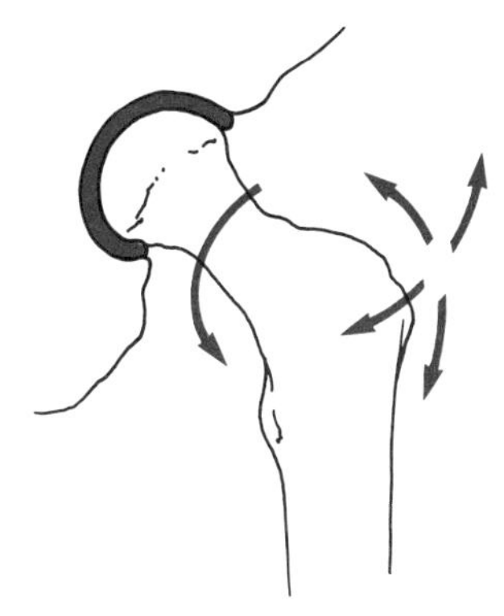

ball-and-socket joint of hip
Very free movement in almost any direction

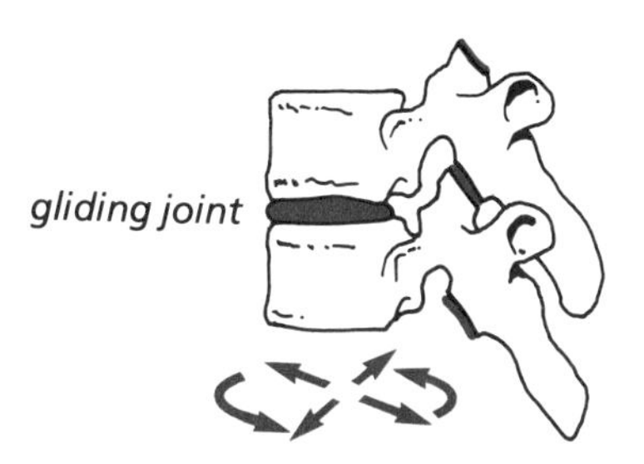

One part slides over another. The amount of movement possible at each vertebra is quite small.

Skeletal Injuries and Disease

Injuries

Fractures and other Injuries

Green-stick fractures usually occur in young children. In this case, the bone does not separate completely, but breaks like a green, sap-filled stick. **Simple fractures** refer to broken bones which do not pierce the skin. **Compound fractures** are breaks in which the ends of the bones push out through the skin. Such breaks are very serious, because of the possibility of infection. (See Figure 4.24.)

Figure 4.24.
Common types of fractures.

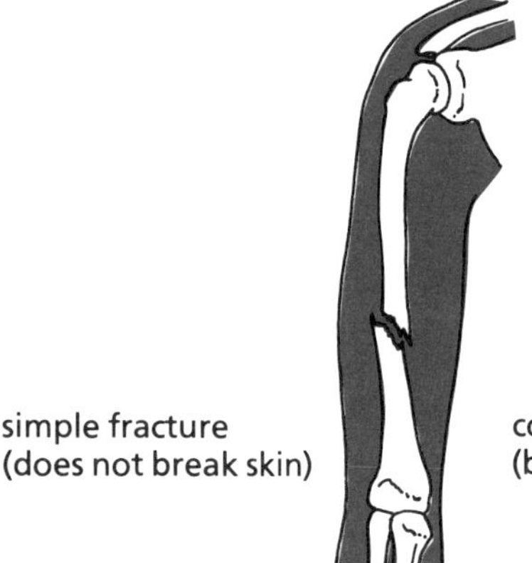

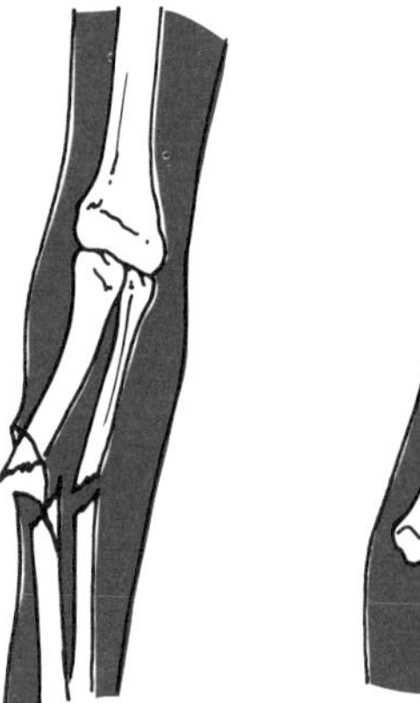

Dislocations occur when the bones of a joint are pulled out of alignment. The ligaments that hold these bones together are stretched, distorted, or torn. Dislocations are thus extremely painful.

A **sprain** results from a temporary separation of the bones, after which the bones return immediately to their normal alignment. The injured area usually swells rapidly and is painful because of the stretching or tearing of ligaments.

Football players, basketball players, and ballet dancers are frequently affected by knee injuries. Sometimes the cartilage on the end of the femur or tibia tears. The joint then becomes very painful and difficult to move. Repair of torn cartilage may require surgical treatment. The amount of synovial fluid on the knee may increase when an injury occurs. The joint then becomes very swollen and painful. This disorder is often referred to as “water on the knee”.

The Healing of Broken Bones

The time that bones take to heal may vary considerably, not only with the size of the bone and the type of fracture, but also depending on the efficiency with which minerals are made available to repair the tissue. The bones of young people heal much more rapidly than those of adults. The bones of old people are very slow to mend.

Sometimes bones do not recover as rapidly as expected. In some exceptional cases, the pieces of bone may fail to rejoin, even after a year. Recently small electrical currents have been used to stimulate the growth of new bone cells. The current also serves to inhibit bacterial growth and thus decrease infection. This treatment has achieved remarkable success in treating injuries.

Steps in the Repair of Broken Bones

STEP 1.
Immediately after a break occurs, blood clots form at the break in the bone.

STEP 2.
After the doctor has "set" the bone and the parts are properly aligned again, these blood clots are absorbed by the body and replaced with new connective tissue called **procallus** ("pro" means *before* or *first*). This substance fills up the spaces in the tissue and absorbs the dead bone cells. It may take from one to eight weeks for this phase to be completed.

STEP 3.
The procallus is gradually transformed into **callus**, a harder substance, which seals the broken ends of the bone, providing some support as new bone is formed. The callus is not nearly as strong as bone, however. To make up for this, it forms a band of thickened material, like a cuff, around the break site.

STEP 4.
Finally new bone tissue replaces the callus. It grows out from the periosteum (a special membrane that nourishes the bone) until the bone finally achieves its original shape and size. To avoid disturbing the action of the procallus and callus during healing, the bone must be immobilized. The doctor, after checking that the parts of the broken bone are correctly aligned, will enclose them in a rigid cast. This prevents the muscles and tendons from pulling the bone out of shape. For

most breaks this also involves immobilizing the joints at either end of the broken bone. During the period that the bone is in a cast, the ligaments tend to lose their elasticity and become stiff, while the muscles lose their strength and become soft. Therefore, after the cast is removed, the limb must be exercised to bring it back to full use.

Disease

The skeleton, like any other part of the body, can be subject to disease. Bones may become infected by an invasion of bacteria as a result of cavities, severe injury, or a heavy blow. The general term for bone infection is **osteomyelitis**.

(os-tee-oe-my-ul-y-tis)
(ar-thry-tis)

Although **arthritis** is often thought of as a disease that affects old people, it has also been diagnosed in very young children. It usually develops between the ages of 25 and 50 and affects the joints. There are many kinds of arthritis, most of which result in swollen joints and are extremely painful. In severe cases, the cartilage separating bones in joints is gradually destroyed and hard bands of calcium fill the spaces.

QUESTIONS FOR REVIEW

SOME WORDS TO KNOW

Match a statement in the column on the left with a suitable term from the list of words on the right. DO NOT WRITE IN THIS BOOK.

1. Joins muscles to bones.
2. Point of contact between two bones.
3. Individual bones that make up the spine.
4. The top bone of the spinal column.
5. The lower jaw.
6. Line formed where two bones have fused together.
7. The vertebrae of the neck.
8. Lubricates joints.
9. Flexible attachment between the bones of a joint.
10. A flexible supporting tissue.

A. radius
B. vertebrae
C. mandible
D. atlas
E. axis
F. suture
G. cervical
H. cartilage
I. lumbar
J. tendon
K. sprain
L. ligament
M. synovial fluid
N. articulation

SOME FACTS TO KNOW

1. What are the functions of the skeletal system?
2. What are the main divisions of the vertebral column?
3. State whether the following bones belong to the axial or appendicular skeleton:
   a) cervical vertebrae,
   b) humerus,
   c) clavicle,
   d) skull,
   e) temporal bone,
   f) mandible,
   g) radius,
   h) sacrum,
   i) atlas.
4. What is the proper anatomical name for the following bones?
   a) shoulder blade,
   b) breast bone,
   c) jaw bone,
   d) spine,
   e) thighbone,
   f) the two bones of the forearm,
   g) collar bone.
5. What organs or structures are protected by the following bones?
   a) vertebrae,
   b) temporal bone,
   c) sternum and ribs,
   d) patella,
   e) pelvis.
6. What is the main function of a joint?
7. Describe the structure of a ball-and-socket joint. Give the functions of the structure that you describe. Use a sketch if you wish.
8. What type of joint permits the following activities?
   a) rotation of the head,
   b) swinging the arm in a circle,
   c) forward bends,
   d) rotating the hands,
   e) bending the elbow,
   f) taking a breath (ribs).
9. Briefly describe where the following are found:
   a) foramen magnum,
   b) fontanel,
   c) maxilla,
   d) radius,
   e) axis,
   f) sternum,
   g) phalanges.
10. What are some of the differences between the male and female skeletons?
11. What is the difference between a simple and a compound fracture?

QUESTIONS FOR RESEARCH

1. Select one of the following topics and prepare a report on it:
   - the job of an ambulance attendant
   - the Red Cross or St. John's Ambulance first aid program
   - healing bones with electricity, a new technique
   - the effect of aging on bones
   - good diets for the young and the old (restrict your research to the effects of diet on bones)
2. With the co-operation of your physical education teacher, build a table of statistics about running and runners. Take measurements of height, leg length, hip width, circumference of thigh muscles, etc. Record the times of each runner's best effort over a standard distance. Make predictions before the runner tries the standard test, about the time a runner will achieve by using his or her body measurements.
3. Interview a person who has arthritis and write a report based on the interview. Ask the person to tell you about the symptoms, the times when the pain is least and when it is most severe. Ask the person about the length of time he or she has suffered from arthritis and what treatments have been received.
4. What first aid treatment should be given for the following common accident problems? a sprained ankle, a broken arm.

Activity 1: THE PROPERTIES OF BONE

Materials

two ribs from a butcher. (Get the longest bones possible.) Alternatively, two chicken leg bones will do. A few small pieces of any uncooked bone. Hydrochloric acid (15%), Bunsen burner, crucible, retort stand, and ring support.

Method

1. Clean as much of the soft tissues from one of the ribs as possible. Immerse the bone in a jar of 15% HCl solution. Soak for at least two days. When the bone becomes pliable, remove it from the acid and wash it thoroughly in running water.

2. Take a small piece of another bone and place it in a crucible and heat strongly in a hot Bunsen flame. Heat until no more smoke comes off. *Record your observations.*
   Note: The first experiment will remove the inorganic materials from the bone, leaving the protein. The second experiment will remove the protein and leave the inorganic material.
   Compare the treated rib with an untreated rib. What differences do you observe? Note the thin covering on the outside of the untreated bone. What is this called? What function does it perform? Try to remove this covering. Why is it so difficult to remove? Does the acid alter the shape of the bone? Does it alter the hardness or flexibility of the bone? Try to tie the bone in a knot! What properties do the inorganic substances give the bone? What are the inorganic substances?
   Does heating alter the shape, hardness, or flexibility of the bone? What is the function of the protein in bones?

Activity 2: DO THE SPINE OR THE LEGS HAVE GREATER INFLUENCE IN DETERMINING OVERALL HEIGHT?

Method

1. Select several students of different height. First measure the total height of each. Then seat them on a flat bench and measure the height of each from the seat to the top of the head. *Record your results.*
2. *What appears to be the major factor in deciding the height of an individual, the length of the spine or the length of the legs? Are the results consistent?*

Activity 3: USING YOUR TEXT, A SKULL, AND THESE DIAGRAMS, ANSWER THE FOLLOWING QUESTIONS IN YOUR NOTEBOOK.

1. Name each of the parts labelled in the diagrams.
2. Give two functions of the frontal bone.
3. What is the name given to the junction between two fused bones?
4. Why is the skull of a baby composed of unfused plates?
5. What function does the occipital bone perform?
6. What is the purpose of the auditory meatus?
7. What bone moves when we chew our food?
8. What bone comes in contact with the occipital condyles?
9. What passes through the foramen magnum?
10. How do the eye sockets help protect the eye?
11. What is the function of the openings at the back of the eyes?
12. Is there any evidence to show that the lower jaw was formed in two halves that later were fused together?
13. Which bones contain sinuses?
14. Which bones contain the tiny bones of the ear?

Fig. 4.25
Using your text, a skull, and these diagrams, answer the following questions.

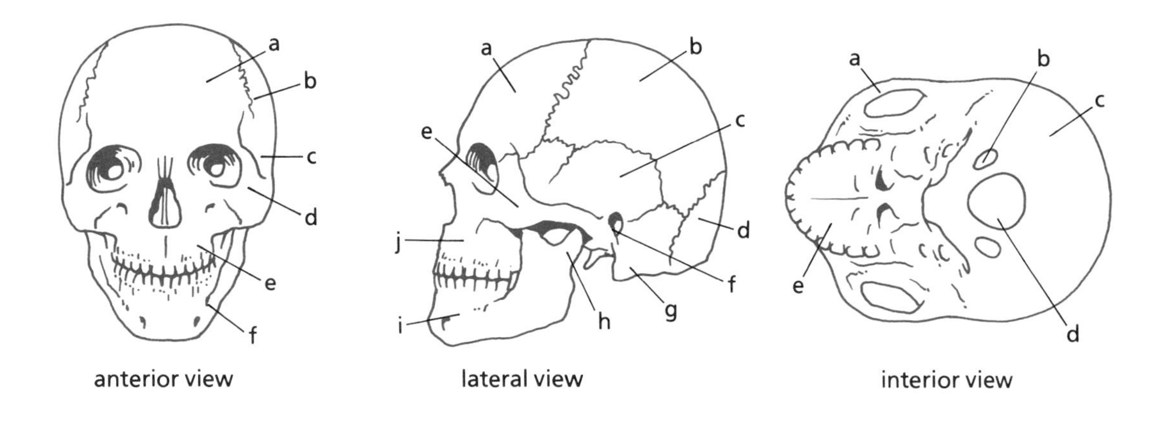

Chapter 5

Muscles

- The Muscular System
- Muscular Contraction and Extension
- Muscle Attachment
- Muscles in Action
- Muscle Disorders

The Muscular System

Animals employ many different methods of locomotion. Some have legs and move swiftly across the land, others have wings for flying, or fins for swimming. Whatever the means of locomotion, it is muscles that provide the moving power. In prehistoric times, people needed speed to catch prey or avoid animal predators. Later people needed strength to hold a plow and till the soil. At the present time both strength and speed are supplied by machines. Now the need is for greater dexterity, especially involving the hands and fingers for jobs such as typing, playing the piano, and operating machines, rather than for strength.

Not all of our muscles are large ones used for locomotion. Some muscles allow us to smile or frown, to wink an eye, swallow food, even to wiggle our ears. Muscles account for about 40 percent of the body mass. There are more than 650 muscles in the human body. All muscles are classified into three types. Although each type has slightly different characteristics at the cell level, they all have the same main function! That is, to contract and, in so doing, perform work.

The three categories are:

1. **Skeletal** or **voluntary muscle**. Most of these muscles are attached to bones and they contract to move the limbs or some group of tissues. Skeletal muscles make up about 35% of a girl's body and about 42% of a boy's body.

2. **Smooth** or **involuntary muscle** is found in the internal organs such as those involved with the digestion. We do not have conscious control over these muscles! They work automatically.

3. **Cardiac muscle** is found only in the heart. This muscle tissue also contracts automatically, without conscious thought. It has tremendous stamina, working without rest for the entire lifetime of the individual. (See Figure 5.1.)

Muscle cells do not increase in number as we grow or when we build up muscles with exercise. The number of cells remains approximately constant, although the size of the cells may increase. A weightlifter or a bodybuilder may have very large muscles indeed. By constant exercise an athlete can cause muscle cells to grow larger; by bringing all the cells

Figure 5.1.
The three types of muscle tissues.

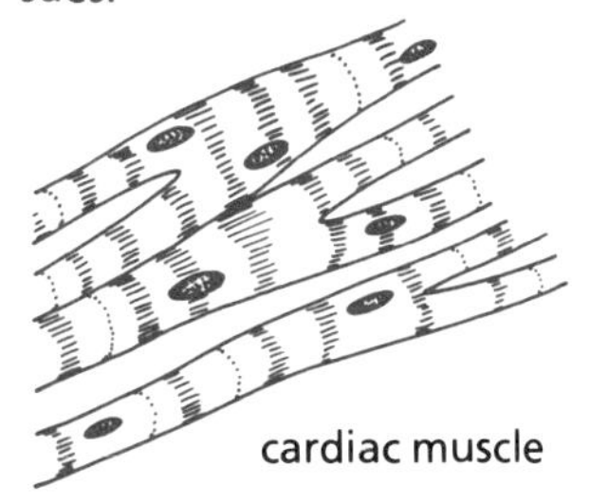

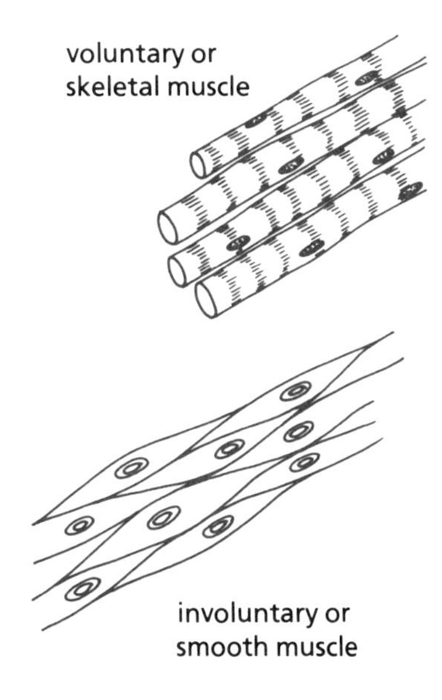

Figure 5.2.
The major muscles of the body.

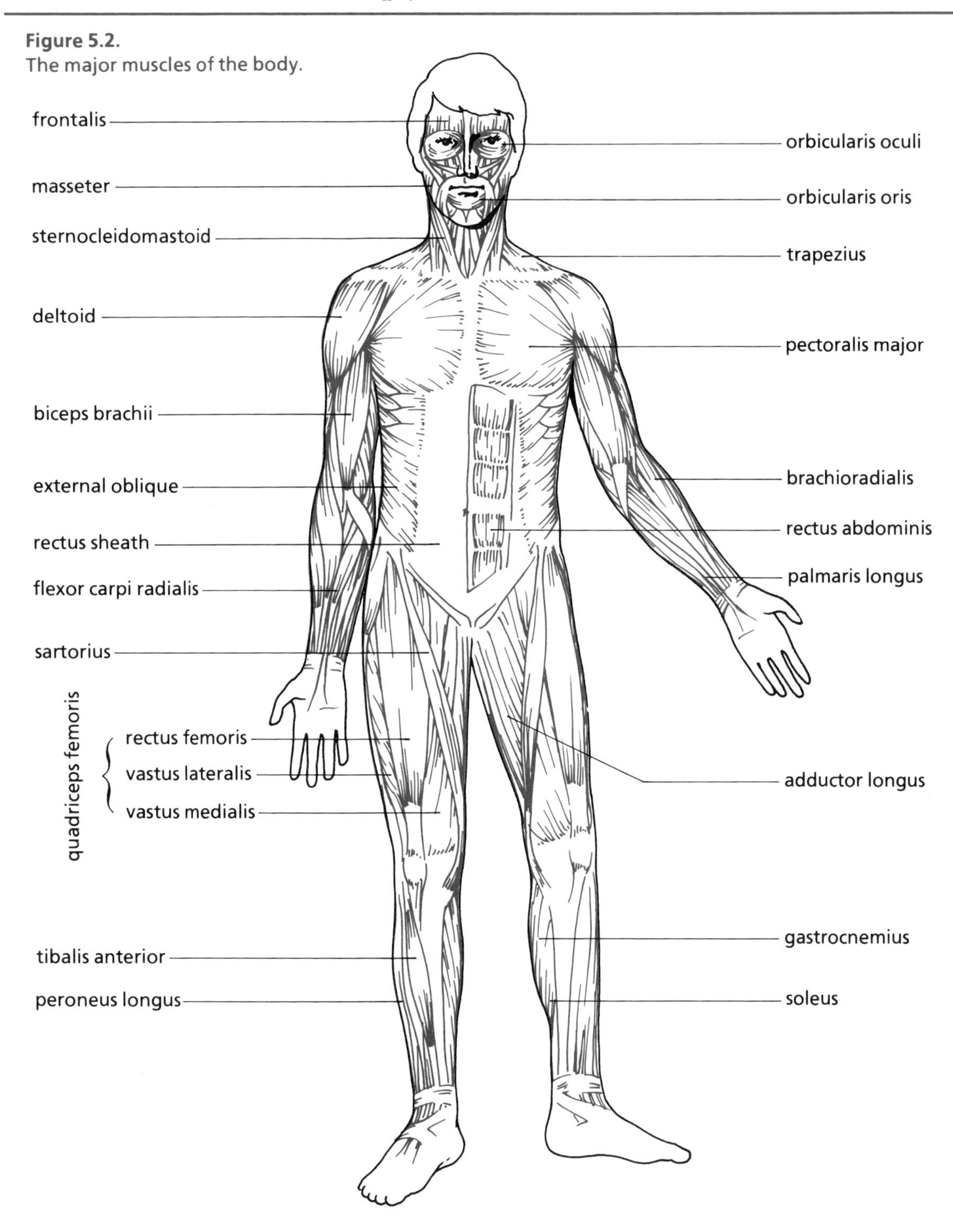

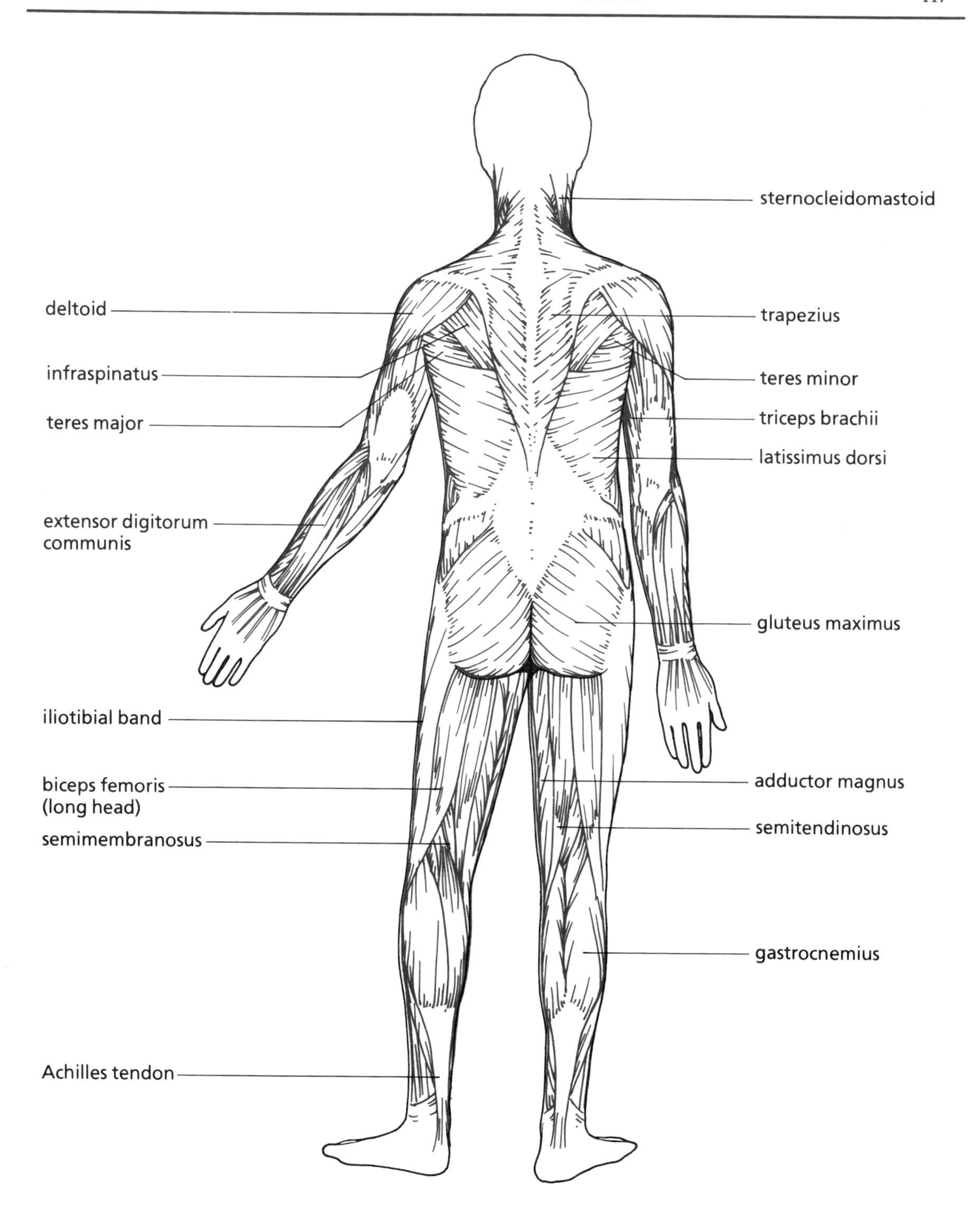
sternocleidomastoid
deltoid
trapezius
infraspinatus
teres minor
teres major
triceps brachii
latissimus dorsi
extensor digitorum
communis
gluteus maximus
iliotibial band
adductor magnus
biceps femoris
(long head)
semimembranosus
semitendinosus
gastrocnemius
Achilles tendon

into play, he or she may be able to perform extraordinary feats of strength, speed, or agility. However, in the absence of exercise the cells gradually shrink in size and lose their effectiveness. If you are ill in bed for some time, parts of the muscle cells reduce in size. You may, therefore, have difficulty in standing or walking until you build up these cells again.

Muscle Contraction and Extension

When muscles are stimulated by a nerve impulse, they contract and accomplish work. Skeletal muscles are arranged in pairs in which one muscle contracts upon stimulation and the opposing muscle extends (relaxes). Muscles thus demonstrate both **contraction** and **extension**. Smooth muscles in organs such as the urinary bladder or the stomach are able to relax and extend the walls of these structures considerably in order to carry or store large quantities of liquid or food.

Muscle cells either contract fully or extend fully. They do not work partially. If you lift a heavy load you simply use more cells than you would for a light load. Cells cannot contract halfway; it is an "all or none" event.

When muscle cells contract, they produce heat. About 80% of the energy used in muscle contraction is converted into heat and "lost" to the body. Sometimes, when we are cold, the body makes rapid contractions of the muscles, just to produce heat and maintain the body temperature. This is known as shivering.

If muscle cells are badly damaged or destroyed they are not easily replaced. Although many tissues of the body such as skin or bone can replace damaged cells quite quickly, nerve and muscle cells are not usually restored.

How Muscle Cells Contract

The cells of skeletal muscle are quite small in diameter, about 10 – 100 μm, but they are very long in comparison with most cells, varying from 3 mm to 7.5 cm in length.

A muscle is made up of several bundles of these cells; these bundles are more usually called fibres. Each of the bundles is enclosed within a sheath (sarcolemma) that is continuous with the tendons at the ends of the muscles.

Inside the sheath there are many nuclei, rather than a single nucleus as is found in most cells. These nuclei are scattered around the perimeter of the fibres. The fibres have characteristic cross-banded markings or **striations**. Figure 5.4 shows how these markings appear under a microscope.

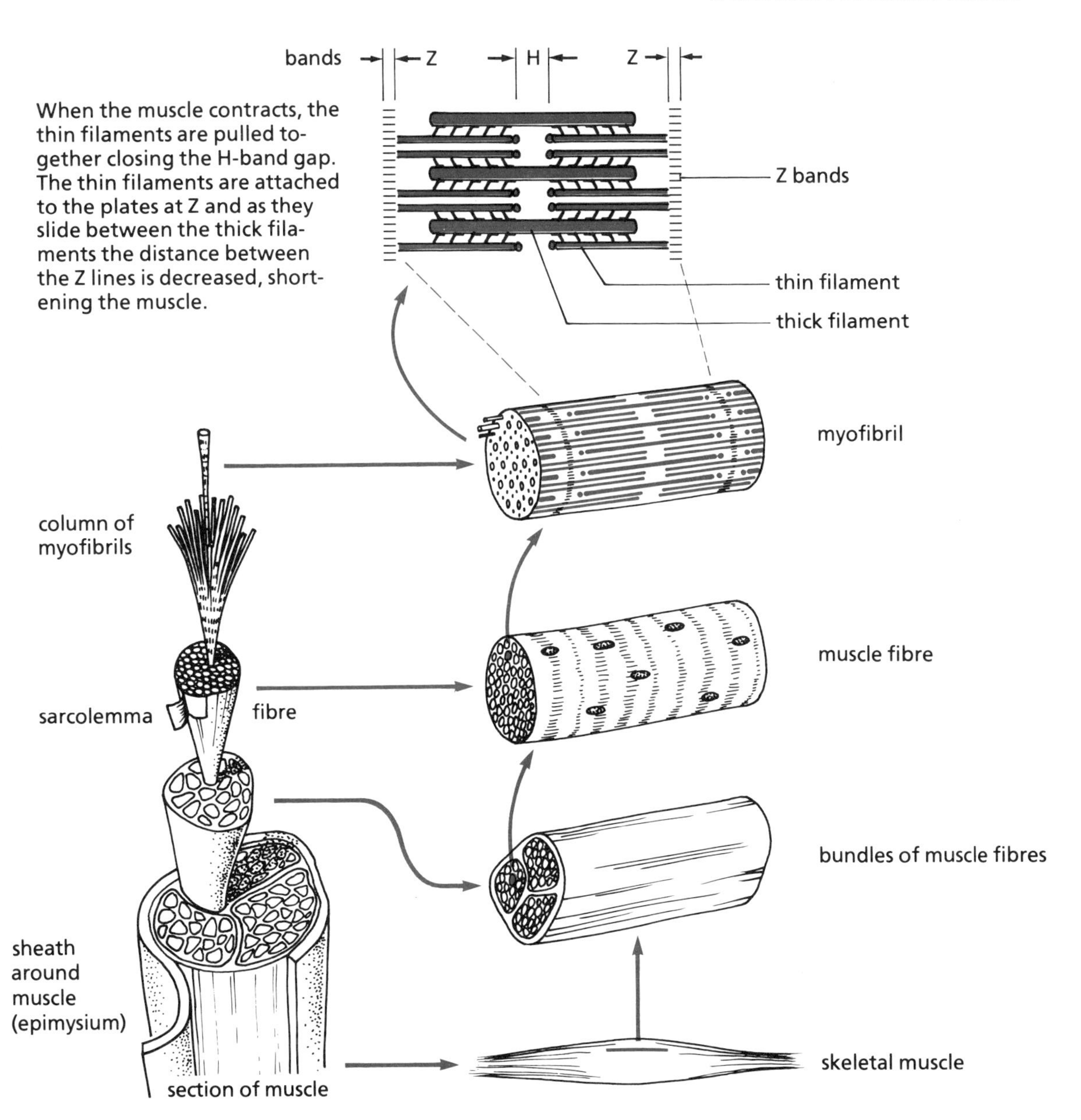

Figure 5.3.
The structure of skeletal muscle.

(my-oe-fy-brils)

Muscle fibres are, in turn, made up of many **myofibrils** bundled together. It is in the myofibrils that the banding or striations are created. The striations, which appear when muscles are viewed through a light microscope, are caused by the presence of many tiny protein strands within the myofibrils. It is these proteins that shorten when the cell contracts. Two types of protein strands are suspended in the muscle cell cytoplasm. The *thick* strands are **myosin** filaments which are surrounded by *thin* **actin** filaments.

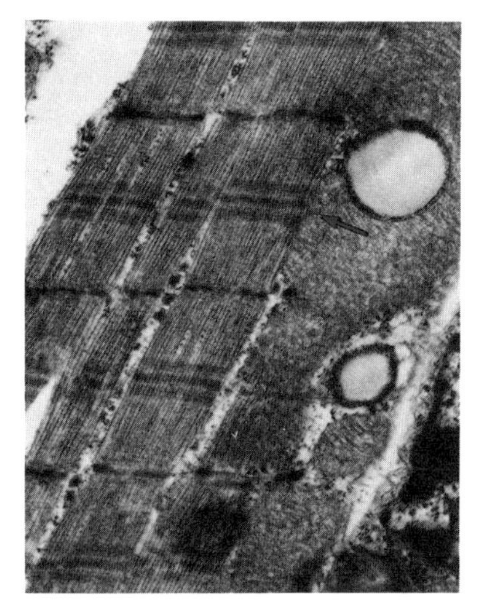

Figure 5.4.
Electronmicrograph of muscle striations.

The thin actin filaments are attached to disclike plates (Z bands) and have heavy club-shaped endings. (See Figure 5.3.) These are the contracting filaments which act to draw the two bands together. Although each group of filaments contracts only a very small distance, the sum of these tiny movements produces a considerable shortening of the whole muscle. The thick filaments have ridged projections along their length; these attach to the thin filaments during contractions. When light shines from behind the myofibrils, it shows the overlapping parallel filaments as bands of light and dark. These appear at right angles to the length of the myofibril. The bands are shown in Figures 5.3 and 5.4.

The muscle cell cytoplasm also contains many substances important to its function, such as glucose, creatine phosphate, and ATP. All of these help to supply energy for the contracting cell.

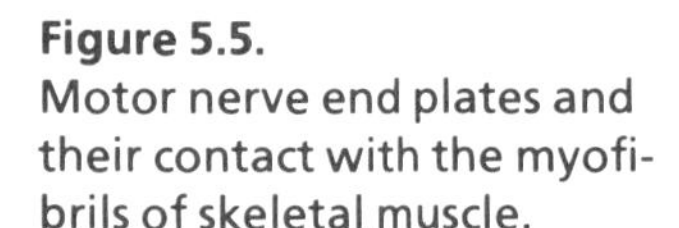

Figure 5.5.
Motor nerve end plates and their contact with the myofibrils of skeletal muscle.

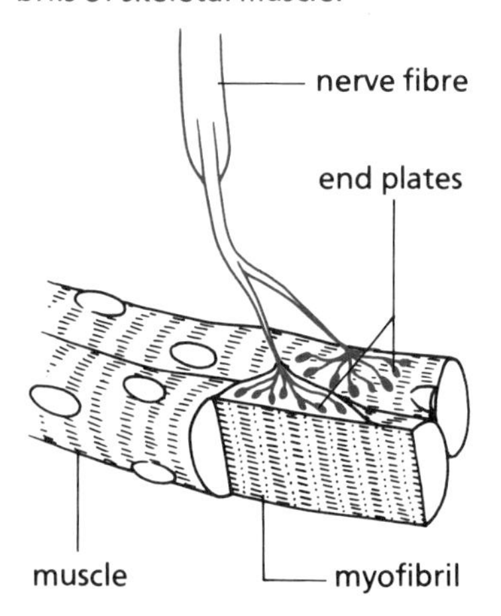

The Stimulation of Muscle Contraction

If you wish to bend your arm to scratch the end of your nose, nerve impulses must deliver the order before the muscle will contract to move the arm. This stimulus (a trigger which causes a reaction) arrives at the muscle along a motor nerve which possesses many tiny extensions. Each extension has a small patch buried in the muscle fibres, these are called motor **end plates**. (See Figure 5.5.) A particular muscle may have from a few to many thousands of end plates. As the impulse arrives at the end plate, it causes the instant release of a chemical (acetylcholine), which passes through the muscle fibre membrane and produces the muscle contraction. An enzyme then quickly destroys this chemical so that the muscle can relax and be free to contract when stimulated again.

Isometric and Isotonic Contraction

Any student who has trained for some athletic event will be familiar with the term **isometric exercise**. In this type of training, muscles are pitted against each other, with no movement involved. Pressing against a wall or hooking the fingers together and trying to pull the hands apart are examples of isometric exercises. Such exercises have been shown to increase strength and muscle size very rapidly. **Isotonic** exercises involve movement, lifting weights or running, for example. (See Figure 5.6.) When muscles become more active, they use more energy and produce more heat. This heat is used to maintain body temperature (homeostasis) by the transfer of this heat to other areas of the body by the flow of blood.

A strong stimulus activates more nerve fibres, which in turn, stimulate more end plates and force more muscle fibres to contract. Our bodies conserve energy by using only a few muscle fibre contractions to lift small objects or make small movements and by employing large numbers of fibre contractions only when we need to move large objects or use great strength.

Many of the muscles in the body employ isometric contractions to support the body or some part of it, such as the

A moves the forearm upward towards the shoulder. (biceps)

B pulls the thigh up towards the waist. (adductus longus)

C pulls the lower leg back towards the thigh. (biceps femoris)

D pulls the upper thigh backwards. (gluteus maximus)

E pulls the lower leg forward into the straight position. (rectus femoris)

F pulls on the heel of the foot, tips the toes downward, and raises the body. This gives the "spring" to the forward thrust of the body. (gastrocnemius)

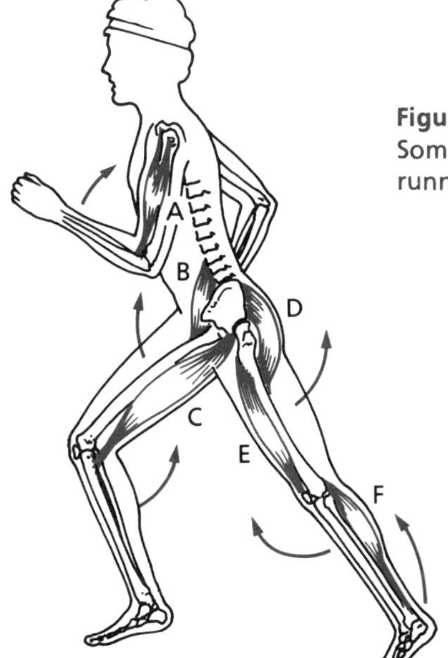

Figure 5.6.
Some of the muscles involved in running.

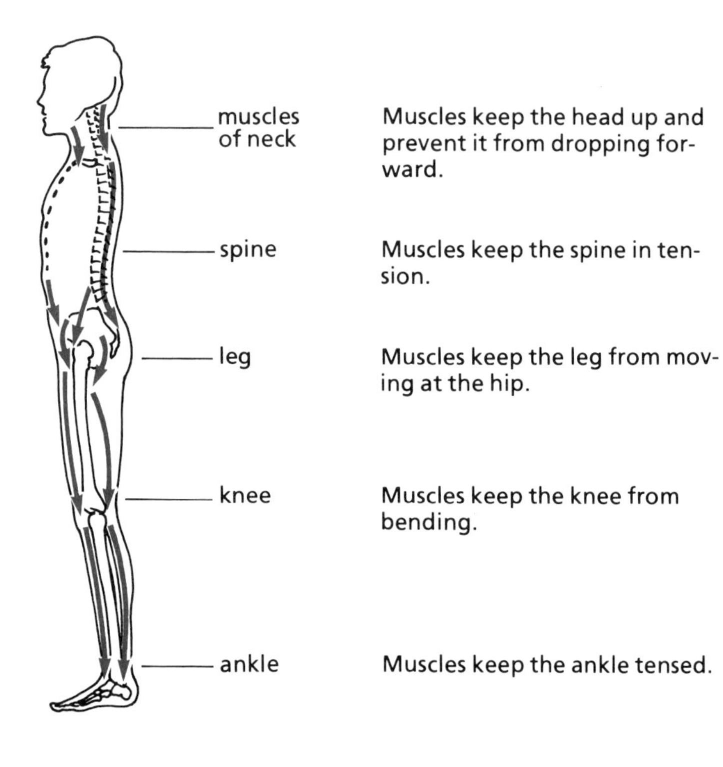

Figure 5.7.
Many antagonistic groups of muscles are required to keep the body upright. These muscles contract and act against each other to provide the necessary support for the body. When a position is held and no movement occurs, it is the result of isometric contraction of the muscles.

head. (See Figure 5.7.) When you are sitting watching television, you might think that you are relaxed, but at least some muscles are working to hold up your head. If you fall asleep, while watching, the head will gradually nod forward, or flop sideways; a hand may slip off your lap and hang down while the body gradually relaxes and slumps in the chair. This is caused by the relaxation of the muscles that were acting isometrically to hold the parts of the body in place. (See Figure 5.8.)

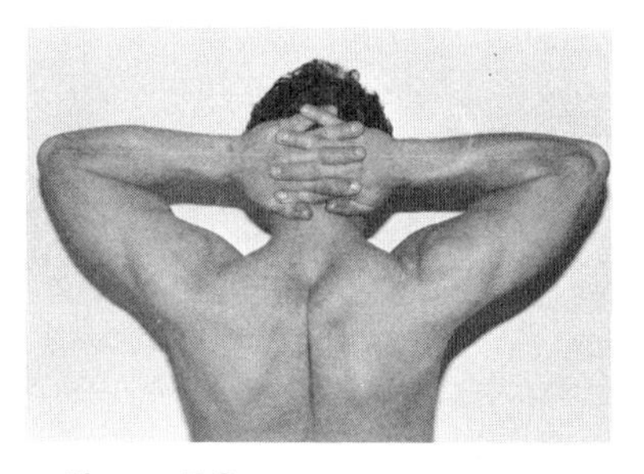

Figure 5.8.
Posterior view showing some of the muscles of the back. Can you identify the trapezius, deltoid, and infraspinatus muscles?

Antagonistic Pairs of Muscles

Many muscles act in pairs. A muscle can only pull (by contracting); it cannot actively push. Thus, once a bone (or other structure) has been moved, movement in the opposite direction can occur only if there is another muscle that can pull the bone in the other direction. For example, if the forearm is moved towards the shoulder (flexion) the biceps muscle contracts; simultaneously, the triceps muscle, which is paired with it, extends. If you wish to move the arm back into a horizontal position, the biceps extends while the triceps contracts to pull the arm down. To hold the arm in a halfway position, both the biceps and triceps contract (isometric contraction) to balance each other. Muscles acting in this way are known as **antagonistic pairs**.(See Figure 5.9.)

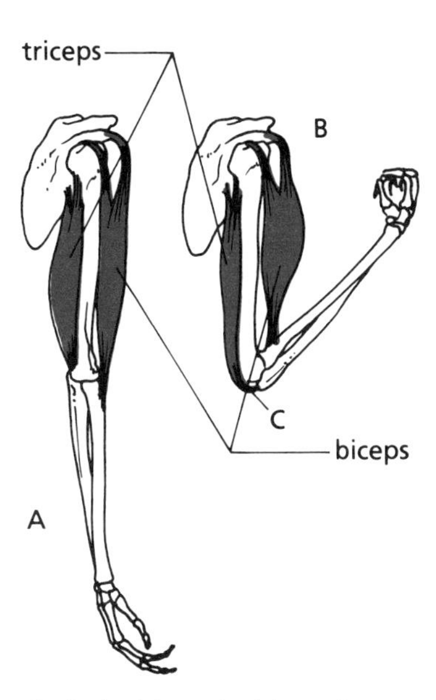

Figure 5.9.
Antagonistic pairs of muscles.

At **A** the biceps is thin and relaxed while the triceps is fat and tensed. This pulls the forearm back. (Extension)

At **B** the biceps is thick and tensed, while the triceps is relaxed. The biceps has shortened and pulled the forearm up. (Flexion)

Note how the tendon of the triceps carries the pull of the muscle around the elbow to attach to the ulna (**C**).

Muscle Attachment

Each muscle is attached at two points. The muscles that move are usually attached by means of tendons. For one bone to move toward another bone, a muscle is required. This muscle will have two attachment points. One end of the muscle must be anchored to a stationary bone, the other to the bone that will move. Then, as this muscle contracts, it will draw the movable bone towards this anchoring structure. The place at which a muscle is attached to the bone is called the **origin** of the muscle. The site of muscle attachment on the movable bone is known as the **insertion**. The insertion of a muscle moves towards the muscle's origin when the muscle contracts. (See Figures 5.10 and 5.11.)

Tendons

Tendons are tough, inelastic bands of connective tissue, which anchor the muscle firmly to bone. They are so strong that sometimes a bone will break before the tendon tears or can be pulled away from the bone. A tendon as thin as a pencil can take a load of several thousand kilograms.

As the tendons are small, they can pass in groups over a joint or attach to very small areas on the bone, areas too small for the muscle itself to find room for attachment. The tendons are tough, but they are subject to wear and tear as they rub across bony surfaces. A group of "bursae", small fluid-filled sacs, are often located between the tendon and the bone against which they rub, to cushion the rubbing action. If the bursae become inflamed (bursitis) they can cause considerable pain. "Housemaid's knee" is caused by inflammation of the bursae of the patella (kneecap). Football players may get bursitis in the shoulder from the strain of throwing the ball, while tennis players may suffer from a similar condition in the elbow.

Figure 5.10.
The biceps and triceps muscles.

Tendons also may become inflamed (tendonitis). This can happen when athletes work out in cold weather without adequate warm clothing, or do vigorous exercise without warm-up exercises.

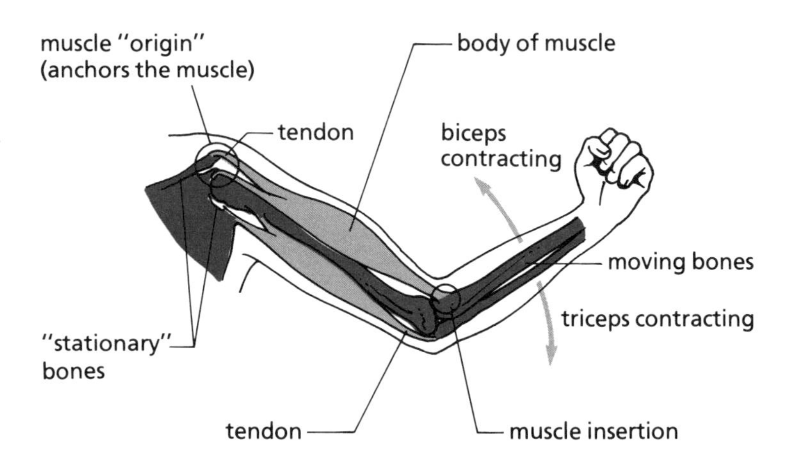

Figure 5.11.
Muscle origin and insertion. The bone in which the muscle is inserted moves toward the bone in which the muscle originates.

Muscles in Action

Bones provide the body with a system of levers and muscles supply the force required to move the bones. You probably have discussed levers in an earlier science course. Figure 5.12 shows some examples of how the body uses the various classes of levers. Sometimes the force exerted is very

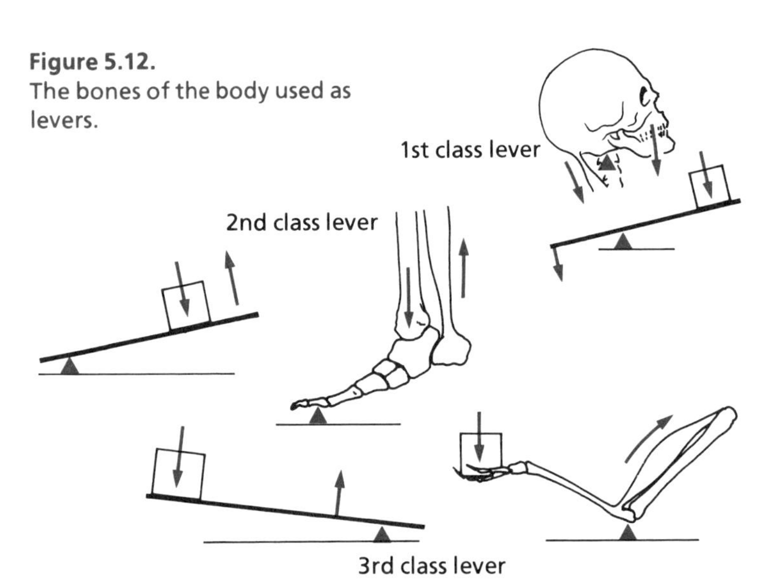

Figure 5.12.
The bones of the body used as levers.

1st CLASS LEVER. The atlas bone on top of the vertebral column is the fulcrum. The mass of the head is pulling down at the front. The muscles at the back of the neck pull the head down at the back, in a seesaw motion.

2nd CLASS LEVER. The ball of the foot acts as the fulcrum. The end of the tibia carries the load and the calf muscle pulls up to raise the body up onto the toes.

3rd CLASS LEVER. The elbow acts as the fulcrum. The load rests on the hand and the biceps muscle pulls the forearm upward, raising the hand and the load.

large indeed. An example will demonstrate how large this force may be. In Figure 5.13, the approximate dimensions of the forearm are given. The elbow acts as a fulcrum (turning point) and a load of 5 kg is placed in the hand, acting at 35 cm from the fulcrum. The biceps muscle is attached at 3 cm from the fulcrum. Let the force exerted by the muscles be xN. The force required to lift 5 kg is 5×10 N.

Clockwise forces = anti-clockwise forces.

Distance from force to fulcrum × force = Distance to load from fulcrum × load.

Then $3 \text{ cm} \times x\text{N} = 35 \text{ cm} \times 50 \text{ N}$

$$3x = \frac{1750 \text{ cm-N}}{3 \text{ cm}}$$

$$x = 583 \text{ N}$$

The biceps muscle then exerts a force of 583 N (58.3 kg) to raise a load of 5 kg!

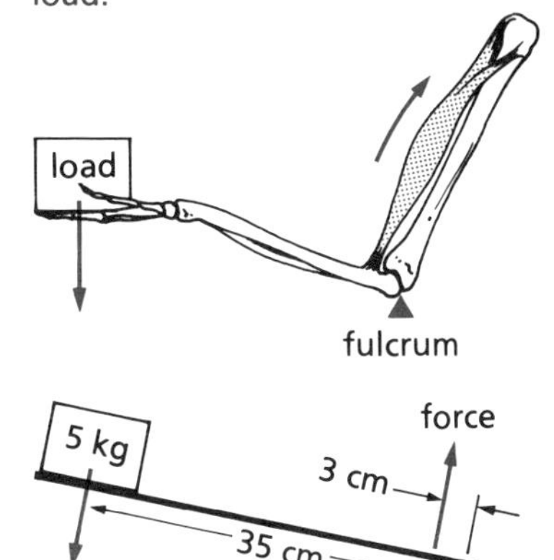

Figure 5.13.
The moments of force involved when the forearm and biceps muscle are used to raise a 5-kg load.

Muscle Fatigue and Energy Needs

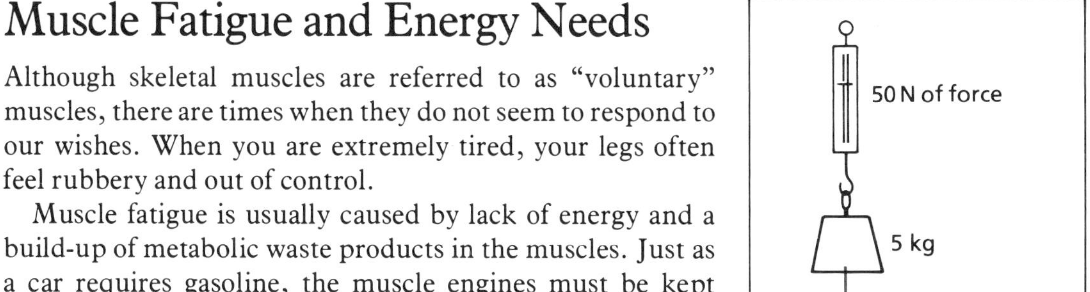

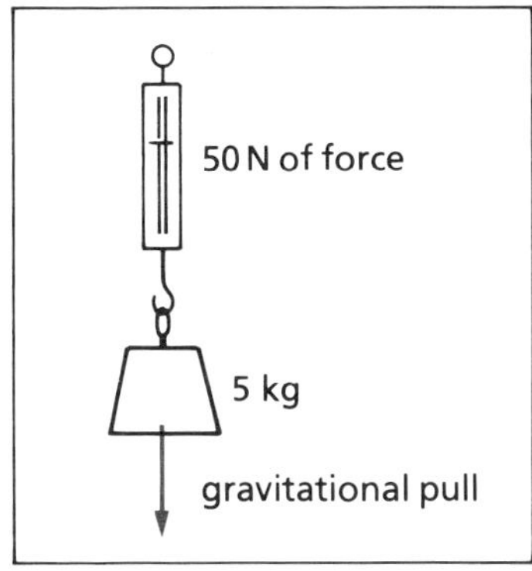

Although skeletal muscles are referred to as "voluntary" muscles, there are times when they do not seem to respond to our wishes. When you are extremely tired, your legs often feel rubbery and out of control.

Muscle fatigue is usually caused by lack of energy and a build-up of metabolic waste products in the muscles. Just as a car requires gasoline, the muscle engines must be kept supplied with energy or they simply come to a halt. The energy used by the cells is carried on ATP (adenosine triphosphate) molecules. The energy transported by the molecule came from respiration reactions that occurred in the mitochondria of the cells. A limited amount of energy is stored in ATP molecules inside each muscle cell. More energy is stored in glycogen, a type of storage sugar molecule. The muscle cells can then convert this to ATP to use as required. The supply of stored energy is quickly used up and the muscles must rely on fresh deliveries from blood circulating through the muscles. During heavy or prolonged exercise, the body uses reserves of energy kept in the liver and delivers this to the muscles. In order to make use of sugar for energy, oxygen is also required. If the body cannot deliver oxygen quickly enough to satisfy the demand, lactic acid (a waste product provided during the chemical breakdown of sugar) is built up and this can cause severe muscle pain. When lactic acid builds up and the amount of available energy declines, the muscles become fatigued.

You may have seen runners with severe cramps at the end of a race. When an athlete stops running, the demand for oxygen decreases and the **oxygen deficit** that has built up is gradually repaid. The cramps then relax until the pain is no longer felt. This is why athletes breathe deeply, gasping for extra air to supply the oxygen needs created by strenuous exercise.

NEW HANDS FOR THE HANDLESS

Thalidomide is a drug that was taken by many women to ease some of the discomforts of pregnancy until its dangerous side-effects were discovered. Many babies were born with seriously deformed limbs, hands, and feet, directly as a result of the drug. For these people, now adults, and for others who have lost a hand, a new invention has made life easier.

Dr. Gustave Gingras and a team of researchers at the Northern Electric Company, using a Russian-produced design, developed an artificial hand that is activated by the body's own electrical currents. To move our own muscles, impulses from nerve endings stimulate the muscles to contract. The new hand works by surface electrodes attached to muscles that pick up these electrical stimuli from the nerve endings.

Dr. Gingras' prosthesis (artificial part) has proportional control. That is, the force of the grip is proportional to the body current that activates it. It allows the person to control how tightly an article is held. For example, if a person sees an easily crushed object, the nerve response "says" it doesn't need much current, thus the electrodes receive less current from the body than if the person sees a massive, solid object. The result is that the "hand" grasps just as hard as it needs to, no harder.

It is hoped that more compact devices with electrodes imbedded beneath the skin will soon be available.

Dr. Gingras has worked for the United Nations in establishing rehabilitation centres in Venezuela and other South American countries. He has worked actively in similar programs for the Red Cross in Morocco and South Vietnam, has taught physical medicine and rehabilitation at the University of Montreal, and directed the Rehabilitation Centre in that city. Dr. Gingras has been called the "father of rehabilitation of the physically handicapped in Canada". (Extracted from an article by Don Wright in the *Mirrored Spectrum*, Ministry of State, Ottawa.)

Muscle Tone and Posture

Muscle tone is the result of sustained contraction, achieved by alternating contractions in different parts of the muscle. If muscle tone is good, the muscle will always be under a small degree of tension. The alternating stimulation of different muscle units prevents tiring and leaves the muscle in constant readiness for action.

Good muscle tone is the difference between a healthy muscle and a flabby one. Unused muscles quickly lose their tone. Regular exercise improves tone and enables us to work harder or longer without tiring. Good circulation and nutrition also contribute to the maintenance of good muscle tone.

Shoulders square and well back – this prevents cramped lungs and promotes good respiration. If the spine is kept in an upright position, backaches and strain on the discs can be avoided. Good posture also prevents the displacement of the soft internal organs. If the body mass is distributed evenly on both hips, as is the case if the spine is straight, the legs are not abused and varicose veins are less likely to develop.

Abdominal, back, and limb muscles that have been exercised regularly are able to hold the body naturally in a healthy position without tiring. Good muscle tone and posture give a lively spring to the step and promote general good health.

Describing Muscle Movement

Although there are many sorts of movements made by joints, only five basic types need concern us here.

Flexion involves the bending of a joint, which decreases the angle between the moving parts. For example, flexion takes place when the forearm is bent towards the shoulder.

(flek-shun)

Extension is the straightening or stretching action that increases the angle between moving parts. It is the opposite of flexion. Extension occurs when the forearm is straightened after being bent at the elbow.

Abduction refers to movements away from the body midline, raising the arms sideways, for instance.

Adduction is the opposite of abduction, as when the arms are lowered and brought towards the midline of the body.

Rotation involves a turning action. Turning of the head occurs when the atlas bone rotates on the axis bone in the neck. (See Figure 5.14.)

Figure 5.14.
Standard terms for the movements of the limbs and joints.

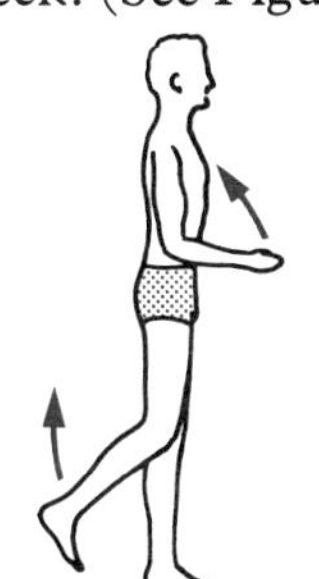

flexion
(moves bones towards each other)

extension
(straightens – moves bones away from each other)

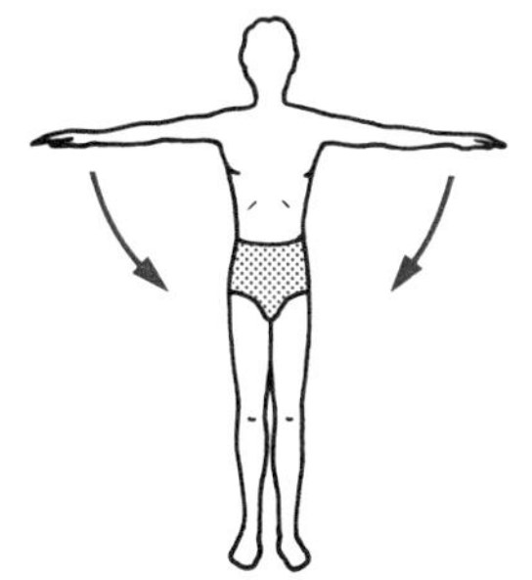

adduction
(towards midline of the body)

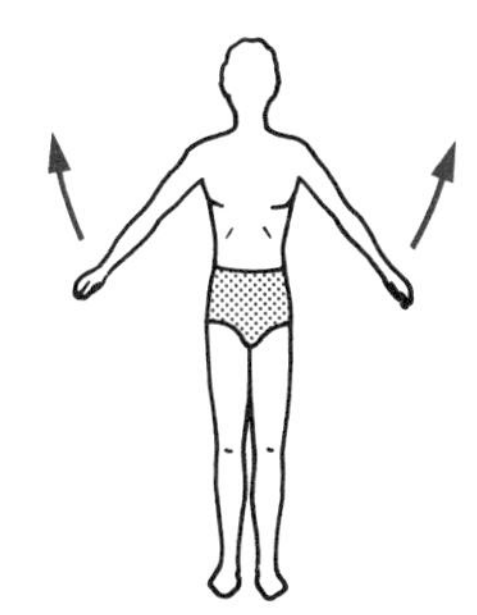

abduction
(away from midline of body)

Muscle Disorders

(dis-troe-fee)

There are several types of **muscular dystrophies**. These are genetic disorders which involve gradual weakening and atrophy of the muscles. Muscular dystrophies cripple the body at different ages depending upon the type of disorder. As yet, no cure is available for any of this group of disorders.

(par-al-i-sis)

Paralysis refers to a condition in which voluntary use of muscles is lost. This disorder is usually of a nervous, rather than muscular, origin. If, for instance, the nerves of the spinal cord are cut at a particular point, the transmission of stimuli to muscles in the lower limbs may be blocked. The person will be paralyzed below whatever point the spinal cord is severed.

Muscle spasm is a very painful, strong contraction of muscles, usually in the limbs. It is often caused by an accumulation of chemicals in the tissues. Massage often helps to relieve the pain.

QUESTIONS FOR REVIEW

SOME WORDS TO KNOW

Match a statement from the column on the left with a suitable term from the column on the right. DO NOT WRITE IN THIS BOOK.

1. Smooth muscles along the intestines.
2. Contraction of muscles which produces movements.
3. Resilient, flexible support material.
4. Attachment of a muscle to a "stationary" or less movable bone.
5. Movement of a limb towards the midline of the body.
6. Movement that involves the bending of a joint, e.g., when the forearm is bent towards the shoulder.
7. Bundles of muscle fibres.
8. Tissues that attach muscle to bone.
9. Muscle contractions which produce no movements.
10. Muscles which are attached to bones.

A. extension
B. flexion
C. voluntary
D. involuntary
E. antagonistic pair
F. isotonic
G. isometric
H. cartilage
I. tendon
J. ligament
K. muscle tone
L. myofibril
M. insertion
N. origin
O. adduction
P. abduction

SOME FACTS TO KNOW

1. What are the three types of muscle? How do they differ?
2. There are very few differences in the structure and function of animal muscles and the muscles of humans. What changes in our culture and way of life are affecting our muscles today?
3. Why are antagonistic pairs of muscles required? Give two examples.
4. Explain the difference between isotonic and isometric contraction in muscles.
5. Explain what is meant by the terms "origin" and "insertion".
6. What are the values of good posture?
7. Muscles are not attached to bone directly, but through tendons. What advantages does this arrangement offer?
8. Explain how muscles are stimulated to contract.
9. Explain how muscle strength is achieved.
10. Make a simple sketch of the inside of a muscle and explain how it contracts.

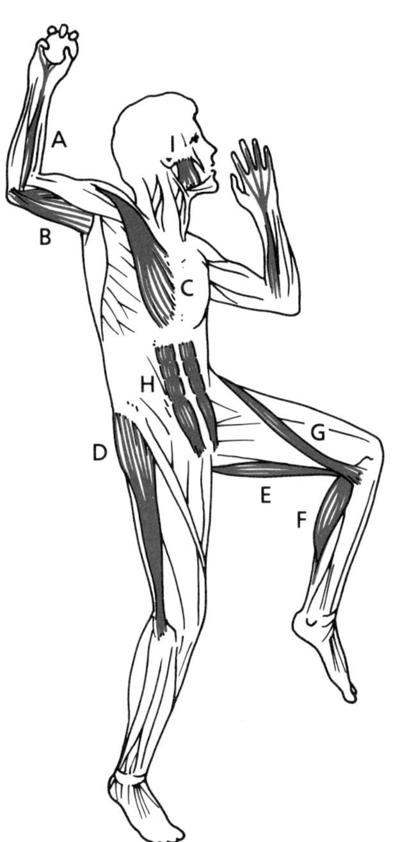

Figure 5.15.
Examine the diagram and determine what movement or movements would result if the muscles indicated by the letters were to contract. (The answers are given on the next page.)

QUESTIONS FOR RESEARCH

1. Select one of the following topics and prepare a report for presentation to the class:
   - physiotherapist
   - masseur
   - trichinosis
   - muscle disorders
   - jogging, its health values
   - problems attributed to jogging
   - posture and modelling
   - muscular dystrophy
   - cramps and growing pains
   - weight lifting and body building
   - occupational therapy
2. Investigate a program of weight lifting. Draw up a list of the exercises used and show which muscles are developed by each exercise. Indicate which exercises are isometric and which are isotonic. What antagonistic pairs of muscles are used in the "hold" positions?
3. Athletics sometimes produce an oxygen debt. What is meant by an oxygen debt? How is it repaid?
4. Investigate the role of calcium in muscle contraction.
5. Find out what causes the muscle cramps suffered by runners at the end of a race.

SOME FACTS TO KNOW

Answers to questions illustrated in Figure 5-15 on the previous page.

A. Flexion of the fingers.
B. Extension of the forearm.
C. Brings the arm down across the chest.
D. Extends the thigh. Flexes the knee.
E. Extends the hip. Flexes the knee.
F. Pulls up on the heel and flexes the foot.
G. Flexes the hip and knee.
H. Bends the trunk forward.
I. Raises the lower jaw.

Activity 1: INVESTIGATING MUSCLES AND TENDONS

Materials

A freshly killed chicken, with feet attached, may be used for a demonstration dissection, or chicken legs can be used for individual student dissections.

Method

This dissection will be arranged by your teacher.

Observations

The dissection will show the following:

- The thigh muscle is made up of many muscles with different functions.
- The muscles are surrounded by a sheath.
- How muscles and tendons are attached. The student should be able to clearly distinguish between the two.
- The flexibility of tendons, their toughness, and that they move in a lubricated sheath.
- Smoothness and design of articulating surfaces.
- The insertions of tendons into bones.
- The antagonistic actions of flexors and extensors should be examined by pulling on opposing muscles or tendons.
- That lean meat is made up primarily of muscle.

Activity 2: EXAMINATION OF MUSCLES AND TENDONS

Method

1. Remove your shoes and stand on one foot. Now raise yourself onto your toes. *Feel the calf muscles and describe what has happened to these muscles. Make a simple sketch of the position of the foot and the lower leg when in the raised position. Sketch in the bones, muscles, and tendons. Describe the movement made by the heel. How are the calf muscles attached to the heel? Feel the Achilles tendon. What has happened to this tendon during the action taken?*
2. *Now point the toes upward, so that the heel is down. Feel the tendon and calf muscles. Feel, also, the muscles at the front of the leg. Describe what you feel and any changes that pointing the toes up has brought about. Feel for the tendons in the front of the foot. How many are there?*

Activity 3: WHAT EFFECT DO LOW TEMPERATURES HAVE ON MUSCLE CONTRACTION AND CONTROL?

Materials
ice cubes, pen

Method

1. *Write your name several times in your notebook.*
2. Now hold your hand in very cold water or hold some ice cubes tightly in your hand for as long as you can. A minute or two should be enough to produce observable results. Now, *write your name in your notebook directly beneath the earlier signatures.*
3. Warm your hand by immersing it in warm water and by massaging it. When it feels warm and loose again, *write your name again in your notebook.*
   Compare the signatures. Describe the differences that are apparent in the signatures and explain what effect the cold temperature had on the muscular strength, control, and co-ordination of your fingers.

Activity 4: WHAT CHANGES TAKE PLACE IN MUSCLES WHEN THEY CONTRACT?

Materials
tape measure, several small weights (2-7 kg; small sand bags work well)

Method Part I

1. Look at Figure 5.11 to discover the points of origin and insertion for the biceps and triceps muscles. Note exactly where the muscles attach to the bones.
2. Measure the length of the biceps and triceps muscles from origin to insertion, while the forearm is in the extended position (completely straight). Flex the forearm until the elbow is at right angles and again measure the lengths of the muscles. Now flex the arm completely until the hand comes close to the shoulder and again measure the lengths of the muscles. *Record your results in a table in your notebook.*

Muscle	*Extended length*	*Flexed 90°*	*Flexed completely*
Biceps			
Triceps			

3. *Subtract the shortest from the longest length for each muscle and determine how much each muscle contracted. Record your results.*
4. Place a tape measure around the upper arm over the biceps muscle and measure the circumference of the arm, while it is in the extended position.
   Flex the arm, tightening the muscle as much as possible, and again measure the circumference. *Record your results.*
5. Select another location on the body and make similar measurements: on the leg, or the distance from the sternum to the pubic bone while bending forward, or some other alternative area. *Determine how much shorter the muscle must get during contraction.*
   Express the distance by which the muscle is shortened during contraction, as a percentage of the overall length of the extended muscle.

Method Part II

1. Refer to the diagram and think of the arm as a lever. Note the point that is the fulcrum, the point of attachment of the "force" (biceps tendon), and use the centre of the hand as the point at which the load is acting.
   Now measure your arm and *record the distances marked A and B. The distances will not be exact, but they will give you a very good idea of the leverage involved. Record the distances and the mass used as a load. Mark them in on a simple arm-lever diagram in your notebook.*
2. *If clockwise forces = Anti-clockwise forces*
   Then Load × distance to load = Force × distance to force.
   $$Load \times A = Force \times B$$
   Find the force that must be exerted by the biceps muscle to lift the load that you have selected.
   Repeat using a heavier load.
   How many times larger is the force than the load?
   When weightlifters are working out with very heavy loads, the force exerted on the tendons can exceed a tonne! This should give you some idea of how strong the tendons and muscles must be, even to lift the mass of the body when we climb up stairs.
   What is meant by mechanical advantage? Does the arm, when used as a lever, have a mechanical advantage or a disadvantage?
   What advantages does the arm obtain by having the biceps muscle attached so close to the fulcrum?

Unit IV

Communication and Control of the body

Chapter 6

The Nervous System

- The Nervous System
- The Brain and its Parts
- Protection of the Brain
- The Spinal Cord
- The Autonomic Nervous System
- Chemicals, Drugs, and the Nervous System

The Nervous System

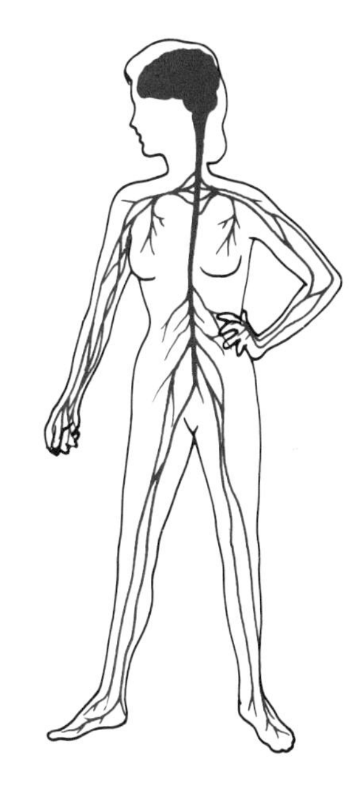

nervous system

As you read this page, hundreds of specialized nerve cells, acting like guards and informants, are receiving and transmitting information to your central nervous system. Even without you being aware of it, everything that is happening in the environment around you, or occurring inside you, is being noted. This information is being sorted, decisions are being made, and responses effected. For example, light entering the eye is being monitored and adjustments made to regulate the intensity of light admitted. Muscles are contracting to permit movements of the eye along this line of print. Other muscles are changing the shape of the lens to bring each word into sharp focus on the retina. Your ears may be registering sounds, while your brain sorts and analyzes the relative priority of each to determine its importance to you. Sensors in many parts of the body may be sending information about the pressure of a piece of clothing, the hardness of the chair, the cramp in your leg, or the itch at the end of your nose. A feeling of hunger may be recognized as your body makes known its need for additional food. Nagging concerns about a future test, appointments to be kept, or a special date for a school dance may be disturbing you. This list is only a small part of the many activities in which your nervous system is engaged, even as you sit and read – a comparatively quiet and inactive occupation.

The many activities of the nervous system can be summarized into four general functions.

1. **Reception** of stimuli and the conduction of impulses to the Central Nervous System.
2. **Interpretation** of the impulses sent to the Central Nervous System, followed by decisions made in the light of remembered experiences.
3. The **sorting** of impulses and the setting of priorities for action upon them. Insignificant information is ignored, while urgent or important information is given priority.
4. The **transmission** of impulses to the motor units which carry out appropriate actions.

Parts of the Nervous System

The nervous system is organized into three major parts:

1. **Central Nervous System** (CNS) is made up of the brain and the spinal cord. It co-ordinates and directs the activities of the body.

(per-if-er-al)

2. **Peripheral Nervous System** (PNS) is made up of the nerves which extend beyond the brain and spinal cord. These nerves bring information in from the sensory and internal organs or carry impulses to effect reactions by the muscles.

(aw-toe-nom-ik)

3. **Autonomic Nervous System** (ANS) controls those parts of the body that act without our thinking about them: the stomach, intestines, and glands, for example. This system helps to prepare the body for emergencies and then returns the body to a normal state after the emergency has passed.

There is only one basic type of nerve cell that transmits all types of impulses. It is called a **neuron**.

The Neuron

Neurons are designed to carry nerve impulses, which are tiny electrical charges, from one point to another. They are living cells, having a nucleus like other cells, but possessing special extensions, the nerve processes (see Figure 6.1.),

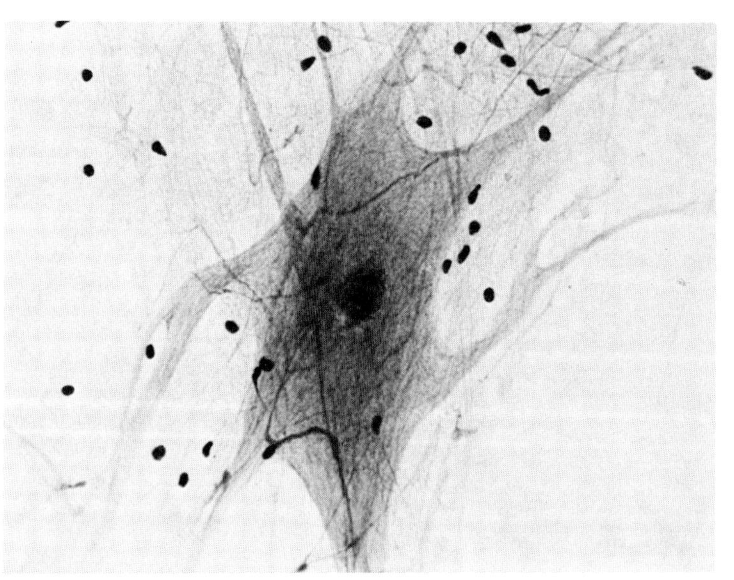

Figure 6.1.
The nucleus, cell body, and extensions of a neuron.

which carry impulses a considerable distance. The **cell body** of a neuron has a special type of cytoplasm inside it called neuroplasm. Neuroplasm flows into the extensions. The shape of the cell body varies. Some are round, while others resemble a diamond or are irregular in shape.

There are two types of nerve processes, **dendrites** and **axons**. Both of these structures contain neuroplasm and are surrounded by a thin membrane. Dendrites are shorter and branch extensively. Their function is to pick up impulses and conduct them *towards* the cell body. Axons carry the impulse *away* from the cell body, passing it on to other neurons or cells. The axon is therefore much longer and less branched. (See Figure 6.2.)

A close examination of what is usually called a "nerve" shows that it is actually many axons bound together, rather like a trunk telephone cable. The axons of many neurons are covered with a white, fatty protein sheath known as **myelin**.

Figure 6.2.
The structure of a motor neuron.

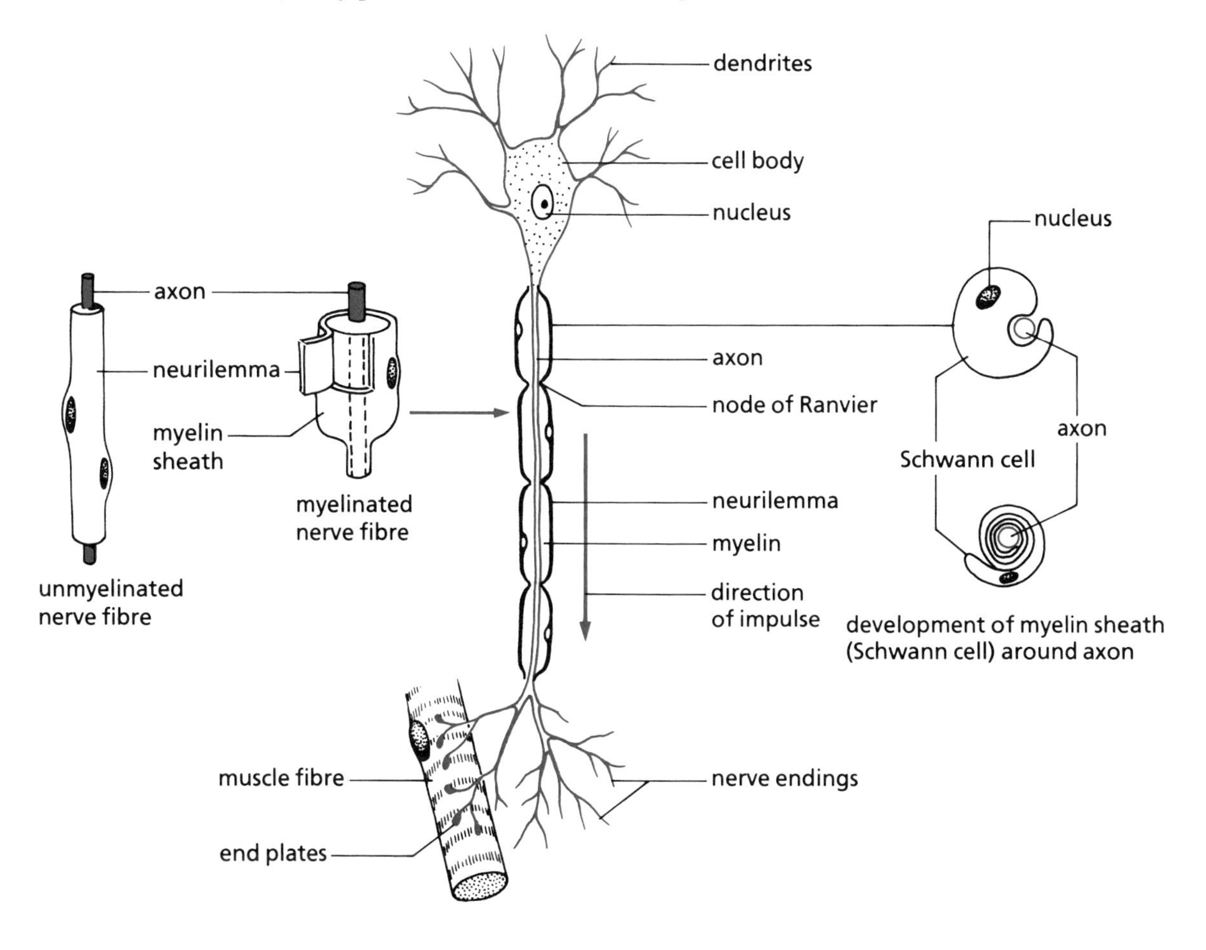

Nerves covered in this way are called myelinated nerves. The main function of this sheath is to insulate the axon, preventing the loss of chemical ions that are present in the nerve fibre. Since these ions are necessary for the transmission of impulses along a nerve cell, the presence of a myelin sheath increases the speed of transmissions. The myelin sheath is formed of flat **Schwann cells**, which are wrapped around the axon like a jellyroll. Between the Schwann cells are small gaps, the **nodes of Ranvier**, which may boost the impulse and help to push it along the fibre. Axons not protected in this way by a myelin sheath are called unmyelinated axons.

(sh-wahn)

(ron-vee-ay)

Myelinated nerve fibres found outside the brain and spinal cord are covered with a delicate membrane known as the **neurilemma**. The function of the neurilemma is to promote the regeneration of damaged axons. Nerves which are not provided with this protection are unable to regenerate. For this reason, damage to the central nervous system is more serious than injury to a nerve of the peripheral system. Within the brain and spinal cord, the myelinated fibres form areas referred to as **white matter**. The cell bodies of the neurons and unmyelinated fibres make up the **gray matter** of the brain and spinal cord.

(nur-i-lem-mah)

Transmission of the Nerve Impulse

The nerve impulse is picked up by the dendrites, passes through the cell body, and is then conducted along the axon to the fine net of nerve endings at the end of the axon. In some locations axons are very long and may reach a length of one metre. Even in such cases the cell body may be no larger than that of other cells.

When the dendrites of a neuron are stimulated by some sensation (the heat of a flame, for instance) a tiny electrical charge is produced. The thin membrane which surrounds the nerve fibre is semipermeable. That is, some molecules can pass through but not others. This membrane can control the passage of certain electrically charged atoms across it, letting them into or out of the cell. Charged atoms are called ions. They have either gained or lost an electron. Chloride ions, for example, have an extra electron and are negatively charged (Cl^-), whereas sodium and potassium ions lack electrons and are therefore positively charged (Na^+, K^+).

When the cell is at rest (has not been stimulated) the membrane allows potassium and chloride ions to move freely across the membrane in either direction. The membrane holds back the sodium ions, which are found in high concentrations in the tissues. These sodium ions build up a positive charge on the outside of the membrane. Inside the membrane there will be a negative charge, because of the greater number of negative than positive ions (because of the Cl^-).

When an impulse is started, the permeability of the membrane changes; sodium ions then flow into the fibre and some potassium ions are pushed out. The result is a reversal of the charge on the membrane at the point where the impulse is passing (during its trip along the fibre). The tiny difference in potential is about 80 mV (80 thousandths of a volt) and only lasts for about 0.001 s. (See Figure 6.3.)

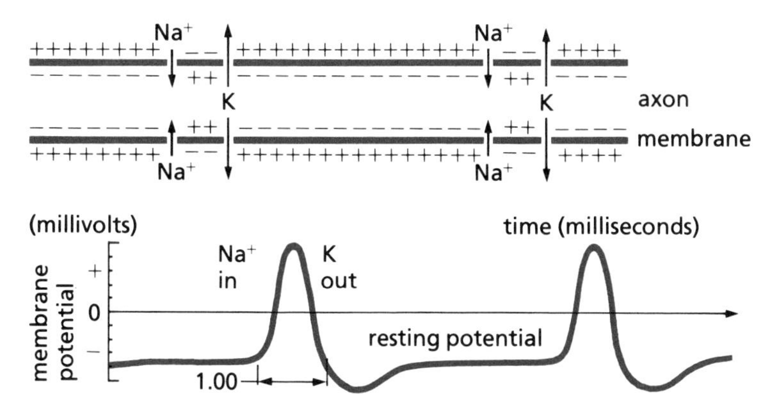

Figure 6.3.
The electrical potential of the membrane changes as the sodium ions move inside the nerve fibre ahead of the impulse. This causes the inner portion of the membrane to become momentarily more positively charged. As the impulse passes, potassium ions migrate outward, restoring the membrane to its original electrical potential.

If electrodes were attached to an unstimulated nerve fibre and the tiny current amplified, the charge would have a negative reading. At the instant the impulse passed the electrode, the needle would jump to show a positive reading and then immediately return to its negative resting position.

The Refractory Period

Immediately after an impulse has passed, there is a short period of time during which the nerve will not respond to a stimulus. It lasts only about two milliseconds and is known as the **refractory period**. If you have looked inside a piano, you will have noticed the row of felt-tipped hammers which strike the strings to produce the notes. If the piano is a good instrument, no matter how fast you repeatedly strike a key, it will repeatedly move the small felt hammer to strike it against the string and produce a note. But the hammer must spring back each time to be ready to hit the string again. The recovery time, the time during which the hammer is not available to strike a note, is similar to the refractory period of a nerve fibre. The fibre must recover from the passing of the previous impulse before it is capable of being charged again.

Reaction Time

The nerve impulse takes some time to travel to its destination in the Central Nervous System and then on to the muscle where it causes a reaction. This is known as **reaction time**. If you have taken driving instruction, you have probably experienced a test for driver reactions to visual or sound stimuli. An emergency can happen at any time while you are driving a car. Perhaps a child runs into the road after a ball. The driver sees the child (stimulus); a message is directed to the CNS; motor neurons are stimulated; and muscles then contract to move the foot operating the brake. The car may travel many metres during this time, as the speed at which an impulse travels is relatively slow – from 1-120 m/s. Alcohol and drugs slow the reaction time drastically. Other factors such as fatigue or pre-occupation may also have considerable adverse effects on the reaction time.

The Size and Strength of the Stimulus

A nerve impulse is self-propagating! Once started, it moves along without outside help. As each point along the nerve fibre membrane changes its permeability, it triggers the next point to change and the impulse sweeps along the fibre. It doesn't stop or fade out, speed up or slow down. A nerve fibre may send 50 to 100 impulses in one second. All the impulses passing along a fibre are of the same strength. If the stimulus is strong, more impulses are sent, but not larger impulses. If the stimulus is too weak, no impulse is sent at all; that is, a receptor must receive a stimulus of sufficient size to be triggered at all. Once this threshold level of stimulus strength is reached, the impulse is sent. A stronger stimulus sends more impulses. This is known as the "all-or-none" theory of nerve transmission.

The Synapse

(sin-aps)

The impulse eventually reaches the end of the axon and crosses to another neuron. The whole nervous system is made up of millions of neuron chains. Each neuron is separate, but connects with others to form these chains. The gap between the end of the axon of one neuron and the dendrite of another neuron is called the **synapse**. (See Figure

6.4.) This gap must be bridged for the impulse to be transmitted farther. When the impulse reaches the end of an axon it causes a chemical called **acetylcholine** to be released, thereby stimulating the dendrites of the next neuron. Immediately after the acetycholine has served its function, another chemical called **cholinesterase** is released at the synapse. Cholinesterase is an enzyme that destroys acetycholine, thus preventing it from stimulating the dendrite continuously.

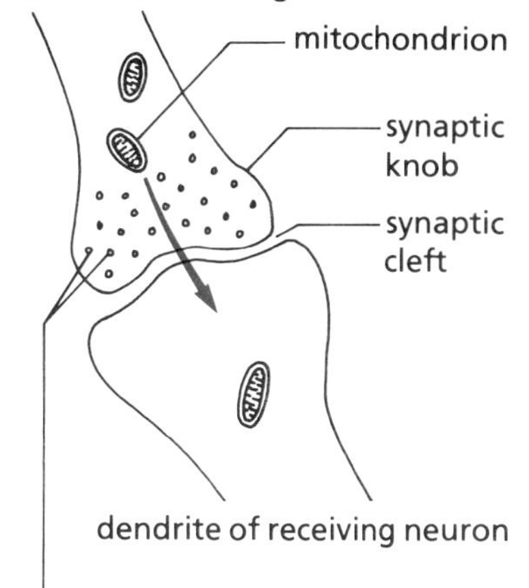

Figure 6.4.
The synapse. A chemical such as acetylcholine is released by the axon of one neuron and carries the impulse across the gap to the dendrites of the next neuron. It then stimulates the production of an impulse in this neuron for further transmission.

The Brain and Its Parts

The brain, with a mass of about 1.7 kg, is estimated to contain about 10 000 000 000 neurons and about another 9 000 000 000 supporting cells. Its size varies with a person's age, sex, and body size. Brain size, like body size, is greater in the average male than in the average female. The greater size, however, is not an indication of greater mental attributes or superiority. The brain continues to grow until we reach about twenty years of age. Reasoning and intellectual ability, however, depend upon pathways, or routes, among the cells of the brain; these pathways continue to develop with use. The brain must be exercised to ensure its fitness and development, just as the muscles require regular exercise to keep them in shape.

The differences in mental capacity between other animals and human beings are related to differences in the size and complexity of certain parts of the brain. The thin outer layer of the cerebrum, called the **cerebral cortex**, is the most significant development of the human brain. The folded (convoluted) arrangement of this structure has increased its functional capacity over that characteristic of a simple unfolded cortex. It is this particularly advanced portion of the human brain that accounts for the complex thinking, reasoning, and learning abilities of people.

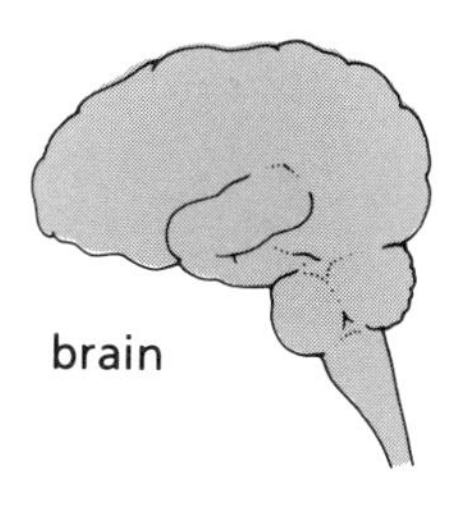

Parts of the Brain

The brain is divided into three distinct parts: the hindbrain, midbrain, and forebrain. In humans the dividing lines between these segments are not immediatey obvious. They become clear, however, if the brains of simpler organisms are examined.

Figure 6.5.
Developmental changes in the brains of some representative animals.

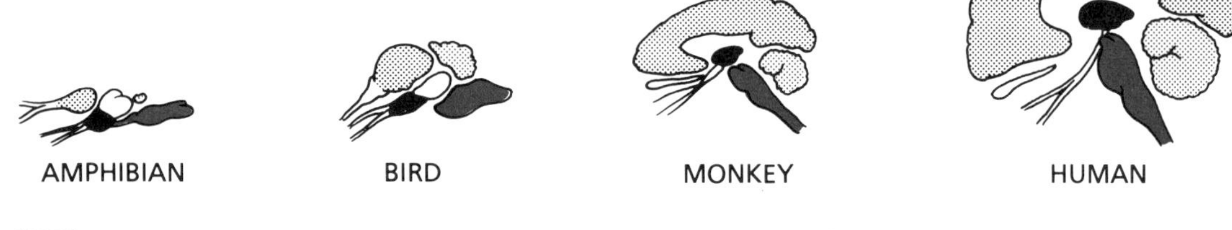

thalamus
medulla
cerebellum
cerebrum

That the brains of all vertebrates have much in common becomes remarkably clear if the embryos of different vertebrates are studied. All have these three parts in almost identical proportions during their early development. (See Figure 6.5.) As embryos increase in size, several changes take place. Although the size of the midbrain decreases in proportion in more advanced animals, the size of the cerebellum and particularly the cerebral hemispheres increases. (See Figure 6.6.) In humans, the forebrain dominates the cranial cavity, and the midbrain is of only limited importance.

Figure 6.6.
Disssection of a sheep's brain.

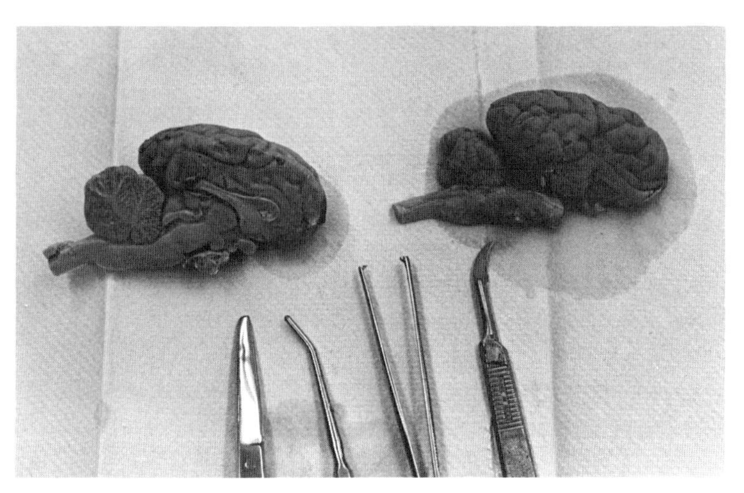

medulla oblongata

The Hindbrain

The hindbrain consists of two major parts, the **medulla oblongata** and the **cerebellum**.

The Medulla Oblongata

The medulla appears as a swollen extension of the spinal cord. Although it is quite small, its functions are vital. Nerve impulses that stimulate the diaphragm and the muscles for breathing originate in the medulla. The heartbeat and regu-

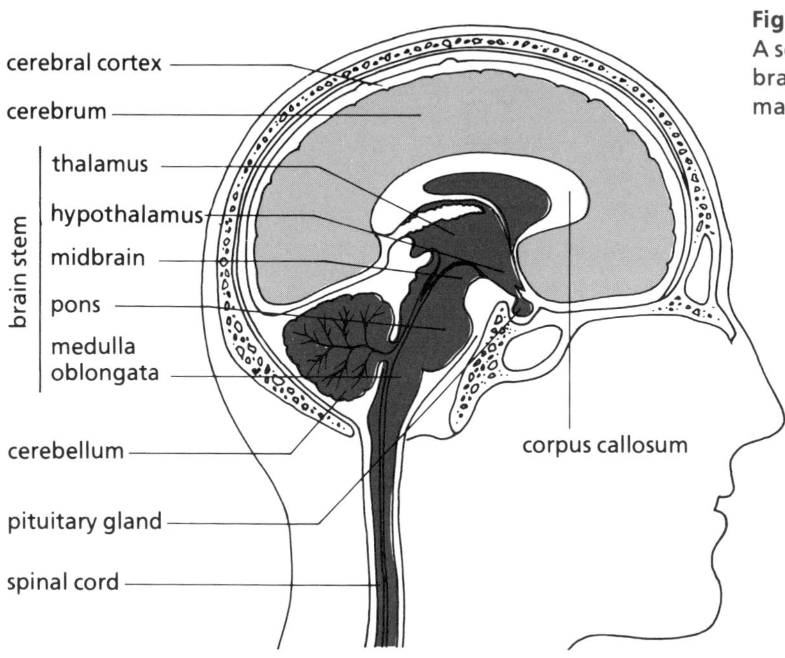

Figure 6.7.
A section through the human brain and head showing the major divisions of the brain.

lation of the diameter of blood vessels are also controlled by this part of the brain. The medulla is like a complex telephone exchange, sorting and relaying incoming and outgoing calls. The medulla provides a pathway for impulses moving from higher parts of the brain to motor nerves and muscles and for impulses travelling to the brain from sensory receptors in various parts of the body. (See Figure 6.7.)

The Cerebellum

The cerebellum is located just above the medulla. It has curved grooves running all over the surface, giving it a furrowed appearance. This part of the brain contains white matter squeezed in between the lobes of gray matter. (See Figure 6.7.)

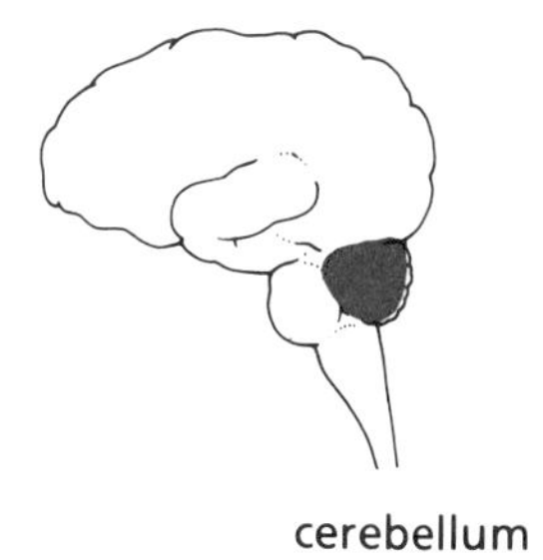

The cerebellum is responsible for balance, co-ordination of movement, and muscle tone. Its function is to organize impulses which originate in the cerebrum and integrate these with the stream of signals coming in from sensory organs. Some very complex muscular sequences are controlled by the cerebellum. For example, a considerable amount of co-ordination is required for a gymnast to make a successful vault. The run up to the boxhorse must be timed so that both feet land together on the springboard. The gymnast must jump up, place the hands upon the box, and somersault the body over the horse. The knees must be flexed, the feet kept together and ready for landing. The final balance and upright position must be controlled so that the gymnast does not fall forward. This complicated series of movements involves dozens of muscles, all programmed to contract in exactly the right sequence.

Although the body may turn and twist and even be upside down, the brain must be aware of the body position and have total control of balance at all times. Co-ordination is produced by an extremely complex organization of the muscles, by the cerebellum, and by the cerebrum, in varying degrees. Pianists, dancers, and skaters must artistically plan the movements that they make. Therefore, thinking and abstraction, memory and emotions may also be very much involved.

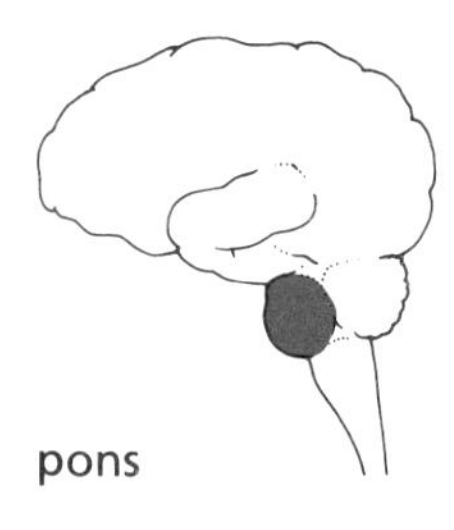
pons

The Pons

The word *pons* means "bridge", which is really a good description of the function of this structure. Lying between the medulla and the midbrain, the **pons** contain fibres which transmit messages from one side of the cerebellum to the other, from the cerebellum to higher centres in the cerebrum and midbrain, as well as between the cerebellum and lower centres, such as the medulla. A few special cranial nerves originate in the pons, which also contains part of the system that controls breathing.

The Midbrain

In terms of size and functions, the midbrain is of only minor importance. It is located just below the centre of the cerebrum and forms part of the brainstem. The midbrain consists of four small spheres of gray matter which act as relay centres for some eye and ear reflexes. Below these spheres of gray matter are some conducting tissues of white matter that connect the higher centres of the cerebrum with the pons, cerebellum, and spinal cord.

The Forebrain

There are two small areas of gray matter squeezed in between the midbrain and the cerebrum; these are the **thalamus** and **hypothalamus**. The hypothalamus is directly connected to the most important gland in the body, the **pituitary**. This gland controls all the other glands in the body and is therefore often called the master gland. The pituitary works in conjunction with the nervous system, the sensory nerves bringing it "feedback" information that aids its regulatory functions.

The hypothalamus controls the autonomic nervous system and the internal organs of the body. It directs the production of special secretions, the activities of the intestinal tract, and the blood pressure, as well as behavioural and emotional responses. Another of its important functions is the regulation of water balance and the control of urine production in the kidneys.

The thalamus forms a sensory relay centre for impulses on their way to the cerebrum. It affects consciousness, temperature, and the degree of awareness of pain.

The Cerebrum

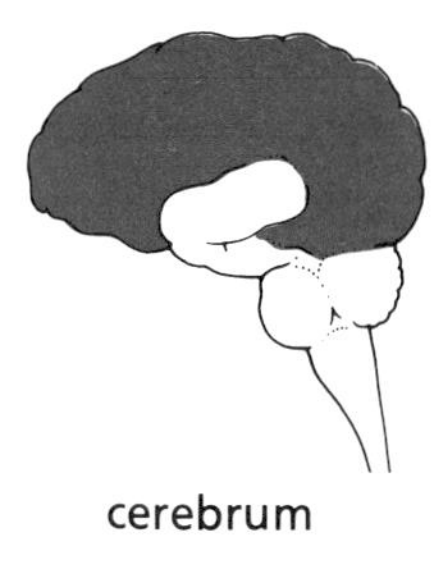

cerebrum

The cerebrum which is the largest part of the human brain, consists of billions of neurons and synapses. It is the highest centre of nervous control and is developed to a far greater degree in humans than in any other animals.

The surface of the cerebrum which is composed of gray matter (2-4 mm thick) is known as the **cerebral cortex**. It spreads like a coat over the surface of the brain. In order to pack in more cells with specialized functions, the coat is thrown into many folds, which increases the surface area. The folds or creases can easily be seen. The deep folds are known as fissures, while other quite shallow folds are called *sulci*. The deepest longitudinal fissure divides the cerebrum almost completely into two halves forming the **cerebral hemispheres**. The cerebral hemispheres are connected internally by the **corpus callosum**, a bundle of fibres which crosses from one hemisphere to the other.

(cor-pus cal-oe-sum)

Each cerebral hemisphere is further divided into four lobes, each of which is also marked by fissures. The most important of these, the Central Fissure, runs at right angles to the axis of the brain from top to bottom. The lobes carry the same names as the bones of the skull that cover them! **Frontal** at the front, **temporal** at the sides, **parietal** at the top and back, and **occipital**, which is quite low at the back of the skull. (See Figure 6.8.)

The frontal lobe has been developed greatly in humans and gives us an advantage over other animals. This lobe controls all voluntary movements: for example, regular walking and running action, arm movements and speech patterns. At the very front of the frontal lobe there is an area that appears to control our basic intelligence and personality.

In addition to the neurons present in the brain, there are also millions of **glial** cells which support and nourish the neurons. Oxygen and nutrients are delivered by a network of blood vessels and capillaries, which spread over the surface of the brain and surround the tightly packed neurons.

Figure 6.8.
The external features of the brain and a midline section.

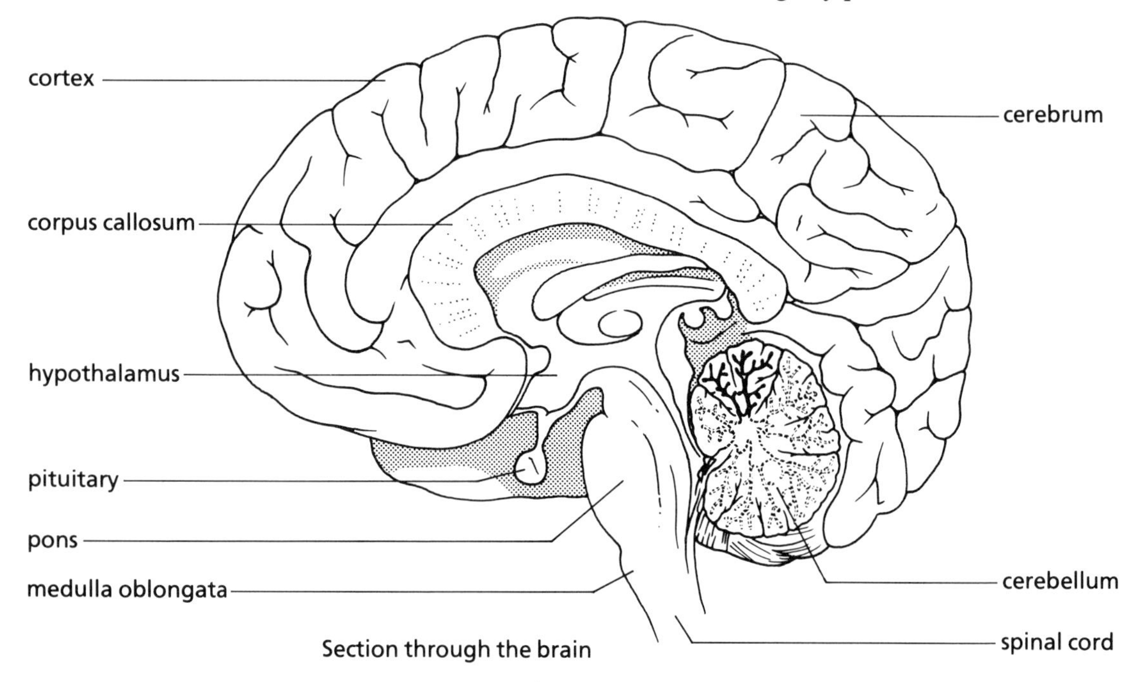

Section through the brain

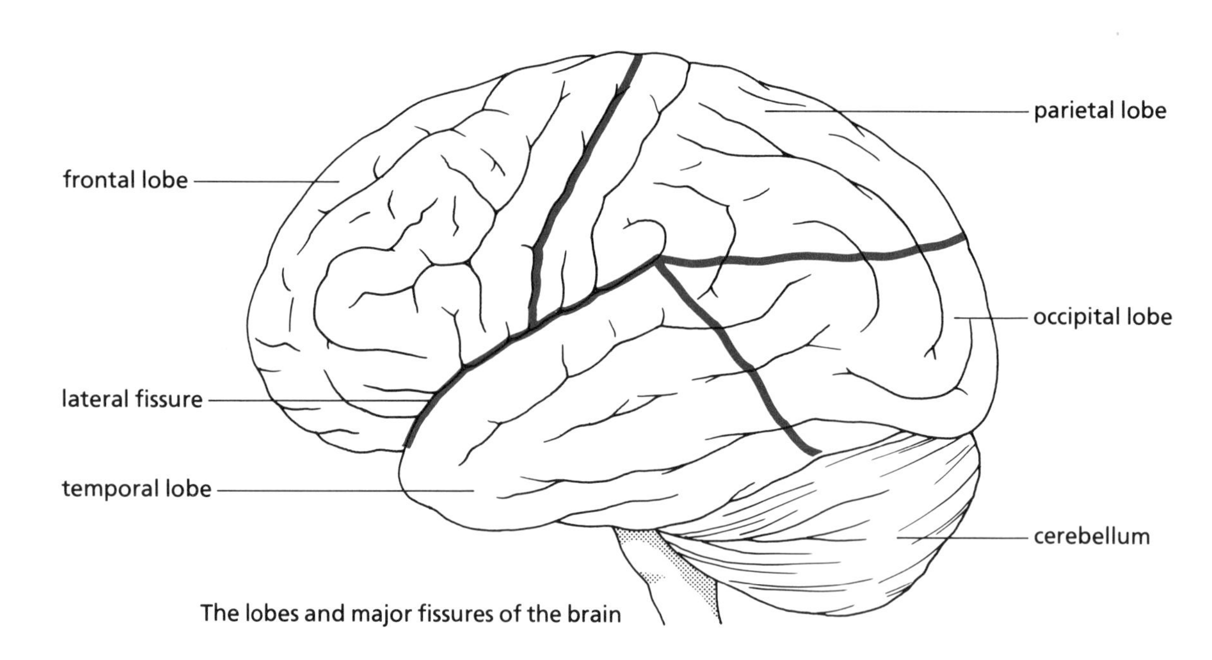

The lobes and major fissures of the brain

LEFT BRAIN-RIGHT BRAIN

The first evidence that the cerebral cortex is responsible for our consciousness came from individuals who suffered from brain injuries. Through working with brain-damaged victims of accidents and strokes, scientists have found that damage to parts of the *left hemisphere* of the brain often result in the inability of a person to speak or perform other sequential skills such as reading, writing, or doing arithmetic. Damage to certain parts of the *right hemisphere* of the brain leads to impairment of ability to determine spatial relationships or to recognize three-dimensional vision patterns such as are used in sketching or reading a map. Observations suggest that those functions described as "reasoned" are better processed in the left hemisphere and those functions considered to be "intuitive" are better processed in the right hemisphere. For example, damage to a specific known area of the left hemisphere (in right-handed people) almost always results in a speech defect.

A familiar example of cerebral specialty is "handedness". Just over 90 percent of all people use their right hands for writing, throwing a baseball, etc. Nerves from most muscles cross over and enter the left side of the brain as well as the right side. Right-handed people use their left hemisphere for many "handed" activities, and left-handed individuals use their right hemisphere for similar tasks. Therefore, injury to the left hemisphere causes greater problems for right-handed people.

The first visual-learning experiments were conducted on animals other than humans (usually cats). The optic nerve in the brain was cut so that the left eye sent impulses to the left hemisphere only and the right eye to the right hemisphere only of the brain. The major carrier of nerve impulses between hemispheres of the brain, the **corpus callosum**, which is located between the right and left hemispheres, was also cut. By using a reward system, an experimental animal was taught to discriminate between a circle and a square with the right eye covered. When the left eye was covered, the animal could not recognize the circle or the square. The animal had "two brains" functioning independent of each other.

Some split-brain humans, those who because of illness or injury have *no* connection between the left and right sides of their brain and, therefore, no crossing-over of impulses, understood with their right brain simple words that were flashed before them by an experimenter. The word "heart" was then flashed so that only part of the word was visible to each visual field. The right brain saw "he", the left saw "art". The patient was given a choice of two printed cards, one with "art" and the other with "he", and was told to point to the correct one with his *left* hand. The subject pointed to "he". Asked to name the word, the subject's left brain "told" him to answer "art".

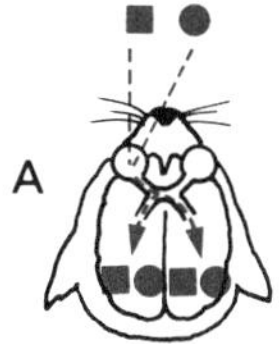

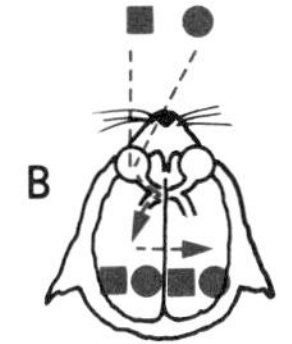

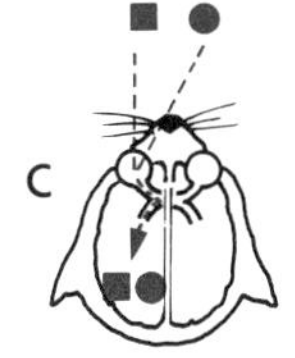

This illustration shows results of "split-brain" experiments with cats. In this case, a researcher taught the cats to press a lever when they saw a circle but not to press it when they saw a square. Even if this training is done with one eye covered, a normal cat can later do this task with either eye covered. If the nerve connections between the eyes (the **optic chiasma**) were cut, the cats were still able to perform the task with either eye. But, when the corpus callosum as well as the optic chiasma were cut, the cats were unable to do the task if the eye patch was switched to the other eye. The researchers concluded that, with no connections between the two sides, they had trained only one half the brain. Illustrations A, B, and C show the results of these experiments. Cat A is normal, Cat B shows a cat with the optic chiasma (nerves between the eyes) severed, and Cat C shows a cat with both the optic chiasma and corpus callosum severed. In Cat C, there are no longer any connections between left and right sides of the brain. When one eye of Cat C is patched, since all connections between the two sides of the brain have been severed, only that side (hemisphere) receives the information the eye sees.

BRAIN MAPS

A young man lies on the operating table of the Montreal Neurological Institute. Part of his brain is exposed and a surgeon is delicately probing the surface of the brain with a tiny electric probe. The young man is fully conscious, although he feels no pain due to the use of local anesthetics. As the doctor probes, the young man experiences a vivid impression in minute detail of a previous experience. He is acutely aware of the people present on that occasion, what was said, and details of the room in which it occurred. As the probing continues at another location, his arm moves or his hand clenches without conscious control, although he knows that it has moved.

Dr. Wilder Penfield, a foremost authority on the human brain, pioneered the mapping of much of the human brain. He determined the areas of the brain responsible for many functions previously not known. One of his particular achievements was isolating the portion of the brain responsible for epilepsy. In people with epileptic seizures, certain regions of the cortex of the brain are damaged. As a result, the electrical charges increase and produce violent reactions or seizures. In addition to mapping the cortex of the brain, and establishing the motor, sensory, and psychological areas present, Dr. Penfield discovered how to remove the cells that caused these violent electrical "storms" within the brain and freed many patients from the distress of epilepsy.

Dr. Penfield has received worldwide recognition for his contributions to medical science. He is a Companion of the Order of Canada and the British Order of Merit, an honour conferred on only 25 people, including such individuals as Winston Churchill, Dwight D. Eisenhower, and Lester B. Pearson.

The Motor Activities of the Cerebral Cortex

If animals or humans are anesthetized and tiny electric currents are then used to stimulate various parts of the brain, it is possible to determine which brain areas are responsible for particular functions. During this process an electrode is placed in contact with some segment of the cerebral cortex; an electrical impulse is delivered; and the response, perhaps the jerking of an arm or leg, is recorded. If the electrode is placed in another location, a facial muscle may contract. Using this procedure little by little, the brain has been "mapped". Should a brain tumor develop or the head be injured, doctors can predict with considerable accuracy which functions of the body will be affected.

The motor functions are controlled by a band of nerve tissues located in front of the central fissure. A separate patch of tissue controls each group of motor muscles. Figure 6.9. shows that some of the body parts that we use frequently and which involve many movements, are allocated a larger portion of the cortex (for example, the hand and mouth). Other muscle groups, which have a limited range of movements, such as the wrist or forehead, are controlled by only small areas of the cortex.

The motor cortex originates voluntary movement of skeletal muscles.

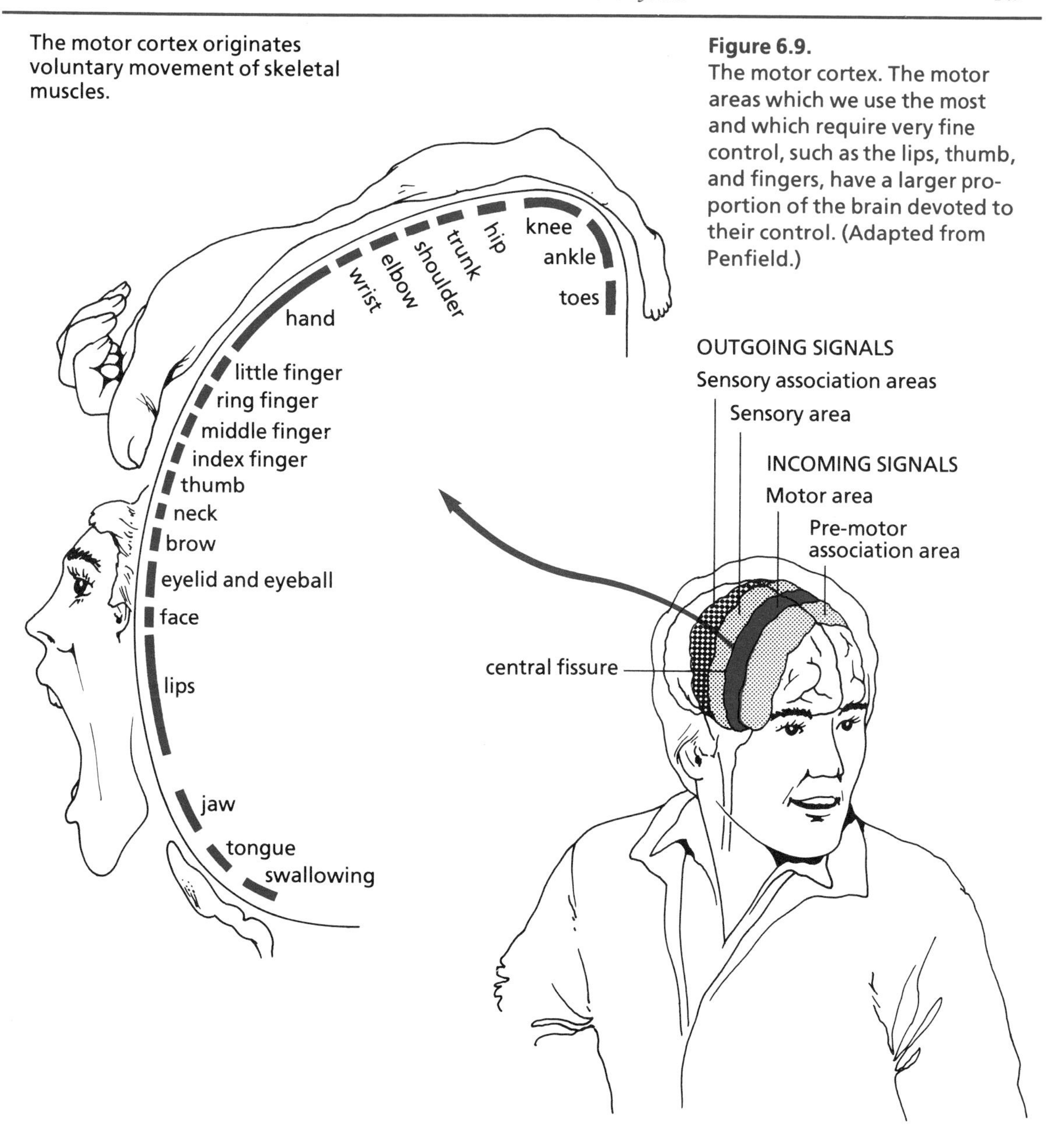

Figure 6.9.
The motor cortex. The motor areas which we use the most and which require very fine control, such as the lips, thumb, and fingers, have a larger proportion of the brain devoted to their control. (Adapted from Penfield.)

The impulses governing voluntary muscular movements are carried by nerve fibres that pass through the brain stem to the medulla. In the medulla, a large number of these nerve fibres cross over to the opposite side and descend through the spinal cord to eventually stimulate the muscles. Thus the majority of impulses arising in the right side of the cerebral cortex control muscles on the left side of the body and vice versa. (See Figure 6.10.)

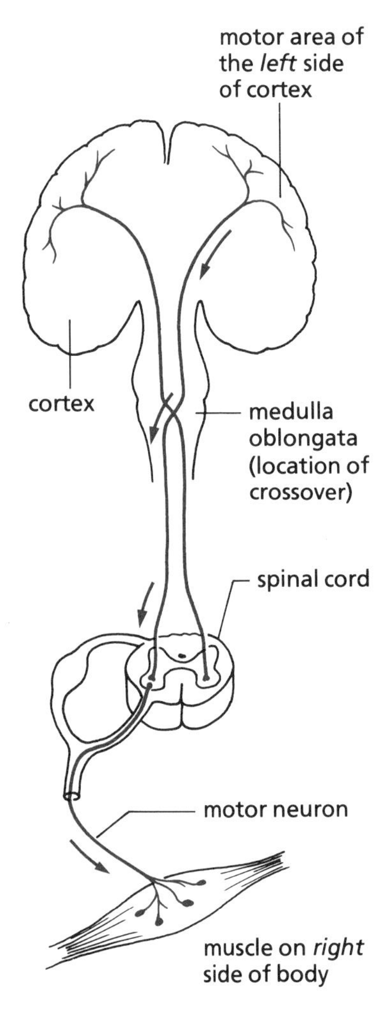

Figure 6.10.
The path of the motor impulses from the left side of the brain to the muscles on the right side of the body.

The Sensory Activities of the Cerebral Cortex

Sensory activities are controlled by cells located behind the central fissure, at the front of the parietal lobe. This area receives all impulses from the sensory organs, including those for touch, pressure, heat, and other sensations. Some of the major sense organs, such as those for sight and hearing, are controlled by special areas that are found respectively in the occipital and temporal lobes.

Co-ordination of several senses is often required, to accurately identify some source, substance, or activity. We may, for example, use smell and colour to identify a chemical. A fire may be heard crackling as smoke is smelled, and the eyes will search for the source. Several senses are often brought into play; each adds further information, enabling the brain to more accurately assess a situation.

This sorting and organizing of sensations takes place in the **association areas** found beside the sensory areas of the brain. Once the information is received it is transferred to the association areas and processed there. The major function of these areas becomes interpretation of the various sensations, redirection of impulses to the motor region for action, or storage of impressions for future reference.

Speech, Appreciation, and Memory

In most people, speech is controlled by a vaguely defined area in the left cerebral hemisphere, while the right hemisphere has charge of our spatial sensations – awareness of our bodies and their particular location in space. If this latter area fails to operate effectively, we have a sense of "floating" or disorientation. Musical and artistic ability also appear to be the responsibility of the right hemisphere.

Perhaps one of the most striking, yet least understood, functions of the brain is its ability to retain, change, or modify ideas, reusing information, some of it received a long time previously. Known as **associative memory**, this capacity makes reasoning, judgment, and moral sense possible. Only a few theories exist to explain how the associative memory operates, and our knowledge about the erasing, or forgetting of information is still very vague.

Short-term memory may enable you to recall what you ate for breakfast yesterday or what clothes you wore, but it is unlikely that you will still remember that information two weeks from now. Long-term memory is really learning; the information it retains is stored for recall at any future time. It is thought that this type of learning may involve changes in the proteins of the brain or at the synapses across certain neural pathways. Reinforcement or repeated events are more readily stored, as are those associated with strong emotional impact, reward, or punishment.

The Inner White Matter of the Brain

Beneath the cerebral cortex lies the white matter, which is composed of myelinated nerve fibres. (See Figure 6.11.) These fibres have three functions. The **commissural** fibres transmit impulses from one hemisphere to the other. The **association** fibres make connections with other fibres in the same hemisphere. The **projection** fibres pass out of the brain to other areas of the Central Nervous System, such as the spinal cord.

Figure 6.11.
A cross-section through the brain showing the location of the white and gray matter. Gray matter is composed of unmyelinated fibres and cell bodies. White matter is formed by myelinated fibres.

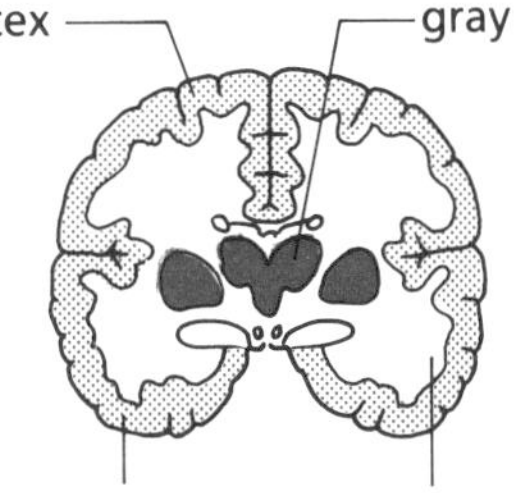

cerebral cortex
core of gray matter
peripheral gray matter
internal white matter

CASE STUDY. Mary, 14 years of age.

Mary was a happy person, friendly with her classmates, and co-operative at school. Mary appeared to be in perfect health, and played on an intramural basketball team. This year, however, Mary's marks had dropped. Teachers and friends had noticed that her attention seemed to wander at times, and she sometimes gazed off into space, oblivious of others around her.

Mary, at the insistence of her parents, made an appointment with her doctor. On the morning of her visit, while brushing her hair, she gave a loud scream and fell to the floor unconscious. Mary's parents ran upstairs and found her lying stretched out on the floor, her arms and legs making rapid, jerking movements. However, in a few minutes, Mary opened her eyes and seemed to be all right again.

When Mary and her mother talked to the physician, he asked some questions about Mary's general health and medical history. He asked when they had first noticed any changes and started to make notes in the file.

Mary told the doctor that sometimes she seemed to "drift off" for a few moments so that when people spoke to her, she could not remember what they had said. The doctor then asked Mary's mother about Mary's appearance when she found her on the floor that morning.

The doctor arranged for Mary to have an EEG (**electroencephalogram**) and gave her a general check up. Eventually, as a result of the tests and reports, the doctor felt confident of her diagnosis and explained to Mary and her mother that Mary had **epilepsy**. The doctor immediately assured them that, with treatment, the disease could be controlled so that Mary would be able to lead the life to which she was accustomed.

The medication prescribed for Mary contained anticonvulsive drugs, but it was almost a year before the right combination of drugs brought Mary's symptoms completely under control. During this period, Mary suffered several seizures and her friends learned to help her and look after her while the attacks lasted. At first, the friends were frightened by the convulsions that Mary experienced. (This was to be expected, as our fears are often due to lack of knowledge and not knowing what to do to help someone.) The friends talked with Mary's parents, who themselves had experienced similar fears until they were given special instructions on how to look after their daughter. They told the friends that even if the attack was severe (a *grand mal* seizure) it usually would last only a few minutes and that a doctor's care would not normally be required. The friends were also told that, during a seizure, an epileptic person is unconscious and feels no pain at all. Even when the body jerks and is →

thrown about, or the person's skin changes colour, or the body goes quite rigid, there is still very little danger to the person, and there is not very much to be concerned about.

To help someone during an epileptic seizure, the following directions should be followed:

1. Keep calm. You cannot stop a seizure once it has started. Let the seizure run its course. Do not try to revive the person.
2. Ease the person to the floor and loosen his or her clothing.
3. Try to prevent the person from striking his or her head or body against any hard, sharp, or hot objects, but do not interfere otherwise with the progress of the seizure.
4. Turn the person on his or her side, so that the saliva may flow freely from the mouth.
5. Do not put anything in the person's mouth.
6. Do not be frightened if the person having a seizure seems to stop breathing momentarily.
7. After the seizure stops and the person is relaxed, allow the victim to sleep or rest if he or she wishes to do so.
8. If the person is a child, the parent or guardian should be notified that a seizure has occurred.
9. After a seizure, most people can carry on as before. If, after resting, however, the person still seems groggy, weak, or confused, it would be better to accompany the person home.
10. If the person undergoes a series of convulsions, with each successive one occurring before the person has fully recovered consciousness, you should immediately seek medical assistance.

What Causes Epilepsy?

The exact mechanism that produces an epileptic seizure is not well understood. In some way, the normal pattern of electrical activity in the brain is disrupted, and a sudden, violent disturbance of electrical waves develops, rather like a short-lived electrical "storm". The brain tissues are very sensitive to changes in amounts of certain chemicals and quite small changes in acid-base conditions. Certain chemicals in the blood can also promote a seizure. Flashing lights and emotional upsets can induce an attack in some persons.

Questions for research and discussion:

1. Is there any evidence that epilepsy is inherited?
2. What proportion of the population suffers from epilepsy?
3. What factors are known to trigger epilepsy?
4. What are the differences between *petit mal* and *grand mal* seizures?
5. Find out more about the treatment of sufferers of epilepsy.
6. At what ages does this disorder usually become evident?

Cranial Nerves

(krae-nee-ul)

The 12 pairs of cranial nerves originate on the underside of the brain. These nerves pass out through tiny holes in the bones of the skull to reach their target organs. They are assigned numbers according to the order in which they leave the brain, starting at the front with the olfactory nerves. Their names describe the organ or the function of the organ to which they are connected. Some of these nerves have sensory functions, others are motor nerves, while some are both motor and sensory nerves. The names, locations, and functions of these nerves are given in Figure 6.12.

Figure 6.12.
The cranial nerves.

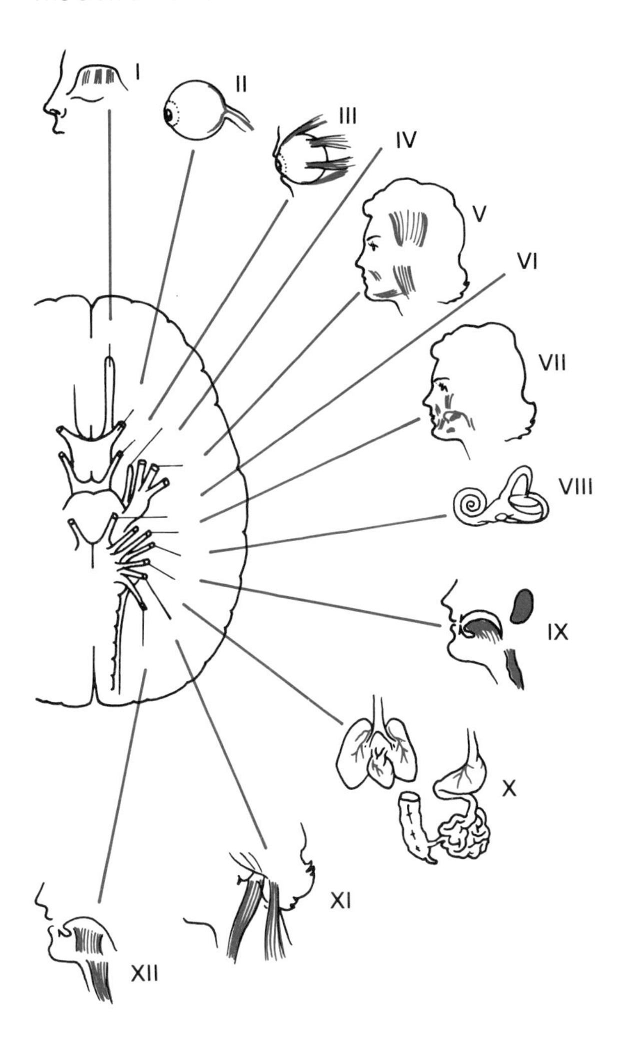

I OLFACTORY
The sense of smell

II OPTIC
The sense of vision

III OCULOMOTOR
Most eye muscles, iris, and ciliary muscle

IV TROCHLEAR
Superior oblique muscle

V TRIGEMINAL
Mastication muscles and sensations from eye, face, teeth, sinuses

VI ABDUCENS
External rectus muscle

VII FACIAL
Facial muscles, salivary glands, taste, and tongue

VIII ACOUSTIC
Cochlear, sense of hearing
Vestibular, sense of balance

IX GLOSSOPHARYNGEAL
Parotid gland, sense of taste, tonsils, pharynx

X VAGUS
Motor nerves to heart, lungs, digestive tract. Sensory from most of the intestinal organs

XI ACCESSORY
Mastoid, rear neck muscles, larnyx, and pharynx

XII HYPOGLOSSAL
Muscles in front of neck and to tongue

Protection of the Brain

The brain is one of the best-protected organs of the body. It is entirely enclosed within the hard bones of the skull. Between the bony case and the actual nerve tissue of the brain are three protective membranes called the **meninges**. (See Figure 6.13.) One of these, closely attached to the inside of the bones of the skull, is a white membrane called the **dura mater**. Covering the brain itself is another membrane called

(men-in-jeez)

Figure 6.13.
Section through the protective coverings around the brain.

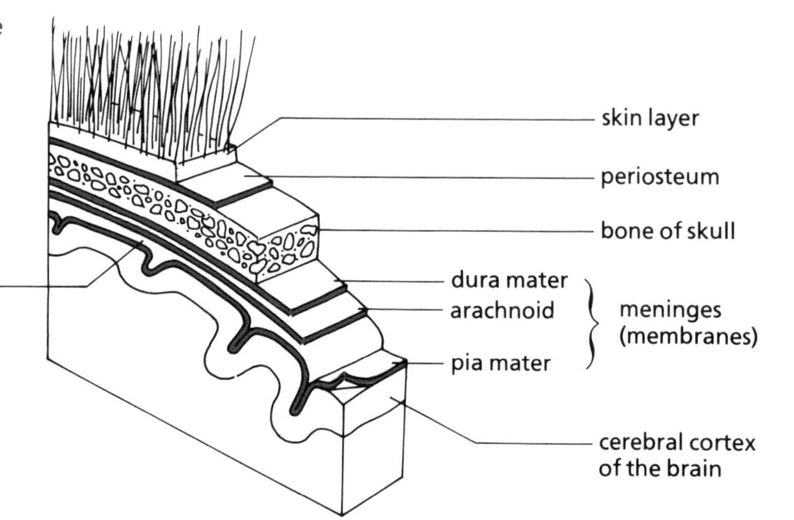

(a-rak-noid)
(sair-ee-broe-spine-al)

the **pia mater**. Between these two membranes is a third known as the **arachnoid**.

The brain is further protected by the **cerebrospinal fluid**, which fills a space between the arachnoid and the dura mater. This fluid bathes the cells and helps to cushion the brain. It carries nutrients to the cells and removes wastes for transfer to the blood stream. This fluid and the membranes around it continue down from the brain to surround the spinal cord, giving it the same protection and nourishment. The cerebrospinal fluid helps to provide a buffer against knocks and bumps to the brain and spinal cord.

The Spinal Cord

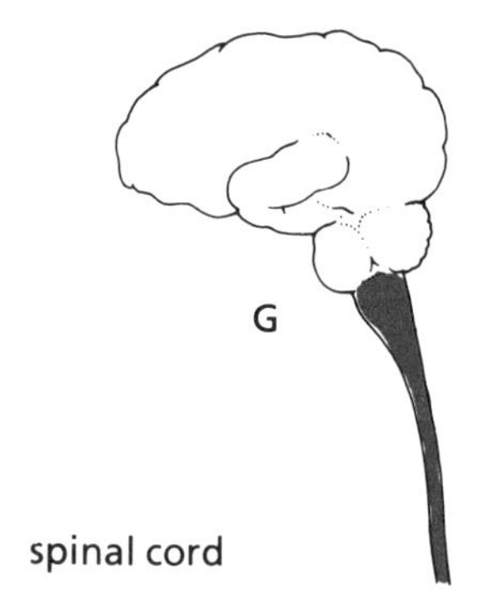

The **spinal cord**, which is part of the brain stem, extends from the foramen magnum, the opening at the base of the skull, down through the vertebral canal, to the level of the disc between the first and second lumbar vertebrae.

Each of the bones of the spine contains a large hole located between the three processes and the centrum. The spinal cord passes through these openings like a thread passing through the holes in a string of beads. As well, a pair of spinal nerves passes through smaller openings in the lateral arms of the 31 vertebrae to connect with small masses of nerve tissue called **ganglia**.

The spinal cord forms a compact cylindrical bundle, with bulges in two regions. One of these is in the neck or cervical region, from which nerves connect with the arms and shoulders. The other lies in the lumbar region, where nerves control the pelvis and legs.

At the level of each vertebra, spinal nerves leave or enter the system to connect with the spinal cord. Sensory nerves carry impulses to the central nervous sytem by way of the **dorsal root** which lies at the back of the cord. Motor impulses are carried away from the CNS by the **ventral root**, which is found at the front of the spinal cord. (See Figure 6.14.)

Figure 6.14
A cross-section through the spinal cord. The gray matter is composed of unmyelinated fibres and cell bodies. The white matter is formed of myelinated fibres.

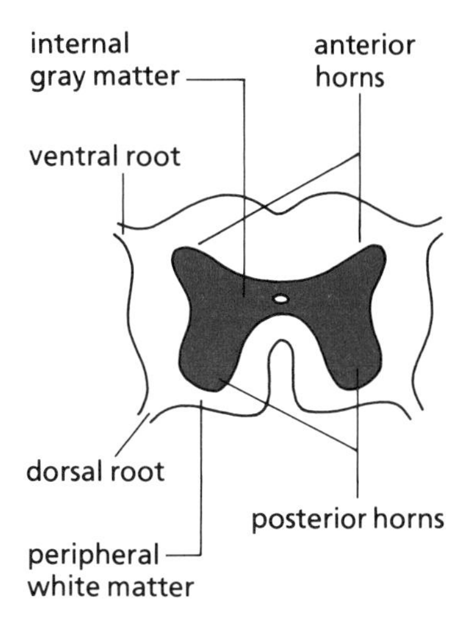

A cross-section of the spinal cord (Figure 6.14) shows a tube of white matter with one deep groove at the back and another at the front. This white area encloses a butterfly- or H-shaped mass of gray matter, in the centre of which is a canal containing cerebrospinal fluid. The central crossbar of the H contains fibres that connect the two halves of the cord. The uprights of the H are called the posterior (dorsal) and anterior (ventral) horns. These connect with the sensory and motor nerves respectively. The white matter around the horns contains hundreds of myelinated nerve fibres, which form bundles of nerves. Some fibres carry sensory messages to the brain, while other bundles carry motor impulses from the brain to the muscles. Additional bundles carry messages up or down to other nerves leaving the cord at segments above or below a particular point along the cord.

The spinal cord has two major functions:

1. It forms a two-way conduction system between the brain and peripheral nerves, both sensory and motor.
2. It controls reflex actions that do not require supervision by higher centres in the brain.

As an example of this second function, when a hot stove burner is touched, the cord will receive a message "hot" and respond reflexively by stimulating motor neurons in the hand and arm to pull away from the hot surface. If the stimulus is strong, indicating greater danger, muscles in the legs may also be sent impulses causing you to jump back; neck muscles may respond by jerking the head around to look at the stove. All of these responses can be dealt with at the reflex level of the spinal cord. Impulses will still be sent to the brain to inform the higher centres about what has taken place and what action or response has been involved.

The Reflex Arc

The reflex arc is the simplest and most basic unit of the whole nervous system. It consists of five parts, which usually involve three neurons (2 or more than 3 are quite common, however). The reflex arc requires:

1. A **receptor**. The receptor recognizes some change in the environment, whether of heat, light, sound, or some other factor. The receptor is stimulated to initiate a nervous impulse.
2. A **sensory neuron**. The sensor, or afferent neuron, conducts the impulse from the receptor to the Central Nervous System.
3. A **central** or **association neuron**. This nerve cell switches the impulse from the sensory "informing" neuron to the "acting" motor neuron. It allows the impulse to be routed into a number of possible pathways.
4. A **motor neuron**. The motor (efferent) neuron carries the impulse to the appropriate organ (usually a muscle) to produce the response.
5. An **affector**. The affector is the muscle or organ that will contract or otherwise respond appropriately to the stimulus. (See Figure 6.15.)

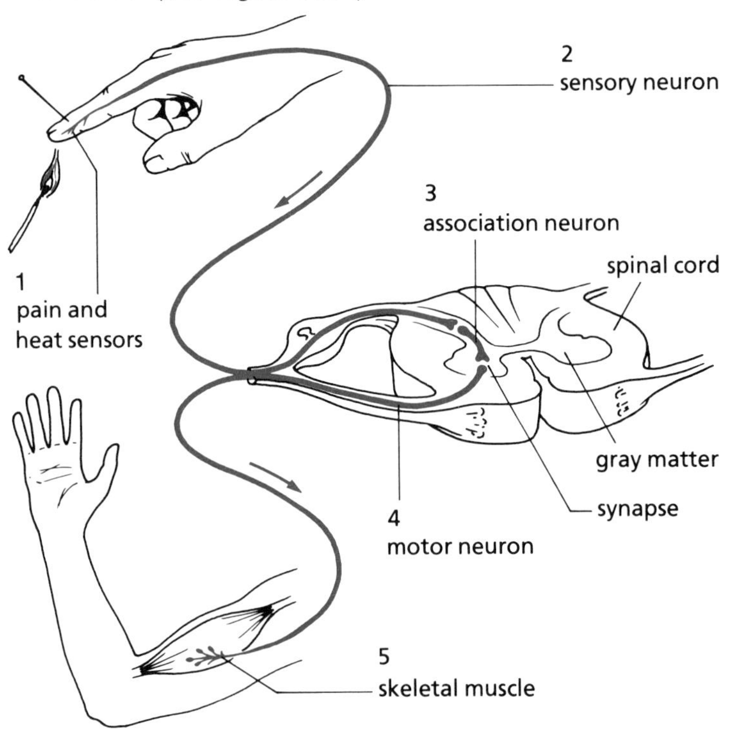

Figure 6.15.
A three-neuron reflex arc. Five parts are necessary: 1) a sensor; 2) a sensory neuron to carry the impluse to the spinal cord; 3) a central association neuron to transfer the impulse to an "action" neuron; 4) the motor neuron to carry the response impulse; 5) a muscle that can respond quickly to effect the avoidance movement.

The Reflex Act

When an affector responds to an impulse in the fashion discussed above it is called a **reflex act**. A reflex act is an automatic or involuntary action, which is always the same when a particular stimulus is involved. The closing of the eye's pupil in response to increased light or the rapid withdrawal of the hand after touching a hot stove, are both examples of the reflex act. Reflex activity is predictable; it provides a rapid reaction protecting the body from harm. Reflex acts occur in a fraction of a second, before a person has time to think consciously about what appropriate action is required.

Acquired Reflexes

Acquired reflexes involve the cerebrum. These are reflexes that have been modified by training or learning, and are conditions called conditioned reflexes. Ivan Pavlov, the Russian physiologist, conditioned animals so that they would respond to certain signals. Dogs usually produce saliva abundantly when food is placed near them (an unconditioned, automatic reflex). If Pavlov rang a bell each time the food was offered, the dogs learned to associate the bell with food. Eventually each time the bell was rung the dogs produced saliva even if no food was present. A conditioned reflex was thus established.

We are constantly modifying our activities, making time-saving, even life-saving, short cuts. The action of our foot on the brake pedal of a car is often the result of a conditioned or learned reflex response to some visual danger signal.

Injury to the Spinal Cord

If dislocation or fracture of a vertebra occurs, damage easily can be caused to the spinal cord. Under normal conditions the cord has plenty of room to accommodate small movements between the vertebrae, without damage to the spinal cord. Recall that the central nervous system is surrounded by three membranes as well as a fluid-filled space. These protective layers cushion and normally prevent injury to the spinal cord. (See Figure 6.16.) However, a severe blow may fully or partially sever the cord. If this happens, a loss of sensation and voluntary control of the motor muscles will occur below the point of injury. Should this occur in the cervical region of the cord, paralysis of almost the entire body results, includ-

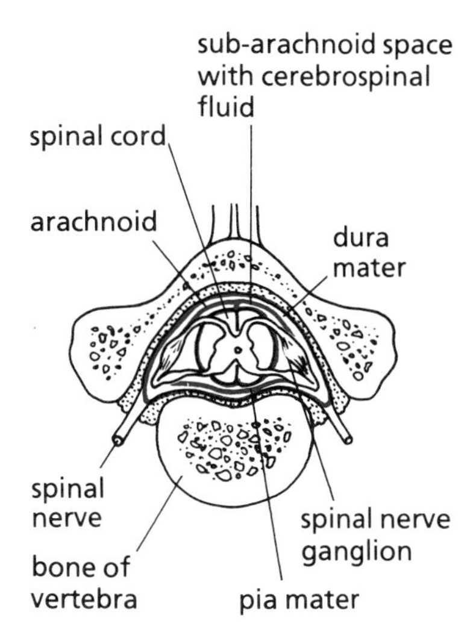

Figure 6.16.
A cross-section of the spinal cord and the vertebra showing the protective meninges of the central nervous system.

ing both the upper and lower limbs. Persons who are afflicted in this way are known as quadraplegics ("quad" means four). If the cord is permanently damaged near the centre of the spine, only the legs will be affected. People with paralysis of the lower limbs are known as paraplegics.

If you are ever called upon to help an accident victim and you have reason to believe that the spine may be damaged, DO NOT MOVE THEM! Get medical assistance from people who have been trained to move such victims to avoid the risk of causing further injury. If you are hurt and lose sensation in the lower part of the body or for some other reason think that your spine may be damaged, don't let anyone move you unless you are sure they are competent to handle victims of a spinal injury. It is very easy for well-intentioned but untrained helpers to do further and permanent damage in such cases.

The Autonomic Nervous System

The name of this system is derived from the same root as the word automatic. This is reasonable because almost all of its functions are performed without thinking. The Autonomic Nervous System controls most of our internal organs and processes: respiration, circulation, digestion, excretion, and reproduction. This system should not, however, be thought of as entirely separate from the central and peripheral nervous systems. The autonomic nervous system is controlled largely by the hypothalamus, which is part of the forebrain. It also involves the medulla, the spinal cord, and the extensive system or peripheral nerves.

The Function of the Autonomic Nervous System

In general, the autonomic system looks after our internal environment, helping to keep the body systems operating normally. It functions to make rapid adjustments in the operation of these systems during an emergency. At such times it acts to speed up the heart and respiratory rates, redirects blood to muscles, and shuts down systems not needed during the emergency. These activities are automatic; we do not have to think about them.

The Ganglia

Ganglia (singular is *ganglion*) appear as bulges or swellings, which are usually located outside the Central Nervous System. These structures are collections of nerve cell bodies.

(gang-glee-ah)

Just outside the spinal cord there are ganglia associated with the spinal nerves leaving the cord at each vertebra. The nerves of the autonomic nervous system are divided into two groups with reference to these ganglia. **Preganglionic fibres** extend from within the CNS to synapses in the ganglia. **Postganglionic fibres** extend from the ganglia to organ muscles or glands.

The Divisions of the Autonomic Nervous System

The autonomic nervous system is divided into two parts: the **parasympathetic** and the **sympathetic** portions. These perform in opposite ways and tend to counteract each other. The sympathetic system prepares the body for emergencies, whereas the parasympathetic system reverses the effect, bringing the body back to its normal condition to conserve energy.

The Sympathetic System

The motor nerves (preganglionic) of this division originate in the spinal cord. They pass out of the cord through the ventral (or anterior) root and connect with other nerve fibres in ganglia just outside the vertebral column. (See Figure 6.17.) The ganglia are arranged in two parallel chains on either side of the spinal cord. These chains of ganglia act as junction boxes where the impulse can be switched into any one of several routes in order to give the best response in an emergency.

If a stressful situation arises, such as when a child runs in front of a car, the body must react rapidly to meet the emergency. Sympathetic nerves send impulses to the heart to speed it up. Other impulses are sent to the skin to shut down the blood vessels there. Blood, containing oxygen and sugar, from the skin is thus diverted to the muscles of the heart and

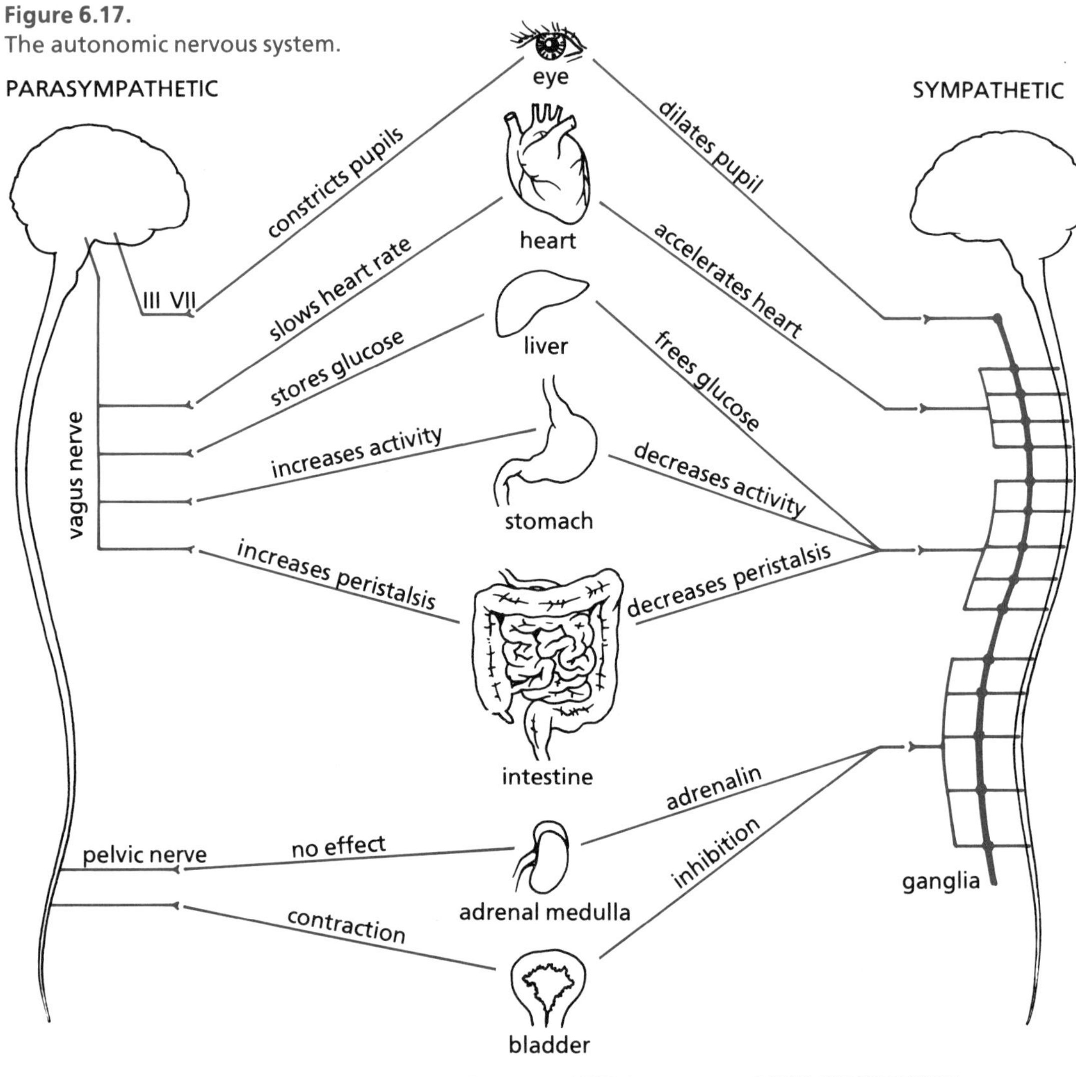

Figure 6.17.
The autonomic nervous system.

limbs to supply their extra needs. Impulses are sent to dilate the pupils of the eye so that maximum peripheral (side) vision is attained. Other impulses sent by the sympathetic system shut down digestive processes for the duration of the emergency and stimulate special endocrine glands, such as the adrenals, to back up the emergency measures already taken. The adrenal glands unlock stores of sugar in the liver to supply muscles that may need extra energy.

The Parasympathetic System

The parasympathetic portion of the autonomic sytem is controlled by nerves orginating in the midbrain, pons, and medulla. Several cranial nerves are involved: the oculomotor nerve which constricts the pupil, the facial and glossopharangeal nerves which affect the salivary glands, and the very important vagus nerve which carries impulses to most of the internal organs. The **vagus** nerve (*vagus* means wanderer) has branches that reach the heart, bronchi, esophagus, stomach, intestines, liver, and pancreas. It helps to control the rate of operation of these organs and returns them to a normal working pace after they have been speeded up or shut down, during an emergency.

In summary, the autonomic nervous system is responsible for the protection and conservation of body resources and the maintenance of normal body functions.

THE ENDORPHINS

The brain secretes many important hormones that spread through the body acting as chemical messengers. In recent years, much has been learned about one class of these brain hormones, called **endorphins**, which behave much like morphine, a painkilling drug. Scientists hope that the use of these substances will be of great benefit in treating severe and long-lasting (chronic) pain, and a variety of emotional problems.

Endorphins are now being synthesized in laboratories by using bacteria whose genes have been spliced.

At the University of Toronto, Dr. Bruce Pomeranz has discovered links between endorphin and the ancient painkilling properties of **acupuncture**. Dr. Pomeranz says, "there is quite a lot of evidence that acupuncture stimulates the production of endorphins and this blocks the transmission of pain messages to the brain." This explains why acupuncture can be used to reduce pain. One researcher recently administered electro-acupuncture to the ears of rats. Following this procedure, examination of the rats' brains revealed substantial depletion of endorphins in three brain regions and a simultaneous increase in endorphin levels in the cerebrospinal fluid. It has been found that during acupuncture brain cells synthesize more endorphins.

Scientists are researching the receptor sites of nerve cells where endorphins have their effect. It is believed that in cases of chronic pain, due to infections in the nervous system, there is little endorphin activity. The pain is due to less endorphin being transmitted to the "pain centre" of the brain. Endorphins have been known to affect the area of the brain which is responsible for our emotions and motivation (expressions of rage, sexual behaviour, and pleasure are found in this region - located in and around the hypothalamus), perhaps producing a state of euphoria.

Tests in which endorphins have been administered to patients with psychological disorders have had mixed results. In treating depressed patients with endorphins, one researcher noted significant improvement for several hours. Other researchers reported no significant change and the condition even worsened when the drug was used on some schizophrenics. It is also possible that some types of endorphins and other neurochemicals produced in the brain could trigger depression and **schizophrenia**. Additionally, the endorphins had the same disadvantages as morphine, producing dependence and addiction.

CASE STUDY. Ross, 19 years of age.

It was Jill's birthday and a group of friends had gathered at Jill's home to celebrate the occasion. Beer, spirits, and other beverages were available and many people were having alcoholic drinks in moderation. Ross, however, was setting a faster pace.

By nine that evening, Ross was talking loudly, dominating the conversation, and rudely interrupting anyone who tried to talk. Some laughed at him as he stumbled against the furniture on his way to the kitchen for another drink. His behaviour was annoying to several other guests.

Between eleven and midnight, Ross fell asleep in a chair in one corner of the room and his friends were content to leave him alone to sleep off the effects of his drinking. Jill's parents were expected home about 1:30 in the morning and Jill wanted her friends to leave before her parents arrived so that she could put the room in order.

Someone woke Ross up and, after a good deal of encouragement, he went out to the kitchen, drank a cup of coffee, and ate a sandwich. His friends asked him if he was all right and he replied "Yes, I'm okay." Indeed, he appeared to be a great deal more capable than before he had fallen asleep. The others left in small groups. Ross was the last to leave. Jill talked with him and, since he seemed to have sobered up, she was only mildly concerned that he would be driving home alone.

Ross does not remember too much about driving home and, as far as he was concerned, he had "no problem". However, a flashing red light and a wailing siren made him pull the car over to the curb. The police officer asked Ross for his driving licence; then as the officer got close enough to smell his breath, he asked Ross to get out of the car. Trying hard to appear relaxed and sober, Ross started to get out, but caught his shoulder on the seat-belt harness and stumbled against the door. After a few more questions, the officer asked Ross to get into the police car. Ross also was asked to lock his own car, which was then left by the roadside to be collected later.

When Ross arrived at the police station, he was asked a number of questions and then was required to write down his name and address. Several tests followed. First, he was asked to walk along a straight line, turn at the end, and walk back. Ross had some difficulty staying on the line and on turning at the end of the line found himself with one foot treading on top of the other. He was asked to touch his finger to his nose and then to pick up some coins. Finally, Ross was asked to take a breathalyzer test and the results showed a blood alcohol content of greater than 0.08% (0.08 g of alcohol in each 100 mL of blood). The officer wrote a report stating the test results and describing various physical observations, such as the condition of Ross' eyes.

It was obvious that Ross' responses were impaired by his use of alcohol. His ability to respond to requests and to control and co-ordinate his movements, keep his balance, or even to write his own name and address clearly, were all seriously affected by the alcohol he had drunk. He should not have been driving a car. The breathalyzer test confirmed the officer's observations. Ross was charged with impaired driving.

When the case came to court, Ross was convicted and fined, and his licence was suspended for three months. The judge warned him that a second offence could result in a two-week jail sentence.

(While 0.08% alcohol is the approved SI Metric way to express the amount of alcohol in the blood at which it is an offense for a person to drive, you should be aware of how this amount is, in practice, expressed. Because of courtroom confusion with decimal point placement, police tests and forensic laboratory results now are given in mg%. The Criminal Code of Canada specifies it is an offense to drive with a blood alcohol level of 80 mg% or higher.)

Questions for research and discussion:

1. How does a breathalyzer machine work?
2. How does alcohol get into the breath?
3. What other samples can be used by the police to determine the amount of alcohol consumed?
4. Find out how reaction time is affected by alcohol.

Chemicals and the Nervous System

The nervous system is easily affected by the presence of chemicals. We have seen, for example, how acetylcholine is released by nerve endings to bridge the synaptic gap. Various other chemicals also affect the nervous sytem, many of them adversely. Some of these chemicals come under the heading of drugs, either medical or non-medical in nature. In some cases they may distort the normal functioning of the nervous system and produce hazardous effects. Among the wide variety of drugs available, three general categories are recognized: stimulants, depressants, and hallucinogens.

Stimulants

Some stimulants are so widely used that we do not think of them as drugs: the caffein in coffee, cola, or tea and the nicotine in tobacco are examples. Other stimulants such as amphetamines have medicinal value and are often prescribed by doctors. These drugs stimulate the sympathetic nervous system, producing reactions such as increased heart rate or release of extra blood sugar. The less potent of these stimulant drugs, such as caffeins, have very mild effects. The amphetamines, such as dexadrine and "speed" (Benzedrine), however, may produce very dramatic results.

When athletes are preparing for competition, they are confronted with special stress. Some experiments have shown that the use of amphetamines may enable the body to draw on special energy reserves and in so doing possibly improve performance. However, it is now recognized that this practice often causes emotional stress to the athlete who then may have problems getting along with teammates or competitors. Amphetamines also increase thirst and sleeplessness and usually reduce the appetite. They thus often cause many new problems while solving others.

Depressants

These drugs act in the opposite way to stimulants. Alcohol, barbiturates, tranquillizers, opiates (opium, heroin, codeine, and methadone) are examples of depressants.

Alcohol reduces the active functioning of the brain and nervous system. It affects the higher centres first, reducing inhibitions. As more alcohol is taken in, lower centres of control are affected, reducing dexterity, reducing visual and auditory recognition, producing slurred speech, and hampering the ability to walk or balance properly. If even higher levels of alcohol are absorbed, unconsciousness and coma may occur.

Barbiturates act in a way similar to alcohol. They affect the nerves in the brain that control sleep. After high doses, unconsciousness may result and breathing may completely stop. A dangerous situation occurs if barbiturates are taken with alcohol. The combined effect of these two drugs can cause accidental death. Barbiturate-alcohol combinations are estimated to cause more than half the accidental suicide deaths in Canada and the United States.

Tranquillizers act like barbiturates in depressing the nervous system, so that stress and tension are not as obvious to the user. Such drugs are of considerable medicinal value for treating patients with mental disorders. These drugs must be prescribed, however, and the effects properly monitored by competent medical staff. Drugs such as heroin are excellent painkillers but have very dangerous side effects. Heroin is extremely addictive, like almost all depressant drugs. These drugs produce an addictive demand by the body for more of the drug, because, in order to be effective, a greater and greater quantity of the drug is required to produce the same results. Even when the drug is no longer needed for the original medical reasons it often is difficult for the patient to stop using it. The body suffers withdrawal symptoms, which can be very pronounced and may cause severe damage to the body or even result in death.

It should also be noted that individual tolerances for depressants, as for many other medical and non-medical drugs, differ greatly from one person to another. Some people have higher tolerance for alcohol just as some people require more or less sleep than others. You cannot assume that because a friend is affected only slightly by a particular drug that your reaction will be the same. For some people even very small amounts of a drug can have disastrous affects.

Glue sniffing and the inhaling of volatile hydrocarbons can produce effects similar to those of alcohol. Such substances are considered far too dangerous for medical use, since complete interruption of respiration and death can result. Some of these compounds cause death by interrupting the impulses that control the contraction of heart muscles.

Hallucinogens

(hal-oo-sin-o-gens)

Mild doses of these drugs distort the visual and auditory senses; they also increase emotional response. When larger doses are taken, hallucinations may be experienced. That is, the subject sees and hears things that are not there. Mescaline, LSD, and some other hallucinogenic drugs are similar to each other in chemical structure. They appear to be closely allied to a naturally occurring chemical called serotonin which is found in the brain. It is possible that some reactions in the brain do not distinguish between serotonin and other similar substances. This then may cause the confused images that a user experiences.

Marijuana

Marijuana is not one pure drug, but is a mixture of more than fifty different reactive ingredients. On the street market the strength of this drug has steadily increased and is now often 20 times as strong as it was a few years ago. Because of the many ingredients, the degree to which these ingredients vary from sample to sample (according to where the plant was grown, the processing of it, and other factors), it is difficult to assign the drug to a specific category. It has some depressant effects and some mind- or perception-distorting effects, like the hallucinogens. The drug does, however, build up in residual amounts in the body. Because of this build-up, quite significant reactions can be experienced after a small dose. The drug also appears to be activated when the body is subjected to stress. Release of endocrine secretions and stimulation of the autonomic nervous system may produce an unexpected "high", as the residual amounts of marijuana in the body are triggered into activity by the stress.

Summary

There is no doubt that the misuse and the abuse of drugs, both medical and non-medical, have produced a large number of individuals who can no longer function normally. Some are dependent on an endless supply of drugs to support their habit. Others have minds so distorted that rational thinking is no longer possible. Some relatively light users are sufficiently depressed, apathetic, and "turned off" that they require one "high" after another. Very little is known about exactly how these drugs affect the neural pathways of the brain to produce the distorted images and "highs" that some seek. Like all systems, the nervous system is delicately balanced; thus, to upset that balance is to risk temporary or permanent damage.

Table 6.1 Drug Guide

DRUG	SHORT-TERM EFFECTS	LONG-TERM EFFECTS
Alcohol Sedative Depressant	• initial relaxation, loss of inhibitions • slowing down of reflexes and reactions, impaired co-ordination • attitude changes, increased risk-taking to point of danger • acute overdose may lead to death	• regular, heavy use increases the possibility of: gastritis, pancreatitis, cirrhosis of the liver, certain cancers of the gastrointestinal tract, heart disease • upon withdrawal following regular use, convulsions and delirium tremens may occur
Amphetamines Stimulants Benzedrine, Dexadrine, Neodrine	• reduced appetite, dilation of pupils • increased energy, alertness, faster breathing • increased heart rate and blood pressure which lead to increased risk of burst blood vessels or heart failure • risk of infection from unsterilized needles if injected WITH LARGER DOSES: • talkativeness, restlessness, excitation • sense of power and superiority • illusions and hallucinations • some frequent users become irritable, aggressive, paranoid, or panicky	• malnutrition • increased chance of infections • psychological dependence • after stopping, there usually follows a long sleep and then depression

DRUG	SHORT-TERM EFFECTS	LONG-TERM EFFECTS
Cannabis Modifier of mood & perception marijuana, hashish, "hash oil"	• a "high" feeling • increased pulse rate • reddening of the eyes • at later stage, user becomes quiet, reflective, and sleepy • impairs short-term memory, logical thinking, and ability to drive a car or perform other complex tasks • with larger doses, perceptions of sound, colour, and other sensations may be sharpened or distorted and thinking becomes slow and confused • in very large doses, the effects are similar to those of LSD and other hallucinogens – confusion, restlessness, excitement, and hallucinations	• a moderate tolerance • possible psychological dependence • loss of drive and interest • marijuana smoke contains 50% more tar than smoke from a high-tar cigarette: with regular use, risk of lung cancer, chronic bronchitis, and other lung diseases increases
LSD (Lysergic Acid Diethylamide)	• initial effects like those of amphetamines • later, distortions of perception – altered colours, shapes, sizes, distances producing exhilaration or anxiety and panic, depending on the user • feelings of panic or of unusual power may lead to behaviour that is dangerous to the user • occasionally, convulsions occur • strong tolerance develops very rapidly and disappears very rapidly	• the long-term medical effects of LSD are not known

DRUG	SHORT-TERM EFFECTS	LONG-TERM EFFECTS
Minor Tranquillizers Valium, Equanil, Vivol, Librium	• calms tension and agitation • muscle relaxation • lessened emotional responses to external stimuli, e.g., pain • reduced alertness • with larger doses, possible impairment of muscle co-ordination, dizziness, low blood pressure, and/or fainting	• physcial dependence • withdrawal reaction (temporary sleep disturbances - abrupt withdrawal leads to anxiety, possible delirium, convulsions, and death)
Sedative Hypnotics Barbiturates (e.g., Amytal, Seconal, Nembutal) Non-Barbiturates (e.g., Placidyl, Dalmane, Doriden)	• small doses relieve anxiety, tension, producing calmness and relaxation • larger dose produces a "high" and slurred speech, staggering, etc. (similar to alcohol) • produces sleep in a quiet setting; otherwise, sleep may not occur • dangerous to drive a car or perform complex tasks • much larger doses produce unconsciousness • acute overdose can result in death	• tolerance and dependence if large doses are used • do not produce completely normal sleep - user may feel tired and irritable after sleeping • upon withdrawal, temporary sleep disturbances occur • abrupt withdrawal leads to anxiety and possible convulsions, delirium, and death
PCP (Phencyclidine) "Angel Dust"	• euphoria - a "high" • faster, shallow breathing • increase in blood pressure and pulse rate • flushing and sweating • lack of muscle co-ordination • numbness of extremities	• possibility of flashbacks • possibility of prolonged anxiety or severe depression

DRUG	SHORT-TERM EFFECTS	LONG-TERM EFFECTS
PCP (Phencyclidine) "Angel Dust"	WITH LARGER DOSES: • a fall in blood pressure, pulse rate, and respiration • nausea, vomiting, blurred vision, rolling movements and watering of the eyes, loss of balance, dizziness, convulsions, coma, and sometimes death • delusions, mental confusion, and amnesia are common	
Opiates Opium, Morphine, Codeine, Heroin	• relief from pain • produces a state of contentment, detachment, and freedom from distressing emotion • large doses create euphoria – a "high" • sometimes nausea and vomiting • acute overdose can result in death • risk of infection from unsterilized needles	• physical and psychological dependence • abrupt withdrawal results in moderate to severe withdrawal syndrome (cramps, diarrhea, running nose, etc.)
Cocaine Stimulant	• same as amphetamines WITH LARGER DOSES: • stronger, more frequent "highs" • bizarre, erratic, sometimes violent behaviour • paranoid psychosis • sometimes a sensation of something crawling under the skin	• strong psychological dependence • destruction of tissues in nose if sniffed • other effects like amphetamines

QUESTIONS FOR REVIEW

SOME WORDS TO KNOW

Match the statement in the column on the left with a suitable term from the list on the right. DO NOT WRITE IN THIS BOOK.

1. Small gap at the junction between two neurons.
2. Carries the nerve impulse away from the cell body.
3. Protective membranes around the CNS.
4. Part of autonomic nervous system that reacts to stress.
5. Carries nerve impulses from receptors to CNS.
6. Involuntary response.
7. Co-ordinates muscular activities, balance, etc.
8. Largest portion of the brain, controls motor and sensory functions, senses, etc.
9. Composed of the brain and spinal cord.
10. Controls autonomic nervous system and the internal organs.

A. hypothalamus
B. dendrites
C. axon
D. Central Nervous System
E. Autonomic Nervous System
F. sensory neuron
G. motor neuron
H. ganglia
I. cerebrum
J. cerebellum
K. meninges
L. reflex
M. synapse
N. sympathetic division

SOME FACTS TO KNOW

1. Draw a neuron and label its various parts. Give a simple explanation of the function of each part.
2. List the major divisions of the brain and give an important function carried out by each part.
3. How is the brain protected?
4. With the aid of a diagram, explain the working of a reflex arc.
5. Which portion of the brain looks after each of the following activities?
   a) arm movements,
   b) vision,
   c) co-ordination,
   d) heart rate,
   e) respiration rate,
   f) hearing,
   g) memory,
   h) sleep.
6. Explain the meaning of the following terms:
   a) synapse,
   b) meninges,
   c) dendrite,
   d) reaction time.
7. What is the difference between white matter and gray matter?
8. What changes are produced in your autonomic nervous system when a child unexpectedly runs into the road in front of a car in which you are riding?
9. If someone says, "I feel nervous", what is meant?
10. Explain how an impulse moves across a synapse.
11. Research one of the drugs listed in this chapter and determine how it affects the body.
12. What are the major characteristics by which drugs can be classified?

QUESTIONS FOR RESEARCH

1. Select one of the following topics and prepare a report on it:
   - psychiatrist
   - neurologist
   - schizophrenia
   - meningitis
   - cerebral palsy
   - the law and the use of drugs (prescribed and/or non-prescribed)
   - the work of the drug addiction clinic
   - phobias
   - stress and the work of Dr. Hans Selye
   - psychologist
   - multiple sclerosis
2. What happens when we sleep? One theory proposes several stages in sleep; investigate this theory.
3. What are biorhythms? Some airlines and truck companies do not allow their pilots or drivers to work when their biorhythms are at low levels. Investigate this theory and try to work out your own biorhythm.
4. What is jet lag? What explanations are there to explain the human biological clock?
5. Find out what is meant by "body language". Do you use it? Can you determine what your friends are thinking from this process?
6. What theories are there to explain memory?
7. What explanations are there for dreams?
8. What are some of the problems encountered when a person is deprived of sleep?
9. Investigate alcohol problems among teenagers.

Activity 1: A COMPARISON OF REACTION TIMES.

Materials

a 30-cm ruler.

Method

1. *Make a table in your notebook for recording your results as shown.*

Reaction distance in cm	
Right hand	*Left hand*

2. Work with a partner for this experiment. One student will hold a ruler vertically by the 30-cm end, between the thumb and index finger. The other student will place his or her arm flat upon the desk with the hand just over the edge. Keep the tips of the fingers about 2 cm apart and level with the end of the ruler. Be prepared to try to catch the ruler when it is released by your partner.
3. The student holding the ruler will give the signal "ready", pause, and then release the ruler. The other partner must catch the ruler allowing it to fall only the shortest distance possible.

4. *Record the distance that the ruler falls, taking the centre of the thumbnail as the measuring point. (Mark the thumbnail.) Repeat the test 5 times, record the results, and then find the average reaction distance achieved. Repeat using the other hand.*
5. *Set up a chart on the chalkboard and record the average results for the class.*

Left-hand average	*Right-hand average*	*Whether right- or left-handed.*

Analyse the results. What generalizations can be made about reaction times? (Relate the distance fallen to the reaction time.) What effect does the dominant hand have on reaction time?

Activity 2: HUMAN REFLEXES

To examine the responses produced when different stimuli are applied to receptors.

The normal reflex response requires three neurons. Each reaction involves the transmission of an impulse from a receptor, through the sensory neuron, to the spinal cord and back by the motor neuron to the muscles that produce the response. The knee reflex involves only two neurons.

Materials
Rubber reflex hammer, feather or strand of cotton thread, penlight or microscope light source.

THE KNEE JERK (patella reflex)

Method 1
Have your partner sit on the edge of the bench so that his or her legs hang freely and do not touch the floor. Feel the position of the patella (kneecap) and locate the tendon that is just below it. Strike the tendon sharply with the edge of the hand, or a rubber reflex hammer. The foreleg should extend involuntarily. Try to determine which muscles causes this upward lift of the leg. The receptors, in this case, are stretch receptors present in the tendon below the knee. These receptors have an important role in helping to maintain our upright position. *Record your observations and explanations.*

Method 2
Repeat the experiment, but this time, ask your partner to clasp his or her hands together and, just before striking the

tendon, have your partner try to pull the hands apart, so that the muscles are strongly tensed. *Record your observations.*

Method 3
Distract your partner's attention by assigning a task, such as counting the number of individual letters in a sentence on a printed page. Once started on this task, again strike the tendon.

Record your observations.
What are the differences between the three experiments? Can you explain these differences? Why do we tightly clench our fists before a fight or under conditions of special stress? What advantages does such action provide? Draw and label the nerve impulse pathway for a knee-jerk reflex.

Activity 3: THE ACHILLES REFLEX

Method
Have your partner kneel on a chair with his or her feet hanging over the edge of the seat. (Remove the shoes.) Push the foot forward and then lightly tap the Achilles tendon. *What happens?*
Which muscles respond? Repeat the test. Is the second reflex stronger or weaker? Explain why this might be so.

Activity 4: THE PUPILLARY REFLEX

Method
Have your partner close his or her eyes for about two minutes. Observe the pupils when you ask your partner to open his or her eyes. *Describe what happens and explain this reflex.* Shine a soft light into the eye for just a few moments. *Observe and record the results.*

Activity 5: THE ACCOMMODATION REFLEX

Method
Have your partner look at a distant object, perhaps across the room or out of the window. Examine the size of the pupil. Now ask your partner to look at a book placed about 20-30 cm in front of his or her face. Examine the pupil size. *Explain any differences in pupil size that you observe.* Repeat the experiment, but this time observe any movements of the eyeballs within their sockets. *Record your observations and explain why this action takes place.*

Activity 6: THE UVULA REFLEX

Method

Have your partner open his or her mouth and with a clean Q-tip, gently touch the uvula (the fleshy, v-shaped mass hanging from the roof of the mouth at the back). What reaction occurs? *Explain this action and its role in swallowing food.*

Activity 7: THE SNEEZING REFLEX

Method

Tickle the lining of your partner's nostrils with a feather or a piece of cotton thread. (Caution: make sure the thread is not inhaled.) *What repsonse is achieved? People often place the forefinger under the nose and press on the upper lip to try to stop a sneeze. Try it while your partner again applies the stimulus. Does this work? Can you explain it?*

Activity 8: A CONDITIONED REFLEX

Method

You have read of Pavlov's experiment on conditioning dog reflexes. Can you design a simple experiment to produce a conditioned reflex in your partner? Here is a suggestion to start you thinking. Shine a light into your partner's eyes at the same time you tap his or her hand. Repeat this 15 to 20 times, then tap the hand but do not shine the light. Observe the response. If there is no result, continue the conditioning and test again. Once you have achieved success, repeat it several times without the light, this conditioned response will quickly be lost. Note how many times the eye responds to a hand tap before the conditioning is lost. Try to invent a conditioned reflex of your own design.

Chapter 7

The Eye and Vision

- Important Senses
- Vision and the Structure of the Eye

Important Senses

We often become so accustomed to our senses and the impressions that they give us, that it is difficult to imagine what life would be like without them. You might get some idea of what your world would be like without senses, if you could imagine yourself alone in a *totally* dark room. The room is soundproofed and designed to eliminate vibrations. It is large enough that you cannot stretch out your arms and touch the walls. There is no furniture within reach and no movement of air. There are no odours and no areas of warm or cold air to direct you. You are shut off from everything and everyone.

Fortunately such a fate is almost impossible – death would occur first. The longest recorded period for which any volunteer has been able to withstand total deprivation of all sensory stimulation is 92 h. (*Guiness Book of Records*: 1962, Lancaster Moor Hospital.) Many people, however, do suffer the loss of one or more of the senses, deafness and blindess being the most common disorders.

Vision and the Structure of the Eye

The eye is, undoubtedly, the most important of our senses. More than 80% of the information we receive comes through our eyes and a larger portion of the brain is devoted to vision than to any of the other senses. The eye responds to light energy, converting stimuli into impulses which are conveyed to the brain for interpretation. It is important to realize that unless impulses from the eye reach the brain, we do not really see. The eye may receive light, the receptors respond and send impulses, but if these impulses do not reach the visual centre of the brain, we are completely blind.

The Cavities That Contain the Eye

The eyes are contained within two hollow depressions in the skull, called the orbital cavities. The bones above the eye as well as the nose and cheek bones, all project forward and offer the eye some protection from large objects, such as basketballs and doorposts. Openings in the bones of the skull at the back of the orbital cavity provide a passageway for the arteries and veins that supply the eye as well as the optic nerve, which carries the nerve impulses to the brain. The orbital cavity also contains several other structures, including the extrinsic muscles which move the eyeball, the tear-producing apparatus, nerves, blood vessels, and some fatty material. This fat helps to absorb shocks and provides a soft seating for the eyeball within the hard bone. The nutrient reserves stored in this fat, may be needed during a severe illness. If such a need arises, the eyes take on a sunken appearance because the reserve of fat has been used.

Muscles of the Eye

The muscles found in the eye may be divided into two groups. The **extrinsic muscles** are found on the outside of the eyeball and serve to move the ball in its socket; the **intrinsic muscles**, which are located inside the eyeball, control the lens and the iris.

(ex-trin-sik)

(in-trin-sik)

The Extrinsic Eye Muscles

Three pairs of extrinsic muscles are attached to the outer coat of the eye. In the chapter on muscles, the idea of opposing or antagonistic pairs of muscles was discussed. One muscle pulls in one direction and another muscle on the opposite side of a structure pulls in the other direction. In this case, one muscle moves the eye upward and another opposing muscle moves the eye downward. The name of each pair of extrinsic muscles is made up of two words, the first of which describes the direction the muscles move the eye and the second of which tells whether the muscles give a straight or oblique pull on the eye.

- The **superior rectus** muscle turns the eye upwards. *Rectus* means straight.
- The **inferior rectus** muscle, which opposes it, turns the eye downward.
- The **medial rectus** is attached to the inner side surface of the eye and turns the eye inward.
- The **lateral rectus** (or external rectus) is attached to the outer side surface of the eye and acts to turn the eye outward, away from the nose.

(rek-tus)

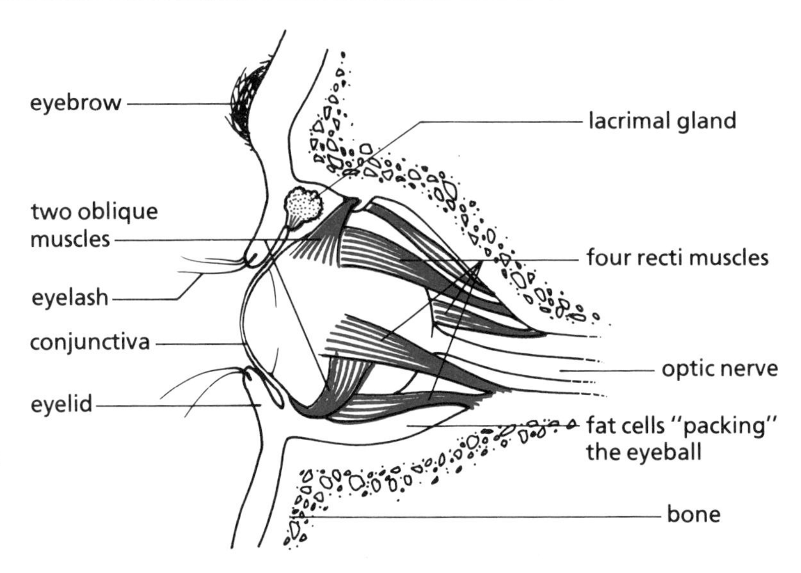

Figure 7.1.
The external features of the eye and the muscles which move the eyeball in its socket.

- The **superior** and **inferior oblique** muscles, with the help of other muscles, serve to rotate the eye (Figure 7.1).

When the eyes are focussed on an object to one side of the body, both will move together in that direticon. If the object is to the left, for example, the left eye will move towards the left side of the face away from the nose, the right eye will move towards the nose. If you hold a pencil at arm's length, then slowly move the pencil in towards your nose, gradually both eyes will turn further and further inward until you appear "cross-eyed". Watch your partner's eyes as he or she performs this exercise. Two nerves (the oculomotor and the abducens) are required to control these complex movements under the direction of special centres in the brain.

Protection of the Eye

Because the eyes are so vital for survival, they are well protected. Above the eyes, on the crests of two bony ridges, are the **eyebrows**. The hairs of the eyebrows help to prevent dust and other falling particles from entering the eye. The ridge and eyebrow also act as a shading device in very bright light. We sometimes extend this ridge by raising the flattened hand above the eyes, to give added protection from very intense sunlight.

The **eyelids** are two very thin folds of skin which close to cover the eye completely, shutting out light or protecting the eyes from damage. The eyelids can be voluntarily controlled, as when we close the eyes or wink, but they may also operate automatically. They can close reflexively, faster than we can think, to protect the eye from a paper pellet or other small object. We "duck" instinctively from any approaching object that the eye perceives as a potential source of danger. This response first requires visual recognition of something moving toward the eye; a message is then sent to the central nervous system which results in a motor message being returned, closing the lids protectively.

The **eyelashes** are attached to the edges of the eyelids. These short hairs are strong enough to remain extended in a gentle curve. They too protect the eyes, forming a fine screen to trap dust and other particles. The hairs are associated with oil producing glands found in the dermis surrounding the hair follicle. This oil helps maintain hair texture.

The Conjunctiva

(con-junk-ti-vah)

An opening in the body must be protected from bacteria. The eye is no exception. A thin transparent membrane called the **conjunctiva** is attached to the edge of the upper lid. It covers the inside of the upper eyelid and then folds down to completely cover the eyeball. The membrane then doubles back to cover the inside of the lower eyelid (See Figure 7.3). It thus completely seals the eye from any bacterial agent trying to enter the body through the eye. The disorder conjunctivitus occurs when this membrane becomes infected and inflamed. One type of conjunctivitis, known as "Pink Eye", is very contagious.

The Lacrimal Glands

Tears are produced by the **lacrimal glands**, located above and to the outer edges of the eye socket. They produce a salty, slightly germicidal fluid, which is composed primarily of water. Secretions of the lacrimal glands flow onto the surface of the eye through several tiny ducts to wash away dust particles and to lubricate the eye. Much of this fluid evaporates; the rest drains into the nasal chamber through the **lacrimal ducts** on the inner edge of the lower eyelids (See Figure 7.2.)

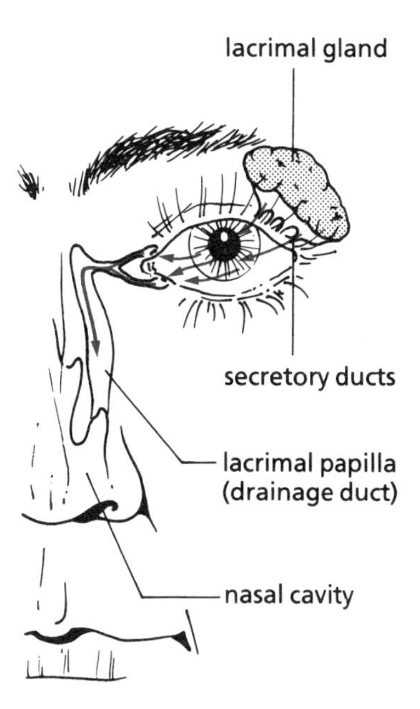

Figure 7.2.
The lacrimal glands and the ducts that drain surplus fluid into the nasal cavity.

During periods of emotional stress, pain, or irritation, the flow of fluid increases under the direction of the autonomic nervous system. If this increased supply is greater than the tear ducts can drain away, the fluid spills over the lids as tears.

The Internal Structures of the Eye

The Iris

The **iris**, which is the coloured portion of the eye, shows much variation, ranging from blue to dark brown and including mixtures of hazel and green. The colours, which result from the presence of various pigments, are determined by genes inherited from our parents.

The iris contains two sets of smooth muscles. These act antagonistically, one set to open the iris, the other set to reduce the opening. The inner muscles are circular; when they contract, they close like a ring, reducing the amount of light entering the eye. (See Figure 7.3.) The outer muscles radiate outward; as they contract they draw the edge of the iris back, increasing the diameter of the opening (dilation) thus allowing more light to enter. If you are familiar with the working of a camera you will recognize the similarity behind the action of the iris and the diaphragm of a camera (See Figure 7.4.)

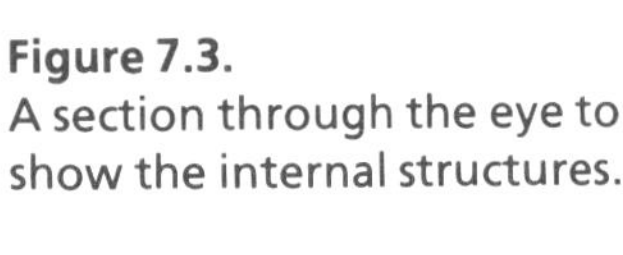

Figure 7.3.
A section through the eye to show the internal structures.

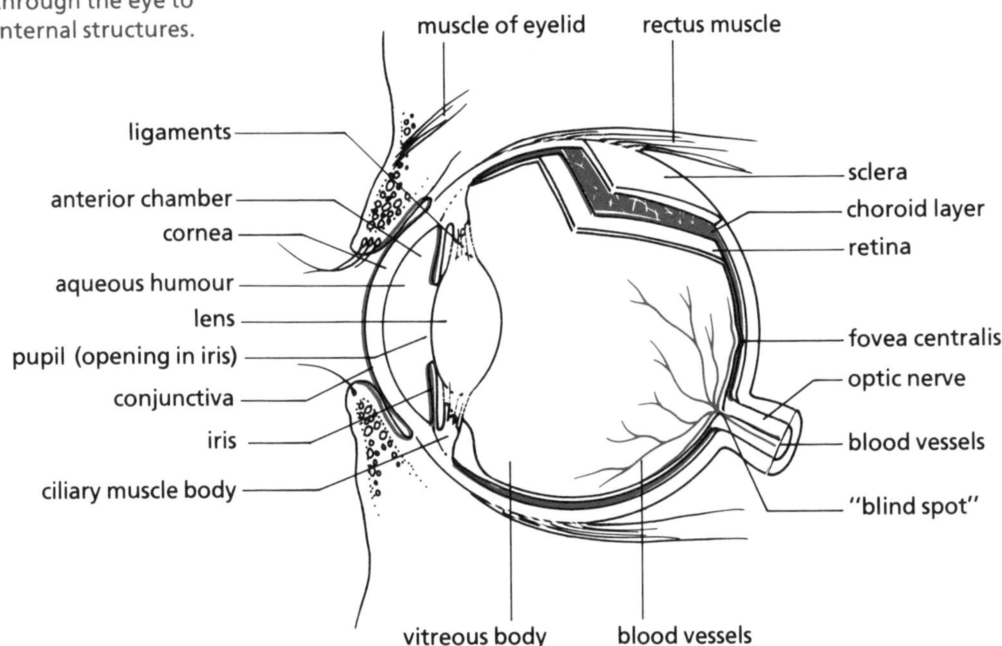

Figure 7.4.
A comparison between the eye and a camera.

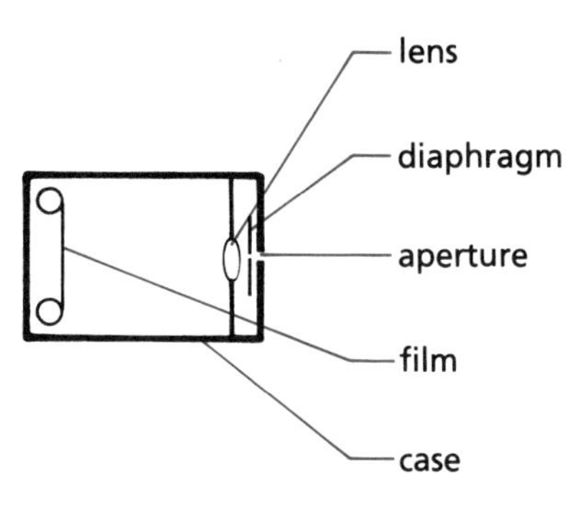

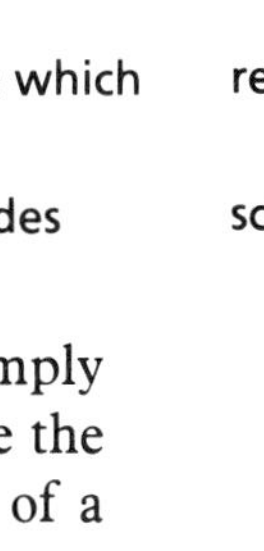

Camera	Function	Eye
lens	Focusses the light rays on a sensitive screen.	lens
diaphragm	Controls the amount of light entering the camera or eye.	iris
aperture	An opening to allow light to enter.	pupil
film	Light-sensitive screen on which the image is focussed.	retina
case	Support structure. Excludes extraneous light.	sclera

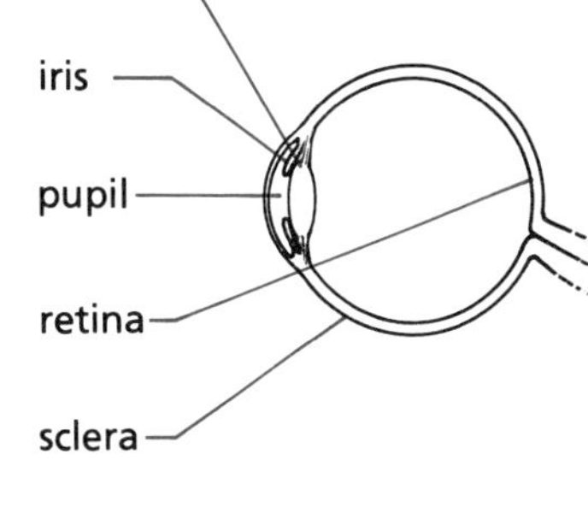

The black centre of the eye, known as the **pupil**, is simply the opening in the iris because, as the eye is dark inside the opening, it shows up as black – like the dark opening of a deep cave or tunnel. The size of the black opening changes as the muscles of the iris react to the change in the amount of light shining into the eye.

The Lens

The **lens** is located immediately behind the opening of the pupil. It is a tansparent disc-shaped structure, which is elastic and therefore capable of changing its shape. The interior is filled with a clear jellylike fluid. The lens is held in place by tiny ligaments that are attached to the ciliary muscles. If these muscles were totally relaxed, the lens would be more spherical in shape. It is, however, held under tension by the suspensory ligaments and is thus pulled into a flattened disc with two convex surfaces. When the ciliary muscles contract they release tension on the ligaments and allow the lens to assume its thickest, most convex shape. When the muscles relax, the ligaments pull the lens into a flattened shape with less convex surfaces. (A comparison of a camera and the eye is shown in Figure 7.4.)

The Layers of the Eyeball

The wall of each eyeball is composed of three layers. The outer coat called the **sclera**, is a white, tough layer forming a case which helps to maintain the shape of the eye. The extrinsic muscles are attached to this layer. At the front of the eye, the sclera becomes transparent and bulges forward to form the **cornea**. The cornea lacks blood vessels and is

(sk-lair-ah)

completely clear; its cells obtain the necessary nutrients from interstitial fluid rather than blood. The cornea contains very sensitive touch receptors.

(core-oid)

The middle layer is the **choroid layer**. It contains many blood vessels and pigmented granules which prevent light from being reflected within the eye – much as the black paint inside a camera prevents stray rays of reflected light from spoiling the film. The inside layer of the eye wall is the **retina**. It lines the interior of the eye and is made up of special light-sensitive neurons. (See Figure 7.3.)

The Retina

(ret-i-nah)

The innermost layer of the eyeball is the **retina**. This coat contains two main types of cells. Next to the choroid lies a layer of *pigmented* cells, while on the inside surface of the eyeball are several layers of different kinds of *nerve* cells. The light-sensitive nerve cells in this inner coat respond to light energy, which enters the eye, and converts it into nerve impulses for transmission to the brain. The other nerve cells in this layer are primarily connector neurons which accept the nerve impulses and carry them to the optic nerve. (See Figure 7.5.) There are two types of light-sensitive cells, the **rods** and the **cones**.

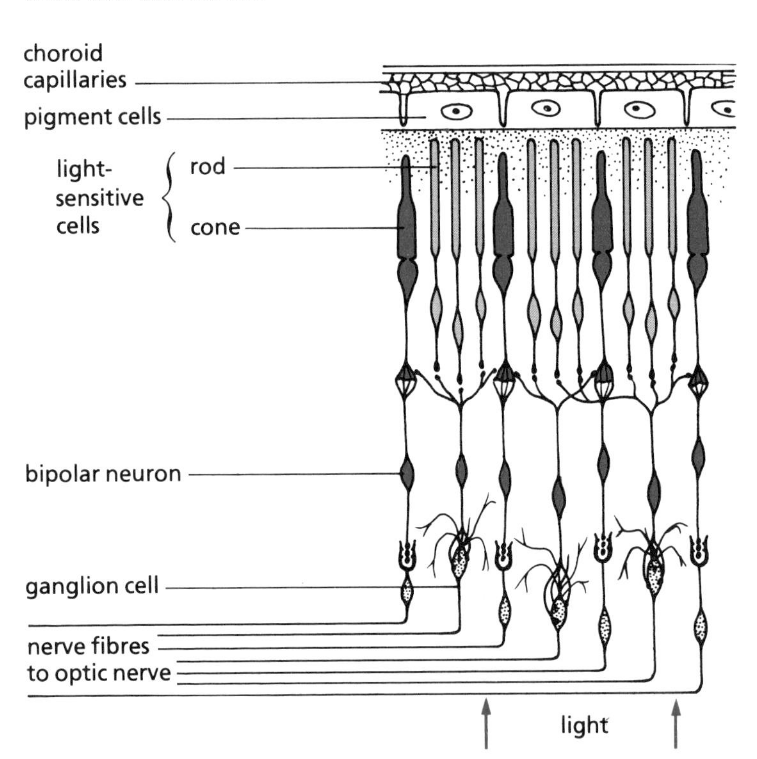

Figure 7.5.
A section through the retina to show the rods and cones, which are the light-sensitive nerve endings of the eye.

The **rods** are sensitive to dim light and are especially useful for night vision. The rods, which contain a substance called **rhodopsin**, require Vitamin A for its production. If the body lacks Vitamin A, a condition called night blindness develops, in which the eye recovers more slowly than usual from changes in light intensity. For example, the oncoming glare of car headlights at night temporarily wipes out the visual capacity; the eyes of a person suffering from night blindness adjust to darkness very slowly. There are more than 125 000 000 rods in each retina; that should give you some idea of how small these neurons are.

WHITE LIGHT AND COLOUR

When white light passes through a triangular glass prism it is separated into six bands of colour: violet, blue, green, yellow, orange, and red. White light is a combination of many wavelengths of light. As white light passes through the prism the various wavelengths are bent unequally and thus cause this rainbow effect. Each colour is, therefore, a name given to a particular wavelength of light. There are some wavelengths beyond the visible range of the eye. Ultraviolet light cannot be seen with the human eye, but its rays cause the skin to tan or burn.

Just as white light can be broken into the colours of which it is composed, the primary light colours (red, green, and blue) can combine to make white light. When these colours are shined toward a central point, the light will be white where they overlap. This colour mixing of the primary light colours is used in producing colour TV. (Don't confuse this with the mixing of primary pigment colours—yellow, blue-green, and magenta—which produces black.)

The most probable theory of how the cones of our eyes distinguish colours relies on the mixing of primary light colours. There are thought to be three types of cones, each of which corresponds with one of the three primary light colours. Evidence for this is still inconclusive.

The spectrum and light mixing.

white light
glass prism
red
orange
yellow
green
blue
violet

When white light strikes a glass prism at a particular angle, it separates the light into the colours of the spectrum.

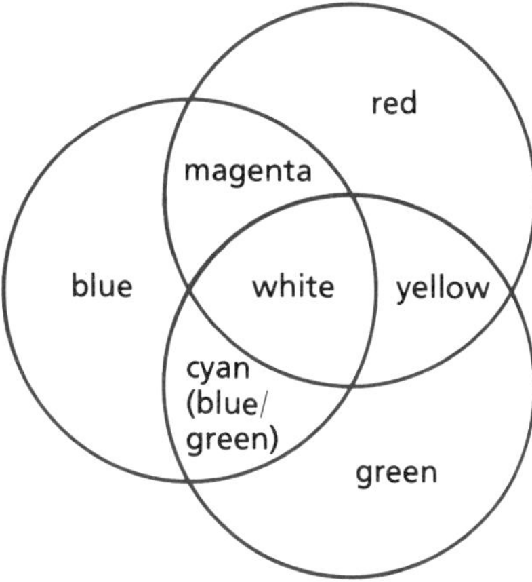

When the primary colours (red, green, and blue) are mixed, other colours are obtained. The principle is used in theatre lighting, colour photography, and in the human eye.

The **cones** of the eye are responsible for detecting colour. There are three kinds of cones, each sensing light of a different wavelength corresponding to one of the three primary colours. The three kinds of colour-sensitive cones are able to recognize many combinations of light wavelengths and enable us to see all the different hues in the world around us. The cones are less numerous than the rods and number about 7-10 000 000 in each eye. Defective cones, which prevent a person from recognizing some combinations of light wavelengths, are usually the result of inherited disorders, such as red-green colour blindness.

Although cones are much less sensitive than rods to low-light intensity, they operate well under bright daylight conditions. Have you noticed your ability to distinguish colour is reduced at night or in a dimly lighted place? Witnesses describing car accidents that occur at night are frequently inaccurate; dark colours, such as reds, browns, and blues, are not easily distinguished at night. After dark, we see mainly greys and have limited ability to distinguish colour differences.

At the back and very close to the centre of the retina is a small yellowish spot in which there are cones but no rods. This spot, called the **fovea centralis**, is an area of special sensitivity which is useful for examinations of details. As you read this page the eyes move in small jumps to keep the print focussed on the fovea. When you look closely at a scale, or some detail of a picture, or the entry point of a sliver into the skin, you use this specialized location on the retina.

(foe-vee-ah)

All the nerve fibres from the rods and cones collect at a point near the fovea and leave the eye as the **optic nerve**, which carries impulses to the brain. At the point where these fibres leave the retina, there are not rods or cones and hence no capacity for vision. This is therefore called the **blind spot** of the retina. (See Activity Number 3.)

The Reflexes of the Eye

There are two pupillary reflexes. The **light reflex** is well known and can easily be demonstrated by shining a small light into the eye and then removing it. The pupil will change in size, growing smaller in light and increasing in diameter as the light dims. This action is caused by the circular and radiating muscles of the iris, as explained earlier in the chapter.

RAPID EYE MOVEMENT (REM) AND DREAMS

During the 24-h day each person undergoes two main types of consciousness, sleep and wakefulness. The body has its own "biological clock" that regulates these 24-h cycles, sometimes called circadian rhythms. Animal experiments indicate that the **pineal gland**, located in the mid-dorsal side of the brain's cortex, is the body's main timekeeping device. **Serotonin**, a nerve transmitting hormone, is built up and stored in the pineal gland. If experimental animals are deprived of serotonin in this centre of the brain, they become permanent insomniacs and die of exhaustion.

Sleep is an essential physiological process. During sleep, the brain fluctuates between two states of awareness, one peaceful and deep, the other resembling a state of wakefulness. Three-quarters of the night is spent in periods of deep relaxation but about every 90 min, four or fives times a night, the brain moves into a second state known as "REM sleep", named for the Rapid Eye Movements that characterize it. About ten minutes after the rapid eye movements cease, the sleeper begins to descend once again through the stages of deeper non-REM sleep. About an hour and a half after the first REM episode the cycle repeats itself. It is during the REM periods, which vary in length from 10–30 min, that a person dreams. At the end of each cycle, the REM episode becomes longer.

Researchers have observed that the more rapidly the eyes flicker, the more intensely the dreamer participates in a dream. The eyes usually scan from side to side but sometimes the movement of the eyes is up-and-down. In one experiment when a dreamer showed the up-and-down eye movements, he was awakened to reveal his dreams. He said, "I saw a boy tossing a ball up and down." Researchers found that when a sleeper is awakened a few minutes after, rather than immediately after, the REM period, only fragments of the dream could be remembered. When a dreamer is awakened rapidly, the dream memories are more detailed than if the person is allowed to be gradually awakened. The REM's are also observed in babies. In fact, the newborn spends about half of the total sleep in REM sleep.

Some people claim that they never dream but most people simply forget their dreams. Psychologists believe that the lack of dreaming may be harmful. In fact, schizophrenics appear to dream less than other people.

Hallucinations may compensate, in some way, for a lack of dreams. For example, a New York disc jockey who stayed awake for 200 h as a publicity stunt claimed that he started seeing a friend's face on the clock dial. REM sleep is apparently a necessary part of the night's rest.

The **accommodation reflex** also results in a change in the size of the pupil but occurs in response to changes in the distance from the object being viewed. If the eyes are fixed on a distant object, then shifted to view a nearby object, the pupils will decrease in size. It has been estimated that the eyes make about 100 000 accommodation adjustments every day.

Adaptation. If you walk into a dark cinema, your eyes require some time to adjust to the dim light, so that you can find your way to a seat. If you go out of a dimly lit cabin into a bright snowy landscape, you will find that your eyes also require time to adjust to the intensity of light. The eye must shift the active reception of light from rods to cones or vice versa and this cannot be effected instantly. This ability is known as **adaptation**.

After-images are recognized in two forms. **Positive** after-images result when the image appears in the same colour as in the original stimulus. If you look at an electric light bulb for a few seconds, then close your eyes, you will find that you can "see" an after-image of the bulb in the same colour.

Negative after-images appear in colours that are complementary to the original colour. If you look at some coloured objects for about 30 s then stare at a white card, the images appear in opposite or complementary colours, in place of the original colours. (Red appears as green.) The after-image effect is believed to be due to the fatigue of the originally stimulated receptors.

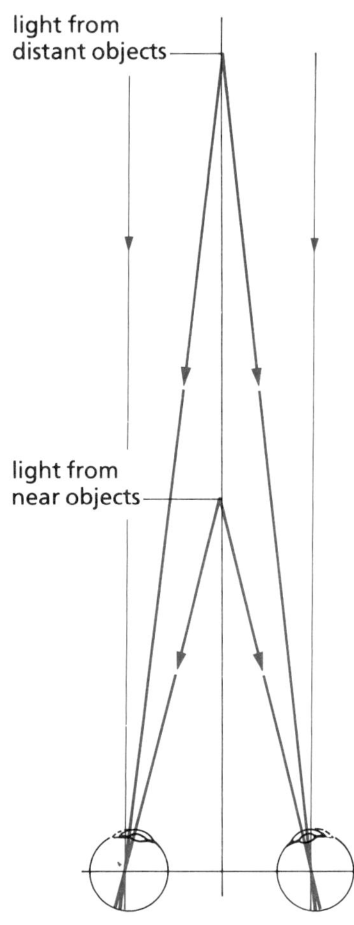

Figure 7.6.
Convergence: In order to see objects very close to us, it is necessary for the eyes to be turned inward or to converge. Objects at a great distance are viewed along parallel sight lines and the eyes are held approximately at right angles to the facial plane of the body.

Binocular Vision

If you look at a photograph taken with a camera it appears flat, in two dimensions only. A skillful photographer can give an impression of depth, but in order to make accurate judgments about distance and depth we require two eyes producing two images. We need **binocular vision**. The eyes turn in their sockets to focus on either near or distant objects. (See Figure 7.6.) The two images that we perceive, one through each eye, must be focussed on areas of the retina of each eye, so that they are superimposed in our brain and are interpreted as one object. If this is not the case, we see two images and have **double vision**, which is similar to the effect of looking cross-eyed.

When you look at a nearby object, your eyes **converge**. If you raise your finger at arm's length and move it toward your nose, while following it with your eyes, your eyes will gradually converge. Eventually as your finger gets close to your nose, you will no longer be able to keep it in focus and you will see "double."

When you close one eye (**monocular vision**), the brain receives only one set of impulses. When both eyes are open, two sets of impulses are sent to the brain. Because the eyes are set a short distance apart, two slightly different images are perceived. Perhaps you have seen stereoscopic slides projected or have used stereoscopic lenses in a geography class to look at **stereoptic** aerial photographs. The two pictures taken from slightly different positions give a very realistic, three-dimensional effect.

Look carefully at Figure 7.7. You will see that nerve fibres from the side of each eye nearest to the nose cross over to the visual centre on the opposite side of the brain. Light striking the lateral side of each retina is carried to visual centres on the same side of the brain. This means that the objects seen out of the right side of each eye are interpreted by the same side of the brain and create a single visual impression, in a way similar to the images of stereoscopic pictures. The recognition and interpretation of these two sets of visual stimuli enables the brain to establish depth and distance relationships.

If you use a telescope or a rifle to sight an object, you must use one eye. The "line of sight" will allow you to see the object, but you will not have a very good indication of exactly how far away the object really is. The other eye would give you a second line of sight. The point at which these two lines meet and cross gives an excellent indication of the relative position of the object. Stand a test tube up in a beaker and place it at arms length. With both eyes open, try to place the pencil in the test tube. Now with one eye closed try the same test. You will note how much more accurate distance perception becomes when two eyes are used rather than only one.

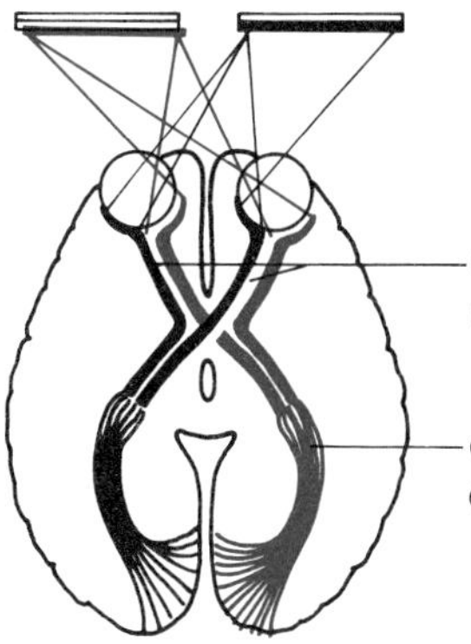

optic nerves

optic chiasma

Figure 7.7.
This diagram illustrates that light coming from a source on the right-hand side (black) will be viewed by portions of the left side of each retina. Nerve impulses from these areas of both eyes are then co-ordinated in the left visual centre of the brain.

Vision Defects

The lens adjusts to allow us to focus on nearby or distant objects. It is "elastic" and can be stretched or relaxed by the ciliary muscles that surround it. When we look at nearby objects the ciliary muscles contract and the lens bulges to become more convex. This causes greater bending of the light rays and brings the image into sharp focus on the retina. To view distant objects the ciliary muscles relax and the lens assumes a flatter, less convex shape, which does not bend the light rays as sharply. (See Figure 7.8.)

The light rays, whether from nearby or distant objects, must be focussed on the retina. If the eyeball loses its round shape and becomes too long, the light rays will focus before they reach the retina, resulting in near-sightedness. Sometimes the eyeball is misshapen so that it is too short and the light rays will focus behind the eyeball, resulting in far-sightedness.

The use of an artificial lens in front of the eye can make adjustments for those small differences and enable people to achieve normal sight. (See Figure 7.8).

Figure 7.8.
The accommodation of the lens and how short- and far-sightedness can be corrected.

The normal eye lens brings the light rays to a sharp focus upside down on the retina at the back of the eye. By changing the thickness, and therefore the curvature, of the lens the image can be brought into focus for both nearby and distant objects.

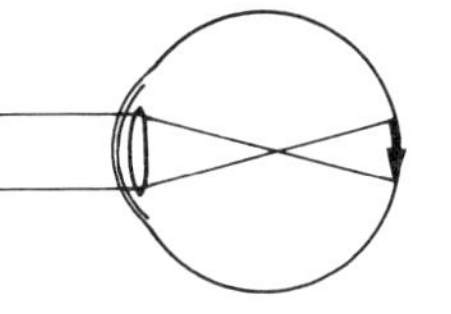

Ciliary muscles relax, tension on the suspensory ligaments increases, and the lens becomes thinner.

Rays from a distant object are parallel (more than 7 m).

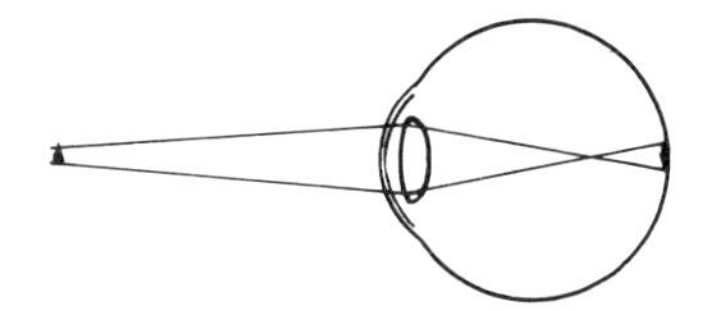

Rays from a nearby object (less than 7 m).

Ciliary muscles contract, the lens becomes more convex.

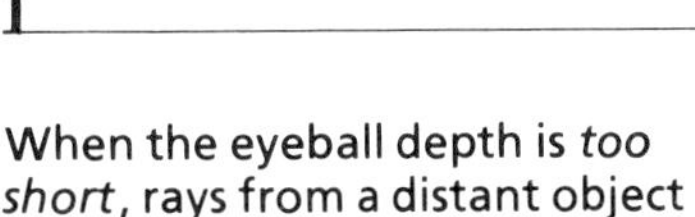

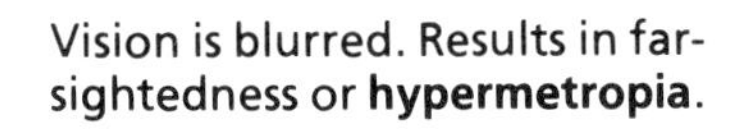

Vision is blurred. Results in far-sightedness or **hypermetropia**.

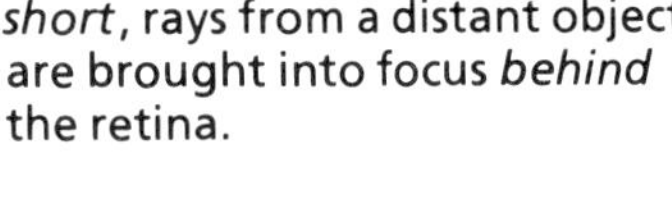

When the eyeball depth is *too short*, rays from a distant object are brought into focus *behind* the retina.

This defect can be corrected using a convex lens.

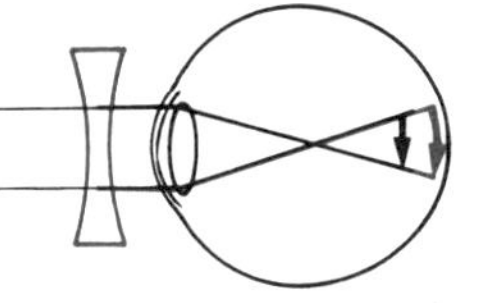

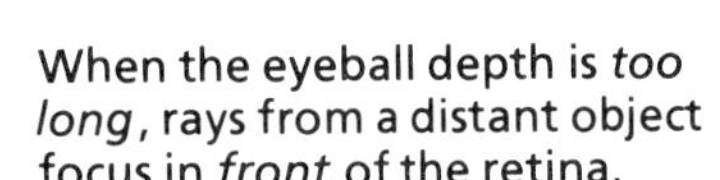

Vision is blurred. Results in short-sightedness or **myopia**. Objects near the eye can be seen clearly.

When the eyeball depth is *too long*, rays from a distant object focus in *front* of the retina.

This defect can be corrected using a concave lens.

Visual Acuity

Until recently visual acuity was measured in feet and referred to as 20/20 vision. With the introduction of metrication, 6/6 vision is gradually being adopted.

Everyone has seen and used a Snellen Eye Chart, which has a series of lines of letters with the largest letters at the top and lines of successively smaller letters in rows beneath. Visual acuity refers to the sharpness with which detail is seen. It is measured by demonstrating the size of type that you can read at a standard distance of six metres. If you are able to read the line of letters that a person of normal vision can read at six metres, you are rated at 6/6. If you have poor visual acuity, you may only be able to read at six metres what a person with normal vision can read at 30 m and you would be rated as 6/30. It is not uncommon to find students with visual acuity better than 6/6. Because there is frequently a difference between the left and right eyes, the test is taken using each eye alone, by covering one while the other is used for reading the chart.

Astigmatisn (a-stig-ma-tizm)

Astigmatism is caused by an uneven curvature of the cornea or lens. Light waves passing through the lens or cornea are distorted, giving an image that may have parts in focus and other parts out of focus. A diagram, like the one shown in Figure 7.9, is used to determine what part of the cornea is affected. A person with astigmatism will see some lines in sharp focus whereas others will be blurred.

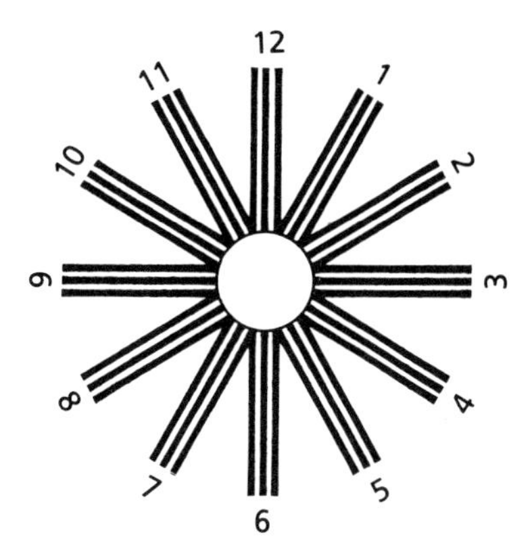

Figure 7.9.
Astigmatism is the blurring of vision caused by an abnormal shape of the cornea or lens.

Colour-blindness

Colour-blindness affects males more frequently than females. About 6–8% of males are colour-blind, whereas only 0.4–0.6% of females have this defect. Total colour-blindness is usually caused by a lack of colour receptors or a defect in the visual centre of the brain. Red-green colour-blindness is more common and may be due to a defect in the colour receptor cones for red and green. Such persons are unable to take jobs where these colours are important, for example, as train engineers or airplane pilots.

QUESTIONS FOR REVIEW

SOME WORDS TO KNOW

Match the description given in the left-hand column with a word shown in the right-hand column. DO NOT WRITE IN THIS BOOK.

1. Controls the amount of light entering the eye.
2. The inner, light-sensitive lining of the eye.
3. Responsible for colour reception in the retina.
4. Changes thickness to focus the image of nearby or distant objects on the retina.
5. A result of distortion of the cornea.
6. A super-sensitive portion of the retina.
7. Muscles that move the eye within its socket.
8. Thin membrane covering the outer surface of the eyeball.
9. Transparent protective coating in front of the eye.
10. The outer coat of the eyeball.

A. iris
B. cornea
C. conjunctiva
D. retina
E. lens
F. rods
G. cones
H. fovea
I. astigmatism
J. blind spot
K. sclera
L. orbit
M. extrinsic muscles
N. intrinsic muscles

SOME FACTS TO KNOW

1. What protection is provided for the eye to keep it from injury?
2. What is the function of the extrinsic eye muscles?
3. What are the three layers of the eyeball? Give the general function of each layer.
4. Draw a diagram of the eye in section and label each of the parts.
5. Muscles are found inside the eye, in the iris, and around the lens. Explain the functions of these muscles.
6. What is the special advantage of having two eyes?
7. Explain the reason for the blind spot on the retina.
8. Explain why an eye might be in perfect condition, with no damage to the eye at all, yet a person could be totally blind.
9. What happens to the eye when a person looks first at a nearby object and then looks at a distant object?
10. When a person is short-sighted what part of the eye is affected? What type of lens is required to correct this defect?

QUESTIONS FOR RESEARCH

1. Select one of the following topics and prepare a report on it:
   - optician
   - cataracts
   - colour-blindness
   - jobs available to the blind
   - eye care
   - ophthalmologist
   - detached retina
   - Braille
   - blind persons and the Olympics for the handicapped
2. Call your local Motor Vehicle Branch and find out what kinds of visual tests and regulations apply to drivers.
3. What effects do the following drugs have on the pupil of the eye?
   a) adrenaline
   b) cocaine;
   c) morphine;
   d) marijuana.
4. What is night blindness? What is visual purple? What has eating carrots got to do with this topic?
5. What is the significance of the Rapid Eye Movement (REM)?
6. Investigate how lasers can be used to help people with eye problems.
7. What are optical illusions? Make a collection of illustrations of this kind of illusion. These can be very interesting and very deceptive. There is a wide variety of these illusions: distorted rooms, reversible figures, impossible objects, and perceptual motion. Try to explain why our minds are confused, by such images, into seeing the illusions.

Activity 1: THE SNELLEN EYE CHART: VISUAL ACUITY

This test determines the sharpness or visual acuity of the eyes.

The chart uses a sequence of black letters of varying size. Beside each line is a number that represents the distance at which that letter can be read by the normal eye. If you read the line marked 6 at a distance of six metres, then you have 6/6 vision in that eye. That is, your vision is within the normal range. If, however, you can only read the line marked 12 at a distance of six metres, then you have 6/12 vision. Many young people can read line 4 at six metres, thus giving them a visual acuity of 6/4; they have better than average vision.

Method
Have your partner stand six metres from the Snellen Eye Chart. Cover one of his or her eyes with a piece of card and determine the smallest line that he or she can read clearly. Repeat using the other eye.
Students who use glasses, should complete the experiment with and without their glasses and *record both results.*

Explain the meaning of visual acuity and the meaning of 6/6 vision. What is the minimum acceptable standard of visual acuity for driving without using glasses?

Activity 2: NEAR POINT ACCOMMODATION

Determine the near point accommodation of the eye.

Materials
metre stick and a pencil

Method
Hold a metre stick just under one of your eyes. Close the other eye. Have your partner slide a pencil along the metre stick towards you, starting at a distance of about 60 cm. Keep the point of the pencil, which should be raised just above the edge of the metric stick, in focus and note the point at which it becomes blurred. Move the pencil away, until the pencil point is sharply in focus again.
Record this distance. Repeat with the other eye.
This test depends on the resiliency of the lens and its ability to change its shape. This ability decreases with age.

Age in Years	Near Point Accommodation
10	7.5 cm
20	9.0
30	11.5
40	17.2
50	52.5
60	83.3

Examine the chart and note the age at which a dramatic change occurs. Have you noticed how middle-aged people have to hold books or papers at arm's length in order to be able to read them? Explain in your notebook, what is meant by near point accommodation, how it can be tested, and your results for each eye. Is one eye stronger than the other? Test your parents, brothers and sisters, or grandparents and see if your results conform to those in the table.
The term "accommodation" is also used to describe the ability of the eye to focus on near and distant objects. What muscles are used for each of the two kinds of accommodation?

Activity 3: DISCOVERY OF THE BLIND SPOT

The nerves that send impulses from the retina to the brain leave the back of the eye and form a thick bundle of nerve fibres, which is called the Optic Nerve. Where these nerves leave the eye, there are no light-sensitive cells and each eye has a "blind spot" at this location. (Refer to the diagram of the eye.)

Method I
Hold your textbook about 45 cm away from your face so that the cross is directly in front of your right eye.

Close your left eye, then slowly move the text towards you. Keep your right eye on the cross; you will be aware of the dot, but do not look at it directly. At some point the dot will disappear, measure the distance from your eye to the book at the point where the dot disappears. Repeat with the other eye. *In your notebook explain what the blind spot is and record your results.*

Method II
Another method of discerning the location of the blind spot is to use the line of letters below.

+ A B C D E F G H I J K L M

Close the right eye. Look at the cross, then slowly read the letters from left to right. At some point the cross will disappear. *Note which letter you have reached when the cross disappears. Is this consistent with the results of others in the class? Did each student keep the book the same distance from the eye?*

Activity 4: COLOUR-BLINDNESS

Determine deficiencies in colour vision.
Colour-blindness is a genetically inherited abnormaility resulting from the deficiency of a particular gene on the X-chromosome. The most common type of disorder is red-green colour-blindness. If the red cones are lacking, wavelengths of red light stimulate the green cones and little differentiation between red and green is possible.

Method
Ishara test (using Ishara Test booklet):
Use the colour-blindness test booklet and follow the directions provided at the front of the book. *Record the numbers that you see and explain the significance of these numbers. The numbers are interpreted in the booklet.*
Holmgren test:
In this test a set of coloured wool strands is used. The subject attempts to match each of the strands with an identical strand in the folder. Mismatching (or hesitation in matching) usually indicates a colour abnormality.

Chapter 8

Other Senses – Hearing, Taste, and Smell

- The Ear and Hearing
- The Structure of the Ear
- The Sense of Taste
- The Sense of Smell

The Ear and Hearing

One dark night, many years ago, I was trying to find my way home through a dense London fog in England that had brought all cars and buses to a halt. I was lost, unable to see even my hand in front of my face, and was feeling my way slowly along the store fronts to obtain some sense of direction. It was deathly quiet, with no traffic moving on the streets. Then I heard the sound of approaching footsteps, coming without any hesitation along the echoing street. It proved to be a blind man that I knew. He took my arm, marched me into the centre of the road and took me home at quite a fast walking pace.

As we walked, he was listening to the echoes of our footsteps against the buildings that lined the London streets. Without sight, his ears had become finely tuned to sounds that I had always ignored. He sensed when we were closer to one side of the street than the other, knew when a side street turned off and, unlike me, knew exactly where we were. After a while, I too, started to notice small differences in the sounds that I heard and could make judgments about which direction we should take. I started to walk with my eyes closed to concentrate harder on the sounds around us.

Humans are very dependent upon sight. Unlike some animals which depend on very acute hearing and a sense of smell to warn them of approaching enemies, our ears and sense of touch take second place. Perhaps you have noticed a dog, head down, with its nose close to the ground following some fresh scent. The dog is often so intent on using its sense of smell that it may fail to see its quarry, even when it is only a few metres ahead.

Hearing is, however, an important part of our early warning system. It is sometimes called the "watch dog" of the senses, because it is the last to relax before we go to sleep and the first to come on duty, even before we are really awake. It is through the ear that we learn to communicate and develop our skills of speech, by listening to word sounds made by others. The ear enables us to judge distance, whether we hear thunder at a distance or the frightening sound of a motor horn close behind us. Also, we can judge the direction and the location of sounds with our ears.

The ear not only recognizes sound vibrations, converting them to nerve impulses for interpretation by the brain, but also functions as the organ of balance. It enables us to recognize our position, whether vertical or horizontal, and to maintain a steady, walking pace without falling.

The Structure of the Ear

The ear is organized into three distinct sections: the **external, middle**, and **inner** ear. (See Figure 8.1.)

The external ear is the part that can be seen on the side of the head. It consists of a flap of skin and cartilage called the **auricle** or **pinna**. The shape of this structure is designed to collect sounds and funnel them into the ear canal. Sometimes, if we have difficulty hearing some quiet sound, we use the hand to "cup" the ear. This extends the effective size of the pinna and increases the amount of sound energy entering the ear. The channel leading into the head from the pinna is the **auditory canal**. The auditory canal is lined with a thin layer of skin that contains fine hairs as well as cells which secrete a waxy substance (cerumen glands). This wax helps to protect the ear drum, although sometimes it can accumulate in large amounts and partially prevent the sound waves from reaching the ear drum.

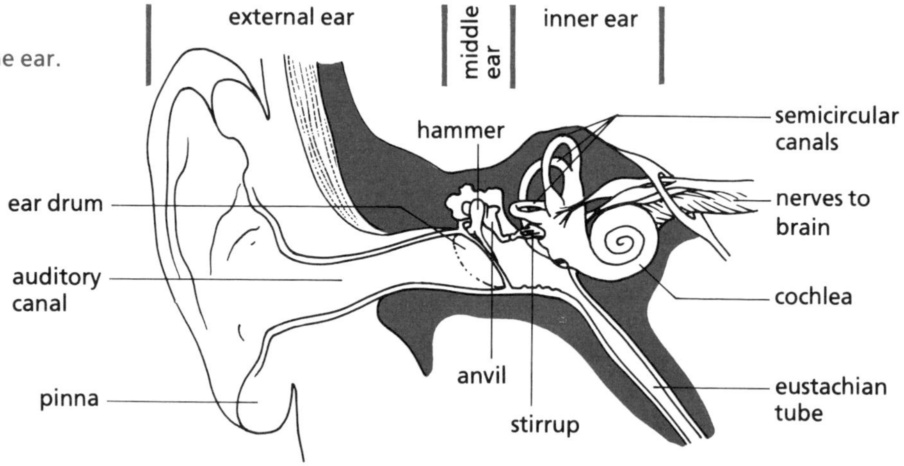

(o-re-kul)

Figure 8.1.
The structure of the ear.

The auditory canal passes through the temporal bone of the skull. The canal is closed at the inner end by a membrane which completely prevents entry into the head. The membrane is known as the **ear drum** or **tympanic membrane** and it separates the external ear from the middle ear.

(tim-pan-ik)

How Sound Travels

Before explaining the operation of the middle ear, we need to review a few simple ideas about what sound is and how it travels. Sound is produced when something vibrates. It may be from a ruler, extended beyond the edge of the desk, which is plucked to produce a sound, or it may be the vibration of a fan or motor. In order for this vibration to be conducted from its source to our ear, it must pass through a medium: something that transmits the energy of the vibrations to the ear, materials such as air, water, or metal. Usually this energy is carried by air molecules, which move away from the source of the sound, like ripples on a pond after a stone has been thrown into the water. The energy of the moving molecules bumping into each other causes them to eventually enter the auditory canal. They then reach and bump up against the tympanic membrane causing it to move back and forth. Perhaps you have seen an apparatus that has five or six steel balls suspended from a frame. If you lift the ball at one end and then let it fall, it strikes the ball next to it, but it is the ball at the far end that jumps away. (See Figure 8.2.) It is in this manner that sound waves are carried from a vibrating source to the ear. The object that vibrates originally to make the sound is represented by the first ball. The molecules of the medium through which the sound travels correspond to the balls in the middle, which appear to be stationery but pass the vibration to the last ball, which then moves. The tympanic membrane receives the sound vibrations and moves back and forth in response, thereby passing the vibrations on to the middle ear.

The tympanic membrane is stretched over the inner end of the auditory canal. On its inner surface, there is a thin mucus membrane, which is continuous with the walls of the middle ear cavity.

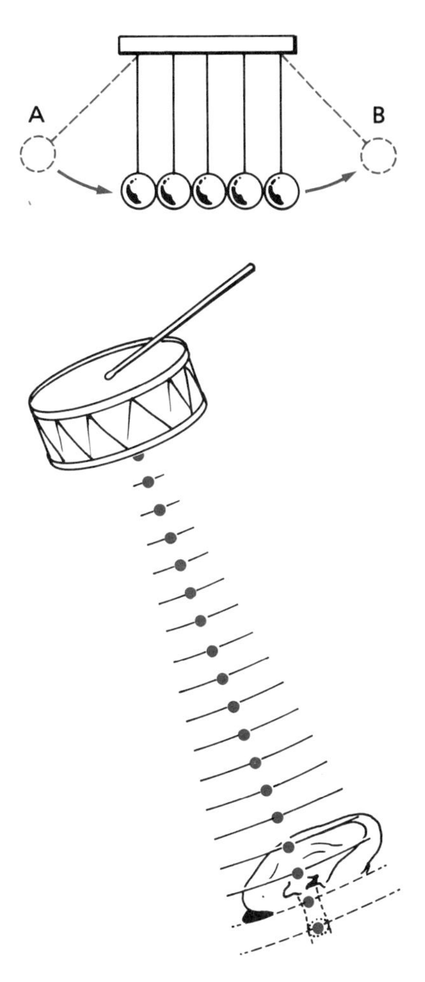

Figure 8.2.
How sound reaches the ear. When the ball at A is allowed to fall, the energy is passed from one to another of the suspended balls until the ball at B, which is free to move, is pushed away from the others. In a similar way, the energy of a vibrating source is transferred to the molecules of the air and passed from one molecule to another until it reaches the ear drum and causes it to vibrate.

The Middle Ear

The chamber of the middle ear is filled with air. If the tympanic membrane is to vibrate freely, the air pressure must be equal on both sides of the membrane. That is, the pressure in the chamber of the middle ear must equal that in the outer ear. You may have experienced a "popping" of the ears while you were riding in a subway train or aircraft, or even during a car ride when you were climbing steep inclines in a hilly part of the country. This popping is caused by an increase or decrease in pressure on the outside of the ear drum, which means that the membrane will bend inward or outward but cannot move freely. To equalize the pressure, you may swallow or yawn. When you swallow, you allow air to move in or out of the **eustachian tube**, which connects the middle ear with the throat where the air pressure is equal to that outside the body and in the outer ear. Swallowing thus tends to equalize the pressure in the middle ear.

(yoo-stae-shun)

Within the middle ear are three tiny bones (ossicles) which are held in place by tiny ligaments. These bones conduct the vibrations from the tympanic membrane across the air space of the middle ear transferring them to the oval window which communicates with the inner ear. They also act to amplify the vibrations.

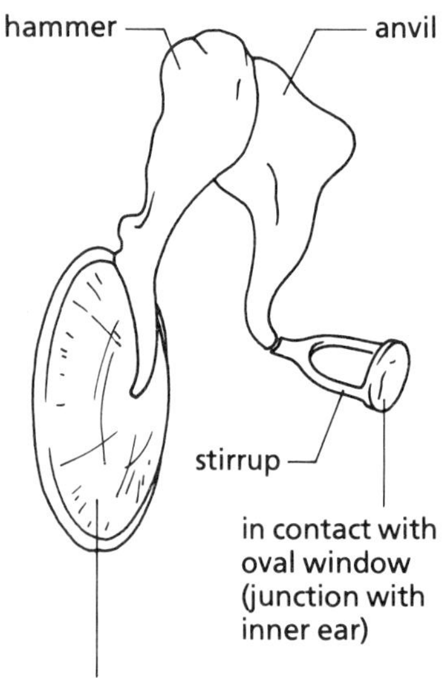

Figure 8.3.
The three bones of the middle ear.

The bones are named according to their shape: the **hammer** (malleus), **anvil** (incus), and **stirrup** (stapes). (See Figure 8.3.) The hammer is attached by the end of its "handle" to the tympanic membrane. As the membrane vibrates, it moves the hammer which in turn moves the anvil. The anvil transmits the amplified vibration to the head of the stirrup. The stirrup rests with its two sidepieces against a small oval membrane called the **oval window**, which separates the middle and inner ear.

Levers are really simple machines and they enable us to use a small force to produce a large movement, or vice versa. The tiny bones of the middle ear use this principle to amplify the sound energy striking the ear drum and increase it twenty times before it strikes the oval window of the inner ear.

The Inner Ear

The inner ear has two functions. Firstly, it converts sound vibrations to nerve impulses, which are transmitted to the brain. Secondly, it aids in the maintenance of balance and recognition of body position. The organs of the inner ear are located in small, cavelike hollows in the skull. In fact, the area where these cavities are found often is called the bony labyrinth (*labyrinth* means maze, a term often used to describe caves). The **cochlea**, which performs the hearing functions of the inner ear, is the organ responsible for transforming vibrations to nerve impulses. The **utricle**, **saccule**, and **semicircular canals** are the organs responsible for maintaining balance and body position.

(yoo-tri-cul)
(sak-you-l)

The organs inside the bony tunnels are filled with **endolymph**. Surrounding them, but separated from them by a thin tube of membranes (membraneous labyrinth), is another fluid, the **perilymph**. (*Peri* means around, *endo* means inside.) (See Figure 8.5.)

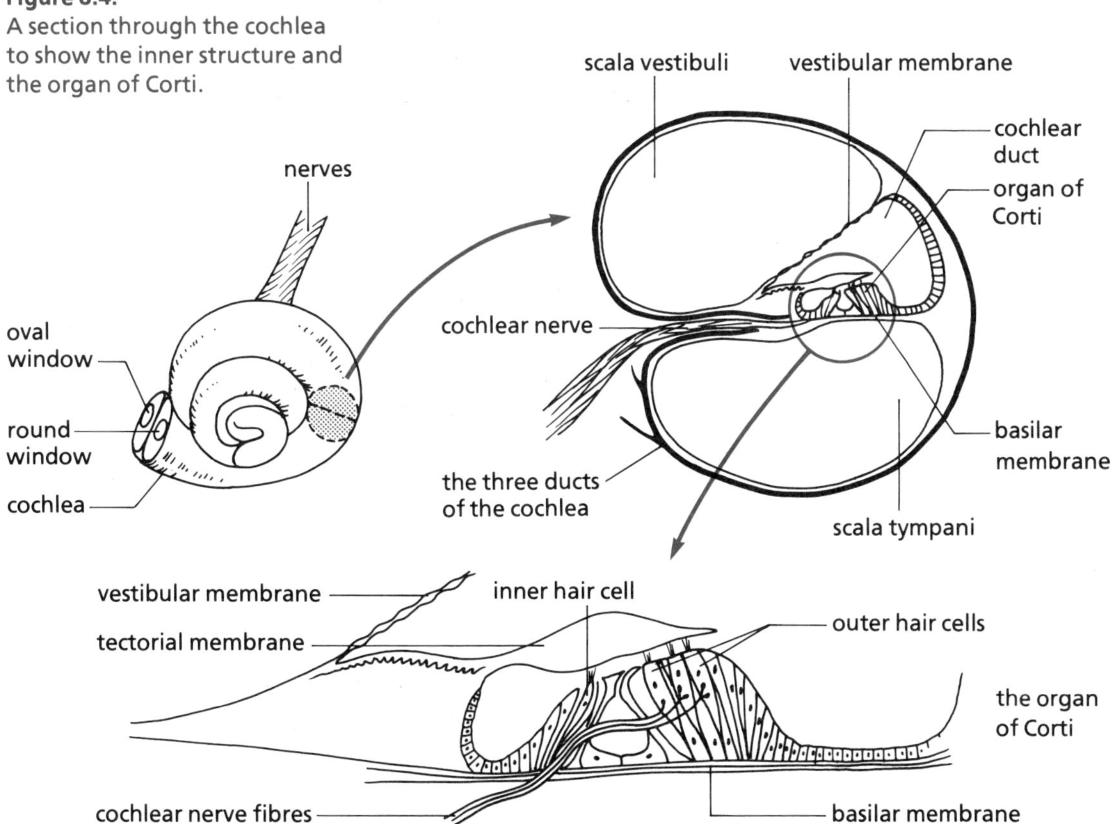

Figure 8.4.
A section through the cochlea to show the inner structure and the organ of Corti.

The Cochlea

The cochlea is shaped like the shell of a snail. It is made up of a tube that winds around a central column, becoming smaller and smaller as it spirals inward. (See Figure 8.4.) The **cochlear duct** lies inside this tube, dividing the cochlea into three parts. Above the cochlear duct is the **scala vestibuli**; below it is the **scala tympani**.

(cok-lee-ah)

The **oval window**, which was described previously, receives the vibrations conducted across the middle ear, is fitted into the scala vestibuli (*vestibule* means entrance). The scala tympani has another membrane-covered opening called

(ska-la)

the **round window**. If we straightened out the spiral tubes of the cochlea it would appear as in Figure 8.5.

As a sound vibration pushes in on the oval window it pushes the fluid on the other side, sending a wave up the scala vestibuli. This fluid wave reaches the end of the cochlea and bounces back, returning along the scala tympani. The movement of the wave is finally absorbed by the round window. As the wave of fluid is pushed along, it affects the membranes on either side of the cochlear duct. This results in stimulation of the **organ of Corti**, which contains the sound receptors.

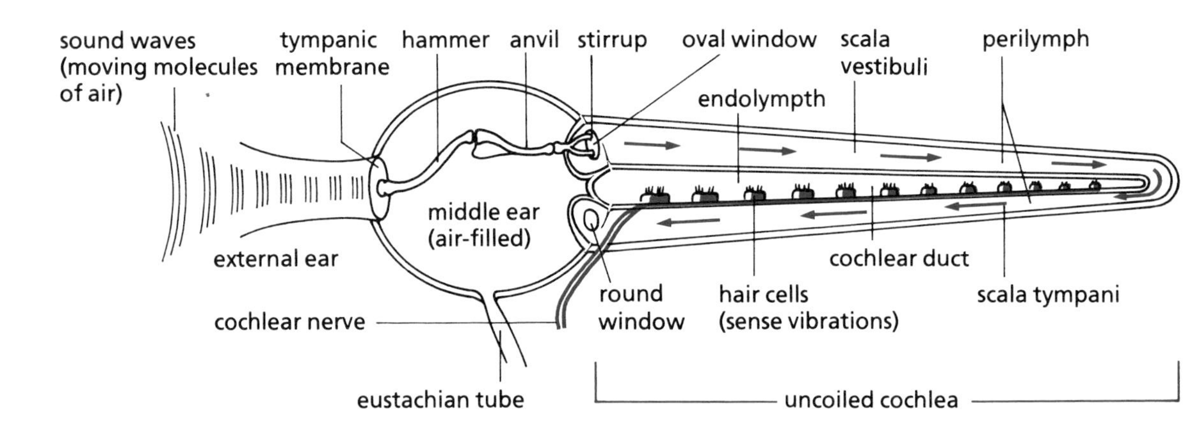

Figure 8.5.
This diagram shows the conduction of vibrations to the oval window of the inner ear. Vibrations of the oval window set the perilymph in motion. This movement is transferred to the endolymph and causes the hair cells of the organ of Corti to move. Impulses are generated in the hair cells and transmitted to the brain via the cochlear nerve.

Organ of Corti

The organ of Corti is located within the cochlea and responds to vibrations in the fluid of the inner ear. (See Figure 8.4.) The cells of this tiny organ possess a number of hairlike protrusions which run the length of the cochlear duct. These hair cells connect with small nerves that eventually join to leave the ear as the auditory nerve. As the hair cells vibrate in response to sounds received by the ear, they rub against a membrane called the tectorial membrane, which lies like a canopy above them. The bending of the hairs which results, stimulates the terminal branches of nerve cells located around the cells of the organ of Corti. The impulses thus initiated are carried to the brain by the auditory nerve. The hairs vary in length and are stimulated by vibrations of different wavelengths. Low notes cause longer hairs to vibrate and high notes cause shorter hairs to vibrate, stimulating the nerve cells to which they are attached. These hairs and nerve cells provide humans with an approximate hearing range of 16 Hz to 20 000 Hz (vibrations per second). (A piano has a range from about 27 Hz to 4186 Hz, from its highest pitched to its lowest pitched notes. Most human speech is pitched at about 1000 Hz.)

Hearing Disorders

The intensity (loudness) of sound depends on the height of the wave produced. It is recorded in decibels (dB). Zero dB is the threshold of audible sound for humans. Ranges in the 100+ dB can be uncomfortable and higher intensities cause physical pain and may damage the ear. (See Figure 8.6.) Technicians working around jet aircraft, or on very noisy machinery, wear special protective ear covers. Musicians, who use powerful electronic amplification, frequently have ear damage that affects their range of hearing as well as their ability to pick up fine overtones, which give quality to the sounds they produce. People who have some auditory impairment frequently increase the volume on their equipment to compensate for their hearing lose and are not aware of the discomfort this brings to others. It has been well established that rock music, with its constant high volume of sound, has already caused irreparable damage to many of its devotees.

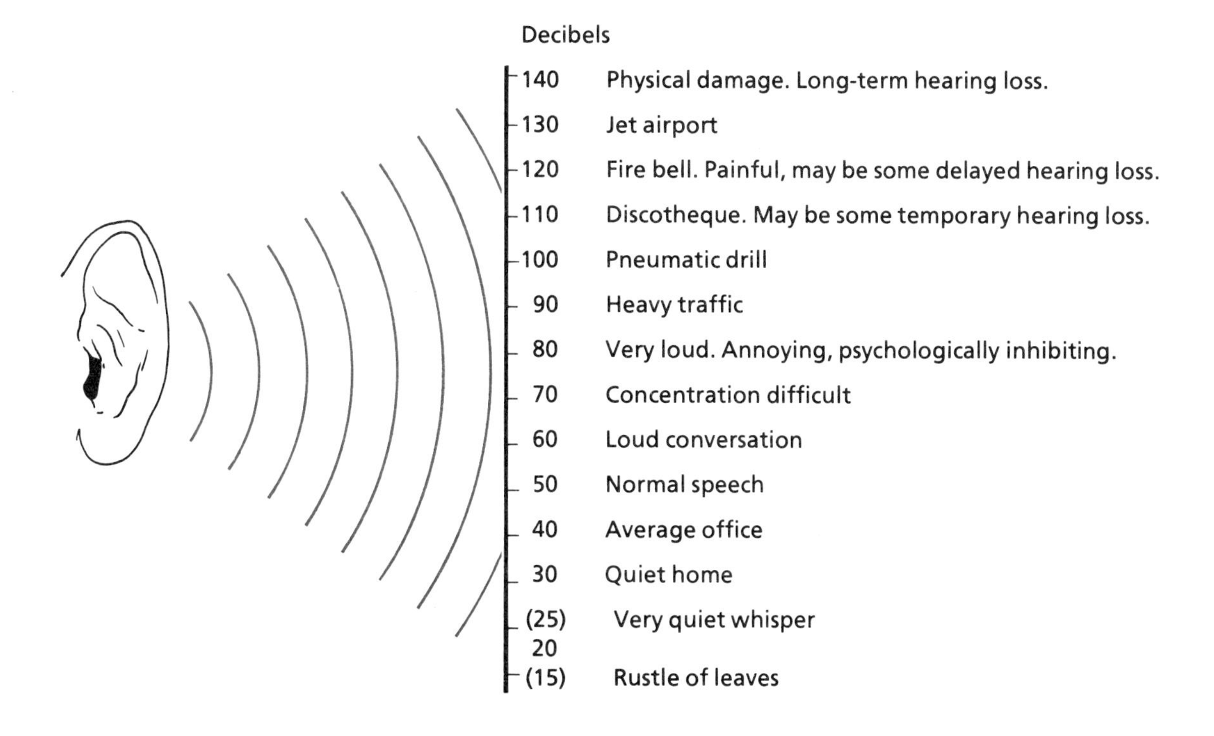

Figure 8.6.
A comparison of some common sound levels experienced in everyday life. Note: The bel is the unit of sound intensity and recognizes the work of Dr. Alexander Graham Bell, the inventor of the telephone and a man who devoted much time and effort to hearing defects.

There are several kinds of deafness. **Conduction deafness** is caused by some interference in the transfer of sound waves to the middle ear. It may be the result of something as simple as a build-up of wax. It also could be caused by damaged or scarred tissue on the tympanic membrane, or by some defect in the action of the bones of the middle ear. **Nerve deafness** is caused by some problem of the nerve cells, either of the sensory cells in the inner ear or the nerve pathway to the brain.

The Organs of Balance

Whereas the cochlea constitutes part of our hearing apparatus, the other structures of the inner ear are responsible for balance.

If you have ever been on a rapidly revolving ride at a fairground, or just spun around quickly, you will know that, for a moment or two after stopping, your sense of balance has been thrown off and you become giddy or disoriented. Two separate organs are responsible for maintaining balance. One is a fluid-filled structure that is divided into two parts called the saccule and the utricle. The other is formed by a combination of three tubes at right angles to one another, called the semicircular canals.

The Utricle and Saccule

Just inside the vestibule of the inner ear, floating in the perilymph, are two membranous sacs, the **utricle** and **saccule**. Each of these is filled with endolymph. (See Figure 8.7.) On the floor of the utricle is a small patch of hair

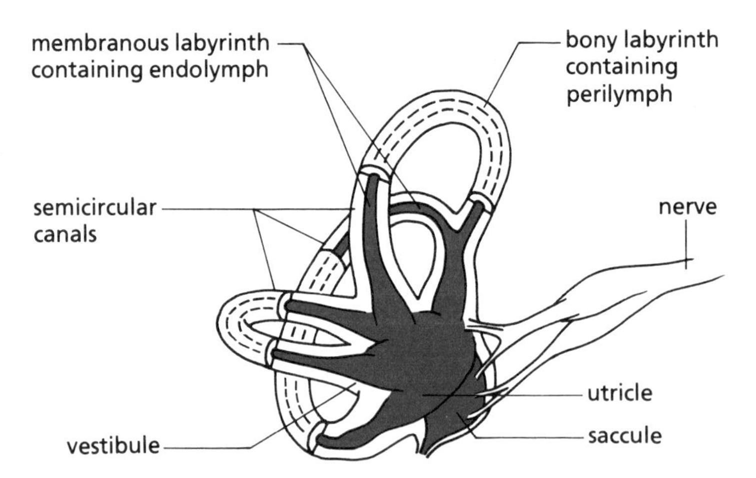

Figure 8.7.
Structures associated with balance in the inner ear. The membranous labryinth is contained within the tubes of the bony labryinth. The vestibule is filled with a fluid called perilymph. "Floating" inside this tube are two membranous sacs called the utricle and the saccule.

cells, and around them is a jellylike semifluid, which contains some tiny particles of calcium carbonate, the **otoliths**. When the head is tilted these particles move down in response to gravity and trigger hair cells to send impulses to the brain indicating the new position of the head. (See Figure 8.8.) The saccule has a similar patch of sensory cells, but it is not known whether these cells respond to head movements or to low-pitched vibrations.

The Semicircular Canals

While the utricle responds to static position, the **semicircular canals** respond to changes in movement such as stopping, starting, or turning. The three tubes of the semicircular canals form part of a circle and each is set at right angles to one another in three different planes. (See Figure 8.9.) There are hair cells present in the semicircular canals and these respond as the endolymph is set in motion.

Figure 8.8.
Inside the saccule are fine hairs which project into a jellylike substance which contains small particles of calcium carbonate (otoliths). When the head is bent forward, gravity affects the otoliths and the gelatin that contains them. This movement stimulates the nerve fibres in the hair cells and a "position" message is sent to the brain for interpretation and action.

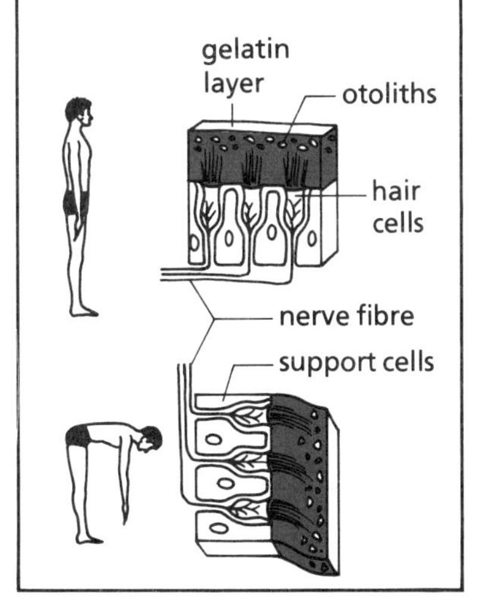

Figure 8.9.
The semicircular canals are in three planes at right angles to one another.

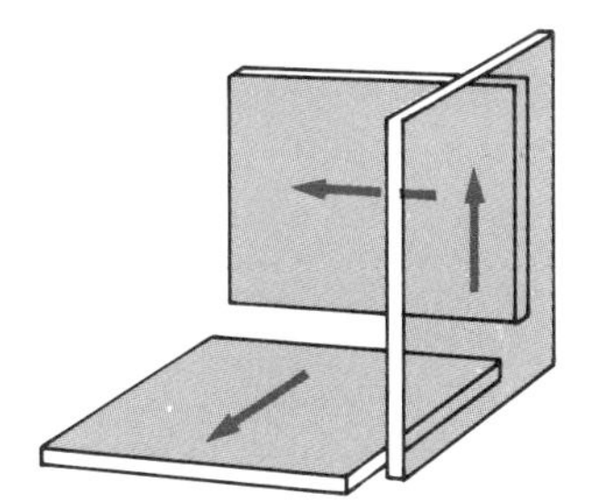

The semicircular canals. Try to imagine this shape carved out like tiny caverns and tunnels in the temporal bones of the skull.

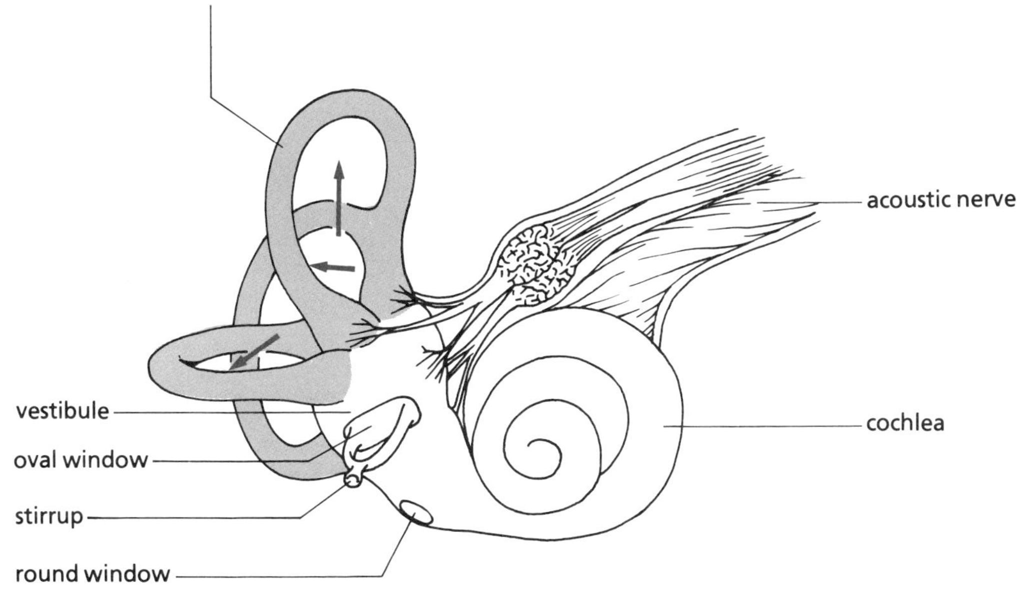

If the head is swung forward and down, the fluid will flow in one of the semicircular canals; another will react if the body is bent to the left or right, and the endolymph in the third tube will flow if the head, or body, is turned on the body axis. The brain recognizes these impulses and sends out motor impulses to correct, or adapt to, the motion by stimulating the appropriate skeletal muscles.

If the flow of impulses is affected by the intake of alcohol, or other substances, the body's reactions are slowed down and the response becomes exaggerated. The body swings too far to one side or another, before the motor neurons cause the skeletal muscles to react. Under the influence of alcohol it becomes difficult to maintain normal balance or to walk a straight line.

The Sense of Taste

If you open your mouth and look into a mirror you can see that the surface of the tongue is covered with tiny bumps (*papillae*). Around these small bumps, buried in the tissues of the tongue, are the **taste buds**. On the top of each taste bud is a small opening or pore, through which solutions enter to stimulate the receptors. Solids cannot be tasted; the substances must first be dissolved. One of the functions of the saliva is to dissolve substances in the mouth, so that they can flow into the taste buds for identification. (See Figure 8.10.)

There are four main tastes: sweet, sour, salt, and bitter. However, the tongue is not equally sensitive to these four tastes in all areas. The back of the tongue is more sensitive to bitter tastes, whereas the tip of the tongue quickly responds

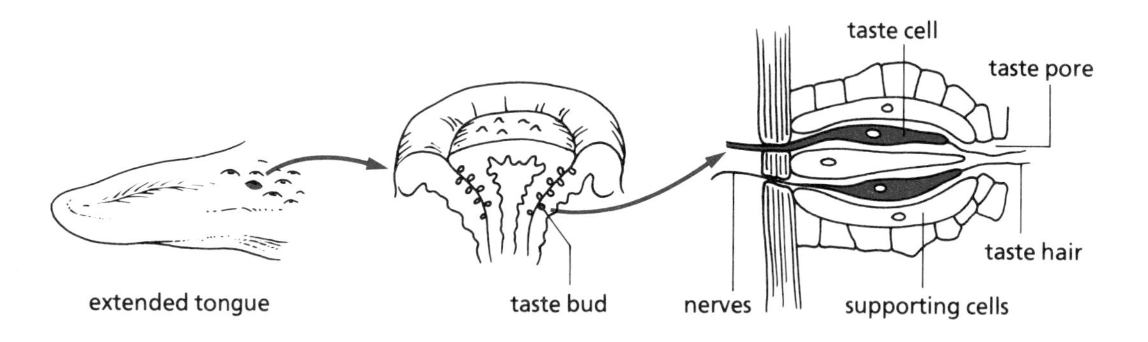

Figure 8.10.
The location and structure of the taste buds.

to sweetness. The sides of the tongue are sensitive to sour tastes while salt is readily recognized on the sides and tip of the tongue. (See Figure 8.11.) Many flavours are combinations of these four taste types.

Taste buds are easily damaged by heat or chemical burns, but they regenerate quickly. Many of the cells found on the tongue have a short life-span and are frequently replaced. Taste buds usually decrease in number with age and the flavour of various foods thus becomes harder to detect in old age. Most people have experienced the taste change that is associated with a bad cold. This lack of sensitivity is not only due to the coating that appears on the tongue, but also the mucus that blocks the small receptors in the nose.

Figure 8.11.
Regions of the tongue where various tastes are most readily recognized. All four tastes can be detected to a limited extent in all parts of the tongue.

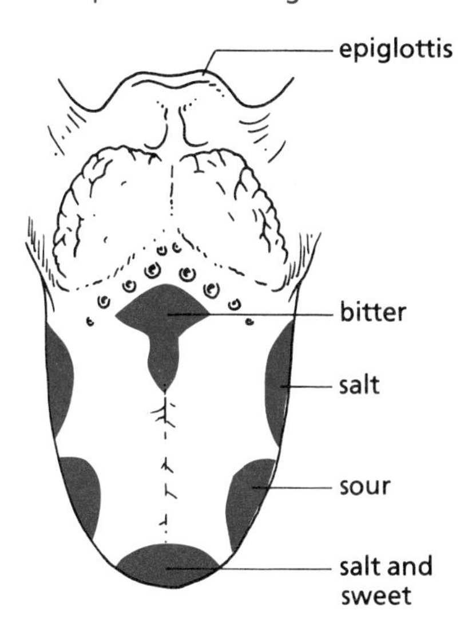

The Sense of Smell

The sense of taste and the sense of smell, while located separately, frequently work together to provide a dual sensation. The sense of smell is not as well developed in humans as it is in many animals. Dogs, for example, can trail an animal's scent many hours after it has passed by. Many animals depend on a keen sense of smell for safety and protection from predators. In humans the eyes have become the dominant sense of the body. Smell is a subtle aid to our taste buds, refining the first impressions. If you taste an orange or a lemon with the nose pinched shut, you will find that your taste impression is quite different compared with that obtained when tasting the fruit with your nostrils open. The texture of food substances also adds to our ability to discriminate between foods that are otherwise quite similar.

Smells are recognized more readily when they are soluble in water or in oils (perfumes are a good example). We become aware of an odour quite quickly, but the cells that recognize the smell rapidly become fatigued. Sometimes we become accustomed to a particular smell that at first we found unpleasant; eventually we are totally unaware of the smell, although it may still be present.

The **olfactory cells**, which are sensitive to odour, are found in a small patch in the upper part of the nasal cavity (See Figure 8.12.) The cells are above the normal stream of air which passes through the nose. As we breathe in we smell many odours. In order to detect faint odours, it is necessary

(ol-fak-tore-ee)

to sniff strongly. When food is in the mouth, odours pass up the passage at the back of the throat (nasopharynx) into the nose and add to the tasting sensation.

The olfactory cells in the nasal cavity are simple columnar cells with sensitive hairlike cilia protruding from them. The cilia are the small sensors. Nerve fibres come from the base of these receptor cells and pass through an opening in the bony roof of the nasal cavity, to reach the olfactory centre in the brain.

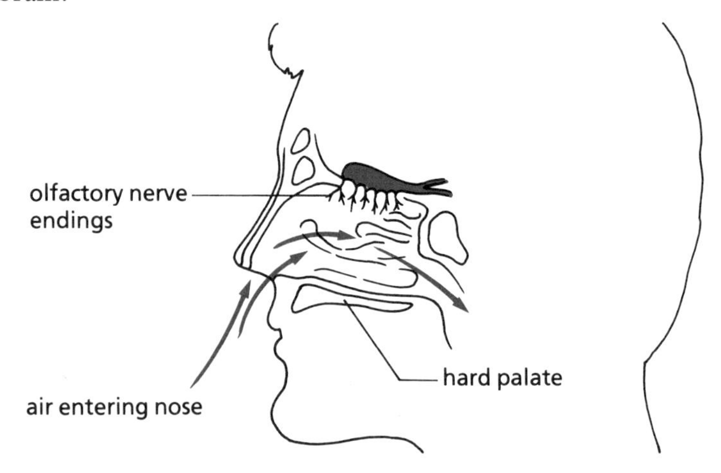

Figure 8.12.
The location of the olfactory nerve endings in the nasal cavity.

QUESTIONS FOR REVIEW

SOME WORDS TO KNOW

Match the descriptions given in the left-hand column with a word shown in the right-hand column. DO NOT WRITE IN THIS BOOK.

1. Converts vibrations to nerve impulses in the inner ear.
2. Snail-shaped structure in the inner ear.
3. Responsible for balance.
4. Fluid found in the inner ear.
5. Permits changes of pressure to take place in the middle ear.
6. Small bones found in the ear.
7. Tube leading to the ear drum.
8. The stirrup rests against this membrane.
9. Membrane that separates the middle and inner ears.
10. Small calcium particles that help to determine the body's position.

A. cochlea
B. eustachian tube
C. anvil
D. auditory canal
E. utricle
F. saccule
G. semicircular canals
H. endolymph
I. oval window
J. round window
K. organ of Corti
L. otoliths
M. ossicles

SOME FACTS TO KNOW

1. What are the functions of the bones of the middle ear?
2. Draw a diagram of the human ear and label the parts.
3. Explain how the sound of a ringing bell reaches the ear drum.
4. What is the range of human hearing? How does this change with age?
5. What effect does prolonged loud music or noise have on the ear?
6. Explain the working of the semicircular canals.
7. Why do we feel "giddy" or disoriented when we suddenly stop after spinning around?
8. What are the four taste sensations of the tongue? Where on the tongue are each of these tastes recognized?
9. Where exactly are the taste buds located? Why is it that we cannot taste dry food?
10. What influence does the sense of smell have on our ability to taste foods in the mouth?

QUESTIONS FOR RESEARCH

1. Select one of the following topics and prepare a report on it:
   - the audiometer
   - ear infections
   - the difference between "deafness" and "being hard of hearing"
   - an interview with a deaf person
   - causes of deafness
   - seasickness or motion sickness
   - sign language and the deaf
   - hearing aids
   - tone deafness
2. Investigate what agencies are involved with helping people with hearing disabilities. How are children with hearing problems taught?
3. How does weightlessness affect astronauts?
4. One recent theory suggests that there are seven different basic types of scent. What are they? Try to design an experiment that could be used by students that would support the theory.
5. What is acoustic fatigue? What types of occupations require protective devices to avoid hearing loss or acoustic fatigue?
6. Why is the sense of smell more pronounced than the sense of taste in humans?
7. The antibiotic streptomycin may be given to treat certain bacterial infections, but it can sometimes cause deafness. How is this possible?
8. What is olfactory fatigue?

Activity 1: SOME INVESTIGATIONS INTO SOUND AND THE SENSE OF HEARING

Method

Obtain a tuning fork. Strike the tuning fork on your knee or against the heel of your hand. Hold it at a reasonable distance from your ear, so that you can hear it easily, then rotate it.

What happens to the volume of the sound? Explain why this happens. Move the sounding tuning fork in a circle around your head and decide where you can hear it best and where you hear it least well. *Explain why this is so.*

Activity 2: AUDITORY ACUITY

A quiet room is required for these experiments. Perhaps a laboratory preparation room can be made available for these tests.

Method
Plug one ear with clear cotton batting. Have your partner sit down with eyes closed. Hold a watch, with a distinct tick, near one ear and move it slowly away from the ear until he or she can no longer hear it. *Measure and record the distance of the watch from the ear.* Now hold the watch beyond the point at which your partner could hear the tick, and bring it closer and closer until he or she can hear the watch again. *Measure and record the distance. Are these distances the same? What might account for any differences? Repeat the experiment testing your partner's other ear. Record your observations.*

Activity 3: THE CONDUCTION OF SOUND THROUGH BONE

Method
Strike the tuning fork and place the point against the hard (bone) surface, just behind your ear. *Describe your impressions.* Strike the fork again, and clamp the point between your teeth. (Don't let the vibrating prongs touch your teeth!) *Explain the difference between bone and air conduction. How is this phenomenon used in some kinds of hearing aids?*

Activity 4: THE ROMBERG TEST–A TEST OF BALANCE

Materials
slide or filmstrip projector

Method

1. Have your partner stand close to the chalkboard and place the projector or lamp so that a sharp shadow of your partner is cast on the board. Have your partner remove his or her

shoes and stand with feet close together. The eyes must be closed or a blindfold used. You will find that the shadow moves as the body sways and you can mark the degree of sway by making a chalk mark at the edge of the shadow.

2. First have your partner stand facing the board and mark the side-to-side sway. Then have your partner turn at right angles and mark the front-to-back sway. *Record this amount of sway in centimetres in each case.*
3. Try having your partner stand on one foot and repeat the experiment. Ask him or her to repeat the test, but with eyes open. *Record your observations. Describe the role of the eyes in maintaining balance. Compare your results with those of other students. How do the results compare?*
   This test is sometimes used to determine the integrity of the dorsal white columns of the spinal cord.

Activity 5: THE SENSITIVITY OF THE TONGUE TO DIFFERENT TASTES

Materials

cotton swabs, Q-tips, 10% sucrose solution, 20% salt solution, 1% acetic acid solution or vinegar, 0.1% quinine solution, distilled water

Method

1. Have your partner stick out his or her tongue. Rinse the tongue with distilled water and dry it with cotton batting.
2. Using a Q-tip, apply a small amount of sugar solution to the tip of the tongue, then try the swab on the sides and back of the tongue. Finally apply the swab to the centre of the tongue. *Grade the taste as: strong, moderate, mild, or negative. Record the sensations experienced by the subject at each location.*
3. Now use each of the other solutions in turn, but do not allow the subject to know which solution you are using. Swab off the tongue with distilled water from time to time to clear residual amounts of the solutions.
   The applications should be in small amounts to a limited area and not spread over the tongue.
4. *Draw a simple map of the tongue indicating which areas of the tongue are most sensitive to each of the four main tastes used in the experiment.*
   Are the four tastes located in different areas of the tongue? Which substances are most readily detected? Which tastes seem to last the longest?

Unit V

How the Body Transports Substances and Defends Itself

Chapter 9

Composition of the Blood

- Characteristics of Blood
- The Plasma
- The Red Blood Cells
- The White Blood Cells
- The Platelets
- Blood Types

Characteristics of Blood

Our senses tell us a few things about blood; it is red, it is sticky, and it has a distinctive taste and smell. It also dries much more quickly than other common liquids. Blood is sometimes called the "life-stream". It flows through the tiny "rivers" and tributaries of the body, carrying the necessities of life - oxygen, nutrients, chemical messengers - to the cells, as well as transporting away the wastes and carbon dioxide that these cells produce.

Most adult men have about five or six litres of blood, while most women have about four or five. Blood is composed of a watery fluid called **plasma**, in which many materials are dissolved. Plasma makes up about 55% of whole blood; the other 45% is made up of solids - red blood cells, white blood cells, and platelets.

Plasma

The blood plasma is a straw-coloured liquid, about 90% water, in which the many substances carried by the blood are dissolved. (See Figure 9.1.) These substances include the following:

Digested food absorbed from the intestines. This includes the simple sugars, fatty acids, and amino acids, which are the end products of digestion. Blood carries these materials from the intestines to the cells, which require these nutrients for energy, repair, or growth.

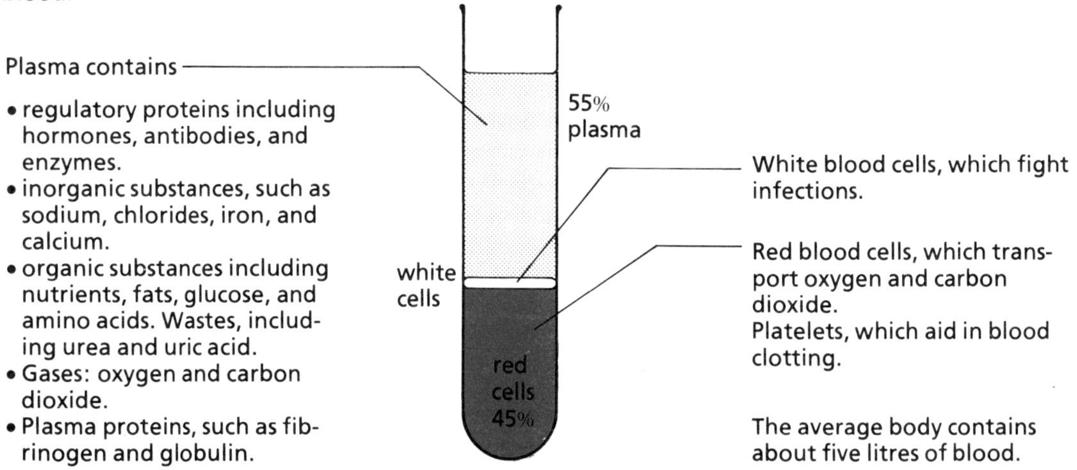

Figure 9.1.
The normal composition of blood.

Plasma contains

- regulatory proteins including hormones, antibodies, and enzymes.
- inorganic substances, such as sodium, chlorides, iron, and calcium.
- organic substances including nutrients, fats, glucose, and amino acids. Wastes, including urea and uric acid.
- Gases: oxygen and carbon dioxide.
- Plasma proteins, such as fibrinogen and globulin.

55%
plasma

white
cells

red
cells
45%

White blood cells, which fight infections.

Red blood cells, which transport oxygen and carbon dioxide.
Platelets, which aid in blood clotting.

The average body contains about five litres of blood.

Mineral Salts and Vitamins, also transported in the same way and to the same locations as digested food.

Urea, which is the waste product formed as a result of cell activities. It is carried to the kidneys for excretion.

Hormones, which are chemical messengers carried from the various endocrine glands, which produce them, to different target organs in the body.

Heat, generated in the muscles and liver. It is transported around the body helping to maintain a steady body temperature.

Carbon dioxide and Oxygen, which may also be dissolved in the blood plasma in small amounts.

Special Proteins, which help to clot blood and antibodies that fight infections and provide immunity against certain communicable diseases. When the fibrinogen (a clotting protein) is removed from the blood plasma, the remaining fluid is called **serum**.

The Red Blood Cells

Red blood cells (erythrocytes) are round cells with a hollow domed depression in each side. If a tennis ball with a small hole in it, letting out air, were pushed in on opposite sides, it would resemble a red blood cell. This shape gives the cell a large surface area with only a small bulk. Its rounded shape helps it pass through the very narrow capillary tubes without getting caught or blocking the bends of these tubes. Red blood cells are so small that if you took a cubic centimetre of blood (about the size of a sugar cube), then divided it into one thousand cubic millimetres, each of these cubic millimetres would contain about 5 500 000 red blood cells! This is the number present in a normal healthy male; a healthy female has about 4 800 000 red blood cells per cubic millimetre of blood.

One very unusual characteristic of red blood cells is that they have no nucleus. They are produced in the marrow of bones, such as the sternum and vertebrae. During the development of these cells a normal nucleus is present, but as the cells mature, the nucleus gradually shrinks and disintegrates.

Red blood cells are required to do a great deal of work and therefore wear out fairly quickly, lasting only about three to four months. Much of the wear and tear occurs on the very thin membrane which contains each cell. During their normal life of about 120 d they travel more than 1100 km, before finally being destroyed by phagocytes (specialized white blood cells). Red blood cells are broken down in the liver and spleen and replaced by new red blood cells. The iron is returned to the bone marrow for use in new cells.

The Function of the Red Blood Cells

(hee-moe-gloe-bin)

Red blood cells carry oxygen from the lungs to all parts of the body as well as transporting carbon dioxide from the cells back to the lungs for release. It is the **hemoglobin** molecules, contained in red blood cells, which enable them to act as carriers. Hemoglobin is composed of a protein called *globin*, a compound with an important iron component. The hemoglobin molecule has 4 attachment sites for oxygen or carbon dioxide molecules. When oxygen associates with hemoglobin, a bright red compound, called **oxyhemoglobin,** is formed. When carbon dioxide joins with the hemoglobin it loses this bright red colour.

Hemoglobin can also unite with carbon monoxide; in fact it attaches itself to this molecule 200 times more strongly than to oxygen. When hemoglobin reaches the lungs, the carbon monoxide cannot be pulled free and the blood is unable to pick up its fresh supply of oxygen.

Figure 9.2.
The blood cells and platelets.

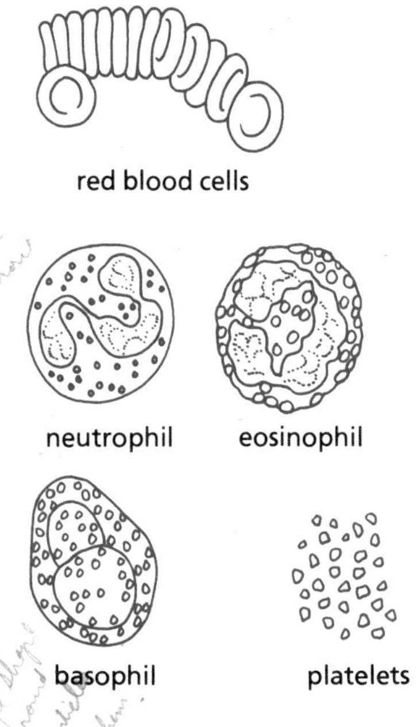

platelets

THE RED BLOOD CELLS

- form about 45% of whole blood
- carry oxygen and carbon dioxide

THE WHITE BLOOD CELLS

The granular white blood cells:

1. Neutrophils make up 65% of the WBC. These cells are actively phagocytic and engulf bacteria and other cells.
2. Eosinophils comprise 2–4% of the WBC. They are moderately phagocytic.
3. Basophils make up 0.5% of the WBC.

The non-granular white blood cells:

1. Lymphocytes make up 20–25% of the WBC. They aid in the development of immunity against infections.
2. Monocytes make up 3–8% of the WBC. The monocytes are actively phagocytic.

THE PLATELETS contain prothrombin.
They are responsible for initiating the clotting sequence.

CASE STUDY. Linda, 17 years of age.

Linda was a likeable seventeen-year-old, with an exuberant personality. One morning, to everyone's surprise, Linda found herself on the floor looking up at a circle of worried faces. Linda had fainted and without any warning that she could remember. Rather shaken by this event, she made an appointment with a physician.

Dr. Wallis questioned her closely about her daily activities and feelings. Had she been feeling tired lately? Did she tire easily? Did she get aches and pains in her muscles? Linda admitted that she had felt rather tired lately and didn't seem to have her old get-up-and-go. However, she had thought it was caused by working rather hard and getting home late. The doctor questioned Linda about her breathing. Was she ever short of breath? Did she have any difficulty in breathing, especially after doing strenuous exercise? Dr. Wallis looked at her fingernails and also noted that Linda looked rather pale.

The next questions were about her blood. Did it seem to take a long time before bleeding stopped, if she cut herself? Did she have a heavy menstrual flow?

When it came to the next questions about her eating habits, Linda was embarrassed. She made excuses about being busy and not having time for breakfast, not liking vegetables, and rarely eating fruit. She said she couldn't stand liver and rarely ate meat except hamburger.

The doctor took a sample of Linda's blood from her arm, to be sent away for analysis, then he questioned her further about her general health and about the medical background of her parents. Dr. Wallis finally gave Linda a complete physical checkup and asked her to return in a few days when the blood analysis report would be available.

When Linda returned to see the physician she was not surprised to hear that she was anemic. Dr. Wallis quickly reassured her that, in her case, it was not one of the more serious forms of anemia and that her problem could be remedied by changes in diet and by taking a prescribed amount of iron tablets.

On a final visit, Dr. Wallis told her to discontinue taking the iron tablets as, with a suitable diet, the supplement of iron was unnecessary.

Questions for research and discussion:

1. Explain the relationship between lack of iron and tiredness or lack of energy.
2. Find out the amount of iron recommended for a person of your age and sex. (See Table 15.8, Chapter 15.)
3. Make a list of all the foods that contain iron. Check off any foods in this list that you have eaten in the last week. What adjustments should you make to your diet?

Anemia

When there are insufficient red blood cells to carry all the oxygen that the body requires, a condition called **anemia** occurs. People who have this condition appear tired and lack the energy to work or play efficiently. They may also catch other illnesses easily. There are several types of anemia, including one quite common type that is caused by insufficient iron in the diet. Because iron is a basic component of hemoglobin, a lack of iron results in insufficient hemoglobin being formed. This reduces the oxygen-carrying capacity of the blood. Whole wheat bread, nuts, raisins, spinach, liver, and other meats are good sources of iron in the diet.

(ah-nee-mee-ah)

The White Blood Cells

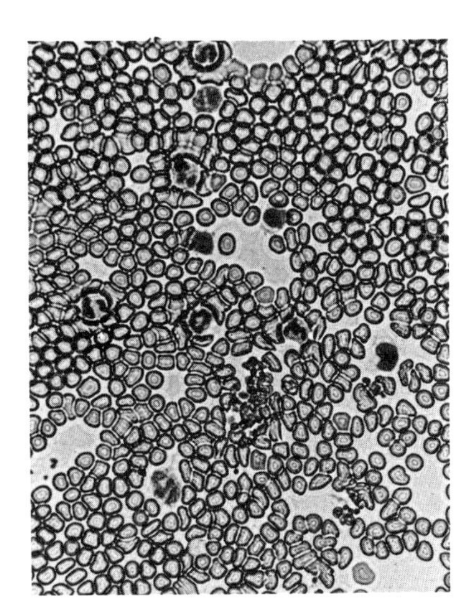

Figure 9.3.
Human blood cells.

White blood cells (**leukocytes**), unlike red blood cells, contain a nucleus. They are colourless and, although they are usually round, possess the ability to change shape. Leukocyte cells are larger than red blood cells but are far less numerous. There is about one white cell for every 600 red blood cells (about 7500/mm^3 of blood). (See Figure 9.3.) Most of the cells are produced in the bone marrow, but some are also made in the lymph tissue, at various sites around the body. They contain no hemoglobin, so are not capable of transporting carbon dioxide or oxygen.

There are several types of white blood cells, differing in shape, size, and the appearance of their nuclei. (See Figure 9.2.) These several types can be conveniently divided into two groups. *Granular* white cells, such as **neutrophils**, **basophils**, and **eosinophils**, have granules in their cytoplasm and possess irregular, lobed nuclei. These cells are produced in the bone marrow. *Non-granular* white cells, such as **lymphocytes** and **monocytes**, are produced in the lymph tissues. They lack granules and have a more regular round nucleus. Usually, granular white cells live for periods of from a few hours to a few days. Non-granular types usually last an average of about 200 d, although some lymphocytes may last for the lifetime of the human who carries them.

All white cells are capable of amoeboid movements; that is, they can change their shape to squeeze through the tiny pores in capillary walls: similar to the way in which the cartoon character, Casper the Friendly Ghost, is able to glide through keyholes. They can also change shape to flow around foreign particles and engulf them (phagocytosis). Some white cells, such as lymphocytes, produce antibodies which react to toxic chemicals or substances produced by bacteria. In general, leukocytes are responsible for fighting infections in the body. Unlike red blood cells, white cells must leave the closed circulatory system and go out into the tissues in order to combat infections. They are subsequently collected and taken up by the lymph system. A more detailed discussion of the role of the lymphocytes and the immune system appears in Chapter 11, which deals with how the body is protected against infections.

White Blood Cell Disorders

Leukemia

Leukemia, a cancerous disease of the blood-forming organs, affects the leukocytes. Although there are several kinds of leukemia, most types involve a considerable increase in the number of white blood cells and a decrease in the number of red blood cells. As the disease progresses, the number of mature white cells decreases and large numbers of immature white cells crowd the blood stream. The body then fails to cope adequately with infections. The reduction in the number of red cells and platelets also decreases the amount of oxygen available, and the body becomes unable to cope with the many internal hemorrhages. Excessive exposure to X-rays and to radioactive elements can cause increased production of white blood cells, and these factors are recognized as among the major determiners of the disease.

(loo-kee-mee-ah)

Mononucleosis

Mononucleosis is not really a disease of the blood but it does result in the production of large numbers of white blood cells. Mononucleosis is most commonly transmitted during a transfer of saliva. It is therefore often known as the "kissing disease". It occurs most commonly among young people and in the more advanced or affluent countries of the world. Mononucleosis is thought to be caused by a particularly stubborn virus, which is not easily eradicated. It causes the production of large numbers of non-granular leukocytes and produces such symptoms as fatigue, swollen glands, fever, and sore throat.

The Platelets

Platelets are much smaller than red blood cells, numbering about 250 000-300 000/mm^3. They are responsible for the initial stages of blood clotting: a process which prevents loss of blood from a wound. If blood did not have a clotting capacity, wounds would continue to bleed until the loss of blood was sufficient to cause death. Platelets are tiny fragments of a specialized larger cell. Each platelet is surrounded by a membrane and filled with **thromboplastin**.

When a wound is made and blood vessels are damaged, the platelets break down and thromboplastin is released into the blood plasma. Two other substances in the plasma, **prothrombin** and **calcium**, react with thromboplastin to form **thrombin**.

(throm-bin)

Another substance, called **fibrinogen**, also becomes involved in this process. Found in the plasma, fibrinogen is converted to **fibrin** when it reacts with thrombin. It is fibrin that forms the threads that block the wound, causing the blood to form a sticky plug which prevents further bleeding. The sequence is a chain reaction which starts when the platelets release thromboplastin. The reaction may be summarized in two steps.

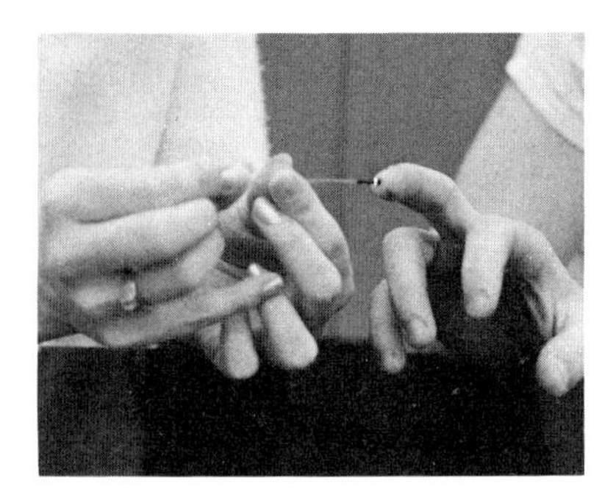

Figure 9.4.
Drawing blood into a capillary tube to determine clotting time.

Step 1. prothrombin + calcium + thromboplastin → thrombin

Step 2. thrombin + fibrinogen → fibrin (clot)

By keeping the starter chemical in containers (platelets) the danger of forming blood clots inside arteries or veins, by mistake, is eliminated.

Clotting Disorders

Hemophilia

Hemophilia is a rare blood disorder that is usually inherited as a sex-linked trait. (See Chapter 20 for details of sex-linkage.) All types of hemophiliac disorders prevent normal blood clotting; the blood may clot either very slowly or not at all. Sex-linked hemophilia is "carried" by the genes of the female, but is an abnormality that occurs more often in the male offspring. Other types of hemophilia affect both males and females.

Blood Types

Blood type is determined by the presence or absence of certain proteins in the red blood cells and plasma. Blood types are the result of genetic effects which are discussed in more detail in Chapter 20. Such proteins in the red blood cells are referred to as **antigens**. If the Antigen A is present, then an individual is said to have type A blood. Type B blood contains the B antigen; AB-type blood contains both A and B antigens. Type O blood has neither of the antigens present.

Substances called **antibodies** may also be present in the plasma. These react with the antigens causing the two types of cells to clump together. The antibodies found in a particular blood type are always the opposite of the antigen. For example, A blood contains the A antigen but the B antibody. B blood has the B antigen and the A antibody. AB blood contains both antigens but neither antibody A nor B, while O blood possesses no antigens but contains antibodies A and B. (See Table 9-1.)

Table 9.1. Human blood types.

TYPE OF BLOOD	ANTIGENS PRESENT IN RED BLOOD CELLS	ANTIBODIES PRESENT IN BLOOD PLASMA
A	A	B
B	B	A
A B	A and B	Neither A nor B
O	Neither A nor B	A and B

If antigens and antibodies of the same type come together, they cause the red cells to clump together or agglutinate. Thus if two samples of blood of the same type are mixed together no clumping occurs, whereas if different blood types are mixed, the cells may clump. If this should happen as a result of incorrect matching of blood during a transfusion, the result could be fatal. This happens very rarely, as some of the recipient's blood is always tested with the donor's blood before being transfused.

Blood Typing

A sample of blood is mixed with a drop of concentrated serum containing A antibodies. Another drop of the sample blood is mixed with a drop of serum containing B antibodies. Whether or not the red blood cells of the sample clump together, will quickly serve to identify the type of blood contained in the sample because clumping would indicate the presence of A or B antigens.

Before a transfusion takes place it must be established that the antigens in the donor's blood will not cause clumping of the recipient's blood, which might block important arteries. In cases of emergency, O blood can be transfused into most people without adverse affects because it contains no antigens. Persons with blood type O are thus referred to as

Table 9.2. Reactions when blood is mixed with serums containing A and B antibodies.

SAMPLE MIXED WITH ANTI A	SAMPLE MIXED WITH ANTI B	ANTIGENS PRESENT IN RED BLOOD CELLS	BLOOD TYPE
No clumping	No clumping	None	O
No clumping	Clumping	B	B
Clumping	No clumping	A	A
Clumping	Clumping	A & B	AB

Table 9.3 The percentage of blood types in several groups of people.

POPULATION	FREQUENCY %			
Blood type	A	B	AB	O
N. American White	41	10	4	45
U.S. Black	26	21	4	49
Inuit	55	5	4	36
Chinese	25	34	10	31

Universal Donors. Persons with group AB blood are called Universal Recipients and can theoretically receive blood from any other group because this blood type lacks the antibody half of the clotting reaction. However, in practice this is not often done and blood from a donor is always cross-matched with a recipient's blood to be sure they are compatible.

The Rh Factor

Not long after the discovery of the different blood types, it was found that another protein was present on the red blood cells in about one out of six people. The presence of this substance caused adverse effects, even if the red blood types of the donor and the recipient had been properly matched and transfused. Most of these problems appeared in women who had given birth to a child, or in people who had previously had a blood transfusion. In 1969 more than 10 000 fetal or newborn deaths were attributed to Rh incompatibility, but now the condition (erythroblastosis fetalis) is almost completely controllable.

Persons who possess this protein are called Rh positive, while those lacking it are called Rh negative. Rh-negative individuals do not automatically contain the Rh antibody; they first have to become sensitized by receiving some of the Rh antigen. Such sensitization can result from a blood transfusion. It may also occur when an Rh-negative mother, while pregnant with an Rh-positive baby, has some of the child's blood seep across the placenta into her own blood stream. When Rh-positive blood enters and thus sensitizes an Rh-negative individual, it stimulates the production of anti-Rh antibodies. Any future association with Rh-positive blood will thus cause a severe reaction. The antibody, once produced in a pregnant woman, is able to travel across the placenta and enter the baby's blood system. If the baby is Rh positive, this may cause destruction of some of its red blood cells.

The Rh factor is only a problem when the mother is Rh negative and both the father and fetus are Rh positive. The first pregnancy in which an Rh-positive fetus develops in an Rh-negative mother does no harm, but once the mother has become sensitized, subsequent pregnancies may involve problems. Tests have now been developed, however, which can be used to determine if the mother has been sensitized. If so, precautions can then be taken to avoid harm to the fetus. It is possible to transfuse the blood of the baby, either during development or immediately after birth.

Immediately after a first delivery, a miscarriage, or an abortion, Rh-negative mothers are now usually immunized with Rhogam, a substance that contains antibodies. These antibodies quickly combine with any fetal Rh-positive antigens and neutralize them before the mother has time to develop her own antibodies. The mother's system will later flush this mixture from her body, through the kidneys, leaving her free of any antibodies that might cause problems in the future.

Blood Donors

Almost anyone who is over 18 and under 65 years of age can become a blood donor. The donor must be free of certain diseases and not be using any drugs or medication. A donor usually gives 500 mL of blood at a time, an amount which is quickly replaced by the body.

QUESTIONS FOR REVIEW

SOME WORDS TO KNOW

Match a statement from the list in the left-hand column with a suitable term from the right-hand column. DO NOT WRITE IN THIS TEXT.

1. A substance required for the clotting of blood.
2. The liquid part of blood.
3. This is present in bright red arterial blood.
4. This contains prothrombin.
5. Produced by glands and carried to other organs in the blood.
6. Pigment containing iron and found in the red blood cells.
7. A condition in which only a limited number of red blood cells is present.
8. Example of a phagocyte.
9. A canceorus disease of the blood-forming organs.
10. Blood cells with no nuclei.

A. hormones
B. red blood cells
C. white blood cells
D. oxygen
E. sensitized
F. plasma
G. platelets
H. hemoglobin
I. anemic
J. monocyte
K. lymphocyte
L. leukemia
M. calcium
N. thrombus
N. antibody
O. antigen

SOME FACTS TO KNOW

1. List the functions of each of the major components found in the blood.
2. What substances are found in the blood plasma?
3. Where are red blood cells and white blood cells produced?
4. What is anemia? What foods can help to avoid an anemic conditon?
5. What causes red blood cells to clump together when samples of type A and type B blood are mixed?
6. Why are persons with blood type O known as "universal donors"?
7. What would happen if type A blood is transfused into a person with AB blood type? Explain.
8. Explain the sequence of events that occurs when blood clots.
9. What does the term sensitized mean when discussing the Rh factor?
10. How does mononucleosis differ from anemia?

QUESTIONS FOR RESEARCH

1. Select one of the following topics and prepare a report on it:
   - laboratory technician (hematology)
   - cholesterol
   - gamma globulin
   - anemia
   - first aid and the pressure points for treating bleeding
   - the Red Cross and the Blood Donor Clinics
   - William Harvey – the discovery of blood circulation
2. Investigate what tests are made for the Rh factor in pregnant women, and what techniques are being used to solve this problem.
3. When an accident victim suffers blood loss, he or she is transfused with plasma rather than whole blood. Why is plasma so effective in meeting the immediate threat to life?
4. Make a visit to a hematology laboratory at a nearby hospital and find out about the many tests that are made on blood.
5. How is blood stored, transported, and prevented from clotting after it has been taken from a donor?

Activity 1: COAGULATION TIME

Method

1. Cleanse the finger with a sterile swab and lance the tip with a sterile Hemolet lancet as previously directed.
2. Place a capillary tube in the peak of the drop, holding it horizontally. *Note the time as the blood starts to be drawn into the tube.*
3. After the tube ceases to draw any further blood, remove it and keep it warm in the palm of the hand. Every 30 s break off a small piece of the tube, about 0.5 cm long. *Keep a check on the time.* Eventually the blood in the tube will coagulate and the glass tube will slide off leaving a thin filament of the coagulated blood in place. *Record the total time for your blood to coagulate. Determine the class average. What factors are responsible for the coagulation of blood?*

Activity 2: PREPARING A BLOOD SMEAR

Method

1. Obtain a drop of blood from the fingertip as previously directed. Place one glass slide flat on the bench and put one small drop of blood on the slide, fairly close to the end of the slide. Hold a second slide at an angle of 45 degrees to the first slide and move it towards the drop of blood. As it touches the blood, the blood will spread sideways across the width of the slide. Draw the glass slide towards the other end of the slide in a pushing motion so that it leaves a thin, wide trail of blood behind. *Note*: The aim is to produce a THIN, wide smear; too much blood will produce poor results.

2. After the slide has dried in the air, cover it liberally with Wright's stain and allow to stand for about two minutes.
3. Rinse the stain with a squeeze bottle of distilled water. Allow the slide to stand for about four or five minutes. Rinse the slide gently and dry the underside with a paper towel.
4. Observe first under low power and then under high power. *Identify red blood cells and the various types of white blood cells. Record which white cells are most common.*

Activity 3: ABO BLOOD-TYPING TECHNIQUE

Note: Most teachers probably use one of the prepared Blood Typing Kits. These I have found work well. However if costs are a problem the following system may be used.

Materials
glass slides, Antigen A and Antigen B serum, box of toothpicks

Method
1. Take a clean glass slide and mark one end "A" and the other "B" with a wax marking pencil.
2. After sterilizing the fingertip surface with 70% alcohol, air dry the surface. Lance the finger. Place one drop of blood on each end of the marked slide.
3. Add one drop of Anti A serum beside the blood at the A end of the slide and a drop of Anti B serum at the B end. Do not allow the dispenser to touch the drop of blood.
4. Using a toothpick, mix the blood and Anti A serum together. Using the other end of the toothpick, mix the Anti B serum and the other drop of blood together. Do not allow the two serums to come into contact with each other.
5. Observe the slide for any signs of clumping. Compare your results with the table below. + = clumping.

Antigen A	Antigen B	
+	–	Type A blood.
–	+	Type B blood.
+	+	Type AB Blood.
–	–	Type O blood.

Compare your results with those of other students. List each student's blood type on the chalkboard and determine the percentage of students with each type of blood. Compare with the tables given in the text. Explain why your blood produced the reactions observed.

Activity 4: Rh-BLOOD TYPING

Refer to the text first for an explanation of this reaction.

Method

1. Take a clean slide and warm it slightly on a low-heat hot plate. Mix a drop of your blood with a drop of Anti Rh or Anti D serum. Use a clean toothpick.
2. Tilt the slide to one side so that the process of clumping, should it occur, is more easily seen. Continue to observe for about two minutes. If in doubt as to whether the reaction has occurred or not, view under a low-power microscope. Often the Rh factor is weaker than that produced by the AB antigens and the reaction is more difficult to detect.
3. *Determine the percentage of students with Rh-positive and Rh-negative blood. Record all results.*

Activity 5: THE HEMOGLOBIN CONTENT OF BLOOD

Materials

Tallquist scale, Hemolet lancets (sterilized), 70% alcohol

Method

1. Swab the end of a finger with an alcohol pad and allow it to air dry. Pierce the fingertip with a Hemolet lancet and wipe away the first drop of blood.
2. Place the second drop of blood on a small piece of filter paper or a square of paper provided in the Tallquist scale.
3. Allow the blood to dry for about four or five minutes and then compare against the colours in the Tallquist scale. Each colour represents a 10% difference in hemoglobin content. Estimate the intermediate percentages. *Record your results. Determine the class average.*

(Other methods are available and are more precise. The directions for using these techniques are supplied with the testing equipment. However, the Tallquist scale was used by many doctors in the not-so-distant past, when house calls were still a common event among doctors.)

Write brief notes to explain the function of hemoglobin and the meaning of anemia.

Chapter 10

The Heart and Circulation of the Blood

- The Circulation of Blood
- The Heart
- The Lymphatic System

The Circulation of Blood

Small, single-celled animals, such as an amoeba or small marine organisms, have no need of a circulatory sytem because food substances may enter, or wastes leave, directly through the cell membrane. Multicellular organisms may have cells isolated from the source of food or waste disposal sites. These organisms require a means of transport so that every cell in the body can receive nourishment and be rid of wastes. The circulatory system provides such a transport system.

The circulatory system of humans is composed of the heart, arteries, arterioles, capillaries, venules, and veins. After leaving the heart, blood flows through these vessels in the sequence in which they are named above.

The Arteries and Other Vessels

An **artery** is a vessel that carries blood *away* from the heart. **Arterioles** are simply small arteries. **Veins** carry blood *to* the heart; **venules** are smaller veins. **Capillaries** link arterioles and venules.

The walls of arteries are thick and contain a layer of muscle. (See Figure 10.1.) When blood is pumped out of the heart it is forced out under pressure. The muscular artery walls are elastic and stretch as this wave of pressure is pushed along. This is the little "bulge" that we feel under our fingers when we take our pulse. The veins also have a muscular coat in the wall, but as the pressure of blood flow is greatly reduced the need for muscle is reduced and this coat is much thinner than that found in arteries.

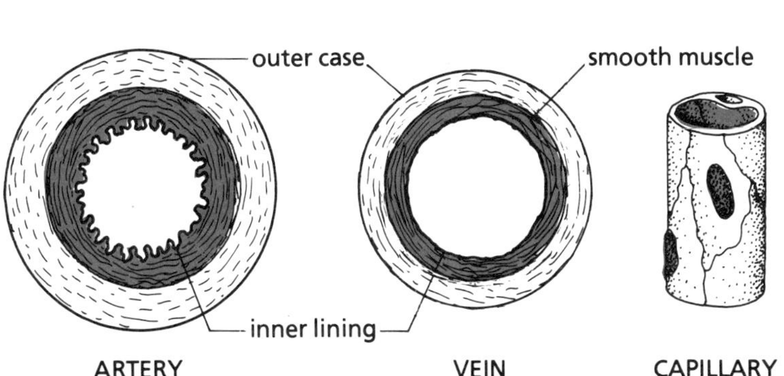

Figure 10.1.
A comparison of the structure of an artery, a vein, and a capillary. Note that the muscle wall in an artery is much thicker than that in a vein. The capillary has been greatly enlarged in order to show detail.

Both arteries and veins have walls far too thick to allow any blood plasma, nutrients, or gases to pass out into the surrounding tissues. The **capillaries**, however, have very thin walls, composed of single layers of endothelial cells. The capillary tube is so small that only one tiny red blood cell can pass through at a time. The blood cells thus pass along the arteries in single file. It is in the capillaries, which link the arterioles to the venules, that the exchange of oxygen, carbon dioxide, nutrients, wastes, and other substances takes place. (See Figure 10.2.)

Figure 10.2.
The action of blood cells and the diffusion of materials between a capillary and the surrounding tissues.

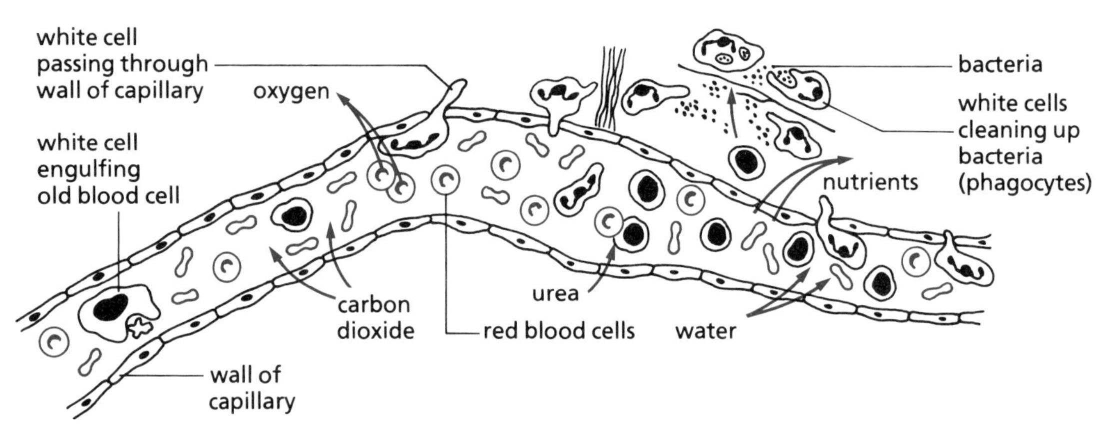

Some of the fluid in the plasma is forced through the capillary walls to bathe the cells of the body tissues. A portion of this plasma re-enters the capillaries. Any additional fluid that is not reabsorbed is later collected and returned via the lymphatic system which is related to, but separate from, the blood circulatory system.

Although the blood is forced into the arteries under pressure, by the time it reaches the capillaries this pressure is very low. (See Figure 10.3.) There must thus be another mechanism for getting blood back to the heart.

Figure 10.3.
Blood pressure in the different vessels of the circulatory system.

As blood leaves the heart it flows "downhill" until the pressure in the capillaries and venous system becomes very low. Blood flow in the capillaries is also very much slower than that in the larger vessels, which allows time for the diffusion of many substances between blood and the surrounding tissues.

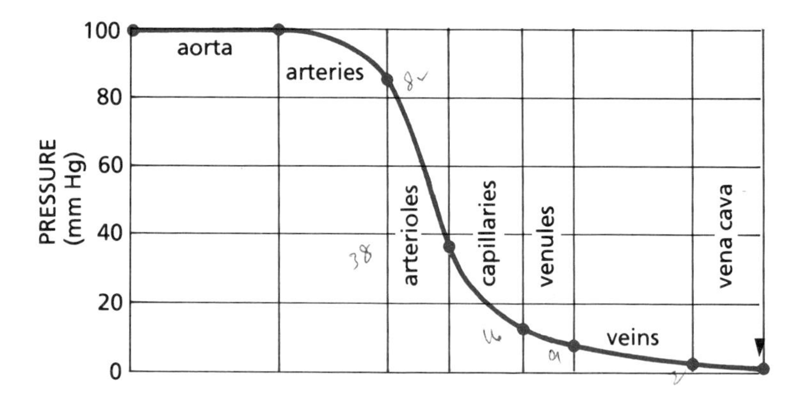

The Veins and Their Valves

If you could see inside a vein you would find many tiny valves, regularly spaced along its length. These valves point in the direction in which the blood is flowing. The valves can be pushed open to allow blood to flow towards the heart, but they close to prevent any back flow (See Figure 10.4.)

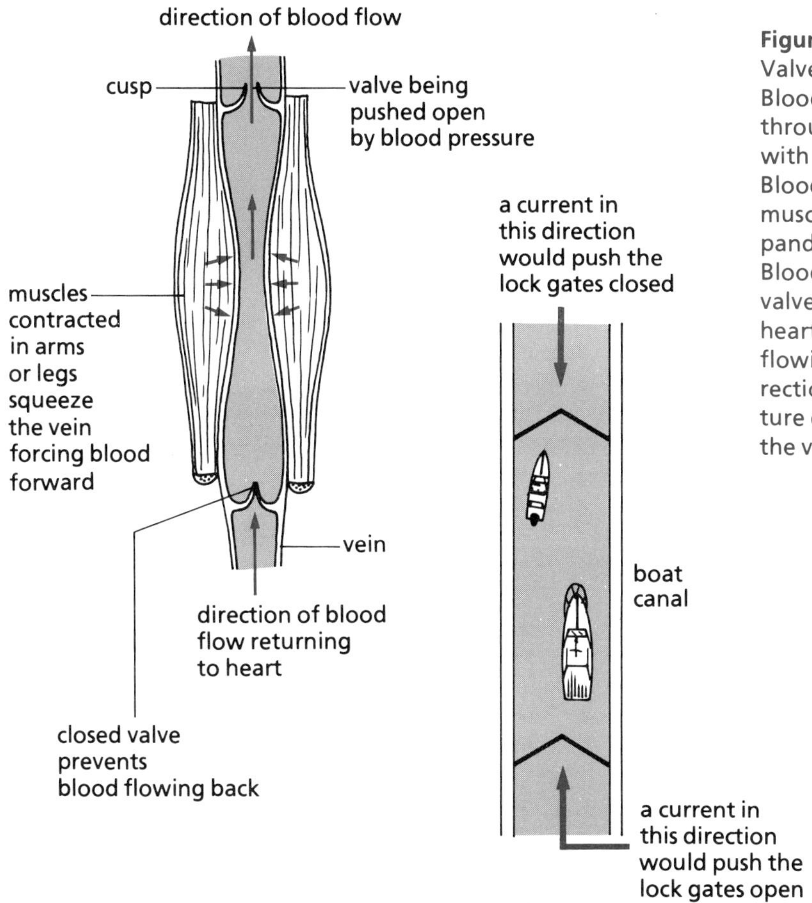

Figure 10.4.
Valve function in the veins. Blood flows back to the heart through veins which are fitted with valves at regular intervals. Blood is forced forward when muscles or nearby arteries expand and squeeze the vein. Blood is pushed through the valves in the direction of the heart, and is prevented from flowing back in the opposite direction by the one-way structure of the valves the next time the vein is squeezed.

Perhaps you have seen lock gates along a boat canal and have noticed that they meet at a slight angle. If the current in the canal is flowing in one direction it tends to push against the gates and keep them closed. If it were to flow in the other direction it would tend to push them open. The valves in the veins act in a similar manner. When the body muscles surrounding these veins contract, they bulge outwards, thereby squeezing the veins and pushing the blood forward through the valves. When the valve flaps move back into place the blood cannot flow back into the vein behind the valve. The blood is thus squeezed back to the heart step by step, by the contraction of the muscles that surround the veins. There is also some suction action on the part of the heart, for as it pushes blood out into the arteries it causes blood returning from the body to be pulled in to replace the pumped blood.

The rate of flow is slower in veins than in arteries, but veins are also more numerous, so that an adequate amount of blood will always be returned to the heart.

If the valves in veins fail to close completely and thus allow the blood to form bulging pools, the condition known as varicose veins results. The veins become swollen because of the increased volume of blood in them. This swelling, particularly in the legs, causes the veins to appear knotted and blue. Varicose veins may cause much pain. The failure of the valves to close is due to a loss of elasticity in the vessels. The veins become stretched and the valve cusps (flaps) fail to meet. Persons who are employed in jobs that require long periods of standing are more prone to this disorder.

Table 10.1 A Comparison of arteries and veins

ARTERY	VEIN
Takes blood away from the heart	Takes blood to the heart
Blood travels in small spurts	Blood travels more smoothly
No valves present	Valves present
Thick muscle walls	Thin muscle walls
Blood rich in oxygen, bright red in colour	Low in oxygen, high in carbon dioxide. Less bright red, more like maroon
Pressure very high	Pressure very low

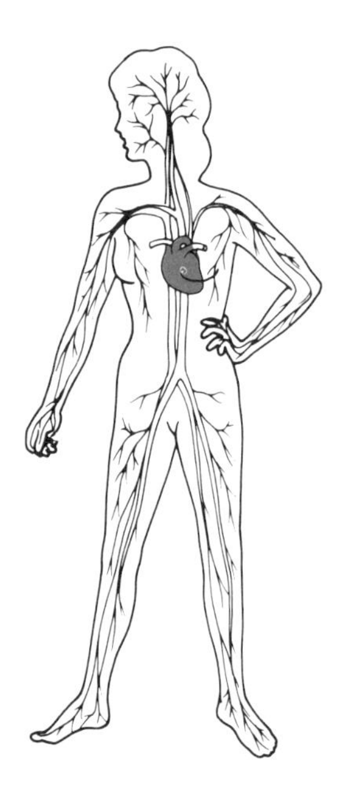

The Heart

The heart is located in the thoracic cavity, well protected by the rib cage. It nestles between the lungs with its lower end slightly towards the left side. The adult heart is about the size of a large fist and has a mass of approximately 300 g. The heart is not rigidly attached to any of the surrounding tissues, but is suspended by the large blood vessels that are attached to it. Because of this, it is able to move loosely in place as it contracts.

Heart Muscle

The heart is not one pump but two separate pumps, each complete in itself. The heart is composed of very special muscle tissue, which is like striated muscle in some ways. (See Figure 10.5.) It has incredible stamina, for it must beat continuously without rest from the early weeks of life (about 18 d after conception) until death. Each cardiac muscle cell is attached to, and pulls against, other cardiac muscle cells during contractions; the muscle is not anchored to bones or other tissues. The muscle cells of the heart have an innate ability to contract and they do so spontaneously. Even fragments of heart muscle placed in a plasma solution will continue to contract rhythmically on their own. If they touch, they have the ability to co-ordinate their contractions into a common rhythm.

Figure 10.5.
The fine structure of heart muscle.

The Parts of the Heart

The right side of the heart is responsible for collecting blood from the body and pumping it to the lungs. (See Figure 10.6, and 10.7.) The blood flows back from the head and arms to this side of the heart by way of a large vein called the **superior vena cava**. It leads into the upper right-hand chamber, the **right atrium** of the heart. (See Figure 10.7.) Blood from the trunk and legs enters this same chamber via the **inferior vena cava**. Both the right and left atria (plural of atrium) are

(ae-tree-um)

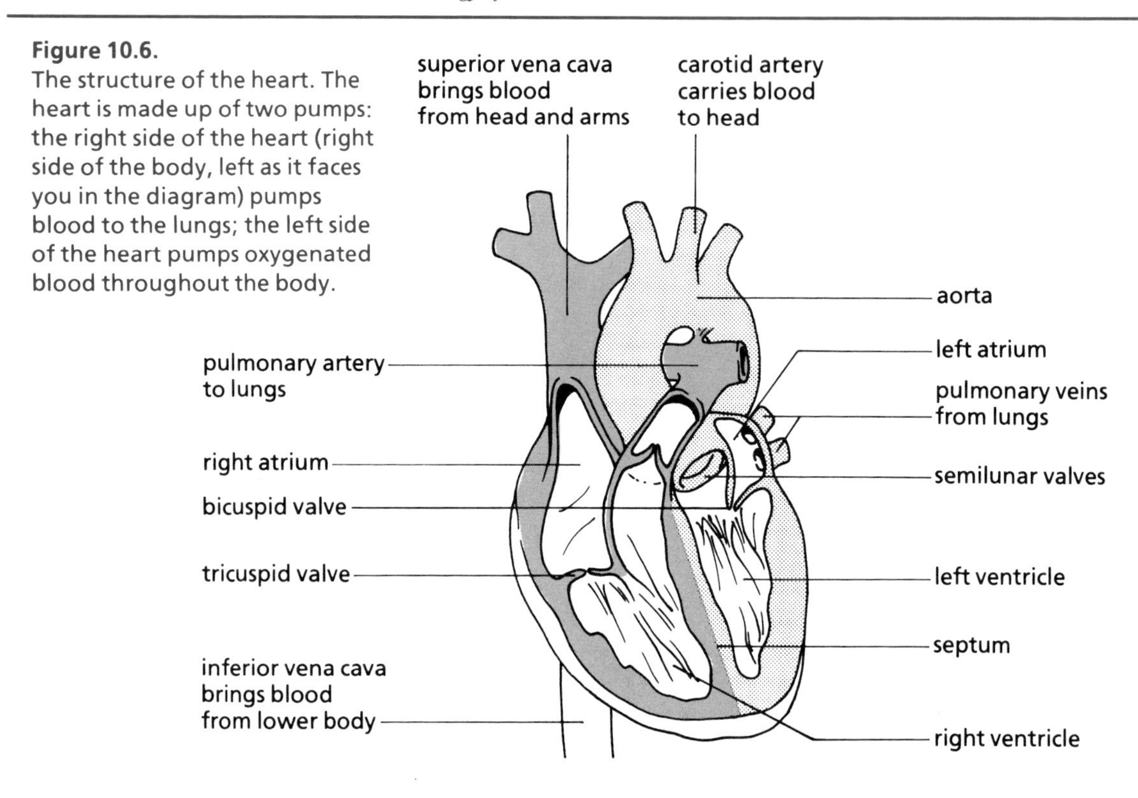

Figure 10.6.
The structure of the heart. The heart is made up of two pumps: the right side of the heart (right side of the body, left as it faces you in the diagram) pumps blood to the lungs; the left side of the heart pumps oxygenated blood throughout the body.

Figure 10.7.
The two pumps of the heart.

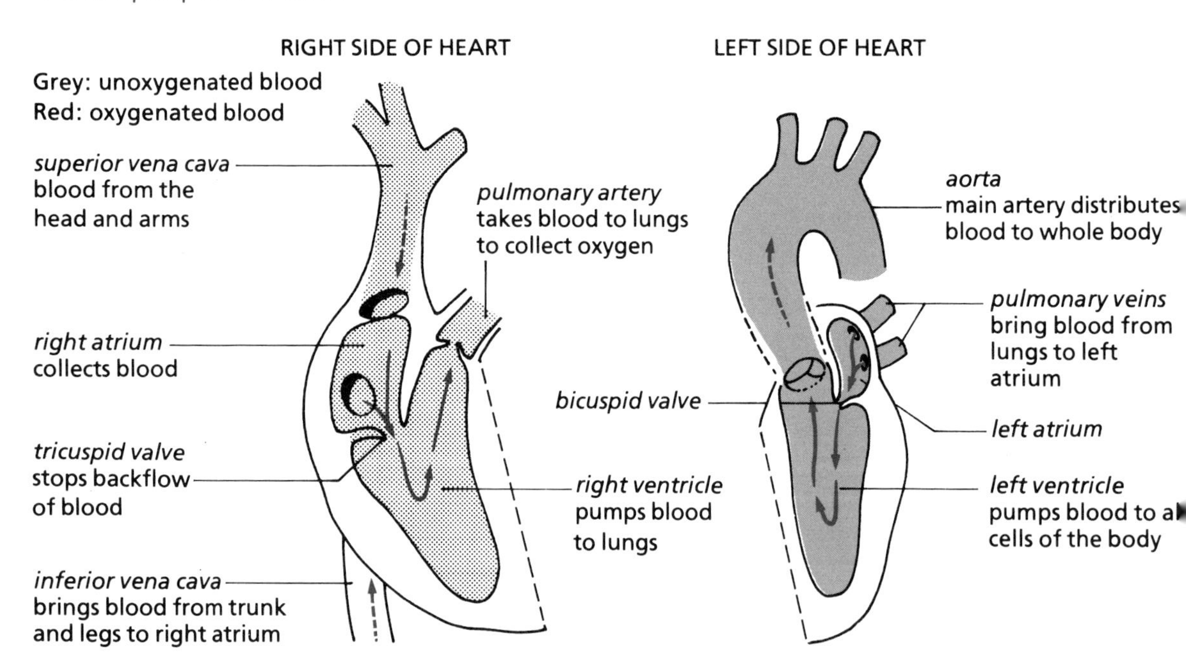

thin-walled chambers which lie above the ventricles. Their main function is to collect blood and pass it into the main contracting vessels, the ventricles.

The **right ventricle** has thicker, more muscular walls than the right atrium and is much larger. It is connected to the right atrium via the **tricuspid valve** which prevents the blood from flowing back into the atrium when the ventricle contracts. When the right ventricle contracts, it pushes blood out through a semilunar valve into the **pulmonary artery**. This vessel carries the blood to the lungs where its load of carbon dioxide wastes is released and a fresh load of oxygen is absorbed. This newly oxygenated blood then flows into the **left atrium** through the **pulmonary veins**. Contractions of this atrium push the blood through the **bicuspid** (or mitral) **valve** into the **left ventricle**. This chamber is the largest and most heavily muscled of the heart. As it contracts it must force the blood to every part of the body, from the brain to the smallest toe. The blood is pushed out from the left ventricle via the aortic semilunar valve, into the largest blood vessel in the body, the **aorta**.

(ven-tri-kul)

(try-cuss-pid)

(pull-mon-air-ee)

The Heart Valves

In the right side of the heart the tricuspid valve connects the atrium and ventricle. (The tricuspid and bicuspid valves are also known as the *atrioventricular* valves, a large word, but quite easy to remember, since it is made up of the names of the two chambers that it separates, the atrium and ventricle). The tricuspid valve is closed while the atrium is filling with blood. When this is completed, the tricuspid valve opens and a wave of contractions starts at the top of the atrium and squeezes the chamber inward and downward, forcing blood through the valve into the right ventricle. The valve then closes. Another wave of contractions starts, this time at the bottom of the ventricle. The blood is now squeezed upward through the semilunar valve, which lies at the entrance of the pulmonary artery leading to the lungs. This process is repeated over and over with each beat of the heart. The same sequence of valve action takes place in the left side of the heart.

The tricuspid and bicuspid valves (the bicuspid is also known as the mitral valve) are composed of flaps of tissue connected by tiny tendons (tendinae) to muscles which pull the valve flaps open. As blood fills the ventricle, these float back into position. The pressure of the blood then holds

THE ARTIFICIAL HEART

Heart disease is a major killer in this country. The use of artificial devices such as pacemakers has prolonged the lives of many people with heart disease. Many believe that a totally artificial heart could be more widely used, and with much more success than a heart transplant. In fact, heart transplant patients generally do not live for very long because the patient's body "rejects" the transplanted tissue. (Only 65% survive one year.)

One problem with the development of an artificial heart is that it must be small enough to fit in the space made available by the removal of the natural heart. Other important factors that must be considered are the ability of the device to provide enough blood for the entire body and to vary the output according to the body's needs–perhaps an electronic, computerized daily schedule could be used. Blood must be pumped gently enough to avoid **hemolysis**–the destruction of the red blood cells.

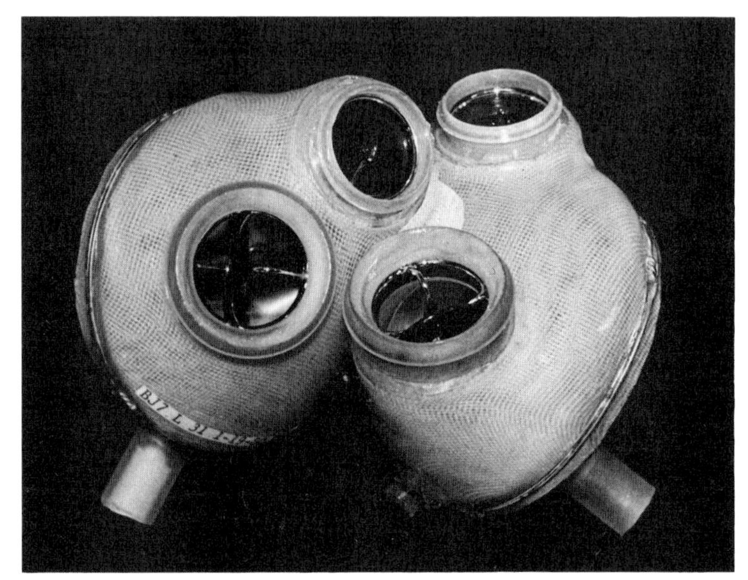

Early experiments with artificial hearts, which began in 1957, met with little success. In 1963, the U.S. National Aeronautics and Space Administration (NASA) helped develop a computer-based control system that regulated the output of an artificial heart according to the various physiological conditions of the natural heart, such as the pressure in the atria.

The accompanying photograph shows the Jarvik-7 heart, named after the designer, Dr. Robert Jarvik, of the University of Utah College of Medicine. This device is intended ultimately for use in humans. The Jarvik-7 was first tested on a calf and work continues with implantation in sheep. The Jarvik-7 implanted in the calf worked for 221 days.

The totally artificial heart for humans will require years of research before it can be safely implanted.

them in place so that there is no flow back into the atrium. The semilunar (half-moon) valves are small flaps of tissue in the arteries that leave the ventricles. These operate by pressure alone; they are not attached to muscles. They also are designed to ensure a one-way flow of blood, preventing backflow into the lower chambers of the heart.

Pulmonary Circulation

The right side of the heart contains venous blood, high in carbon dioxide and low in oxygen. (See Figure 10.7.) Such blood is usually coloured blue in diagrams. The left side contains arterial blood, bright red, rich in oxygen, and low in carbon dioxide. Usually arteries carry bright-red, oxygenated blood and veins the dull-red, unoxygenated blood. However, in the heart there are two exceptions. The pulmonary artery carrying blood to the lungs for oxygen contains the dull, carbon-dioxide rich, venous blood; whereas the pulmonary veins carry oxygen rich, bright-red, blood from the lungs to the heart. Arteries carry blood away from the heart regardless of the oxygen content of the blood. (See Figures 10.8 and 10.9.)

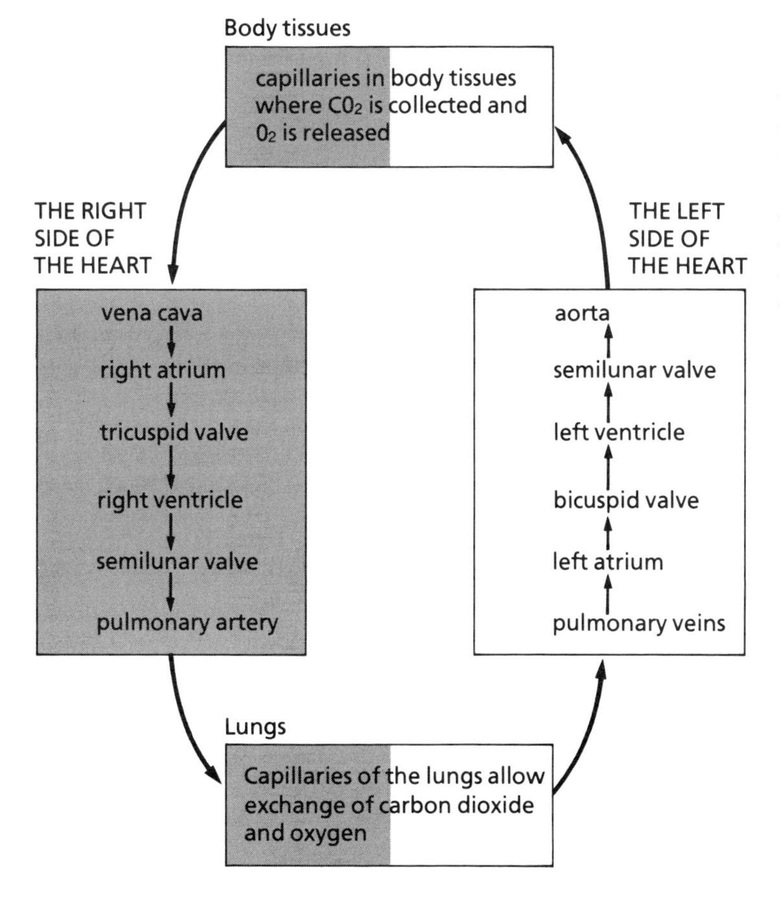

Figure 10.8.
The sequence of vessels and valves through which the blood flows. Although only the exchange of gases is explained in this illustration, it is important to realize that nutrients and wastes are also exchanged in the body tissues.

Nervous Control of the Heart

In the wall of the right atrium, near the entrance of the superior vena cava, is a bundle of nerve tissue called the **sinoatrial node** or SA node (See Figure 10.10.) It is here that nervous stimuli originate to control the rate of the heartbeat. For this reason the SA node is often called the **pacemaker**. When this stimulus is released, it starts a wave of contrac-

(sy-noe-ae-tree-al)

Figure 10.9.
The major arteries and veins of the body.

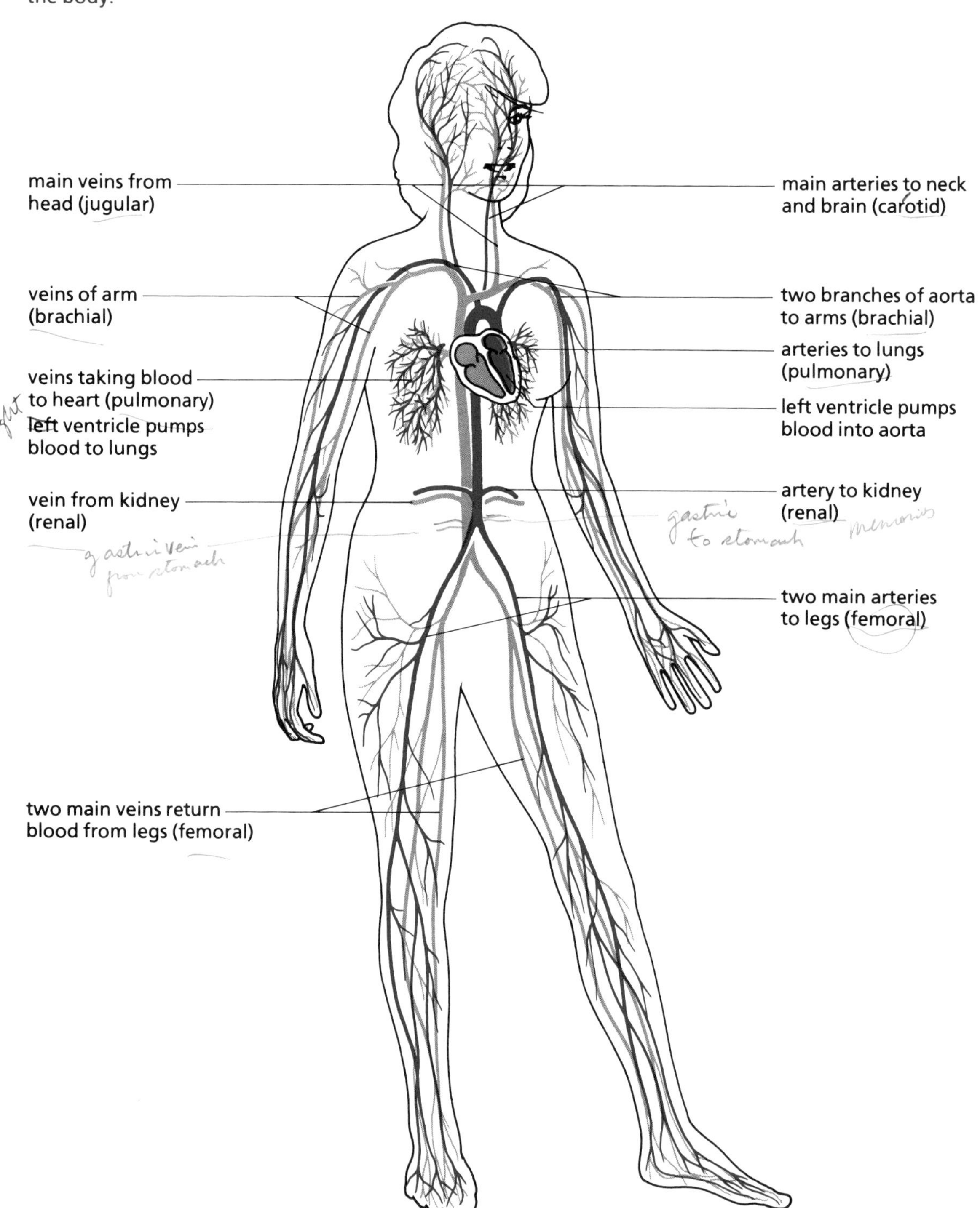

tions in the cardiac muscle fibres, which spreads outward over the right and left atria. The **atrioventricular node** is another small mass of nerve tissue located in the wall between the two sides of the heart, approximately level with the junction of the atria and ventricles. From this AV node, a bundle of nerves, called the Bundle of His, extends out to reach both the ventricles. There a network of nerves spreads out to start contractions from the bottom of the ventricle. These nerve fibres are known as **Purkinje fibres**.

Figure 10.10.
The nodes and nerve-conducting pathways of the heart.

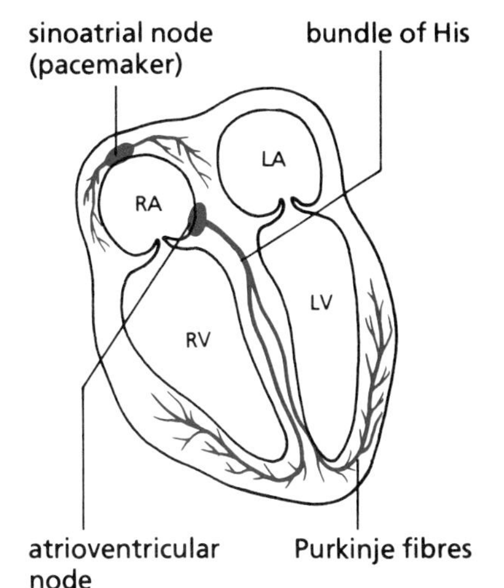

(purr-kin-gee)

As we have seen, the heart has two unique characteristics: it beats automatically and it beats rhythmically in a continuous manner. The adult male heart beats on average 75 to 80 times a minute while sitting. The female heart rate is about 10 beats more per minute than this. The heart contracts like this, only occasionally missing a beat, for an individual's entire life (See Figure 10.11.)

The Heartbeat Sounds

The stage during which the ventricles are contracting is called **systole**. The period during which the heart is relaxed and the ventricles are filling with blood is known as **diastole**. By using a stethoscope the sounds made by the heart can be amplified and thus heard. There are two distinct sounds. The first, a "lub", is the loudest and longest. The second is a quieter and shorter "dub" sound. The sounds are rhythmically repeated in a "lub-dub", "lub-dub", "lub-dub" pattern.

(sis-toe-lee)
(dy-a-stoe-lee)

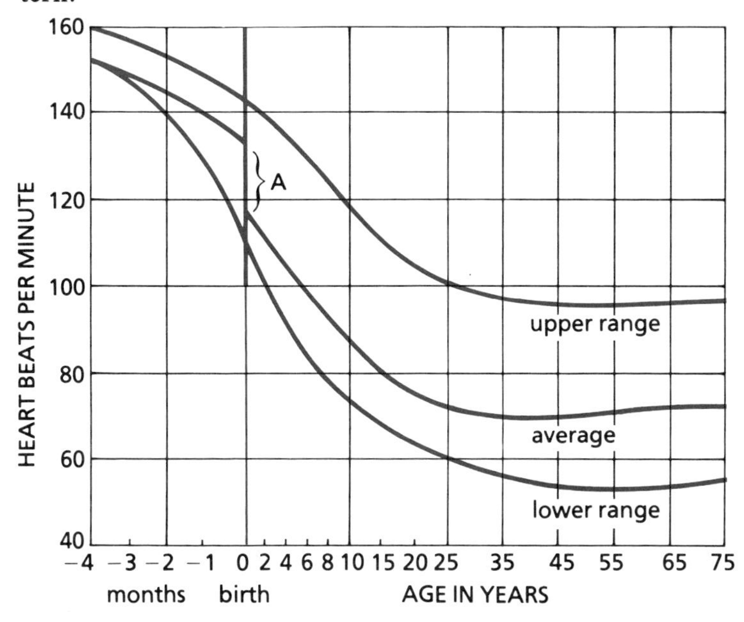

Figure 10.11.
The relationship between heart rate and age. The upper and lower lines represent the limits within which the normal heart rate may be expected to fall. There are considerable differences between individuals, the figures are approximate only. Note the jump in heart rate that occurs at the time of birth indicated at A.

The first sound (the "lub") is made by the sudden closing of the valves between the atria and ventricles and the opening of the semilunar valves. The second sound (the "dub") is made by the closing of the semilunar valves and the opening of the valves between the atria and ventricles. (See Figure 10.12).

CASE STUDY. Bob Bedford, aged 40 years.

Snow had been falling steadily and a gusty wind had curled the snow into banked drifts along the wall beside the driveway. In spite of the cold, the winter sunshine made the chore of shovelling the snow a not unattractive task.

Bob Bedford was nearly 40 years old. He enjoyed his work as an accountant and most of his evenings were spent following his passion for reading books about sports and watching sport programs on T.V.

Bob started removing the snow, shovelling the fresh white powder up onto the already high piles along the driveway. The task did not seem too strenuous and in half an hour he had it cleared and with a satisfied feeling he went inside to relax.

While struggling with the zippers of this rubber overboots, Bob suddenly felt an intense, agonizing pain in his chest. He pressed his hand tightly against his chest to ease the pain, leaning back against the wall to keep himself from falling.

The pain was so intense he could not call for help, but slowly slipped down the wall to the floor. The pain seemed to be spreading into his neck, and his arm also began to hurt. He had a cold feeling of fear and despair.

Bob's wife called out, asking what he was doing, and when she received no reply, came out to see where he was. She cried out as she saw Bob lying on the floor and ran over and knelt beside him. Bob was unable to speak. She quickly loosened his shirt and straightened his legs, which were cramped up under him. She pulled off her jacket and pushed it under his head and then ran to telephone for an ambulance.

Bob remembers little of what happened then. His chief recollection was of several people bending over him, one attaching some wire leads to his body, another placing a mask over his mouth and nose. Later he realized that the nursing staff had placed a needle in his arm and attached an intravenous bottle which also contained his medications.

The doctor found that his pulse rate was rapid and his blood pressure was low. He also administered a medication into Bob's arm to kill the pain. Bob's lips were blue and his skin dull gray in colour. At this point, Bob was moved to the **coronary** care unit, where he could be kept under constant observation.

The muscles of Bob's heart, suddenly required to do extra, unaccustomed work, had been damaged. As the heart cannot be immobilized while it heals, it is necessary to reduce its work load until it has had time to mend. Thus, rest is prescribed.

The damage to the heart muscles causes it to work less efficiently and it tries to compensate by speeding up. However, this does not help, for the rapid pumping rate does not allow time for the damaged heart to adequately fill with blood, and so the quantity of blood present in the arteries is reduced and the pressure drops.

The main purpose of the care and therapy prescribed for Bob was to gain some time while the heart healed. With good care, the damaged muscles will repair. Although the muscles will never be as strong as they were before the attack, Bob's chances of returning to work and enjoying a nearly-normal life are quite good.

Questions for research and discussion:

1. Find out how the risk of having a heart attack can be reduced.
2. What factors in our style of living today contribute to heart attacks?
3. What can you do to help someone who has suffered a heart attack, after they return home from the hospital? (Don't forget mental as well as physical needs.)

Figure 10.12.
A dual stethoscope is being used in this illustration which allows two persons to hear the heart beat at the same time.

Factors Affecting Heart Rate

The rate at which the heart pumps depends on several factors. It will vary with the amount of activity or muscle effort in which an individual is engaged. You can easily take your own pulse, either by placing your finger on the carotid artery on the side of the neck or by placing the tips of your fingers on the inside of the wrist, where the radial artery lies close to the surface. (See Figure 10.13.) If you lie down relaxed, most of your muscles are inactive and your heart rate will be accordingly slow. If you sit, some muscles are required to actively support the upper part of your body and head. The heart will thus beat faster. When standing, muscles in the lower body, as well as the trunk and head, must work and the heart will beat even faster. If you run or exercise, the heart will automatically adjust to speed up in response to the increased activity.

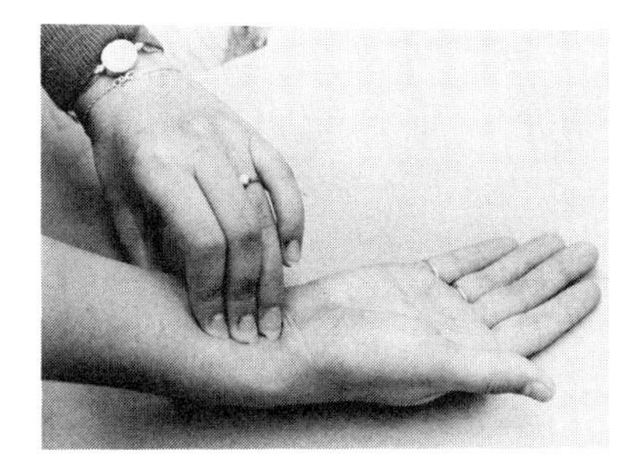

Figure 10.13.
Taking the radial pulse.

Emotions, such as fear, excitement, shock, and tension, will all affect the heart rate. If you use a pulse monitor, which has an audible beep for each beat, you can easily hear the heart rate change as you move from lying down to sitting or standing up.

Chemicals can also change the heart rate. Carbon monoxide will slow the heart and excess oxygen will speed it up. Drugs, such as nicotine and alcohol, as well as hormones (e.g., adrenalin) secreted by the body, will also affect the heart rate. Adrenalin increases both the force of the beat and the rate at which the heart beats. Thyroxin, a hormone from the thyroid gland, increases overall body metabolism, including that of the heart.

A CANADIAN HEART SURGEON

How can the heart of a patient be operated on without stopping it? If it stops, of course, the patient dies. The answer is that the heart must be slowed down enough to permit work on it but not so much that it stops sending blood to all parts of the body. In the early 1950's Dr. Wilfred Bigelow developed a technique that lowers the body temperature and slows down the metabolic processes of the body, while surgeons work to make vital repairs to the damaged heart. The process Dr. Bigelow developed was called **hypothermia**, which simply means lowering the body temperature. Hypothermia is usually induced by immersing the patient in an ice bath, or by wrapping the patient in a rubber blanket containing a circulating refrigerant. Muscle relaxants are given to the patient to prevent shivering, which is an automatic response by the body to maintain its normal temperature.

While developing the hypothermia technique, Dr. Bigelow used groundhogs for many of his experiments. These animals normally hibernate, which means that they are well adapted to sustained low temperatures. However, many of the animals died during the experiments because of a breakdown in the co-ordination of the heart action. Trying to solve this problem led to the concept of an artificial co-ordination system, the pacemaker.

The first pacemaker was made by Dr. Callaghan, then working at the Banting Institute in Toronto, and Jack Hopps, of the National Research Council in Ottawa. These early efforts were refined by other researchers and now pacemakers are implanted in at least 40 000 people each year throughout the world.

Dr. Bigelow also did pioneer work on micro-circulation. He discovered how damaged blood cells block the tiny capillary vessels after shock or special stress. Since blood passes from all arteries to veins through these tiny vessels, this work was very important.

The Volume of the Blood Pumped by the Heart

When skeletal muscle fibres are stretched, the heart pumps more forcefully. During exercise, the ventricles increase in size and stronger contraction is used to force the blood to the heart and throughout the body. Cardiac output is therefore increased. Normally it is the amount of blood returning to the heart that regulates cardiac output. As a result of certain disorders or during excessive bleeding, the stroke volume falls and cardiac output is reduced. The body responds to this by trying to pump faster, increasing the rate of contraction.

Regulation of the Heart Rate

If nothing occurred to influence the pacemaker it would produce a constant steady beat, regardless of the body requirements. There are, however, a number of special reflexes that change this pace, according to the body demands. The stimuli for these changes come from both the parasympathetic and the sympathetic divisons of the autonomic nervous system, centred in the medulla. Sympathetic nerves act to quicken the heartbeat, while parasympathetic stimulation slows the rate. There are many chemical receptors in the blood vessels; some of these are sensitive to oxygen and others are sensitive to carbon dioxide. These receptors keep the autonomic nervous system informed of changes in oxygen content of blood.

The Electrocardiogram

Your entire heartbeat takes less than a second to complete. If electrodes are attached to the body at certain places, tiny pulses of current, present when different muscles of the heart contract, can be picked up and fed into a machine. This device then displays them on a screen or moving chart called an electrocardiogram. The normal pattern seen on an electrocardiogram is shown in Figure 10.14. When arteries

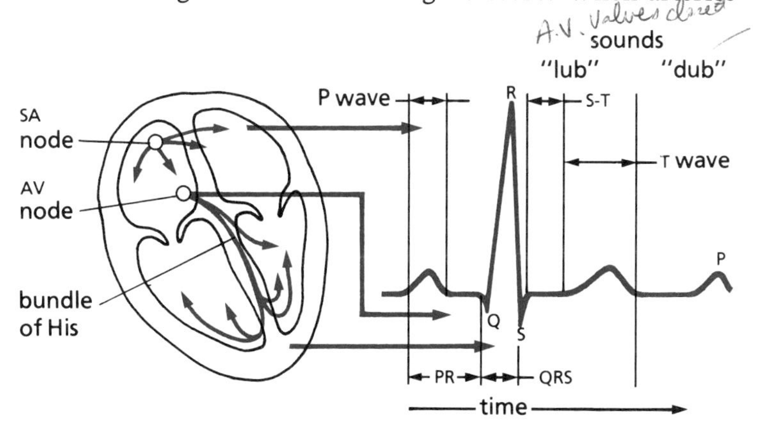

Figure 10.14.
The electrocardiogram (ECG) records the small electrial changes that occur in the muscles of the heart when they contract. A normal pattern is shown above. The P wave shows the passing of the impulse from the SA node through the atria. The PR portion shows the contraction of the atria and the passing of the impulse to the fibres in the ventricles. The QRS portion indicates the passing of the impulse through the ventricles and the ST section is the contraction of the ventricles. The T to P wave is the final relaxing of the ventricles.

supplying the heart muscles become partially blocked, this pattern changes. Damage to the heart and even some changes caused by chemicals can be detected with this machine. Doctors also use these records to make comparisons, similar to using "before and after" photographs. In this way they can see the progress being made by a heart attack victim or any unusual pattern changes in a previously normal heart.

Blood Pressure

If the blood is to reach the hands and feet, the brain, and every part of the body, it must be pumped out of the heart under very considerable pressure. The highest pressure occurs in the aorta, the great vessel that carries oxygenated blood away from the heart. As the blood passes into smaller vessels and the distance from the heat becomes greater, the pressure becomes greatly reduced.

The pressure in any vessel varies as a result of five major factors:

1. *The amount of blood*. If as a result of some injury a person has lost a lot of blood, the pressure in the system drops because of the decrease in volume.
2. *The heart rate*. The faster the heart pumps blood, the greater the pressure which is built up. The pressure falls as the heart rate decreases, especially during rest or sleep.
3. *The size of the arteries*. When the arteries dilate (become bigger in diameter), the volume of the vessels increases and the pressure falls. If the arteries constrict, pressure is built up because of the extra resistance to blood flow.
4. *Elasticity.* The walls of the arteries must be flexible and elastic. They must be able to expand as a surge of blood is forced out of the heart, and then relax after the surge has passed. If they cannot stretch in this way, they are described as hardened. Hardening of the arteries is a condition common in older people. When the arteries do not expand to accommodate blood flow the blood pressure in the system is increased.
5. *The viscosity of the blood*. Viscosity refers to the thickness of the blood: thick, sticky fluids flow less readily than thin watery liquids. The balance between the number of red cells and the amount of plasma present is one factor that controls the viscosity of the blood.

The SI unit for pressure is the pascal (Pa)
Blood pressure, like that of all liquids, should be measured in kilopascals (kPa). Since all sphygmomanometers in current use are still calibrated in millimetres of mercury (mm of Hg), we have retained that unit of measurement.

To convert millimetres of mercury to kilopascals, please note that a pressure of 100 mm of mercury = 13.3 kPa.

The usual blood pressure for young adults is 120/70 or 115/75 mm Hg. (See Figure 10.15.) The numbers refer to the pressure in millimetres of mercury. The numerator of this fraction represents the highest pressure generated when the ventricles contract. It is known as **systolic pressure** or systole. The denominator shows the **diastolic pressure** recorded when the ventricles relax and the elastic walls of the arteries offer the least resistance.

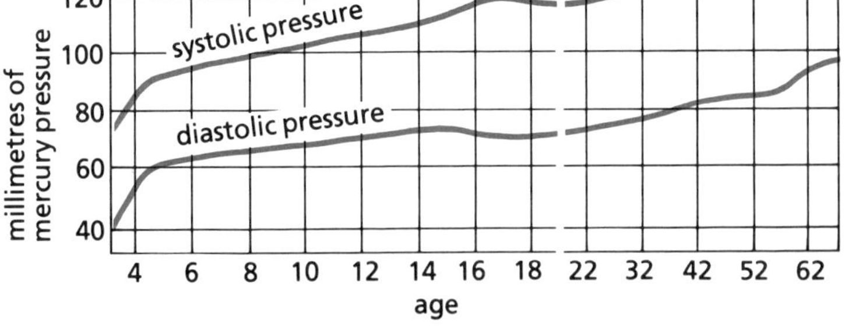

Figure 10.15.
The relationship between blood pressure and age. The lines show *systolic* and *diastolic* pressures that are average for each age group. Variations beyond these limits do not necessarily indicate a disorder.

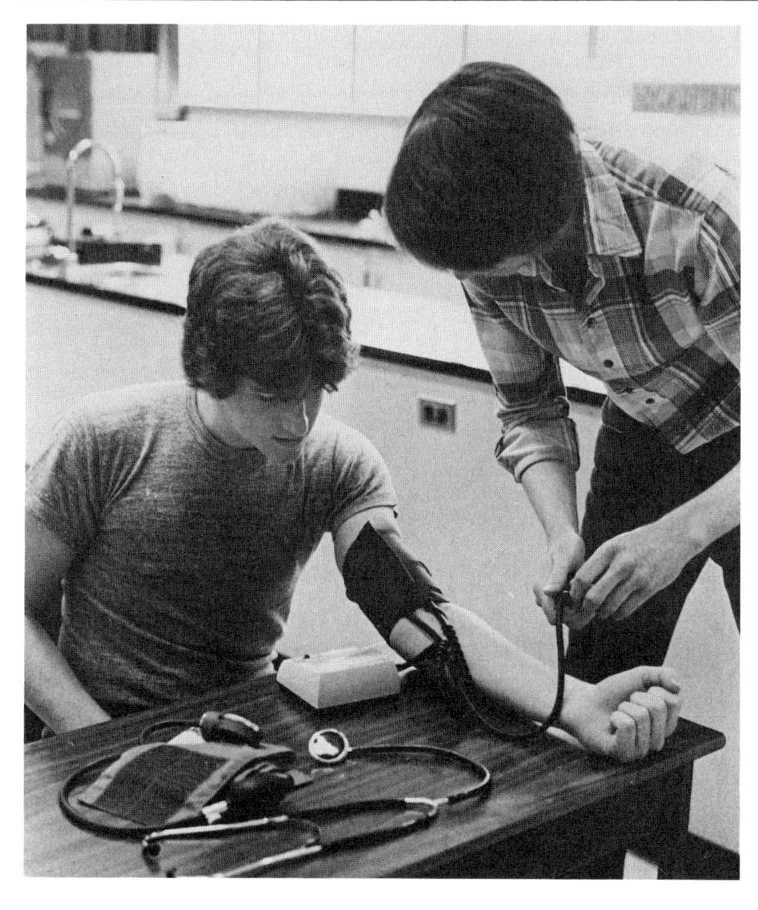

Figure 10.16.
Taking blood pressure reading with the sphygmomanometer.

The pressure is taken with a **sphygmomanometer** (See Figure 10.16.) This instrument uses an inflatable cuff to prevent blood passing through the artery of the arm. By using a stethoscope and listening for the sound made by the blood trying to force its way through the constricted artery, the blood pressure can be determined. The cuff is inflated to create a pressure about normal in the arm artery (about 160 mm Hg), and it is then gradually lowered. At the point when the blood just manages to squeeze through the constricted artery, the systole pressure is determined. When the sounds disappear it is a signal that the blood can pass through the constricted artery easily, because the heart is relaxed between contractions. At this time the diastole pressure is determined.

(sfig-moe-man-om-eter)

Hypertension occurs when the arterial blood pressure is significantly above average. This usually involves a *sustained* systolic pressure above 140 mm Hg or a *sustained* diastolic pressure above 90 mm Hg. Overweight people may have considerable fatty deposits within their blood vessels. These deposits decrease the internal diameter of the blood in them. This heightens the risk of heart attack. Anything that abnormally overloads the heart, such as stress and nervous tension, can increase the risk of coronary heart disease, although not all the factors are well understood.

The greatest dangers for people with high blood pressure are cerebral hemorrhage or heart failure. A cerebral hemorrhage occurs when the pressure becomes too great for some small brain arteries to withstand. If an artery feeding a part of the brain has a weak wall, it may burst, thereby preventing that part of the brain from receiving its supply of oxygenated blood. These brain cells may die causing the patient to lose control of some related part of the body. (Speech and limb movements are commonly affected by circulatory disorders of this kind.) This is the brain accident that produces a stroke.

In several chapters, reference has been made to the need for homeostasis or balance of all processes in the body. Monitoring blood pressure and taking prompt action when it is too high or too low is important for assuring that some processes remain homeostatic.

Interstitial Fluid

All of the cells and tissues of the body must be continuously bathed by fluids. These fluids enable nutrients to pass from the capillaries across the spaces between the cells to reach the cell membranes. They also prevent the cells from drying out.

(in-ter-stish-al)

The fluid that surrounds the cells is called **interstitial fluid** (*interstitial* refers to spaces between things). It is very abundant, making up about 15% of the body mass. You have probably seen this fluid, as the clear, colourless liquid that appears when you graze the skin or break a blister.

Interstitial fluid is very similar to the plasma of the blood. It contains approximately the same substances as plasma but in different proportions. The differences in these substances are due to the movement of materials and exchanges that take place between cells, interstitial fluids, and the blood plasma.

The blood is laden with substances needed by the cells. Not every cell, however, is in direct contact with a capillary. Interstitial fluid bridges this gap carrying materials from capillaries to the isolated cells. Some of this fluid then diffuses back into the blood stream, but much of it enters small tubes which are part of the lymphatic system. (See Figure 10.17.)

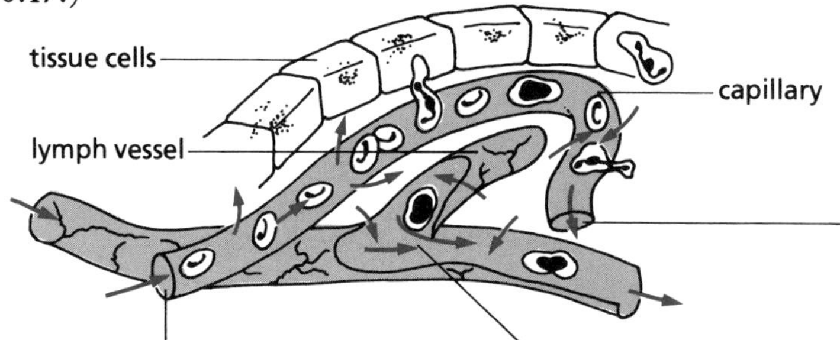

Arterial end of capillary, slightly higher pressure and higher concentration of nutrients. Plasma flows out of the capillary.

Excess fluid (lymph) is picked up by tiny closed-end lymph vessels and drain towards the lymphatic ducts.

Venous end of the capillary. Lower pressure and concentration of nutrients. Fluid tends to flow into the capillary.

Figure 10.17.
The exchange of plasma fluids between the capillaries and the interstitial spaces of the tissues and the collection of excess fluid by the lymph vessels.

The Lymphatic System

(lim-fat-ik)

The lymphatic system has several functions, the most important of which is return of the interstitial fluid to the blood circulatory system. It forms an extensive network of vessels, but, unlike arteries and veins which interconnect, the lymphatic tubes function in one direction only, controlled by tiny valves within the vessels. The lymphatic system has no pump, such as that found in the heart, but the larger vessels are fitted with valves similar to those found in the veins. Skeletal muscles and other active tissues surrounding the lymph vessels massage the tubes, pushing the fluids along towards the major collecting vessels. The fluid that enters the tubes is know as **lymph** and it has much in common with both the blood plasma and the interstitial fluid.

One of the most important features of this system is the presence of white blood cells. Red blood cells cannot leave the capillaries, but white blood cells do leave these small vessels using amoeboid movement to squeeze out into the interstitial spaces. Bathed in interstitial fluid they then go to work engulfing any bacteria or foreign proteins that they find, thus helping to combat infections. The material they collect is broken down and digested, usually by the cell itself. Sometimes several cells will combine to surround larger particles. These cells then move back through the walls of the lymph vessels and are carried along by the lymph to the lymph nodes. (See Figure 10.18.)

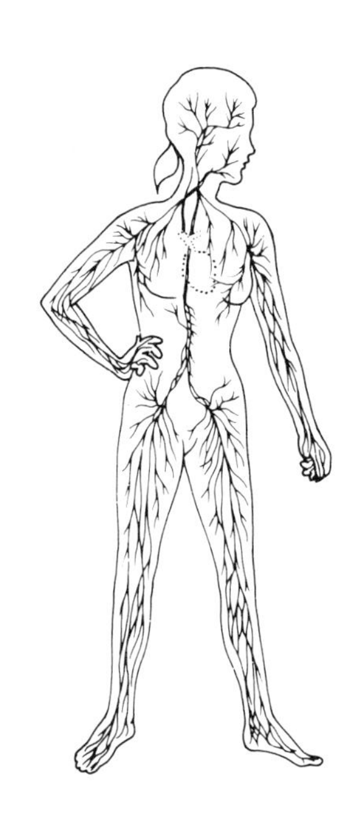

A BOMB TO KILL CANCER CELLS

When cancer takes hold in a particular site in the body, the cancer cells multiply rapidly and crowd out the normal cells around them. To remove the cancer cells by surgery involves many risks. One of these risks is that some cancer cells will be freed and washed into the lymphatic system, then carried to new locations in the body where they can set up new cancer sites. If the cancer cells could be killed in place without removing them, some of the risks could be reduced.

Since radiation was known to destroy cells, it was thought that, if an accurate beam of radiation could be aimed at a cancer site, these cancer cells might be destroyed without damaging normal tissues.

Dr. Harold Johns, a physicist then at the University of Saskatchewan, developed the first machine to be used for radiation therapy. The first machine was known as the Cobalt Bomb. Cobalt is an element 300 times more effective in radiation therapy than radium and about 6000 times cheaper. The first

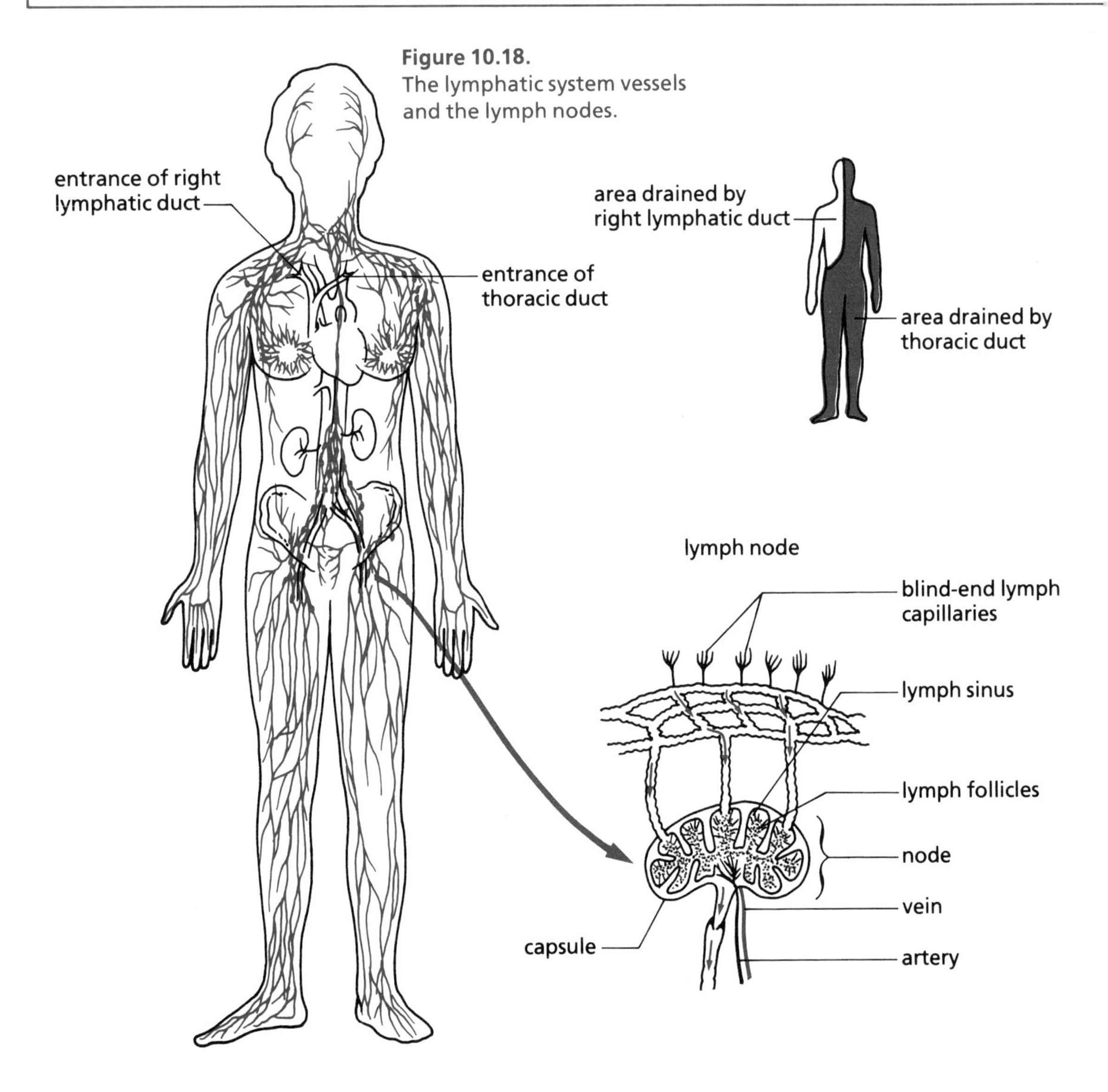

Figure 10.18.
The lymphatic system vessels and the lymph nodes.

discs of cobalt, each not much larger than a .25¢ piece, were produced at Chalk River, Ontario, and the machines to hold the cobalt were also designed and built in Canada.

Radiation from the cobalt source penetrates the cancer cells and prevents them from functioning and multiplying. Healthy cells replace the cancer cells and repair the damage.

The first two Cobalt Bombs went into operation in 1951 and these became the models for later machines around the world. Atomic Energy of Canada Limited makes more than half the cobalt therapy units in the world.

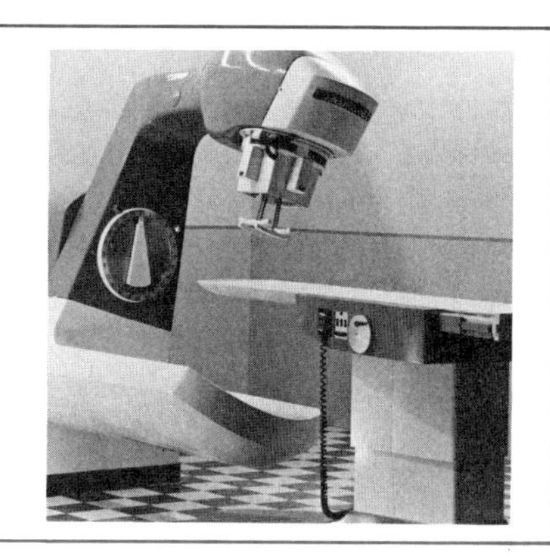

The Lymph Nodes

The **lymph nodes** are small, round structures found scattered along the medium-sized and large lymphatic pathways. The nodes are concentrated in four major areas; the groin, the neck, the abdomen, and the armpit.

(limf)

Non-granular white cells are produced in these nodes. The nodes also contain a permanent number of phagocytic cells, which help to filter out any other particles or infectious agents in the lymph. These cells may also trap cancer cells that have entered the lymphatic system.

The network of small lymph vessels gradually leads into larger and larger collecting vessels until finally they converge into either the **great thoracic duct** or the **right lymphatic duct**. The great thoracic duct empties into the venous system at the junction of the left **jugular** and **subclavian** veins (veins from the neck and arms). The right lymphatic duct opens into the junction of the right jugular and subclavian veins. Figure 10.18 illustrates the lymphatic system of the body.

The **tonsils** consist of three lymph organs located at the back of the mouth, at the base of the tongue, and in the upper pharynx. The pharyngeal tonsils are known as adenoids. They consist of a mass of lymphatic and germinal tissue, which produces white blood cells. These glands filter the lymph and help to supply the body with a continuous supply of leukocytes.

The **spleen** is a soft organ, purplish in colour, located behind the left side of the stomach. It also functions primarily to produce lymphocytes. The spleen also collects old red blood cells, filtering out and breaking down those that have become damaged or worn.

QUESTIONS FOR REVIEW

SOME WORDS TO KNOW

Match a statement from the column on the left with a suitable term from the column on the right. DO NOT WRITE IN THIS BOOK.

1. Blood flows out of this chamber into the aorta.
2. This chamber carries blood to the lungs.
3. The pressure created when the ventricles contract.
4. The pacemaker.
5. These vessels carry blood away from the heart.
6. Blood vessels where the exchange of materials takes place.
7. Valve found between the right atrium and right ventricle.
8. Returns interstitial fluid to the veins.
9. Found around cells in the tissues.
10. Separates and disposes of old red blood cells.

A. right ventricle
B. left ventricle
C. right atrium
D. left atrium
E. The SA node
F. The AV node
G. spleen
H. aorta
I. vena cava
J. arteries
K. veins
L. capillaries
M. tricuspid valve
N. bicuspid valve
O. systole
P. diastole
Q. interstitial fluid

SOME FACTS TO KNOW

1. Compare arteries, veins, and capillaries. Note the differences in both structure and function.
2. Draw a simple diagram of the heart and indicate with arrows the direction in which blood flows through each of the chambers and vessels.
3. Explain how the heart rate and the contraction of the heart muscles are controlled by nerves.
4. What factors affect the pulse rate? Describe the effects of exercise on the heart and circulation.
5. a) What is the normal heart rate for a person of your age and sex?
   b) What is the significance of recovery rate of the pulse after exercise?
6. How does the blood get back to the heart through the veins when the pressure in these vessels is so low?
7. What are some of the factors contributing to heart disease?
8. Define any six of the following terms.
   a) anemia,
   b) leukemia,
   c) hemophilia,
   d) systole,
   e) diastole,
   f) thrombus,
   g) lymph,
   h) interstitial fluid.
9. Briefly describe the main functions of the lymphatic system.
10. What is the function of the lymph nodes?

QUESTIONS FOR RESEARCH

1. Select one of the following topics and prepare a report on it:
   - causes of heart attack
   - heart transplants and tissue rejection
   - blue babies
   - arteriosclerosis
   - the ECG machine
   - fibrillation of the heart
   - taking care of your heart
   - stroke victims (aneurism)
   - high blood presure (hypertension)
   - the CPR program in first aid
   - the intensive care unit
2. What are the latest findings on cholesterol and its formation in the blood vessels?
3. Investigate the many innovations and artificial parts that can be used in the heart: pacemakers, valves, etc.
4. One in five people who reach sixty years of age dies of a heart disorder. Often these disorders are termed by the layman "heart attacks". Research a list of the more common terms associated with heart disease and prepare a short explanation of each. Cardiac bypass, myocardial infarction, congenital heart disease, angina, are among the more common terms used.
5. What is a lymphoma? Find out where they are found and what is known about their causes and treatment.

Activity 1: DEMONSTRATE THE VALVES PRESENT IN THE VEINS OF THE ARM

Method 1

1. Have your partner roll up the sleeve on the right arm to above the elbow.
2. Leave the arm hanging by the side for a minute or two until the veins of the arm become distended. Then have your partner rest his or her arm on the desk top and note the large blue veins showing on the surface of the inner arm.
3. Press your index finger against one of these veins near the elbow and stop circulation through the vessel. Very little pressure is needed. Maintain the pressure, but gently stroke the vein above the pressure point in the direction of the heart with the other index finger.
4. The blue line will soon disappear over a short portion of the vessel. Release the pressure and observe the vein refill. *Record your observations so far.*
5. Apply the same pressure, but this time stroke the vessel below the pressure point and away from the point. Does the blue line of the vein disappear? *Look up the relevant information on the valves found inside the veins and explain this series of observations. Why are these valves necessary?*

Method 2

1. Hold your right hand above your head and allow the left hand to hang at your side. After about twenty seconds put both hands out in front of you at shoulder level. *Note the colour of your hands and the size of the veins on the backs of your hands. Record and explain your observations.*
2. Have your partner lie down and observe the size and colour of the veins in the neck, hands, and feet. Raise and lower your partner's arms and note any changes in the vessels. *Record your observations.*

Activity 2: PULSE RATE

The pulse is an indication of the rate at which the heart is beating. It varies widely with amount of exercise, emotional state, and general health. It also varies between sexes.

Method

1. Take your pulse either at the radial artery or at the carotid artery in the neck.
   If the radial artery is used, place the fingers firmly over the radial artery along the inside of the wrist, below the line of the thumb. This will press the artery against the bone and the pulse can readily be felt.
2. Having found your pulse, count the number of beats in one minute in each of the following positions. At rest (lying down), sitting, and standing. Now run in place for one minute; make sure that you lift your knees and work hard. Take your pulse immediately following the exercise. Count the number of beats in fifteen seconds and multiply your answer by 4 to get the beats per minute. After resting for exactly two minutes, again take your pulse and *record all your results in a suitable table in your notebook.*

Record your observations

	At rest	*Sitting*	*Standing*	*1 min exercise*	*Recovery rate*
Males					
Class average					
Females					
Class average					

Compare the results for girls and boys. Compare these results of students who exercise regularly with those who do not. Compare the results of non-smokers and smokers. A chart of all the results on the chalkboard may enable you to draw some other conclusions. Are there major differences in their results? Look especially at the recovery rates.

Note: Your heart changes pace rapidly, so take the pulse immediately after the exercise. What other factors may affect these results?

Activity 3: BLOOD PRESSURE

The blood pressure cuff or sphygmomanometer is a valuable diagnostic aid for the doctor. An explanation of how this instrument works appears earlier in the chapter. There are now several varieties available through biological supply houses; all give excellent results if the directions are followed properly.

Materials

stethoscope and sphygmomanometer

Method

1. Wrap the cuff around your partner's arm snugly, just above the elbow. If there is a white ring or dot on the cuff make sure that it is just over the brachial artery.
2. Place the stethoscope just under the rim of the cuff over the brachial artery. Inflate the cuff, first closing the valve on the rubber bulb. Inflate to a reading of about 150 mm, by squeezing the rubber bulb.
3. Slowly release the air by opening the valve just a little. Listen carefully for the first sounds of a pulse.
4. As soon as you hear the first quiet sounds of a pulse, *record the* systolic *pressure.* Continue to allow the air to escape and to listen to the sounds of the pulse. At the point when you can no longer hear the sounds of a pulse, *record the* diastolic *blood pressure.*
5. *Record the blood pressures in the following positions or after the activity suggested.*

At rest lying down	*At rest sitting down*	*At rest standing*	*After exercise still standing*

6. *Keep the results for girls and boys separate. Average the results for each sex. Compare your results of those who exercise regularly with those who do not. Compare results of non-smokers with smokers. What generalities, if any, can you disover? Consider any other factors that may influence these results.*

Activity 4: HEART SOUNDS

The sounds produced by the heart are caused mainly by the vibrations produced when the heart valves close and the blood bounces against the walls of the ventricles or blood vessels. The *stethoscope* makes it very much easier to hear these sounds. Some microphones can also be used to magnify the sounds for discussion by the class. There are two major sounds that can be heard.

The First Sound is produced when the valves between the atria and ventricles close and the semilunar valves open. This sound has a lower pitch and makes a "lub" sound. It occurs at the beginning of systole.

The Second Sound occurs at the end of systole. This sound is produced by the closure of the semilunar valves and the opening of the valves between the chambers. The pressure in the arteries is higher and the pitch is also somewhat higher. The sound is a short "dub" sound.

Method

1. Swab the ear pieces of the stethoscope with 70% alcohol. Look at the sketch showing the locations where the sounds are best heard. (See Figure 10.19.) Listen to your partner's heart, using the four areas suggested by the diagram.
   How many distinct sounds can you hear? State the causes of each of the heart sounds that you can hear.
   Which sound persists for the longest time? Did you notice any missed beats? What variations in sound occur between the different areas?

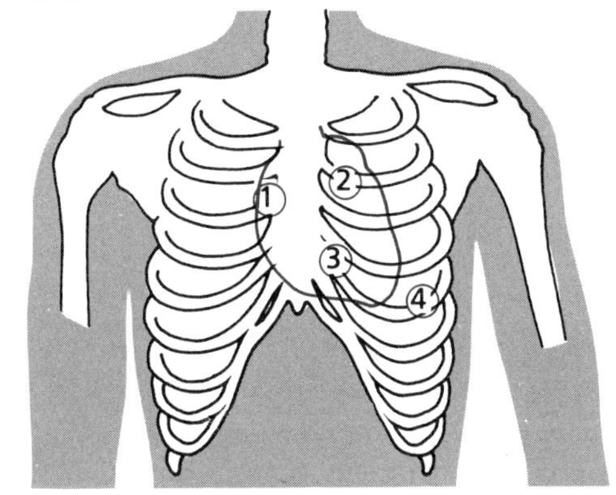

Figure 10.19.

Areas where the heart sounds *can be heard* most clearly.

1. The semilunar valve into the aorta.
2. The pulmonary semilunar valve.
3. The tricuspid valve.
4. The bicuspid valve.

2. *Listen to your pulse in the radial and carotid regions. Describe what you hear.*
3. Run on the spot for a minute and then listen to your heart again. *What differences are there? Describe the differences in your notebook.*

Chapter 11

The Body's Defences against Disease

- An Excellent Host
- Pathogens
- Body Defences
- Drugs that Fight Infection

An Excellent Host

Throughout our lifetime, we are under almost constant attack from a wide variety of invaders, attempting to pierce the body's defences and establish a new home within the human body. For many of these microscopic organisms, the human body makes an excellent host. It possesses, among other things, the three most important conditions necessary for their survival and multiplication. The human body has warmth, moisture, and a rich variety of excellent food sources. (See Figure 11.1.)

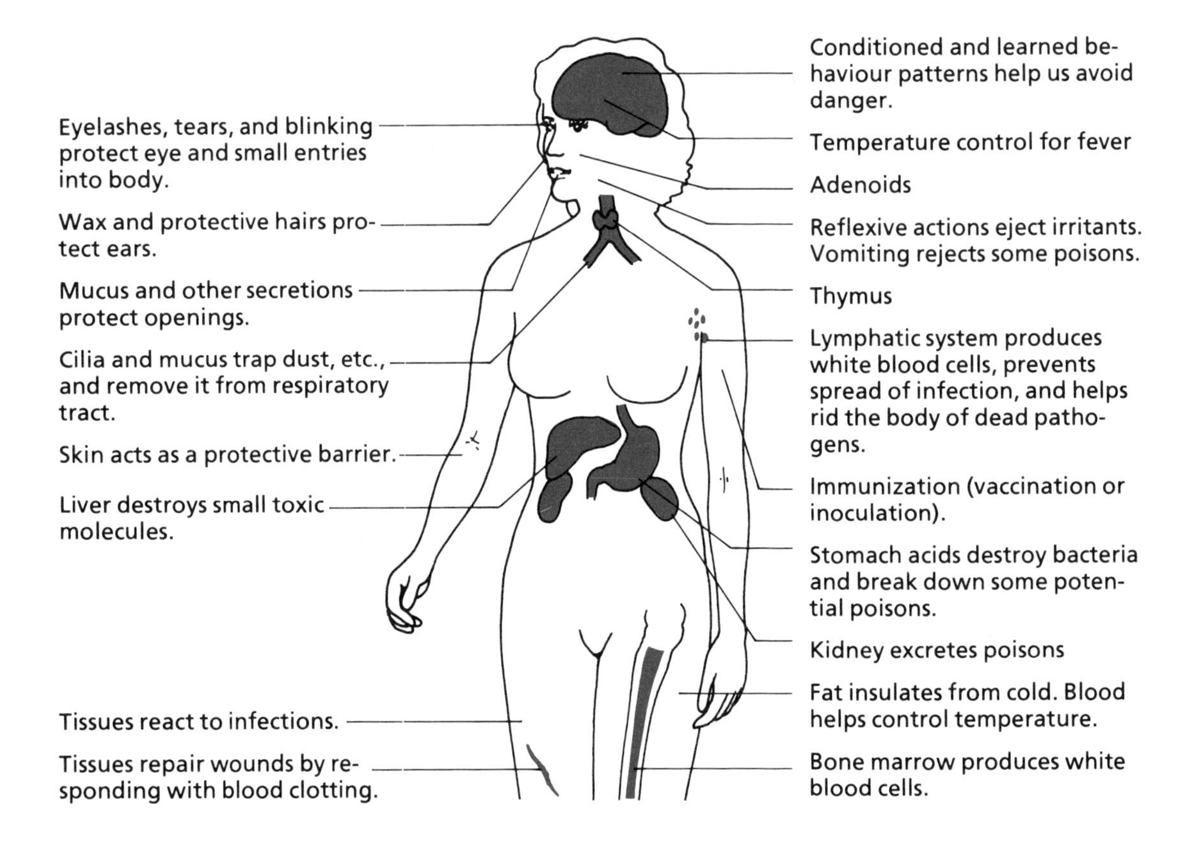

Figure 11.1. The defences of the body against disease.

Pathogens

Any micro-organism that can harm the body, or produces a disease within the body, is called a **pathogen**. This does not mean that all microscopic organisms are pathogens. Our body contains many small organisms, especially in the mouth and digestive tract, which do us no harm at all. Some, such as the bacteria that live in the large intestine, actually help us; certain essential vitamins are formed during the process by which the bacteria break down waste products that are present in the intestine.

Once the pathogens (germs) have entered the body, they may establish an infection and produce a number of **symptoms**. Symptoms are changes in the body that take place as a result of the infection. Symptoms vary with the disease in question, but common examples are fever, headaches, localized pain, or changes in skin colour.

(sim-tum)

Pathogens may enter the body through natural openings such as the nose and mouth, or through wounds and cuts which have broken the outer barrier of the skin. After they have entered the body, they may produce a local infection near the site of entry or they may travel throughout the blood stream causing infection in almost any tissue or organ of the body.

Types of Pathogens

All known pathogens may be classified into 6 categories:

1. **Viruses** are the simplest form of life, being not much more than complex molecules.
2. **Bacteria** are single-celled organisms, found in several shapes.
3. **Rickettsiae** are much smaller in size than bacteria and are found inside other cells.
4. **Fungi** are plantlike organisms, which may combine to form multicellular organisms.
5. **Protozoa** are single-celled organisms, some of which have the ability to move on their own.
6. **Parasitic worms** are multicellular animals. Many of them also live in other animal hosts.

Viruses

Viruses are the smallest of all the pathogens. They cause such diseases as colds, influenza, poliomyelitis, and measles. Viruses are so small that one million of them could fit within a bacterium. When viewed under an electron microscope they appear to have a central core of genetic material surrounded by a protein case. (See Figure 11.2.)

Viruses are the simplest form of life known. They are

considered to be alive because they are able to reproduce. To survive, however, they must live and reproduce within the cells of some other organism. Once inside, they take over control of the cell and use the cell's protein to manufacture more viruses similar to themselves.

When a virus invades a cell, only the genetic material of the virus enters the cell; its protein coat remains outside. It is the injected nuclear material that takes over the host cell and starts producing very large numbers of new viruses from the cell contents.

There are very few drugs that can control viruses. Often a doctor prescribes medication that helps to relieve the symptoms of an infection (bring down the fever, for example), but it has very little effect on the viruses themselves.

Figure 11.2.
Virus reproduction. A virus, known as bacteriophage (phage), is capable of entering a bacterium. By using its DNA it is able to overcome its host and reproduce many identical viruses from the basic structural materials of the bacterium.

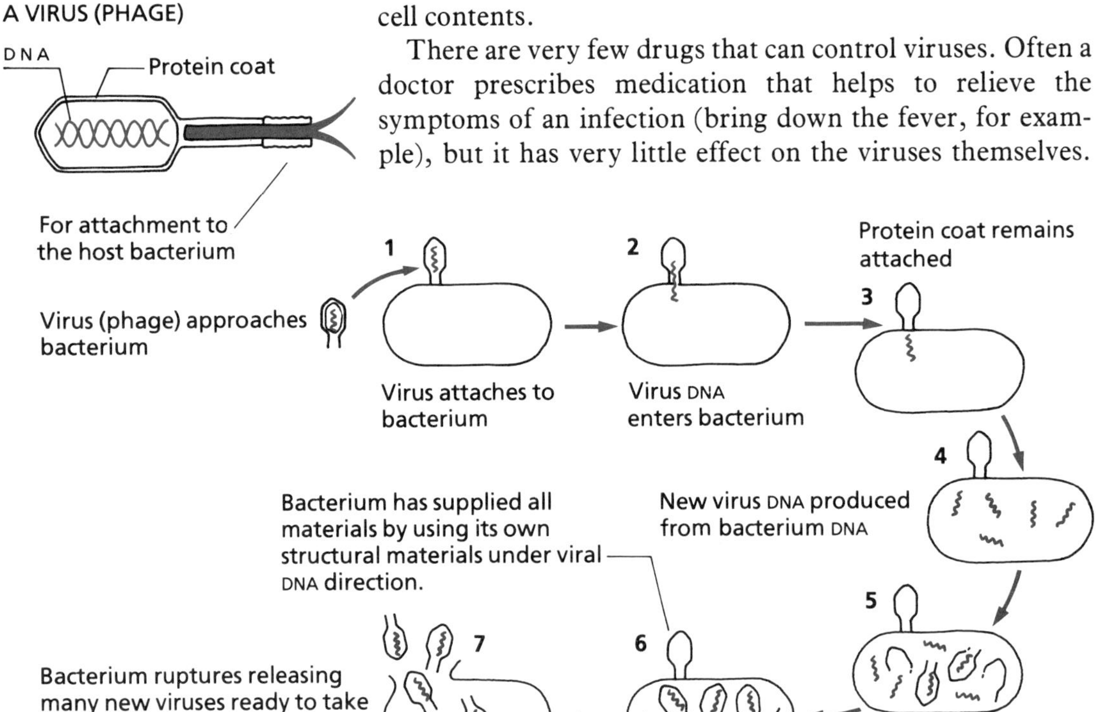

Bacteria

Bacteria are simple, single-celled organisms which reproduce rapidly by dividing. Under the right conditions, they may divide every twenty minutes. Bacteria are small enough that many thousands could fit into the space occupied by the period at the end of this sentence.

Many bacteria that are not pathogenic are found on every surface we touch and in every item of food we eat. It is estimated that pasteurized milk contains more than 20 000 000 living bacteria per litre.

Bacteria do not live inside cells, but produce poisonous enzymes and **toxins** (poisons) that may seriously affect the body. Some of these toxins destroy body cells in a localized area, providing food for an increasing number of bacteria. Some toxins may be carried into the blood stream and affect an entirely different part of the body. Although there are over 1500 known species of bacteria, only 100 of these are pathogenic in humans.

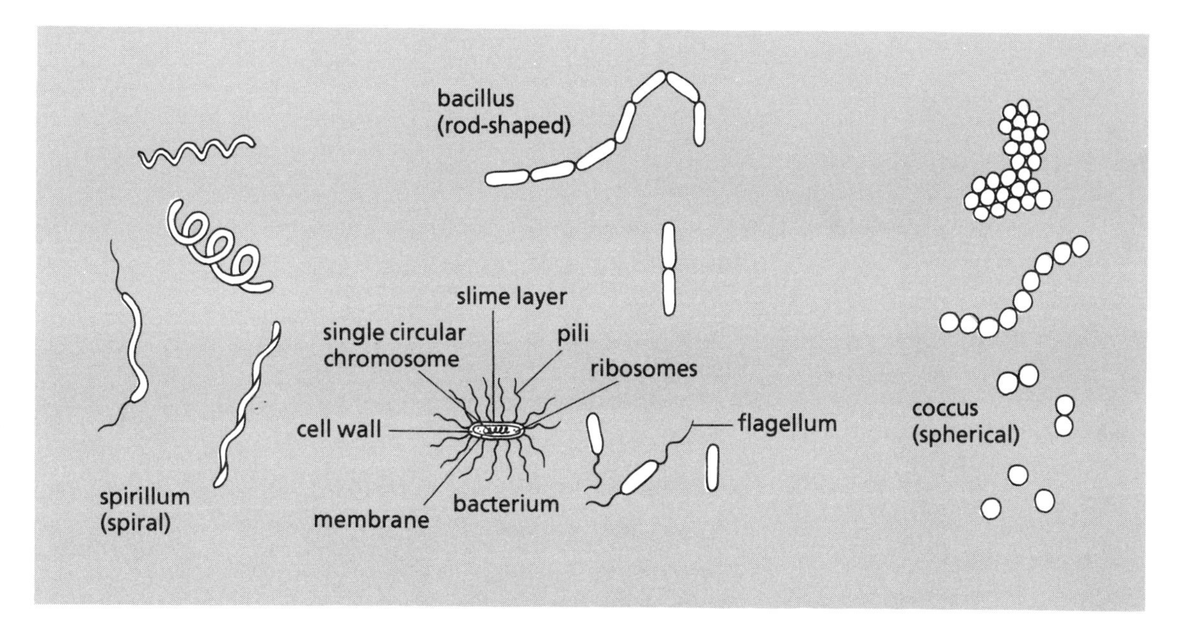

Figure 11.3.
The shape and structure of bacteria. Bacteria do not contain nuclear membranes or well-organized cell organelles. The cell membrane is covered by a tough outer coat. Bacteria have the capacity to reproduce very rapidly.

Bacteria are found in three basic shapes (See Figure 11.3 and 11.4) and may occur singly or in pairs, in clusters or in filamentlike strands. Bacteria that are rod-shaped are known as *bacilli*. Spherical-shaped bacteria are called **cocci**. Spiral-shaped bacteria are the **spirilla**.

Examples of diseases caused by bacteria are: gonorrhea, diphtheria, strep throat, tuberculosis, boils, and meningitis.

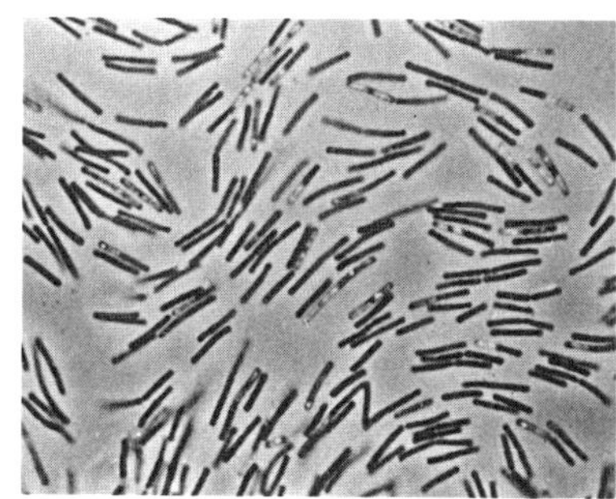

Figure 11.4.
Rod-shaped bacillus bacteria.

Rickettsiae

Rickettsiae are plantlike organisms, rather similar to bacteria in appearance, but very much smaller in size. These organisms cannot exist outside of living tissue; they are true parasites. The rickettsiae cause several very serious diseases in man, such as Rocky Mountain spotted fever and typhus. Because these organisms must always live inside other organisms, they are usually transmitted by bites from ticks, fleas, and lice.

Fungi

The fungi are primitive organisms that do not make their own food by photosynthesis, but feed off other organisms; they are thus parasites. The spores (reproductive cells) of these plants are present in the air and are commonly found in damp places. Athlete's foot, ringworm, and some respiratory disorders are the most common problems produced by these pathogens. Only a few types of fungi are responsible for human diseases.

Protozoa

These are simple single-celled organisms, larger than bacteria but still so small that they can only be seen under a microscope. Most of the diseases produced by these pathogens are associated with tropical climates and many of them are transmitted by water. Amoebic dysentery, malaria, and sleeping sickness are examples. In some of these diseases, such as malaria, the pathogen may remain dormant in the body for a long time, then without warning induce a new bout of the disease.

Parasitic Worms

Parasitic worms vary in size from tiny species, the size of a pinhead, to tapeworms, which can be several metres in length. Many are found in other host animals and are transmitted to us in the foods we eat. (See Figure 11.5.) They are not micro-organisms but they are usually microscopic in size. In North America and Europe, the process of inspecting meats, especially beef and pork, has greatly reduced the incidence of disease caused by such organisms.

Transmission of Diseases

Diseases caused by pathogens are known as **infectious diseases**. If an infectious disease can be transmitted to other people, it is called a **contagious disease**. Diseases can be transmitted in several different ways.

Direct Contact

This method involves actual contact with the pathogen or the host carrying the pathogen. Touching the skin, kissing, sexual intercourse, or the spray resulting from a sneeze or cough may transfer the pathogen from one person to another.

Figure 11.5.
Transmission of a parasitic tapeworm through an intermediate animal host.

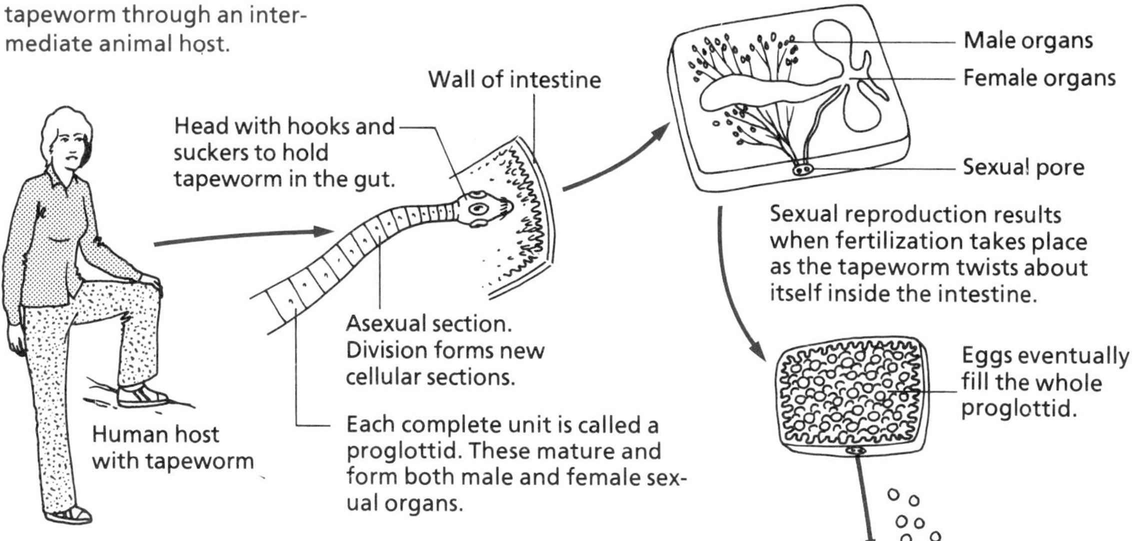

Eggs are released and pass into the feces. When inadequate methods of sewage disposal are present, eggs may enter food of animal host.

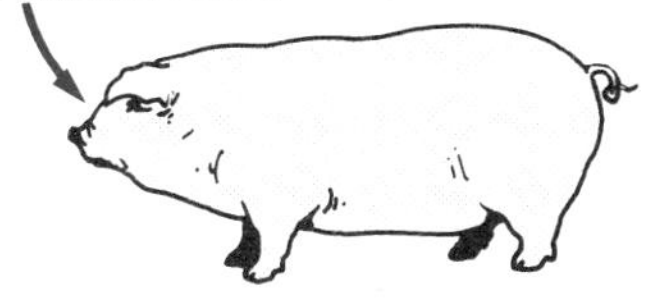

Eggs develop inside pig and enter the muscles. If meat is inadequately cooked and not inspected before sale, the tapeworm may be present in human food.

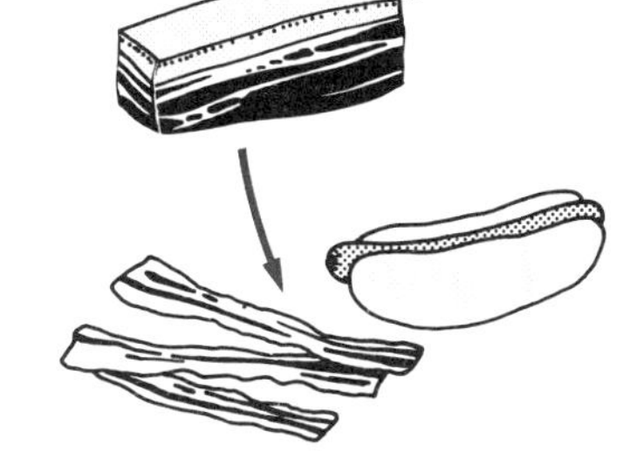

Eggs or tiny worms may be carried into new host in undercooked meats.

Indirect Contact

This method of transmission occurs when some object or substance conveys the pathogen from the person with the disease to some uninfected person. Water, food, towels, clothing, bedding, hypodermic needles, cutlery, or dishes may be involved. The survival of pathogens carried on these objects depends on the warmth and dryness of the objects and time involved in transmission.

Airborne Transmission

Pathogens are sometimes carried by air currents and breathed into the body. They may be spread by spitting, excrement, or other body discharges that dry and form dust containing pathogenic spores. This dust may eventually be carried by air currents to uninfected individuals.

Vectors

Insects are the chief agents of this form of disease transfer. Flies, mosquitoes, ticks, lice, or other animals may carry the pathogens on their feet, mouth parts, or inside their bodies. In some instances, the insect vector may bite a diseased host and ingest the pathogens. It may then bite another individual injecting the pathogens into this new host. Rabies may be transferred by this means or by contact with the saliva of an infected animal. Transfer of disease by vectors usually takes place during a predictable sequence of events.

Body Defences

The skin is frequently referred to as the "first line of defence". While it remains unbroken, the skin forms a very reliable and effective barrier to infection. Inside the openings of the body, the epidermis is lined with cells that secrete mucous. This adds another defensive barrier as it helps to trap any pathogens that may enter. Cilia, which line the respiratory pathways, act to clear out bacteria that might otherwise reproduce in the moist warmth of the lungs. The cilia move back and forth, creating tiny currents which sweep particles out with each breath of exhaled air. The eyes are protected by a mildly antibacterial lubricant that washes away any dust or bacteria that may find its way onto the surface of the eyeball.

Cellular Defences: White Blood Cells

If an infection becomes established in the body, there is an immediate response to the pathogens involved by an army of white blood cells. As the infection develops, a characteristic swelling and redness develops as well as fever and discomfort. The redness and swelling are caused by dilation of the blood vessels around the site of infection. Such dilation

Figure 11.6.
Inflammation. A response of the body to infection.

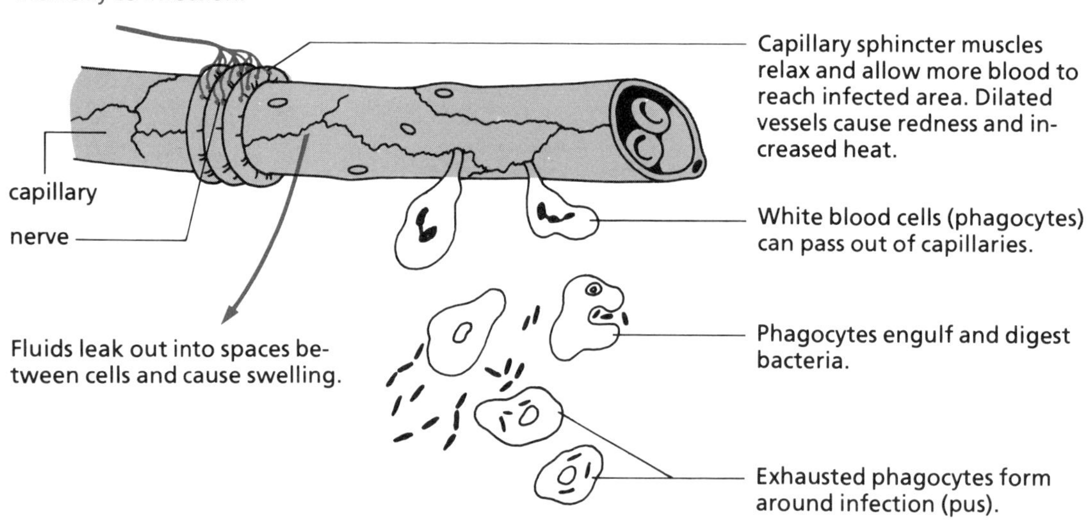

brings extra numbers of white blood cells to the area. (See Figure 11.6.)

Normally, the number of white cells is about 7000 to 10 000/mm^3 of blood. When an infection is established, a signal is given to increase this white cell count. Initially, immature white cells are released into the blood. Then the bone marrow and lymph nodes start to manufacture more white cells, both to provide a larger army of infection fighters and to replace damaged cells. Some white cells are able to surround and engulf foreign particles or pathogens and destroy them. Such white cells are known as **phagocytes**.

Histamines

Histamines and certain other chemicals in the blood activate the smooth muscles of the blood vessels causing them to expand and increase vessel diameter as much as ten times. This provides fluids and white cells with faster and more efficient passage into the tissues. It also raises the temperature around the infected area. This fever increases the activity of the white blood cells, while the extra blood and lymph fluid flush away debris and toxic substances into the lymphatic system for disposal.

Antigens and Antibodies

Antigens are special chemicals, usually protein or carbohydrate molecules, that are found adhering to pathogenic organisms. Lymphocytes (a type of white blood cell) "recognize" these antigens and are stimulated to produce special molecules called **antibodies** which react with them.

Each antigen has a different shape, against which specific antibodies fit like pieces of a jigsaw puzzle. For each antigen there is only one antibody, which recognizes its particular shape. When the appropriate antibody reacts with an antigen, it may cause the pathogens to stick together in large masses, thus immobilizing them. Alternatively, it may help phagocytes recognize the pathogens more easily and thus destroy them more rapidly.

Antibodies act to destroy the pathogens that have produced a disease. They can also provide immunity to disease, thereby destroying pathogens before they have a chance to become established in the body.

How Antibodies Work

If a pathogen enters the body, it causes the body to produce large numbers of antibodies to prevent or minimize the infection. This "antigen-antibody" reaction involves several chemical possibilities:

1. Antibodies may neutralize the antigen, so that it has no effect.
2. Antibodies may immobilize the antigen-containing organisms stopping them from moving, while the white blood cells engulf and destroy them.
3. Antibodies may break down the cell wall of the antigen-containing pathogen thereby killing it.
4. Antibodies may attach themselves to the pathogen thereby attracting white blood cells and increasing the chance of the pathogen being destroyed.
5. Antibodies attack and kill some types of pathogens.

It should be noted that the antigen-antibody reaction is one of the major difficulties encountered during organ transplants. The antibodies recognize that the transplanted organ is composed of proteins that do not belong to the body; they are recognized by the host's antibodies as foreign proteins.

Figure 11.7.
Immunization and the production of antibodies.

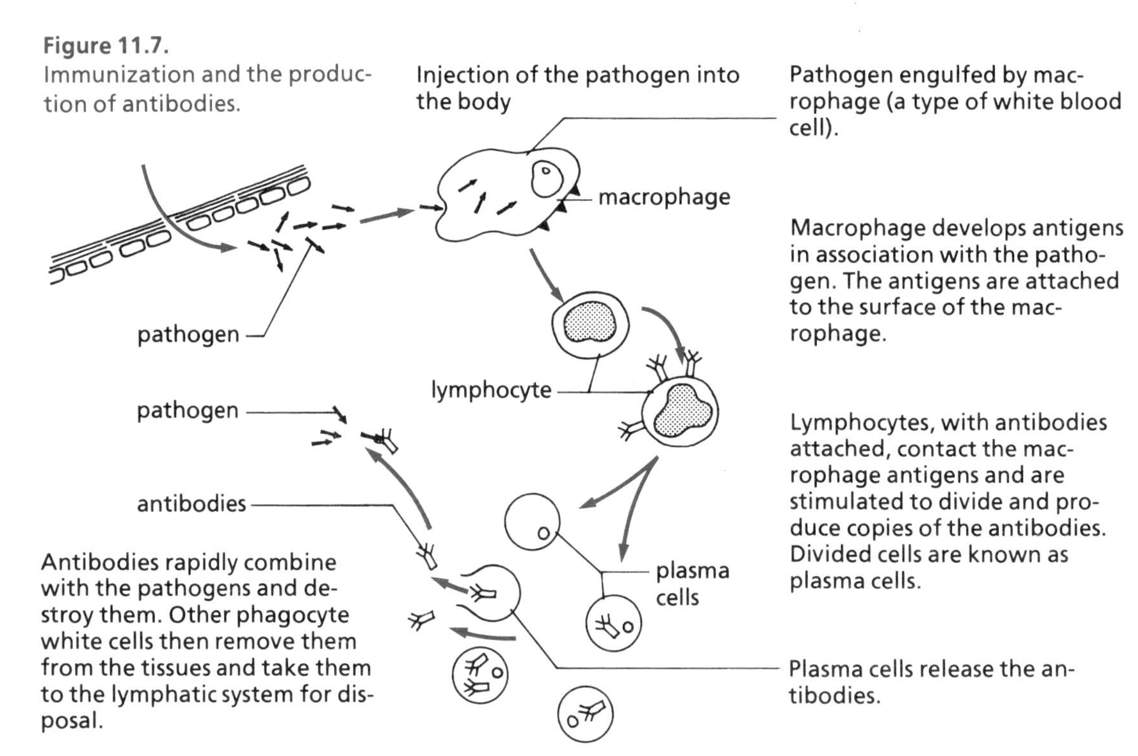

As a result, the antibodies go to work to destroy the cells and cause rejection of the transplanted organ. (See Figure 11.7.)

Natural Immunity

Newborn babies possess a natural immunity to those diseases for which the mother has acquired immunity. The antibodies built up in the mother's body cross the placenta and enter the baby's blood, providing immunity that lasts for several months. The baby may obtain from its mother other antibodies given during breast-feeding.

If a person contracts a specific disease, the body builds up antibodies against the disease. Sometimes the attack may be so mild that the individual does not even realize that he or she has had the disease. These antibodies remain in the blood and respond quickly to further infection by pathogens of the same type. Such persons have a built-up **immunity** to a disease as a result of exposure to a specific pathogen.

Artificial or Acquired Immunity

Vaccines are prepared from very weak or dead pathogens, which will usually produce only a very mild reaction when injected into the body. Once the pathogen is introduced, the body starts to build up antibodies against it. Any future invasion of this disease will be met by these antibodies which will prevent an infection from developing.

Immunity may also be acquired by injecting antibodies from the blood serum of animals or other humans that are already immune to the disease. These measures are very temporary as the antibodies soon break down. This is because without antigens, there is no mechanism within the recipient to initiate the production of his or her own antibodies.

Immunization is now available for many serious diseases, such as typhoid fever, typhus, yellow fever, rabies, polio, and cholera. Persons travelling to many countries around the world are required to be immunized for many of these diseases and health certificates are checked by immigration officers at border points. There are now immunization programs in North America for most of the common childhood diseases, such as diptheria, whooping cough, poliomyelitis, measles, mumps, and German measles. Some vaccines require additional "booster" shots periodically to rebuild the quantity of antibodies and maintain effective immunity. (See Table 11.1.)

Table 11.1. Immunization. A schedule of recommended vaccines to protect you against a number of potentially dangerous diseases.

2 months	First immunization. Four vaccines in one against diphtheria, whooping cough, tetanus, and polio.
4 months and 6 months	Two more injections against diphtheria, whooping cough, tetanus, and polio.
12 months	One injection of a combined vaccine against measles, mumps, and rubella (German measles).
16 to 18 months	First booster dose of vaccine against diphtheria, whooping cough, tetanus, and polio.
4 to 6 a	Second booster dose against diphtheria, whooping cough, tetanus, and polio.
11 to 12 a	Third booster dose against diphtheria, tetanus, and polio.
16 to 18 a	Fourth booster dose against diphtheria, tetanus, and polio.
Adults	Polio vaccine every five years. Tetanus vaccine booster every ten years. Women planning to have children should have their immunization updated, including rubella immunization, before becoming pregnant. Where children have not been immunized from early childhood, a modified immunization program should be decided upon, in consultation with the doctor or local health unit.

Ontario's recommended immunization schedule. By permission. The Ontario Ministry of Health.

Allergies

In the last few decades, the number of substances or particles to which individuals may become allergic has increased enormously. In every classroom there are students allergic to pollen, dust, fruits, milk, additives, insect stings, cosmetics, fibres, penicillin, as well as a host of other substances. Part of this allergic sensitivity is inherited from our parents. We must also, however, be exposed at some time to an allergy-producing substance present in the environment; and modern technology is constantly producing new chemical products which cause reactions in some people.

There is usually no noticeable reaction the first time a person is exposed to an **allergen** (substance capable of producing an allergic reaction). However, the body recognizes the presence of the antigen associated with the allergen, and a reaction occurs in the lymph tissue cells. The lymph tissues produce antibodies which become attached to the lymph **mast cells**. Inside the mast cells there are histamines that

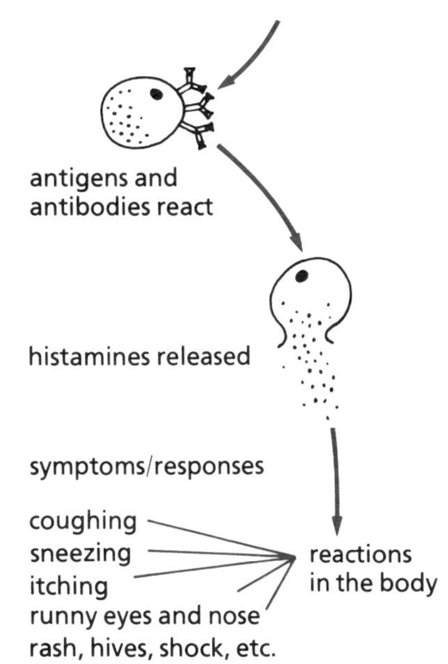

Figure 11.8.
Allergy responses to environmental factors.

cause various reactions in the body. (See Figure 11.8.) The first exposure thus produces the antibodies, but it is not until the second exposure that the antigens rapidly associate with the antibodies on the mast cells. These mast cells, containing the histamines, then rupture and the histamines spread out into the blood stream.

The reactions produced by histamines vary from sneezes, coughs, runny eyes, itching, and rashes, to more severe symptoms, such as breathing difficulties. In very severe cases, the person suffering from the allergy may go into shock and even die.

Allergies can usually be controlled by injections or tablets containing antihistamines. There are tests in which a small amount of a particular substance is introduced into the skin. If the person is allergic to the substance (for example, dust or a type of food), the skin swells and becomes red, thus identifying an allergy.

INTERFERON – THE VIRUS FIGHTER

Interferon is a substance produced by the bodies of humans and other living organisms when their cells are attacked by viruses.

When a virus invades a body cell, instead of continuing to turn out the normal proteins needed to sustain itself and other parts of the body, the cell may start to produce more viruses. The viral DNA duplicates to form identical viral DNA's and, thus, more viruses, using the cell substances to carry out this process. The body cell breaks apart and dies. The millions of viruses so produced attack other healthy cells and repeat the process. Fortunately for us, the initial viral infection somehow triggers the first infected cell to produce a substance called **interferon**. Interferon passes out of the infected cell and lines the cell membrane of healthy cells. These healthy cells respond by producing "antiviral proteins" against that specific virus. So healthy cells can now become immune to that type of viral infection.

Interferon is produced in humans in white blood cells. Unfortunately, it takes 45 000 L of blood to produce only 400 mg of interferon. Researchers are now able to splice the genes of bacteria to produce a new species of bacteria which will produce an interferon closely resembling human interferon.

Interferon has been used to treat patients with such varied problems as breast cancer, certain allergies, and the common cold. Research is ongoing, as much more still needs to be learned about this very interesting natural fighter of viral infections.

Drugs That Fight Infections

When medications are taken to help the body fight an infection, the drug may act to either kill the pathogens completely or slow down their activity so that the body can combat them more effectively. When a doctor prescribes a drug, the directions will usually state that the medication must be taken regularly until all the tablets are completely used up. Some people are tempted to discontinue the drug once the symptoms have eased, thinking that the drug is no longer necessary. This is unwise. The drug may have reduced the activity of the pathogen, but the body's defences may not have had time to destroy all the bacteria. The pathogens may then build up again and overcome the body's defences. If we take the complete course of medication prescribed, it is unlikely that any pathogens will survive. Failure to kill all of the bacteria will allow the most drug-resistant ones to escape. These bacteria might be or become immune to the drug. If these immune pathogens then reproduce, use of that drug in the future will be far less effective because more and more bacteria will have this drug-resistance.

Antibiotics

In the last few decades, the number of **antibiotics** available has increased enormously. Many antibiotics are produced by organisms that live in the soil or grow as moulds on decomposing plants or foods. Antibiotics compete with the pathogens and work to effectively prevent the pathogens from reproducing. This gives the body time to destroy the invading bacteria. Since the most useful effect of an antibiotic is to reduce the rate of reproduction of the pathogen, it should be administered during the early stages of the infection. If a sufficient amount of the antibiotic is used, it may kill the bacteria.

Table 11.2. Some major types of diseases, their cause and prevention.

TYPE OF DISEASE	GENERAL CAUSES OF THE DISEASE	PREVENTING THE DISEASE
Deficiency diseases	Lack of proteins, minerals, vitamins, etc., in the diet.	Education about a balanced diet. Help to areas of poverty and famine. →

Endocrine disorders	Over- or under-secretion of hormones due to growths, enlargement or atrophy of the glands.	Good medical attention and regular check-ups. Can be controlled with medication in most cases.
Infectious diseases	Pathogens such as bacteria, etc. Contact directly or indirectly with persons carrying the disease. Poor hygiene, contaminated water, etc.	Good public health, water, and sewage services. Inoculations and immunization. Good personal hygiene. Avoid contact with infected perosns.
Cancer	Chemical and other carcinogenic hazards. These produce rapidly dividing abnormal cells.	Elimination or reduction of known carcinogen hazards (tobacco, etc.). Early diagnosis by regular check-ups (breast examination, Pap tests, etc.)
Degenerative diseases	Due to old age and the wearing out of tissues and organs. May affect hearing, vision, joints, etc.	Healthy lifestyle and regular exercise. Good diet during earlier years. Care.
Mental disorders	Some may be congenital or inherited. Some due to stress, chemical factors, and environment.	Education of parents. Reduction of stress; learning to handle stress. Understanding of child and adult needs.
Accidents and injuries	Carelessness. Working or driving when tired or when reflexes are slow. Taking risks.	Awareness of hazards. Safety education. Good working habits and adherence to safety rules.
Occupational disorders	Jobs where dust, fibres, chemicals, heat, stress, etc., are common. Special hazard exposure to asbestos, radiation, etc. Emotional and stress jobs such as air traffic controllers, executives, etc.	Protective clothing, shields, ear protectors, helmets, masks, goggles, etc. Stress control, planning, rest, methods of relaxation.
Habitual diseases	Smoking, alcoholism, drugs, excess food.	Understanding the risks involved, selecting alternatives, avoidance. Not starting habits that are difficult to control.
Congenital diseases	Usually an infection or illness during pregnancy which affects the unborn child.	Pre-natal care and supervision of the mother during pregnancy. Avoidance of contact with contagious diseases. Good personal health habits and diet.

QUESTIONS FOR REVIEW

SOME WORDS TO KNOW

Match a statement from the column on the left with a suitable term in the column on the right. DO NOT WRITE IN THIS BOOK.

1. A cell capable of surrounding and engulfing a bacterium.
2. Disease-causing organisms.
3. A disease that can be transmitted from one person to another.
4. Molecules that react to antigens.
5. Resistance to a particular disease.
6. A preparation of dead or weak pathogens.
7. A response to environmental factors, such as pollen, dust, or fibres.
8. A chemical substance released by cells that have been infected by a virus.
9. A substance that inhibits the growth of bacteria.
10. A poisonous substance.

A. pathogen
B. bacterium
C. virus
D. parasite
E. toxin
F. vector
G. phagocyte
H. histamine
I. allergy
J. interferon
K. contagious
L. immunity
M. vaccine
N. antibiotic
O. antigens
P. antibodies

SOME FACTS TO KNOW

1. Describe the differences between viruses and bacteria.
2. Give five ways in which the body protects itself against infections.
3. What happens when an infected site becomes inflamed?
4. What is the difference between natural and artificial immunity?
5. How do vaccines protect the body from a particular disease?
6. Give four examples of substances that cause allergies. How can allergies be controlled?
7. Briefly describe how various diseases may be transmitted.
8. Give two ways in which white cells help to protect the body from infections.
9. Define the following terms:
   a) contagious diseases,
   b) infectious diseases,
   c) antibiotic,
   d) symptoms,
   e) pathogen.
10. a) How can we protect ourselves from colds?
    b) Of what value are most cold remedies?

QUESTIONS FOR RESEARCH

1. Select one of the following topics and prepare a report on it:
   - allergist
   - antibiotics
   - homogenization of milk
   - infection control in hospitals
   - herbicides or insecticides and their effect on humans
   - transplant rejection
   - salmonella and botulism
   - common childhood diseases and their symptoms
   - methods of preserving foods
   - rabies
   - pathology laboratory assistant
   - pest control
   - food allergies
   - immune response
2. What measures are taken in your school to prevent the spread of infections? You might investigate cafeteria dishwashing methods, how tables are wiped after use, washroom disinfection, changing areas and showers, garbage disposal methods.
3. In what ways does the body prepare itself to combat viruses? How does interferon help in this process?
4. What does the term "drug resistance" mean? You may need to check with your local Public Health Office and obtain a list of possibly drug-resistant bacteria.
5. Find out what the inoculation schedule is for your province. Make a personal chart of the inoculations that you have had and whether they are up to date. Also list any diseases that you have had.

Activity 1: WHERE CAN BACTERIA BE FOUND?

Materials

sterile agar plates, scotch tape, glass-marking pencils

Method

Note: Keep the petri dishes containing the agar closed until you are ready to inoculate them.

1. Turn the closed petri dish upside down and mark off in four quarters with a large cross. Number each of the 4 quarters.
2. Take a piece of scotch tape, with one end folded over to give you a non-sticky part to hold onto. Press the sticky part of the tape against any surface that you wish to test for the presence of bacteria: the desk, a ruler, your hand, the sink, or a cafeteria table.
3. Quickly tip up one side of the petri dish lid and place the sticky portion of the tape against the surface of the agar in one of the four quarters. Smooth the tape onto the surface and then remove the tape and quickly re-cover the dish.
4. *Record in your notebook the surface tested and the number of the part of the dish where you placed the tape.*
   Repeat this process using a fresh piece of tape for each surface tested.
5. Place the petri dish (marked with your name) in the incubator for three or four days at about 37-40°C.

6. Remove the plates each day and look for any changes. Do not open the cover; view the contents through the transparent top. *Count the number of colonies and anything else that you observe, and record.*
7. *Make a suitable chart to contain your results.*

Quadrant #	Source of bacteria	Number of types of colony present				Total number of colonies			
		Day 1	2	3	4	1	2	3	4

Questions

1. *Did you find bacteria in all of the places that you tested?*
2. *How can you tell different bacterial colonies apart?*
3. *You probably found some other things growing on the agar. How can you tell the difference between moulds, fungi, yeasts, and bacteria?*
4. *What conditions are present to make these organisms grow so well?*

Activity 2: COUNTING BACTERIA IN MILK

How many bacteria are there in a litre of milk? There are many millions, far too many to count. When large numbers are involved we need to use a sampling and dilution technique. In this experiment we must make the assumption that, if we dilute the sample so that the colonies are widely spread, each colony develops from a single bacterium.

Materials

(Note: All the glassware and petri dishes must be sterilized before use.)

bottle of unsterilized milk from a farmer or dairy,
4 sterile pipettes (capacity 1 or 10 mL graduated in 1 mL steps),
1 empty sterile flask,
3 sterile flasks containing 99 mL of sterile distilled water each,
3 petri dishes,
flask of warm, liquid lactose nutrient agar

Method

1. Pour some of the milk into a sterile flask, stopper it, and shake it well to make sure the bacteria are evenly distributed throughout the liquid.
2. Use a pipette and transfer 1 mL of milk to a flask containing 99 mL of water. Shake well for about half a minute. Label the flask "A". Take a 1-mL sample of this diluted milk and transfer it to a sterile petri dish. Mark this "A" also.
3. Use another clean pipette and transfer 1 mL of milk from flask A to flask B, which contains 99 mL of water. Shake well. Transfer 1 mL of this solution to another petri dish marked B.
4. Make another dilution by taking 1 mL from flask B and adding it to the third flask (C) containing 99 mL of water. Shake well. Take 1 mL from this flask and add it to a third petri dish (C).
5. Pour about 15 or 20 mL of agar into each of the dishes and swirl them around gently until the milk and agar are mixed. The agar should be cool but liquid.
6. Incubate the plates in a warm oven at 37°C for 48 h. *Examine the plates after* 24 *and* 48 h *and record the number of bacteria colonies present on the plates. Record your results in a suitable table.*

Discussion and Questions

a. *Each time you took a* 1-mL *sample and added it to a* 99-mL *amount of water you diluted the milk one hundred times. You did this three times. What was the final dilution factor?*

b. *The decrease in the number of bacterial colonies each time you diluted the milk should be obvious from your results. Take the number of colonies in your final dilution and multiply this number by the dilution factor. This will tell you how many bacteria there were in* 1 mL *of the original undiluted milk sample. How many bacteria would there be in a litre of this milk?*

c. *Are the majority of the bacteria present on the plates pathogenic or non-pathogenic?*

d. *Why must milk be pasteurized? Look up the meaning of this term and find out the temperatures involved.*

e. *Design an experiment of your own to determine some question about the bacteria present in milk. For example: How long must milk be boiled to kill all the bacteria present? or How does refrigeration prevent milk from going sour?*

Activity 3: EFFECTS OF ANTIBIOTICS ON THE GROWTH OF *BACILLUS SUBTILIS* AND *ESCHERICHIA COLI*

Materials

a distinfectant (such as Lysol), cultures of *B. subtilis*, *E. coli*, sterile nutrient agar plates, glass-marking pencil, forceps, 70% alcohol, inoculating loop, millimetre ruler, Bunsen burner, and commercial antibiotic discs containing tetracycline, penicillin, streptomycin, erythromycin, etc. (Some discs have dispensers)

Method

Put your initial and the date on the lower side of the plate. Divide lower side of nutrient agar plates into four sections with a marking pencil. Streak the entire plate with *B. subtilis* or *E. coli*. Using a sterilized forceps, place one disc in each section of the plate. Cover and incubate for 48 h at 37-40°C.

Questions

1. *What do the clear areas around each disc indicate?*
2. *Which of the two organisms is more sensitive to the chemicals in general?*
3. *Which would be the best drug to control* E. coli? B. subtilis?
4. *What are antibiotics?*

Unit VI

How the Body Exchanges Oxygen and Carbon Dioxide

Chapter 12
Breathing

- The Need for Oxygen
- The Air-conducting Structures
- The Mechanism of Breathing
- Smoking

The Need for Oxygen

If you have ever taken a first aid course or Red Cross course in swimming, you will have been taught that the first concern of a person giving first aid is to see that the victim is breathing. This is a more important consideration than stopping bleeding or attending to broken bones. The body must constantly receive an adequate supply of oxygen, for if the brain is without oxygen for even four minutes, it may be permanently damaged.

If a machine is to operate, it must be supplied with energy; cars need gasoline, other machines need electrical energy, and humans must have chemical energy, which is obtained from food, especially sugar. If you take a small spoonful of sugar, place it in a crucible, and heat it over a Bunsen burner, it will begin to turn brown as it melts. After a time, it will ignite and burn fiercely. This combustion process, like another type of burning, requires oxygen. For this reason, in order to get sugar to release its stored energy for use in the body, the body must have an adequate supply of oxygen. It is the function of the respiratory system to bring oxygen into the body, and to provide a site where it can be transferred into the blood stream. The circulatory system will then transport it to all the cells of the body where energy reactions take place. The respiratory system also provides a means by which the waste gas of metabolism, carbon dioxide, can be eliminated from the body.

The Air-conducting Structures

With the exception of the **alveoli**, all of the structures in the respiratory system have walls that are too thick to allow diffusion of air into the blood stream. The organs involved are the nasal cavity and sinuses, pharynx, larynx, bronchi, bronchioles, and alveoli.

(al-vee-oe-ly)

The Nasal Cavity

The nasal cavity is separated from the mouth cavity below it by a bony platform which forms the hard palate of the

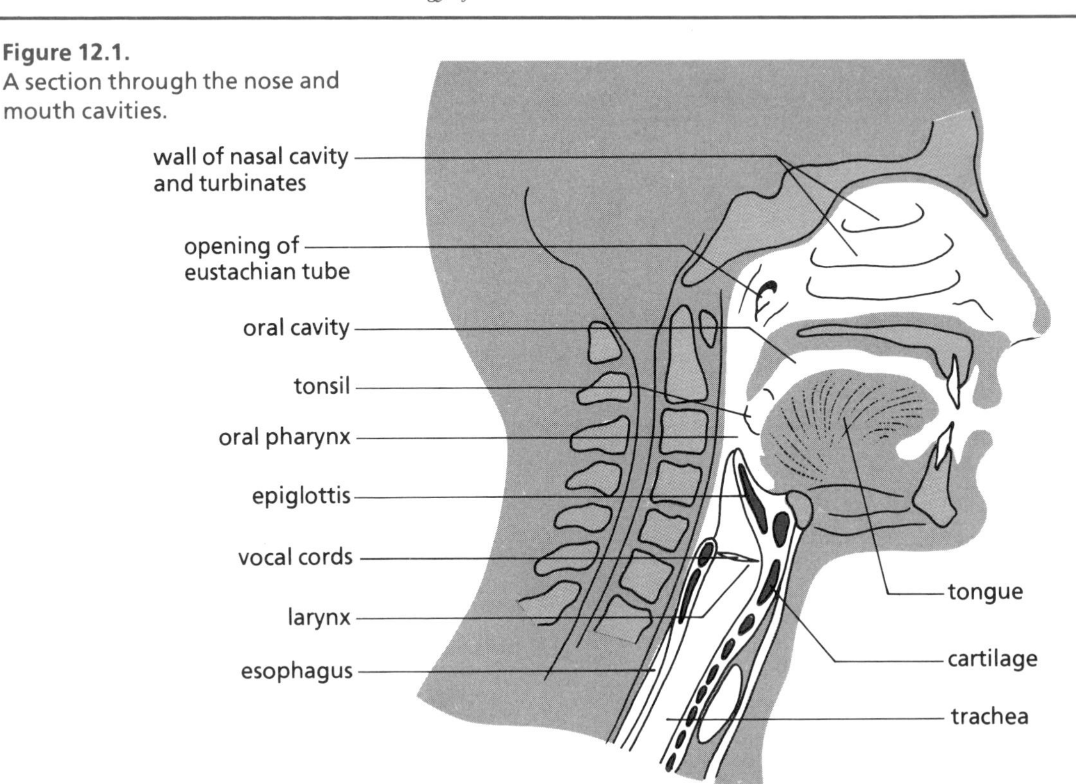

Figure 12.1.
A section through the nose and mouth cavities.

mouth. Extending into the nasal chamber, like sagging shelves, are the **turbinate bones** (See Figure 12.1.), which help to increase the amount of surface area in the nose. In the nasal cavity air is warmed, moistened, and cleaned before it is passed down into the lungs. The air is warmed by contact with the many surfaces of the cavity, for the lining tissues are well supplied with vessels filled with warm blood. The greater the surface area that is in contact with the air, the more efficiently the warming is accomplished. This is the value of the turbinate bones. On a cold day, the blood vessels in the skin covering the turbinate bones are capable of raising the temperature of air from freezing to almost body temperature by the time the air has entered the respiratory passages. The warmth and moisture present in air which has passed through the nasal cavities is easily seen when we exhale on a cold day, as our breath quickly condenses in the cold air. Moisture is supplied from secretions of the epithelial tissue in the nostrils. The cells of these tissues also produce mucus which, together with many small hairs, aids in trapping dust and other fine particles in the air.

The Pharynx

(fair-inks)

The pharynx forms a tube common to both the respiratory and digestive systems. It starts at the back of the nasal cavity and extends down to the larynx (voice box). The upper part of the **nasal pharynx** is covered with ciliated epithelial cells that trap the fine particles in the air. It also contains the **tonsils** and **adenoids**, which consist of a mass of lymphoid tissues. Sometimes, especially in young children, they become infected and enlarged and may interfere with breathing or swallowing. The individual then breathes primarily through the mouth. This prevents the nasal chamber from cleaning, moistening, and warming the air that enters. In some severe cases of enlarged tonsils it is considered worthwhile to remove them.

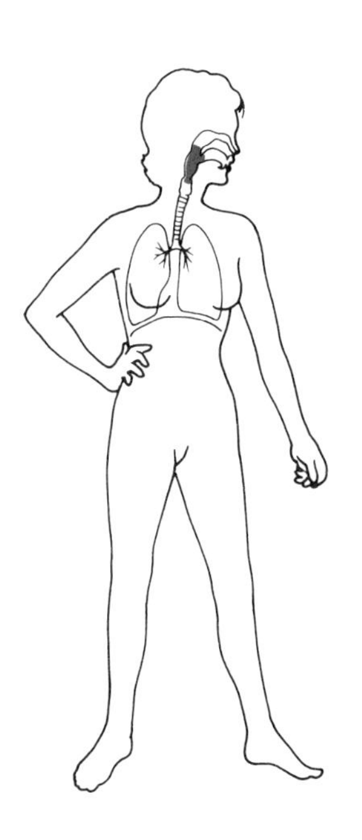

The second portion of the pharynx, the **oral pharynx**, lies behind the mouth cavity and forms a passageway for both food and air. Its walls are lined with epithelial cells which can stand up to the rough wear and tear of foods passing through. This tough lining extends into the last portion of the pharynx, which divides into two tubes, one carrying food, the other air.

The Larynx and Epiglottis

(lair-inks)

The larynx is a boxlike structure located at the opening to the respiratory passageway. It is formed by several pairs of cartilage. The largest of these, the **thyroid cartilage**, forms the framework of the Adam's apple. Above this is a leaf-shaped flap that forms the **epiglottis**. The epiglottis seals the opening into the respiratory tract, thereby preventing the passage of food into the lungs while swallowing. The larynx contains two flaps of cartilage controlled by muscles called the **vocal cords**. (See Figure 12.2.) As we speak, air passes out of the lungs and through the larynx. It causes these cords to vibrate and produce sounds, in much the same way as we can make a blade of grass produce sounds when we hold it tightly between our thumbs and blow on its edge. The pitch of sounds made by the vocal cords can be changed by tightening or loosening the muscles that hold the cords in place. The thickness and length of the cord also affect the quality and pitch of the sounds produced. As the sounds pass through the mouth, they are moulded by the shape of the mouth and the position of the tongue into words that we can recognize.

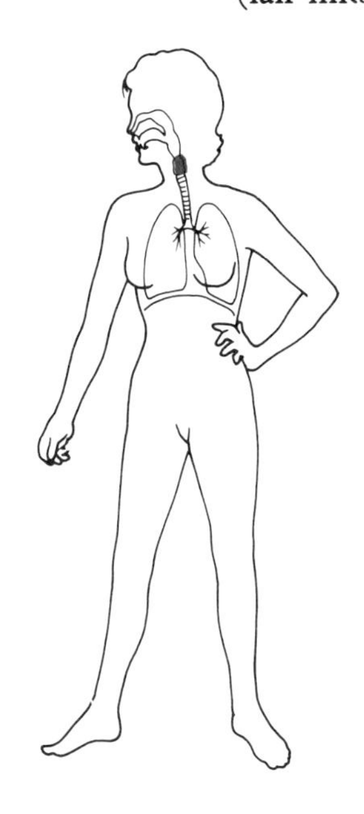

Figure 12.2.
The larynx, trachea, and epiglottis.

epiglottis
This softer part rests against the esophagus.
vocal cords
thyroid cartilage
cricoid cartilage
air
rings of cartilage
Rings of cartilage which prevent the trachea from collapsing if the air pressure should change significantly (anterior view).
trachea

The Trachea and Bronchi

(trae-kee-ah)

(brong-ky)

The **trachea** is a tube, about 12 cm in length, which extends from the larynx into the chest cavity, where it divides into the right and left **bronchi**. The trachea is constructed of smooth muscle in which C-shaped rings of cartilage are embedded. (See Figure 12.2.) The rings are not quite complete at the back where the trachea is loosely attached to the **esophagus**. When food is swallowed, the esophagus, which has walls made of muscle, expands against the trachea. The rings serve primarily to prevent the trachea from collapsing if there are changes of pressure in the tube. The rings thus ensure that the trachea is always open. The right and left bronchi are also supported by small rings of cartilage. These tubes, in turn, branch into smaller tubes, forming what is often called the bronchial tree. The smallest branches are called the **bronchioles**, which lack cartilage. They become smaller and smaller in diameter and more numerous, spreading throughout the entire lung tissue. (See Figure 12.3.)

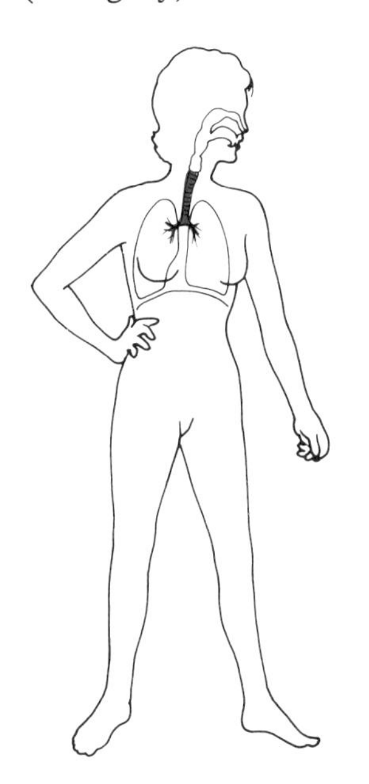

The bronchial tree terminates with the **alveolar ducts** which lead into tiny chambers where gas exchange takes place. These chambers, or air sacs, are called the **alveoli**. (See Figure 12.4.) (See Table 12.1.)

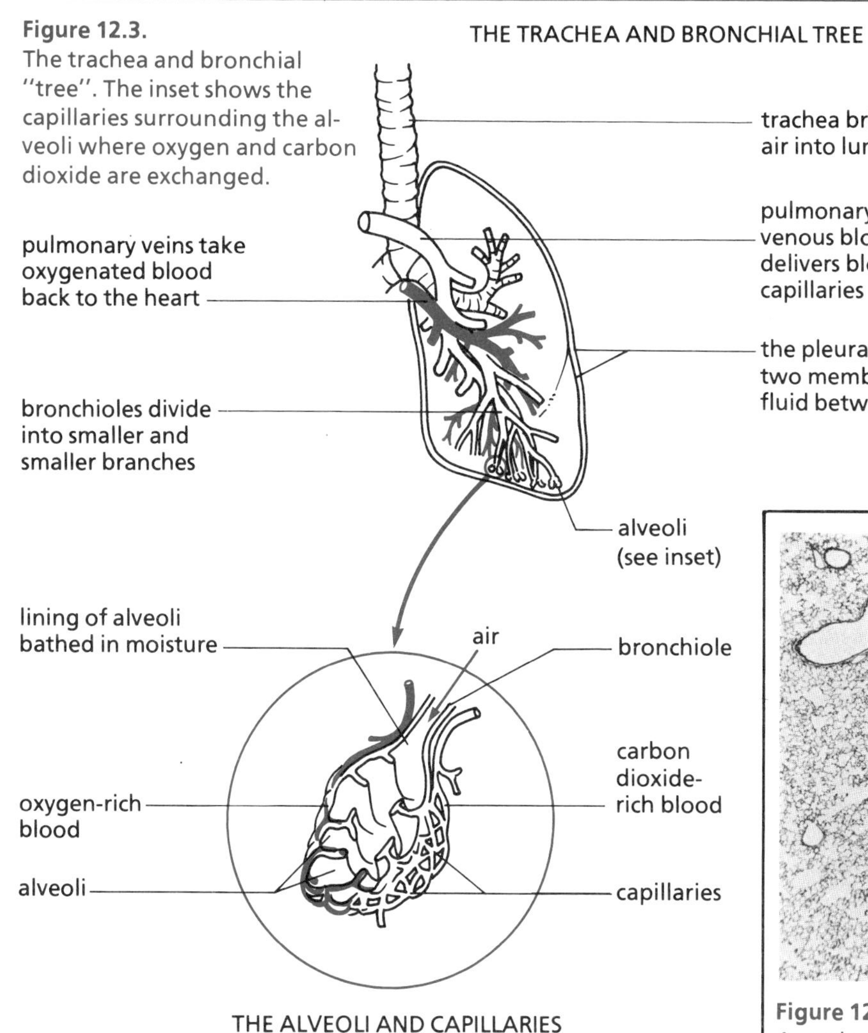

Figure 12.3.
The trachea and bronchial "tree". The inset shows the capillaries surrounding the alveoli where oxygen and carbon dioxide are exchanged.

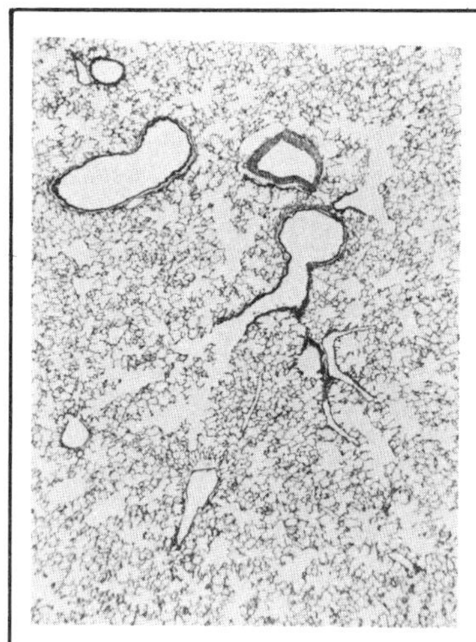

Figure 12.4.
A section of a normal lung showing the bronchioles and tiny alveoli.

Table 12.1
Comparison of the numbers and size of airway tubes in the respiratory system.

Organ	Number of Tubes	Diameter (mm)	Total cross-sectional area in cm^2
Trachea	1	18-25	2.5
Bronchi	2	12	2.3
Small Bronchi	1 020	1.3	13.4
Bronchioles	262 000	0.5	534
Alveolar Ducts	4 200 000	0.4	5580
Alveoli	300 000 000	0.2	50-70 m^2

The branching structure of the bronchial tree makes it very effective. As the number of tubes increases, although the diameter of the tubes decreases, the total cross-sectional area available also increases. This enables a large volume of air to be rapidly dispersed into numerous branching tubes and then transferred into the blood stream.

The Lungs

The lungs are two cone-shaped organs moulded into the form provided by the thoracic cavity. They are well protected by the surrounding ribs, sternum, and spine. The base of each lung lies in contact with the diaphragm and the top of each lung reaches just above the clavicles.

The Pleura

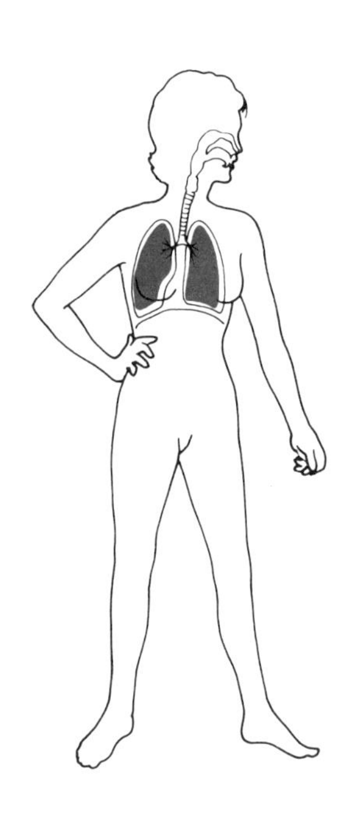

The lungs are contained within the **pleura**, two membranous sacs which surround the lungs. The outer membrane (**parietal pleura**) lines the inner surface of the chest wall and covers the upper surface of the diaphragm. The inner membrane (**visceral pleura**) adheres to the surface of the lungs. These two membranes are so close together that only a very thin film of fluid separates them.

The pleura help to isolate each lung, and the film of fluid has a lubricating function. It reduces friction produced when the lungs move against the walls of the thoracic cavity. Because two smooth surfaces adhere closely together when there is a film of moisture between them, when the rib cage expands it pulls the lung wall with it. If you take two glass slides, wet them, and place the moistened surfaces together, you will find that it is quite difficult to separate them. This same action "glues" the lungs to the walls of the rib cage.

You may have heard of someone who had a collapsed lung. Such a condition occurs when this seal is broken and air gets between the two membranes of the pleura. Inflammation of the pleura is known as **pleurisy** and may be caused by pneumonia, tuberculosis, or influenza. When the membranes become inflamed, breathing becomes difficult and painful. Coughing, fever, and rapid, shallow breathing are common symptoms.

The right lung is divided into three lobes and fills most of the right side of the thorax. The left lung has only 2 lobes and shares some of its space with the heart.

The Mechanism of Breathing

For air to enter the lungs, two basic actions must occur, both of which have the effect of increasing the volume of the thoracic cavity. The **diaphragm** is a thin, dome-shaped sheet of muscle, approximately level with the bottom of the ribs, which is stretched across the bottom of the thoracic cavity separating it from the abdominal cavity. This sheet of muscle is curved upward in the middle, like an upside-down saucer. When we breathe in, the sheet is pulled downward, which tends to flatten it out, thereby making the cavity above it larger in volume. The second action causes the rib cage to move upward and outward. This results from contraction of the internal **intercostal muscles** which lie between the ribs.

(dy-ah-fram)

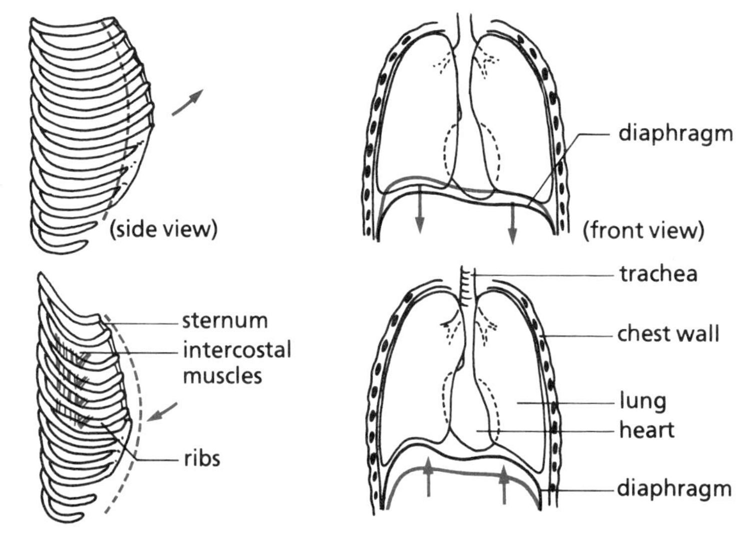

Figure 12.5.
The mechanics of breathing.

Inspiration. The ribs move upward and outward; the diaphragm moves downward and flattens. Both of these actions increase the volume of the chest cavity and decrease the pressure. As a result, air rushes in to equalize the pressure.

Expiration. The ribs and diaphragm return to their former relaxed positions. This decreases the volume of the chest cavity and increases the pressure. Air is forced out.

(See Figure 12.5.) Thus these two actions increase both the vertical dimensions of the thorax and the back-to-front (and side-to-side) measurements of the chest. Since the chest is a closed cavity, with no outside opening, this increase in its volume causes a drop in the air pressure inside it. As the pressure drops below atmospheric pressure, air from the outside rushes into the lungs to equalize this pressure. At the start of an intake of breath, air moves rapidly into the lungs. Then, as the two pressures become equalized, the air flows in more slowly. As described, the process of breathing in (inspiration) requires that muscles actively contract. Breathing out (or expiration) requires no muscle contraction; it is the result of muscle relaxation.

Relaxation of the diaphragm causes its belllike shape to be regained, and it resumes pushing upward against the base of the lungs, increasing the air pressure inside them. The internal intercostal muscles, which moved the ribs upward and outward, now relax and the ribs move downward and inward, pressing against the walls of the lungs, also increasing the internal air pressure. (See Figure 12.5.) This results in air being pushed out of the lungs until the internal and external pressures are equalized once more. The ability of the lungs to expand and contract as the pressure changes is due to the elastic, flexible nature of the lung tissues.

Lung Capacity

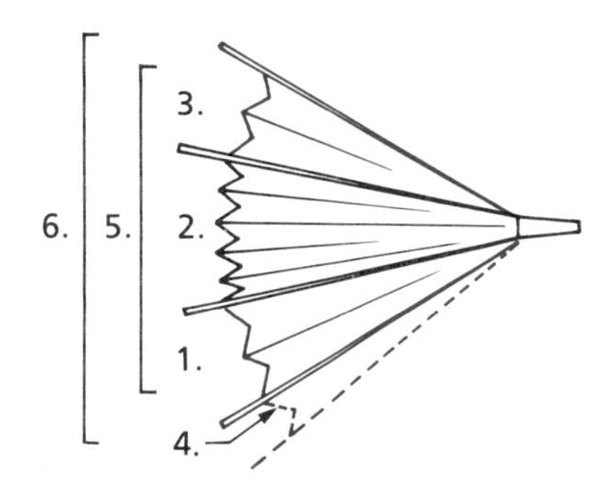

1. Inspiratory reserve volume
2. Tidal volume
3. Expiratory reserve volume
4. Residual volume
5. Vital capacity
6. Total lung capacity

The action of the lungs is rather like a bellows. When we make small movements with the handles of a bellows, only small amounts of air are drawn in and pumped out. When we are resting only a small portion of each lung is used. When we exercise vigorously, we require more air, and larger portions of the lungs are used. There is always a small amount of air left in the lungs even after a forced expiration.

The average number of breaths per minute varies between 14 and 20 for a healthy adult. The amount of air moved by a normal individual breathing while at rest is called the **tidal volume**. This is only a portion of the potential lung capacity. If you breathe in and out normally, then, at the end of a normal exhalation, forcibly push out as much extra air as you can, the air you remove is called the **expiratory reserve volume**. Similarly, the amount of extra air you can forcefully pull in at the end of a normal inhalation fills the **inspiratory reserve volume**. These three volumes together make up the **vital capacity** of the lungs. No matter how hard you try to push air out of the lungs, there will always be a small amount left in the spaces and tubes. This is called the **residual air capacity**. (See Figure 12.6.)

For an average adult person the capacities are as follows:

Expiratory Reserve Volume	1500 cm^3
Tidal Volume	500 cm^3
Inspiratory Reserve Volume	2000 cm^3
Vital Capacity	4000 cm^3

The average residual air capacity is about 1500 cm^3, making the total capacity of the lungs about 5500 cm^3. The vital capacity of each individual will vary according to an individual's size, build, sex, physical condition, and age. The vital capacity can vary among adults from 1500 cm^3 to 7000 cm^3. For adults in good physical condition, the average vital capacity for males is about 4500 cm^3 and for females about 3100 cm^3 (See Figure 12.7.)

Figure 12.6.
Lung volumes and capacities.

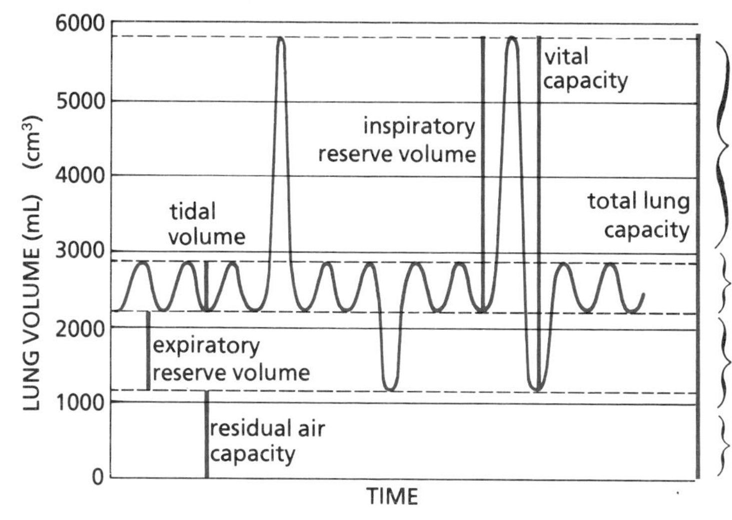

Figure 12.7.
Two types of classroom respirometers used to determine lung capacity.

The Exchange of Gases

The bronchioles end, as we have aready discussed, in air sacs called alveoli. Each tiny alveolus is surrounded by a network of capillaries. It is here that the exchange of gases takes place; oxygen and carbon dioxide pass through the walls of the alveoli to enter, or leave, the capillaries of the circulatory system. This exchange can take place because the walls of the air sacs are extremey thin and moist. The gases dissolve into a layer of moisture that lines the membranes of the alveoli.

The Exchange of Carbon Dioxide and Oxygen in the Lungs

At this point, the partnership between the respiratory system and the circulatory system comes into effect. The respiratory system has brought oxygen molecules to the exchange site in the alveoli. (See Figure 12.8.) Now oxygen will cross the membranes, enter the blood stream and be transported to the cells and tissues, which require oxygen for their activities.

You will remember that diffusion was defined as a movement of molecules from an area of high concentration to an area of low concentration. The air we breathe into the alveoli is made up of nearly 80% nitrogen, which the body cannot use and therefore it is exhaled. About 18% - 20% of the air is oxygen, which is a much higher concentration of oxygen than that found in the blood. (See Figure 12.9.) Due to these differences in concentration, the oxygen diffuses from an area of high concentration—in the alveoli-to an area of lower concentration-in the blood of the capillaries. (The blood arriving at this point will be high in carbon dioxide but have very little oxygen.)

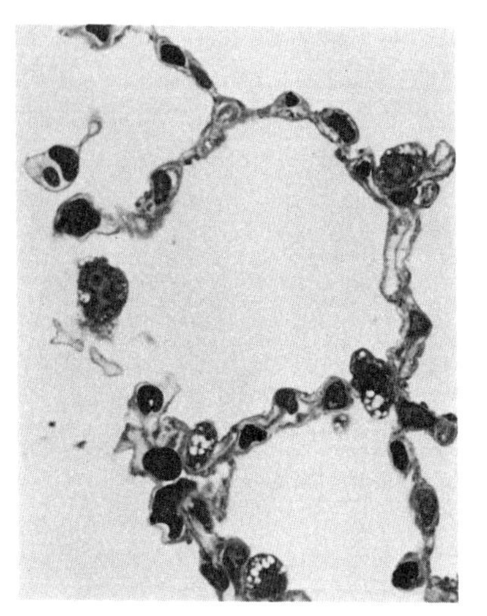

Figure 12.8. Thin-walled alveoli. The dark bodies are red blood cells. Some white cells with U-shaped nuclei can also be seen.

Figure 12.9.
The composition of inhaled and exhaled air.

Inhaled air

500 mL Atmospheric air	
oxygen	21 %
nitrogen	78 %
carbon dioxide	0.04%

Other gases in small amounts.

150 mL of air fills the spaces in the tubes of the respiratory system. The composition does not change as it is not in contact with respiratory surfaces.

350 mL of air reaches the alveoli. Here it diffuses into the moisture on the respiratory surfaces and the O_2 passes into the blood, while CO_2 is taken up from the blood.

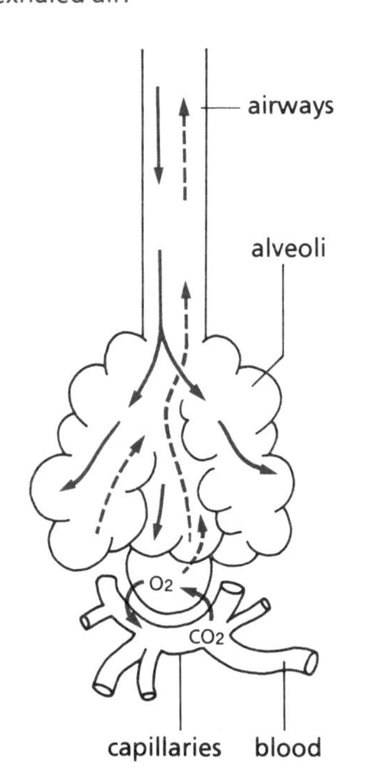

Exhaled air

Expired air	
oxygen	16%
nitrogen	78%
carbon dioxide	5%

Other gases in small amounts.

Some of the "dead air" from the tubes will be mixed with the air from the alveoli.

Alveolar air. This has been in contact with the alveoli. In the alveoli the air has a composition of;

Oxygen	14.5%
Nitrogen	80 %
Carbon dioxide	5.5%

The oxygen must be dissolved to cross the membrane from the alveoli to the capillaries. The walls of the alveoli are bathed in a film of moisture in which the oxygen gas becomes dissolved. As the dissolved oxygen molecules cross into the capillaries they pass into, and mix with the blood plasma, the watery part of the blood. The plasma could not possibly, however, transport all the oxygen that the body needs. Therefore, the oxygen in blood plasma is quickly picked up by **hemoglobin** molecules contained in red blood cells. Each hemoglobin molecule has four sites to which the oxygen molecules can be attached. Usually, when the hemoglobin molecules leave the lung area, 99% of these carrying sites are filled with oxygen.

(hee-moe-gloe-bin)

The Transport of Gases in the Blood

The blood, now rich in oxygen, is rapidly carried back to the left side of the heart and then forcefully pumped into the large arteries and arterioles. These vessels are all thick-walled; and until the blood reaches the thin-walled capillaries in the tissues of the muscles, brain, or other body parts, it cannot diffuse into the cells where it is needed. In capillaries the blood is moving slowly enough, and the vessel walls are thin enough to permit such diffusion.

The Exchange of Gases in the Cells and Tissues

The concentration of oxygen in the capillaries is high. Outside, in the tissues, oxygen is in low concentration, because it has been used up by the activities of the surrounding cells. Therefore, there is movement from a high concentration to a low one as the oxygen molecules diffuse out of the capillaries into the tissues around them. These same tissues, although they are low in oxygen, contain high concentrations of carbon dioxide, which is the waste gas of cellular metabolism. Carbon dioxide, therefore, diffuses into the blood stream where the concentration of this gas is low. Here is it picked up and attached to vacant carrier sites on hemoglobin molecules that have just emptied their oxygen into the tissues. It is then carried back to the lungs, where it diffuses into the alveoli and is finally exhaled (See Figure 12.10.)

Some carbon dioxide is also carried in the plasma of the blood as carbonic acid and by ions in the plasma that form certain salts called bicarbonates. Carbonic acid is produced by combining carbon dioxide and water ($CO_2 + H_2O \rightarrow H_2CO_3$).

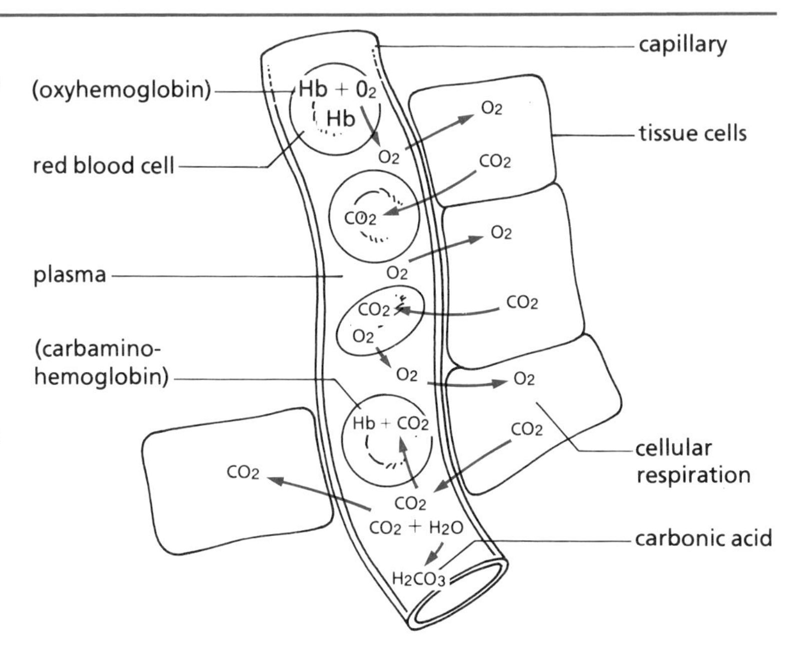

Figure 12.10.
The exchange of gases between the capillaries and surrounding cells.

The oxygen is carried to the tissues attached to the hemoglobin molecules. This produces a high oxygen concentration inside the capillaries, compared to a low pressure inside the tissue cells. The CO_2 level inside the cells of the tissues is high, while the CO_2 level is low inside the capillary. CO_2 diffuses out to take the place of O_2 on the carrier molecules of hemoglobin. CO_2 also combines with water to form carbonic acid and is carried in the blood plasma.

As described, the carrier sites on the hemoglobin molecule that are used to transport carbon dioxide, are also used for oxygen. If they are all full of carbon dioxide, oxygen cannot be picked up and transported. It is, therefore, just as important for carbon dioxide to be cleared from the body as it is for tissues to be supplied with oxygen. Both oxygen and carbon dioxide become only loosely attached to hemoglobin and are easily freed when the pull of diffusion concentrations is exerted.

Carbon monoxide, however, which is a colourless, odourless, and poisonous gas, clings very tightly to hemoglobin molecules and will not let go when oxygen requires these sites. Carbon monoxide molecules are also picked up at a faster rate than oxygen would normally be absorbed.

Control of the Amount of Oxygen Delivered

During a basketball game or a race, the body demands a great increase in oxygen supply. This demand by the body for more oxygen can be met in two ways:

1. Pumping the blood faster and supplying more blood-carrying oxygen.
2. Changing the concentration of oxygen that the blood is carrying.

You will have noticed that when you exercise the **rate** at which you breathe (the number of breaths per minute) and the **depth** of breathing (how large a breath you take) changes. This helps to exchange the air in the lungs more efficiently, bringing more oxygen in contact with the exchange sites in the alveoli. At the same time, the heart rate increases and speeds the extra supplies along to the tissues requiring oxygen.

Nervous Control of Breathing

Whenever we increase our activity rate we breathe faster. What causes these changes? We don't consciously think "I'm going to run. I must now start breathing faster". The reaction is automatic.

As we engage in some strenuous activity, the rate of cell metabolism increases and the waste products of metabolism, including carbon dioxide, begin to accumulate. Thus the cells, and consequently the blood, show an increased carbon dioxide content. Although a low concentration of oxygen and a high concentration of carbon dioxide both act to stimulate breathing, the carbon dioxide level has the more important effect. Carbon dioxide levels in the blood passing through the brain are monitored by the respiratory control centre in the central nervous system. This respiratory centre is located in the medulla oblongata and the pons. It normally operates with a regular rhythm, but is greatly influenced by information passed to it from various parts of the body. A constant stream of impulses regarding the oxygen needs of the body and the carbon dioxide concentrations present in the tissues are processed by the CNS. The respiratory centre responds to these signals by increasing or decreasing the breathing rate to meet the body's demands.

When high carbon dioxide levels are recognized, the cells in the medulla send out nerve impulses to the diaphragm and intercostal muscles to speed up their action. This causes a more rapid exchange of carbon dioxide and oxygen and gradually returns the concentration levels of each to normal.

The respiratory rate varies considerably from one person to another. The average resting respiratory rate for adults is between 14 and 20 breaths per minute. Newborn babies may take 40 breaths per minute, slowing to about 30 breaths per minute at around one year.

CASE STUDY. David, 17 years of age.

Exams were over and the party was planned at a cottage on Sandy Lake. Everyone was in high spirits and excited by the prospect of summer fun ahead.

The water was warm and inviting, and everyone was anxious to swim, water ski, and sail. All the friends could swim, although not all were excellent swimmers. As the afternoon progressed, the wind dropped and the lake became flat and calm. Two of the girls had been sailing and their boat was drifting, not far from shore. David and Al decided to swim out to talk to them.

Distances over water can be very hard to judge, and David made a poor assessment of how far the boat was from shore. When they were about halfway to the sailboat, David realized that he was in trouble. He looked behind and the shore seemed a long way off, so he struggled on. His breath was coming in short gasps and his tired arms were flailing without producing much progress. Panic rose in his throat and he called to Al, who was by now well ahead and failed to hear his cries. The girls were watching the sails and shouting to Jim.

David started to scream and stopped swimming. He sank below the lake surface, gulping in mouthfuls of water as he tried to shout. He came sputtering to the surface, gasping for air, then started to sink before he could scream again. David didn't recall much more until he found himself in the boat with the worried faces of his friends bending over him.

Al was a fast swimmer and had taken lifesaving instruction from the Red Cross. He had searched and found David beneath the water and brought him to the surface. He then gave David mouth-to-mouth resuscitation while holding his unconscious body until help arrived. The two girls had paddled the sailboat over to the pair in the water as quickly as they could and then hauled them into the boat.

All three friends took turns giving David artificial respiration until he started to breathe normally again and regained consciousness. David was in shock as they paddled the boat to shore. On shore the parents who owned the cottage took over, first treating David for shock and then taking him into hospital for a medical checkup.

There is no doubt that his friend's knowledge of lifesaving and his swimming ability saved David's life.

Rescue breathing. Mouth-to-mouth artificial respiration.

1 **Use rescue breathing** when persons have stopped breathing as a result of:

- drowning
- choking
- suffocation
- excessive drugs
- electric shock
- heart attack
- gas poisoning
- smoke inhalation

Quickly remove the victim from the cause or remove the cause from the victim.

2 **Start immediately**. The sooner you start, the greater the chance of success. Apply rescue breathing anywhere:

- on a dock
- on the beach
- in a boat
- from a boat
- standing in water
- kneeling in water
- on the ground
- in a car
- on a hydro pole
- in a chair
- on a bed
- on the street

- Send someone for medical aid.

3 Open airway by lifting neck with one hand and tilting the head back with the other hand.

4 Pinch nostrils to prevent air leakage. Maintain open airway by keeping the neck elevated.

5 Seal your mouth tightly around the victim's mouth and blow in. The victim's chest should rise.

6 Remove mouth. Release nostrils. Listen for air escaping from lungs. Watch for chest to fall.

Repeat last three steps twelve to fifteen times per minute

Continue until medical help arrives or breathing is restored.

Questions for research and discussion

1. Consider the account and list the positive actions that were taken to help the victim. Some steps in helping a drowning victim were not mentioned. Include these in your list.
2. What did the parents probably do to minimize the effects of shock after David reached the shore?
3. What effect does cold water have in such cases (refer to index under *hypothermia*).

NOTE:
Time is all-important in such cases, as the following statistics demonstrate:

Time breathing stopped, in minutes	Chances of recovery
1	98 out of 100
5	25 out of 100
10	1 out of 100
11	1 out of 1 000
12	1 out of 100 000

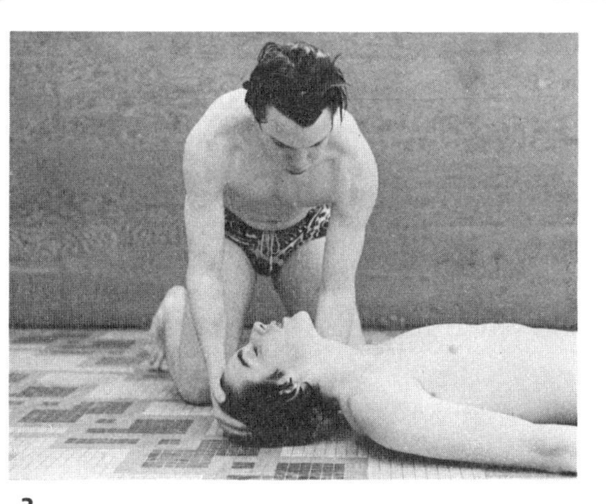

3

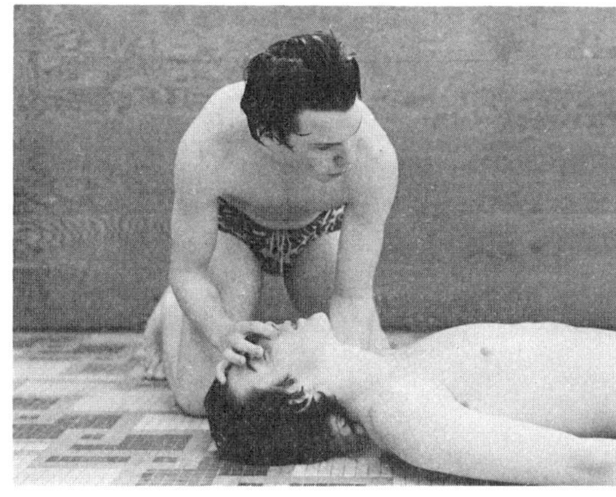

4

5

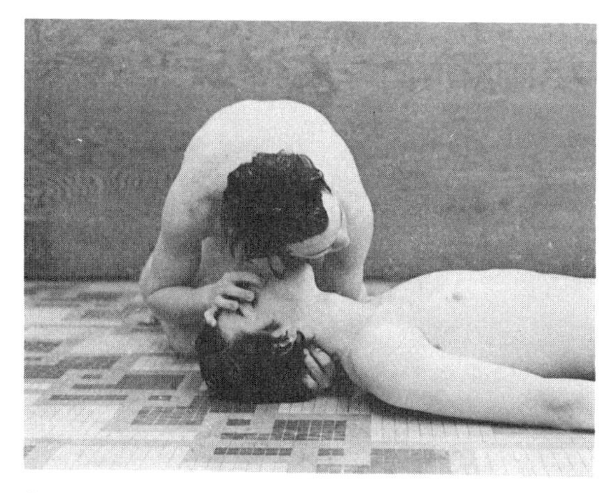

6

Smoking

Tobacco is big business. Anyone who smokes a pack of cigarettes a day will contribute about $10,000 during their lifetime to the treasury of the tobacco industry which has a total income exceeding $8 billion a year. Tobacco companies spend more than $100 million annually on promotional campaigns. Terry Fox raised more than $22 million dollars to fight cancer, a fraction of the advertising budget spent to promote a proven cancer-inducing habit!

Smoking also affects non-smokers because a smoke-filled environment has a measurable effect on everyone in the room. Non-smokers must breathe in smoke-laden air and this affects their heart rate and blood pressure, as well as the amount of carbon monoxide and carbon dioxide in their blood. Smoke also causes eye irritation, headaches, sore throats, coughs, and allergic reactions in some people.

The smoke from a cigarette left in an ashtray will give off approximately twice the tar and nicotine that a smoker puffs into his or her lungs while smoking a whole cigarette. The children of smokers are ill more often than the children of non-smokers, particularly with diseases that affect the respiratory organs.

The Effects of Tobacco Smoke

More than 1000 known chemical compounds have been identified in tobacco, including more than 200 toxins. Some of these compounds are found in the smoke, some remain in the ashes, still others are formed during combustion. It is the composition of the smoke that enters the lungs that is the chief cause of concern.

The major compounds found in cigarette smoke are tars, nicotine, phenols, carbon monoxide, hydrocarbons, arsenic, and more than 15 other agents known to cause cancer (carcinogens). The effect of these compounds will vary according to whether the smoke is inhaled or not. Usually pipe and cigar smoke is too strong and acrid to inhale, but cigarette smoke is quite commonly inhaled.

As smoke passes through the bronchi, many chemical substances are deposited on the bronchial walls and it is in these locations that cancer frequently develops. As we have already seen, the walls of the respiratory passageways are lined with cilia. Several substances contained in smoke irritate the ciliated cells and nicotine can actually paralyze cilia, preventing them from cleaning air as it enters the lungs.

Lung Disease and Smoking

When cells are exposed to constant irritation over a long period of time, carcinogens produce changes in the cells which cause them to multiply at an unusually rapid rate. These cells are abnormal and obstruct the work of the normal cells around them. (See Figure 12.11.)

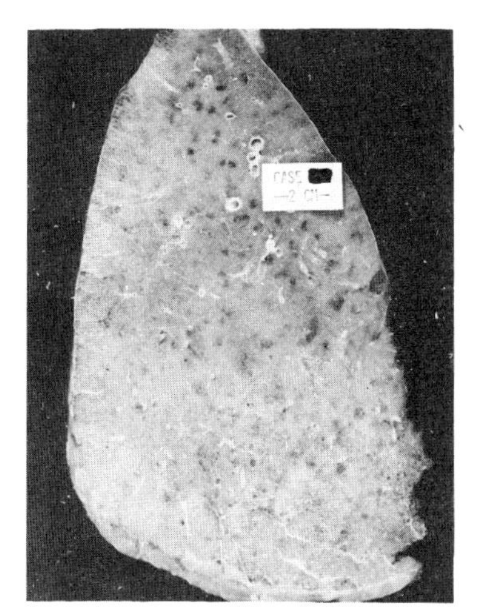

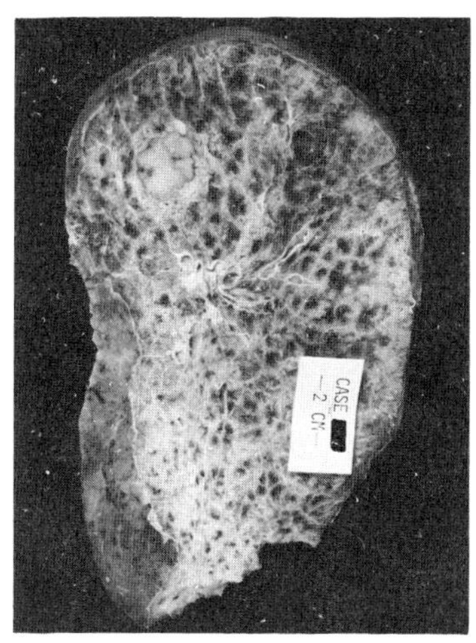

Figure 12.11.
A comparison between a normal lung (left) and a cancerous lung (right). The tumour is seen as a smooth grey body. The dark patches are greatly enlarged air spaces caused by emphysema.

An increased incidence of cancer deaths among smokers has been demonstrated repeatedly by many studies. Each packet of cigarettes must, by law, carry a warning that smoking may be injurious to health.

Cancer is not the only disease caused, or aggravated by smoking. Bronchitis, emphysema, ulcers, cirrhosis of the liver, and heart disease are all examples of disorders that have been attributed either directly or indirectly to smoking.

Smoking During Pregnancy

Women who smoke while they are pregnant risk affecting both their own life and that of the unborn baby. More stillbirths, premature babies, and spontaneous abortions occur when pregnant women smoke than when they don't. In addition the babies of smoking mothers are, on average, of lower weight at birth than those born to women who do not smoke. Such babies also have a higher incidence of neurological damage.

QUESTIONS FOR REVIEW

SOME WORDS TO KNOW

Match the statement in the left-hand column with one of the words in the right-hand column. DO NOT WRITE IN THIS BOOK.

1. Tiny sacs where gases are exchanged.
2. Muscles that raise the ribs.
3. Tubes that branch off from the trachea into the lungs.
4. A membrane which surrounds each of the lungs.
5. The amount of air filling the lungs (not including the residual air).
6. The part of the brain which controls the rate of respiration.
7. Tubes which connect the nasal and oral cavitites with the trachea.
8. A sheet of muscle which separates the thoracic and abdominal cavities.
9. The vocal cords.
10. The gas which has the major influence on controlling the rate of respiration.

A. diaphragm
B. lungs
C. trachea
D. bronchioles
E. bronchi
F. alveoli
G. pharynx
H. intercostals
I. sinuses
J. pleura
K. vital capacity
L. reserve volume
M. oxygen
N. carbon dioxide
O. medulla
P. cerebrum

SOME FACTS TO KNOW

1. List the structures, in the correct sequence, through which a molecule of air would pass from the point of entry into the respiratory system until it reaches the alveoli.
2. In point form, explain the mechanical process of breathing in air.
3. Explain the terms: inspiratory reserve volume, expiratory reserve volume, tidal volume, vital capacity.
4. Sketch two or three alveoli and their associated blood vessels.
5. What is the functional value of the pleura?
6. What processes take place in the nasal chamber?
7. Briefly describe the structure and function of the larynx.
8. In what ways does smoking affect the respiratory and circulatory systems of the body?
9. Briefly outline what you think are the rights of smokers and non-smokers.

QUESTIONS FOR RESEARCH

1. Select one of the following topics and prepare a report on it:
   - The Canadian Lung Association or a similar organization
   - emphysema
   - rights of a non-smoker
   - chronic bronchitis
   - hay fever
   - air pollution and respiratory diseases
   - pneumonia
   - asthma
   - lung cancer (lip, tongue, or larynx)
   - the heart-lung machine
   - smoking and pregnancy
2. There are a number of reflexive responses associated with the respiratory system and related organs. What happens when we yawn, sigh, cough, sneeze, hiccough?
3. Smoke reacts in the lungs with asbestos. The risk of cancer in people who smoke where asbestos is present is very much higher than normal. Make a list of common carcinogenic materials in the environment and find out how they affect the human body.
4. Find out what is considered a good recovery rate for the pulse after a standardized activity such as the Step Test. Make a survey of your class to determine who is "fit" when measured by this testing process. Comment on how accurate this system is when comparing results produced by the athletes or the more sedentary students in the class.

Activity 1: HOW DOES THE AIR ENTERING THE LUNGS DIFFER FROM THE AIR LEAVING THE LUNGS?

Materials

thermometer, glass plate, test tube, limewater, and a straw. Alcohol for sterilizing the thermometers.

Method

1. Using the thermometer, first find the air temperature of the room. *Record this result.*
2. Place the thermometer between your teeth or lips about 2-3 cm into your mouth. Do not allow the thermometer to touch the tongue. Breathe *out* rapidly so that the air passes over the bulb of the thermometer and determine the temperature of the exhaled air. *Record this result, then, by subtraction, find out the increase in temperature caused by the air being warmed in the lungs.*
   Note: You must read the thermometer immediately after it is removed from the mouth. The thermometer should be cleaned with alcohol before being placed in the mouth.
3. Breathe out onto a glass plate. *Note what collects on the surface of the glass and explain what you see.*
4. Take a test tube of limewater. *Gently* blow through the straw into the limewater. If you blow too hard it will splash out of

the test tube. The contents of the tube will turn "milky" as a white precipitate of calcium carbonate forms. *What gas is tested with this experiment?*

5. *Summarize your results.*

Questions

1. *What three changes between the composition of exhaled air and inhaled air have these tests demonstrated?*
2. *What structures in the respiratory system are responsible for warming and adding moisture to the air entering the lungs?*
3. *Where does the exchange of carbon dioxide and oxygen take place?*

Actvity 2: THE RATE OF RESPIRATION

Method

1. Count the number of breaths that your partner takes per minute while sitting quietly. Try to make the count while your partner is not aware of your observation. You might try to watch the chest movements while he or she is reading or writing. *Why is this necessary?*
2. Have your partner exercise vigorously for at least two minutes, then count the number of breaths taken immediately after exercise.
3. *Record your results and compare the two rates. Make a list of the respiratory results of the class. Put a check (√) mark beside the results of students who play on teams or exercise regularly. Place an asterisk (*) beside the name of students who smoke regularly. What do the results suggest concerning respiration rates and exercise and/or smoking?*

Activity 3: MEASURING THE CAPACITY OF THE LUNGS

Materials

spirometer

Method

A. TIDAL VOLUME

Tidal volume is the amount of air inhaled or exhaled during normal, quiet breathing.

Sit by the spirometer and breathe quite normally for about a minute. After taking a normal breath, place the mouthpiece between your lips and exhale into the tube. Try to be natural and not force the air out in any way. *Repeat this three times and average the results. Record the number of litres of air in your notebook. Tidal volume = TV.*

B. EXPIRATORY RESERVE VOLUME

Expiratory reserve volume is the amount of air that can be forcibly exhaled after a normal exhalation. Stand up, breathe normally for a minute or two. After a normal exhalation, place the tube in your mouth and forcibly remove all the extra air that you can. *Repeat this three times and average your results. Record this as ERV, or expiratory reserve volume.*

C. INSPIRATORY RESERVE VOLUME

Inspiratory reserve volume is the amount of air that can be forcefully inhaled after you have already taken in your normal breath of air.

Stand up and breathe normally for a minute or so. Now breathe in as deeply as you can. Breathe out into the mouthpiece slowly as you would normally, but do not forcibly remove any of the extra air. *You can find the IRV by subtracting your TV from this result. Record your IRV.*

D. VITAL CAPACITY

Vital capacity is the amount of air that can be forcibly removed from the lungs after the largest possible inspired breath.

Stand up and breathe normally as before. Now breathe in as deeply as possible. Place the tube to your mouth and exhale as hard and long as you can to try to empty all the air from your lungs. *Repeat three times and average your results. Record your VC.*

A cross-check can be made by adding:

$$IRV + ERV + TV = VC.$$

Analyzing your results.

You should now have a table showing the different volumes. Compare your scores with those of other students. If you have completed any of the other experiments such as respiration rates, thoracic index, etc., you might try to make a large chart on the chalkboard and see if there is any correlation in the results. Try to compare students by sex and size. (Mass, posture, and the types of activity in which the student is involved also affect the results.)

Check to see if there is any correlation in the results you obtain between smoking and vital capacity, or between exercise programs and vital capacity.

If some students have very low scores, enquire if they have any temporary respiratory infections or suffer from such problems as asthma.

Activity 4: WHAT EFFECT DOES SMOKING HAVE ON PULSE RATE, RESPIRATION RATE, AND BLOOD PRESSURE?

Materials

sphygmomanometer, thermometer, cigarettes

Method

1. Select a person who normally smokes, but has not had a cigarette between waking and the start of the experiment. The person should also avoid other stimulants such as tea or coffee until after the experiment.
   Seat the subject comfortably and take the resting pulse rate, respiration rate, blood pressure, and skin temperature of the hands. Holding the bulb of an ordinary classroom thermometer between the thumb and forefingers or rolled into the palm of the hand will give a reasonable reading.
2. *Record these results.*
3. The subject now lights a cigarette, still sitting and smoking at the normal rate.
4. Immediately the cigarette is finished, take the pulse rate, respiration rate, blood pressure, and hand temperature. Record these results. Allow five minutes to elapse and repeat the readings. If time allows, continue to take the readings at five-minute intervals. *Record all results.*

Questions

1. *What effect does smoking have on the factors tested?*
2. *Account for the change of temperature in the hands.*
3. *What part of the inhaled smoke caused the changes in the heart and respiratory rates?*

Activity 5: EXAMINE THE EFFECTIVENESS OF A FILTER IN A CIGARETTE

Materials

smoking apparatus and cigarettes

Method

1. Take a filter-tip cigarette and using the smoking machine, determine the amount of tars present in 3 cm of the cigarette. Mark the filter paper from the machine and set this evidence aside.
   Carefully cut off the filter tip from a cigarette of the same brand. Mark the 4 cm point on the cigarette and repeat the experiment. Mark the filter paper from the smoking

machine "No filter". *Compare these two stains. Does the filter tip effectively reduce the amount of tars taken into the lungs?*

2. Take the filter tip that you removed from the cigarette in the last experiment and tape it to the end of another cigarette of the same brand. This cigarette will have the same amount of tobacco in it, but twice the amount of filtering material. *Mark the results of this test "Filter × 2". Is all the tar removed or is an even longer filter required?* Try to determine how long the filter would have to be in order to have a tar-free cigarette. Think about the experiment as you work and decide if you are using adequate controls. Decide if the results are honest comparisons. (For example, remember to mark them so that equal amounts of tobacco are smoked in each test.) *Record all your answers and then write a statement about the effectiveness of the filters used in cigarettes.*

Activity 6: DEMONSTRATE THE SUBSTANCES CONTAINED IN CIGARETTE SMOKE

Materials

paper tissue or white handerkchief, cigarettes

Method

1. Have a smoker draw in a mouthful of smoke from a cigarette but do not allow him or her to inhale the smoke. Have the smoker breathe this smoke out through a paper tissue held over the mouth. Be sure that none of the smoke escapes and that all the smoke passes through the tissue.
2. Now ask the smoker to take a second draw on the cigarette and inhale the smoke. Have the student hold the smoke in the lungs for a moment or two and then exhale through another part of the paper tissue. *Compare the two stains.*
3. The difference in the two stains indicates the amount of tars and particulate matter retained in the lungs after one inhalation of smoke.

Activity 7: COMPARISON OF DIFFERENT BRANDS OF CIGARETTE AND THE EFFICIENCY OF THE FILTERS

Materials

smoking machine (Eduquip) or lab-built equipment as illustrated (Figure 12-14), filter paper, different brands of cigarettes.

Method

1. Place a cigarette in the holder, light the cigarette, and pump the bulb evenly and slowly until the cigarette is completely smoked. (As cigarettes vary in length and in the length of the filters that they contain, it is better to mark off a set distance on each type of cigarette before you start: about 4 cm, for instance.)
2. Remove the filter paper from the holder and mark it with the brand of cigarette used. Place another brand of cigarette and another filter paper in the smoking machine and repeat the experiment.
3. *After you have tested several brands, compare the results and rate the brands according to the amount of tars trapped by the filter papers.*

Activity 8: WHICH HALF OF THE CIGARETTE HAS THE MOST TARS: THE FIRST OR THE SECOND HALF OF THE CIGARETTE?

Materials

smoking machine and cigarettes

Method

1. Set up the smoking machine as before. Mark off a distance of 4 cm from the end of the cigarette opposite to the filter-tip end. Then divide this distance into two and place another mark so that there are two equal divisions.
2. Fit the cigarette into the holder and smoke it down to the first 2-cm mark. Snip off the cigarette as soon as it reaches this point. Remove the filter paper and mark it "First half". Place another filter paper in the machine, relight the cigarette, and smoke the second half of the cigarette.
3. Remove the second filter paper. Mark it "Second half" and compare the two stains. *Which half of the cigarette contains the most tars? What should smokers do about this item of information?*

Unit VII

How the Body Obtains Energy and Material for Growth

Chapter 13

The Digestive System

- Process of Digestion
- Absorption of Nutrients

The Process of Digestion

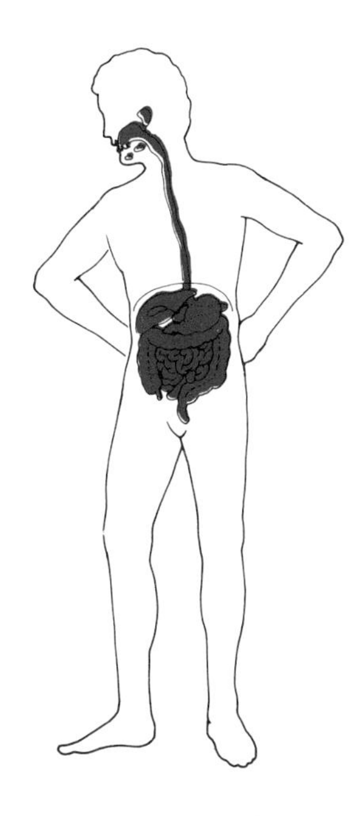

The process of digestion involves the breaking down of our food into molecule-sized particles that will dissolve. Digestive activity may be divided into three parts. The first stage is **physical** digestion, the mechanical process of breaking down food into smaller pieces. This starts on our plates when we use a knife and fork. It continues when we use our teeth to bite, grind, and separate the food fibres. The second stage involves **chemical** digestion; by the action of enzymes, these small food pieces are broken down into even smaller molecular-sized particles. The third stage involves **absorption** of these molecules by the body.

The digestive system functions both to break down food and to absorb it. This third stage is of obvious importance. Food, broken down in the digestive tract, serves no purpose if the nutrients contained in it do not eventually reach the cells. These nutrients must pass out of the digestive tube, pass through the walls of the tract, and pass into the thousands of capillaries which line the small intestine. In order to do so, the molecules that are the end products of physical and chemical digestion must be small enough to go into solution and pass through the tiny pores in the cell membranes. The blood that flows thorough these capillaries will then transport the nutrients to all the cells of the body, providing energy for cell processes and materials for the growth and repair of tissues.

The pores in cell membranes are extremely small, even submicroscopic. They are only large enough to accept the very smallest molecules. Therefore the end products of digestion must also be very small. These molecules must pass through several membrane barriers before arriving in the cells where they will be used. Reducing the food we eat to molecules small enough for absorption is the function of the digestive system. (See Figure 13.1.)

Digestion in the Mouth

Several different digestive processes begin in the mouth. The most obvious of these involves the action of the teeth.

The Teeth

If you look at your teeth in a mirror, you will see that they vary in shape and size. At the front are the **incisors**, four on

Figure 13.1.
The major components of food and the final products of digestion after being broken down into smaller sub-units by the action of enzymes.

COMPONENTS OF FOOD	INTERMEDIATE PRODUCTS OF DIGESTION	FINAL PRODUCTS OF DIGESTION
Carbohydrates	Disaccharides: sucrose, maltose	Monosaccharides: glucose, fructose
Fats	Emulsification into smaller droplets	Glycerol and fatty acids
Proteins	Short chains of amino acids	Individual amino acids
Minerals and Vitamins	Require no digestion, absorbed directly	

the bottom and four on the top. These are chisel-shaped teeth, which are excellent for biting or cutting food. On either side of the incisors are the **canines** (also called cuspids). These are more pointed in shape and are useful for tearing or shredding food. In humans, the teeth located posterior to the canines are almost square in shape. They are known as **premolars** and **molars**. These teeth, which are flattened on the upper surface, are used for grinding and chewing food, especially tough, fibrous foods, such as meat.

With rare exceptions, each of us receives two sets of teeth during our lifetime. The first set, the **deciduous** or milk teeth, starts to appear at about six months of age. They then erupt at the rate of roughly one or two each month, until a full first set of twenty teeth is present.

A permanent tooth appears to replace each deciduous tooth as it falls out between six and thirteen years of age. This permanent set includes three additional teeth on each side of the jaw, which come in only once. A normal adult thus has 32 teeth: 8 incisors, 4 canines, 8 premolars, and 12 molars. (See Figure 13.2.)

Structure of Teeth

The most obvious part of the tooth is the **crown**, which is the part visible above the gum. It is covered with **enamel**, the hardest substance in the body. Unfortunately, the enamel cannot be replaced if it gets chipped or worn away. It protects the tooth by providing it with a hard biting surface. If it does get chipped or worn away by bacterial acids the **dentine** is exposed. Dentine is the hard, bonelike material that forms

(den-teen)

Biting Tearing Grinding surfaces

(Wisdom tooth)

2 incisors 1 canine 2 premolars 3 molars

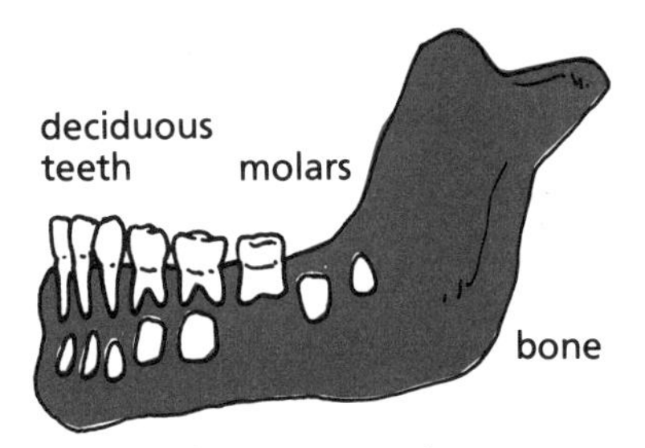

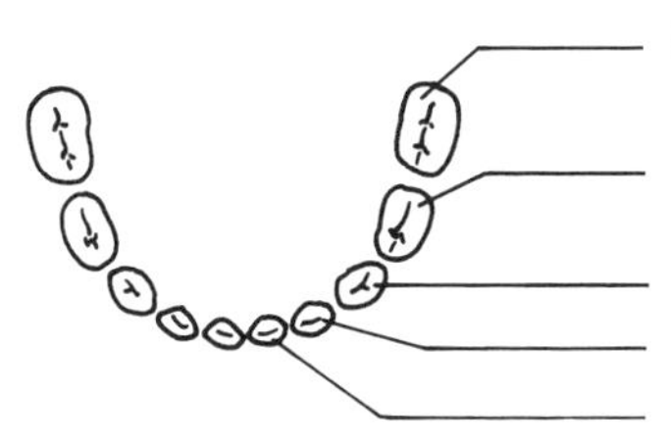

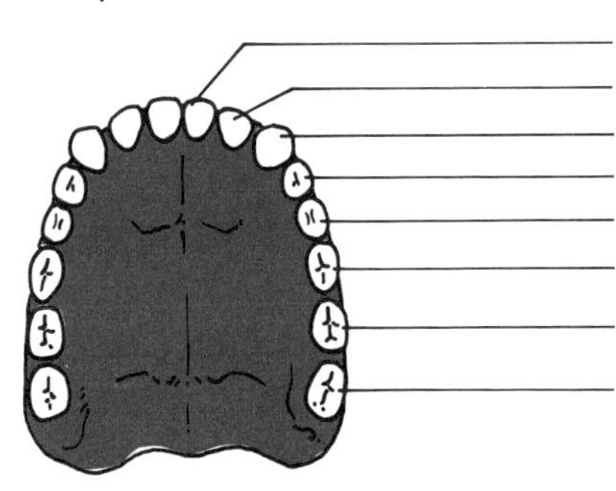

Figure 13.2.
The teeth and the average time of eruption through the gums.

the greater part of the tooth. It has, however, poor resistance to abrasion or damage, food or bacterial acids. Sometimes a tooth does decay; if the inner **pulp** is reached the nerves and blood vessels of the tooth become exposed. These nerves react strongly to chemicals or direct contact and make us painfully aware that something is wrong and that the tooth needs the attention of a dentist.

The dentine is penetrated by minute tubes which radiate out from the pulpy interior of the tooth. Dentine is sensitive to touch, temperature, acids, and sugars. Fine fibres in these small tubes transmit information about the type of sensation received to the nerves in the pulp cavity.

The **root** of the tooth is held in place by fibrous tissue that forms a firm connection between the tooth and the jawbone. The root is covered with a thin layer of **cementum**, a bonelike material which helps hold the tooth in place. The

Figure 13.3.
Longitudinal section through a molar tooth that illustrates the internal structure.

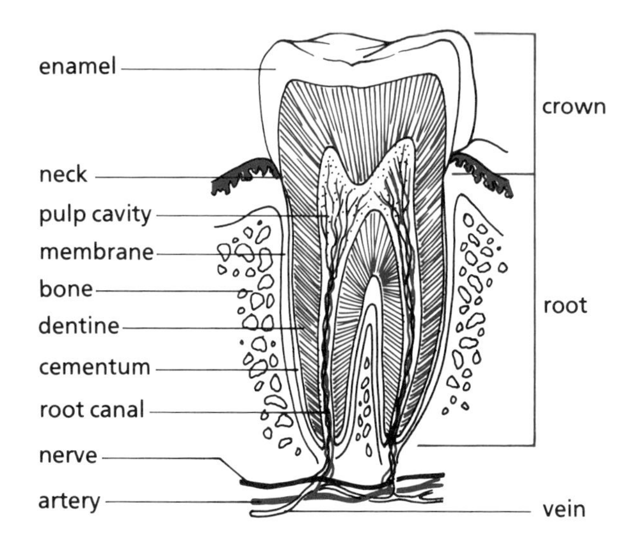

cementum extends up past the gumline, thereby also helping to protect the dentine. (See Figure 13.3.)

The Tongue

The tongue is a very necessary part of the chewing process, for it helps to position and reposition the food on the molars for chewing. It also aids in mixing the food and saliva. When the food is soft, the tongue rolls it into a ball (bolus) and moves it to the very back of the tongue, preparing it to pass into the pharynx at the start of the swallowing process. (See Figure 13.4.)

Figure 13.4.
The tongue has a rough, irregular surface produced by three different types of raised papillae. Taste buds, small specialized nerve cells which detect dissolved chemicals, are found in the furrows around the papillae.

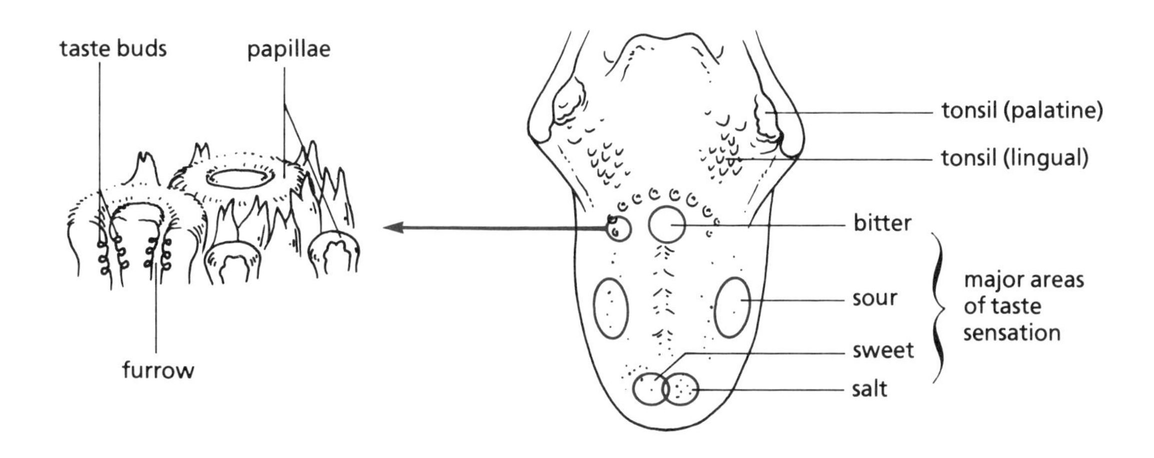

The tongue is an enormously versatile organ. It is attached not at the back of the mouth as you might imagine, if you stick out your tongue, but to the floor of the mouth. It can be twisted and turned, rolled or extended.

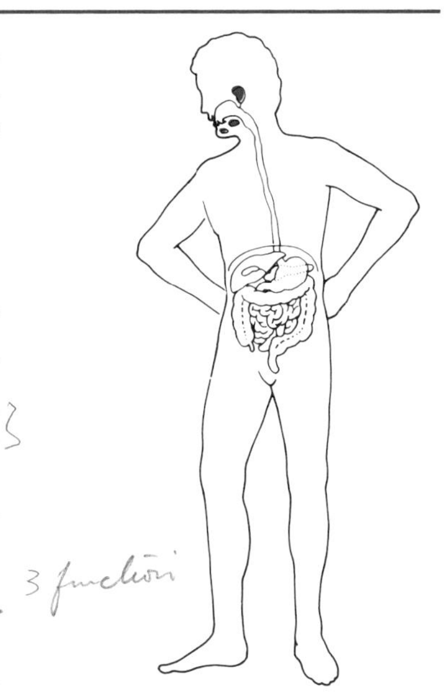

The Salivary Glands

Sometimes, when watching a commercial for particularly delicious-looking food on television or smelling a roast of meat cooking in the oven, our mouths fill with "water", or **saliva** as it is properly called. Description of food, a whiff of a particular smell, or a visual image can all stimulate the production of saliva. However, it flows, to some extent, all the time. Saliva is very slightly acid; of the approximately 1000 mL produced every day, 99% is water.

Saliva performs several useful functions. Firstly, it moistens dry food. When eating a cracker, for instance, it can help to bind loose crumbs together into a smooth ball, so that they will not stray into the respiratory system and cause you to cough. The saliva softens food as it is swallowed so that the rough edges of a potato chip will not scratch the walls of the esophagus. Saliva also contains an enzyme that helps to break down starch, which is made up of chains of small sugar molecules. The saliva's enzyme **amylase** starts the chemical digestion of starchy carbohydrates in the mouth.

Saliva is produced by three pairs of glands which lie outside the oral cavity and empty their secretions into the mouth by means of short ducts. (See Figure 13.5.) One pair of saliva glands is found at each side of the mouth cavity, slightly in front of and below the ear. These are the **parotid** glands. The other glands, the **sublingual** and **submaxillary** glands, are found in the floor of the mouth. Glands such as these, which empty their secretions into a chamber rather than into the blood stream, are called **exocrine** glands.

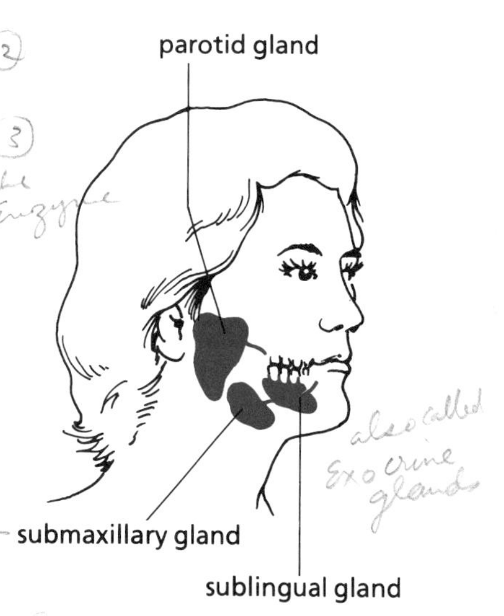

Figure 13.5.
The location of the salivary glands. The glands are found in pairs, one on each side of the mouth cavity.

Other Structures in the Mouth

Feel the roof of your mouth with your finger or the tip of your tongue: the front part is hard (the hard palate), but farther back is a soft part. By using a mirror to examine your mouth or by looking into the mouth of a friend, you can see that this soft palate is, at one point, formed into a piece of tissue that hangs down. This is the **uvula**. (See Figure 13.6.) Look carefully at Figure 13.6 and you will see that the hard and soft palates separate two spaces, the nasal chamber and

(oo-view-la)

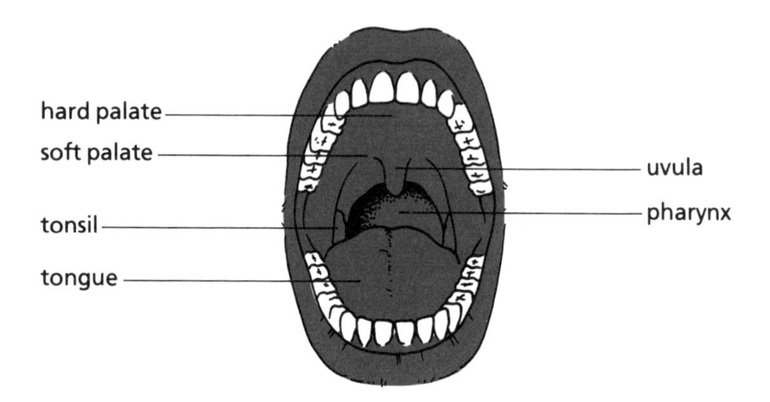

Figure 13.6.
The mouth showing the location of the teeth, uvula, and tonsils.

the mouth cavity. The palate is thus rather like a ceiling that separates the downstairs room from one above. You can see in the diagram that the tube leading down from the nasal chamber connects with the tube leading out of the back of the mouth. Most of us have had the unpleasant experience of having liquid go up this tube into the nose, when someone made us laugh unexpectedly while we were drinking. The tubes from the mouth and the nose come together at the **pharynx**. Below this junction, leading down from the pharynx are two more tubes. The trachea, which leads air into the lungs, and the **esophagus**, which carries the food into the digestive tract.

(fair-inks)

Swallowing

To ensure that food and drink pass directly into the esophagus and not into the lungs, where they would cause us to choke or cough, special structures are present. When a bolus of swallowed food is moved to the back of the tongue, the soft palate moves downward to partially seal off the nasal passage. At the same time the **epiglottis** (See Figure 13.7)

(ep-i-glot-is)

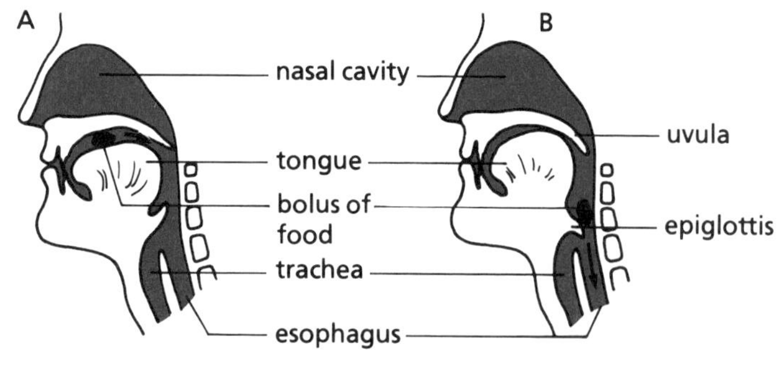

Figure 13.7.
The swallowing process. At A the uvula has moved to close the opening into the nasal chamber and to open the entrance to the pharynx. At B the epiglottis has moved to protect the entrance to the trachea and to direct food into the esophagus.

closes the opening into the respiratory system (the trachea) thereby preventing the food from going into the lungs. Muscular contractions of the pharynx help to pass the food down into the esophagus. The movement of food from the tongue into the pharynx is under voluntary control. The second stage, involving the epiglottis and the movement of the food into the esophagus, is involuntary.

The Esophagus (ee-sof-a-gus)

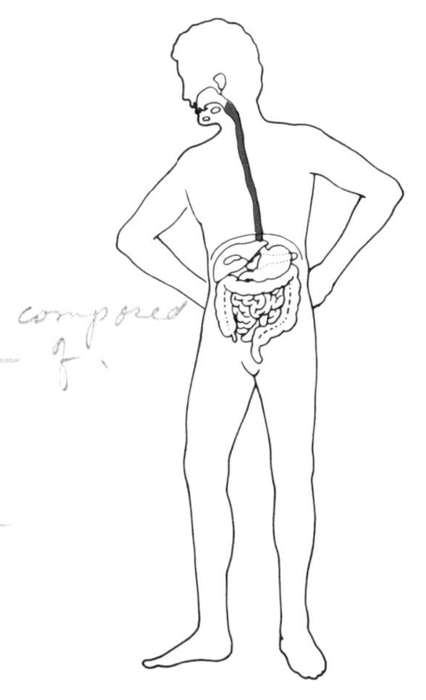

The esophagus is a flexible tube, about 25 cm in length, which leads from the pharynx to the stomach. It passes through the neck, the thoracic cavity, and the diaphragm to reach the stomach. Like most of the digestive tract, it is composed of four layers. On the inside of the tube is a thick lining covered with a film of slippery mucus, which helps the food to pass easily. The second layer contains many glands, which produce this mucus, as well as blood vessels and nerves. External to this, there are two layers of muscle. One layer is arranged in a circular fashion and passes around the tube. The other is longitudinal, running the length of the tube. Finally, there is a thin sheet of connective tissue around the outside, which helps to anchor the esophagus to the surrounding tissues.

A bolus of food is moved down the esophagus by **peristaltic** action. This is achieved by rhythmic contractions of the circular and longitudinal muscles. (See Figure 13.8.) The esophagus eventually passes through the diaphragm and enters the abdominal cavity where the major organs of the digestive systems are found (See Figure 13.9.)

(pair-i-stal-tik)

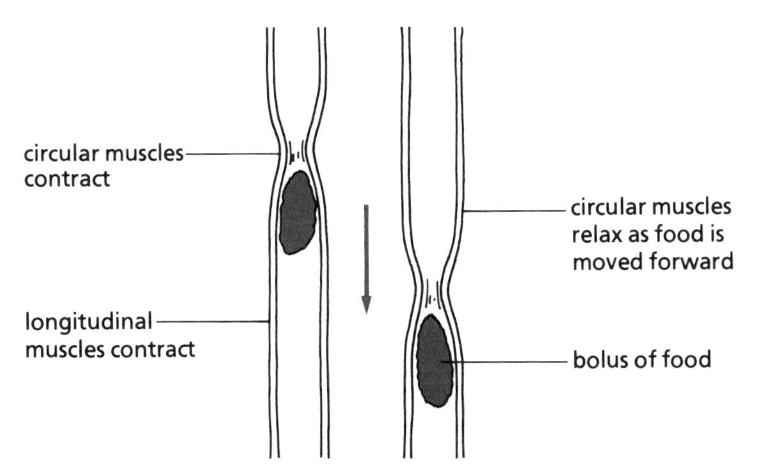

Figure 13.8.
Peristalsis moves the bolus of food along the digestive tube. The circular muscles contract behind the bolus of food and the longitudinal muscles contract in front of the food.

Figure 13.9.
The digestive organs in the normal position and in an extended view.

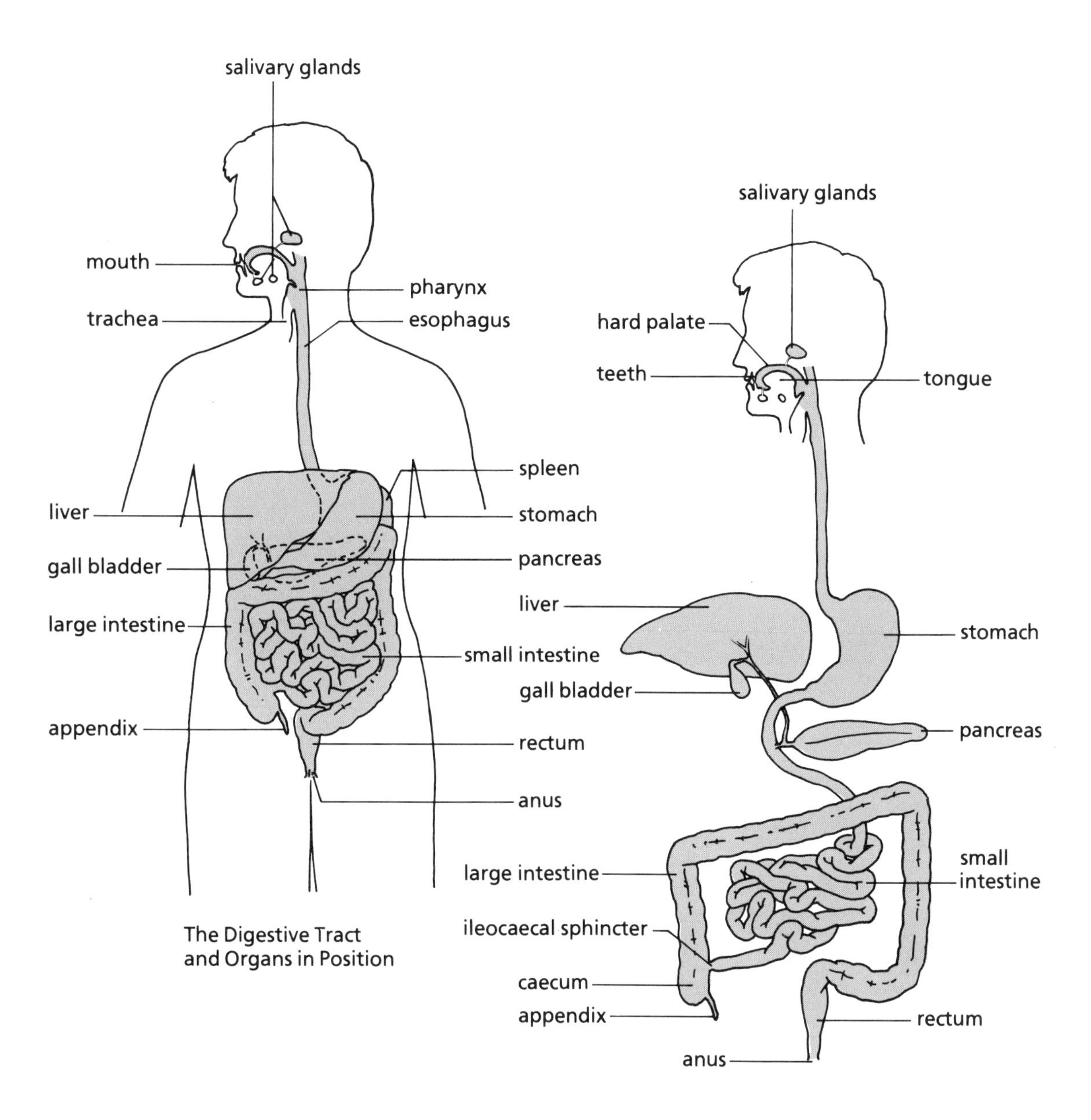

CASE STUDY. Mr. Parsons, 34 years of age.

The job was getting to be too much for Mr. Parsons. Sales had spiralled down into a slump. He had recently lost an excellent assistant to a rival firm. Money problems in financing the extension to the plant had not been solved in the way that he had expected. The constant round of business meetings, business lunches, and overtime made him feel perpetually tired.

Mr. Parsons was 34–young for his senior executive position, but he had achieved a successful career by long hours and hard work. Mr. Parsons had always been aware of having a "sensitive" stomach, and he always carried some antacid tablets to ease his discomfort.

Lately, the stomach pains seemed to be getting worse. The pains came about an hour before meal times, but disappeared shortly after he had eaten. Before he went to bed, the pains started again, but he could usually ease them by drinking a glass of milk. However, today the pain was different. He felt sick and didn't want to eat. His skin was sweaty and he looked pale.

Mr. Parsons decided to go home early and, once there, he dropped onto a chair feeling quite ill. His wife was concerned and, against Mr. Parsons's wishes, phoned their family physician, who asked them to meet her at the emergency entrance of the hospital immediately.

When Dr. Barker examined the patient, she noted that although Mr. Parsons felt sweaty, his hands and feet were cold and he was very pale. His pulse rate at 130 was well above normal and his blood pressure was low at 80/65. Dr. Barker gently pressed on the abdomen, but it seemed soft and normal. The doctor asked about the colour of the feces that Mr. Parsons had passed and made a note that a stool analysis must be taken. The colour of the stool had become much darker than usual and was almost black that morning.

The doctor also noted Mr. Parsons' history of using antacid tablets and the relief the patient gained from eating. She diagnosed a bleeding ulcer and prescribed immediate treatment.

Mr. Parsons received a blood transfusion to replace some of the lost blood and to help restore his blood pressure. The doctor also prescribed certain drugs to lessen his pain and discomfort. **Barium X-rays**, taken later, confirmed the doctor's diagnosis and Mr. Parsons was given some dietary advice for the future.

Once an ulcer has formed, the stomach acids irritate the ulcer site each time they are released. Food in the stomach helps to absorb the acid and give temporary relief. Antacids help to neutralize the acid condition and also ease the pain.

Bleeding from the ulcer into the digestive tract causes the feces to become much darker in colour, and it may become almost completely black.

Diet control, with smaller and more frequent meals and restriction of coffee and alcohol which stimulate the gastric secretions, is necessary. Cigarettes tend to retard the healing process and so these also should not be used.

The objective of the treatment given to Mr. Parsons was to relieve the pain and promote the healing of the ulcer.

Mr. Parsons had great difficulty in adapting to the changes the doctor prescribed. These included more regular working hours and meal times and a reduction in work pressures. It was not long before Mr. Parsons went back to his old work habits, and two years later he had a second and more serious ulcer, which did require surgery.

Questions for research and discussion:

1. Find out what other kinds of ulcers there are.
2. What diet recommendations do doctors make for patients suffering from ulcers?
3. How common are ulcers? What are the most common factors believed to initiate ulcers?
4. If you know someone well who has an ulcer, get an interview and write up a case study indicating the causes, symptoms, treatment, and dietary advice that person has received.

The Stomach

(s-fink-ter)

At the junction of the esophagus and the stomach there is located a ring of muscle called a **sphincter**. This ring of muscle can pinch the esophagus, acting like the drawstring on a purse to control the passage of materials into the stomach. The sphincter at the entrance of the stomach is called the **cardiac sphincter**.

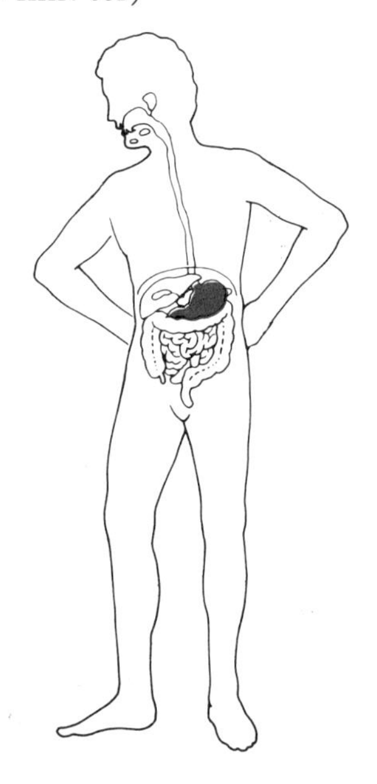

The stomach is located immediately beneath the diaphragm, towards th left side of the abdomen. It forms a muscular bag that stretches as it fills with food. Its walls are made up of four layers in a manner similar to those of the esophagus, but with a few important differences. The inner wall of the stomach is folded into many ridges or wrinkles called **rugae**. (See Figure 13.10.) At the bottom of these rugae are gastric pits or glands, through the openings of which the gastric juices are secreted. The muscle layer of the stomach has three divisions instead of two. One runs the length of the stomach, one circles the stomach, and the other stretches diagonally around it. These layers contract rhythmically to mix the food contents with the gastric secretions.

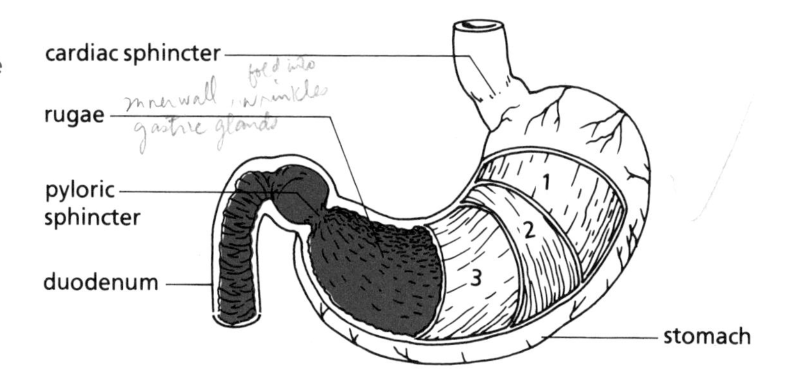

Figure 13.10.
A cut-away section to show the structure of the stomach.

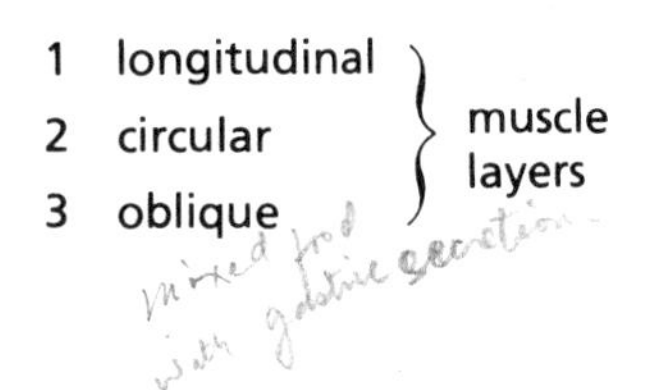

Secretions of the Stomach

The **gastric juice** is supplied by approximately 35 000 000 gastric glands. It contains the enzymes **pepsinogen** and **rennin**, as well as a supply of **hydrochloric acid**. This acid changes the pH or acid-base balance of the stomach to allow the enzymes to work efficiently while breaking down proteins. It also helps to kill any bacteria that may have entered the digestive tract with the food. Small quantities of **lipase**, an enzyme which helps to break down fats, and rennin, a milk-clotting agent which is especially important for young children, are also produced in the stomach.

(ly-pase)

In the cells that produce these gastric juices, carbon dioxide and water combine with salt to form hydrochloric acid and sodium hydrogen chloride ($NaHCO_3$).

$$CO_2 + H_2O + NaCl \rightarrow HCl + NaHCO_3$$

The hydrochloric acid formed in this way reacts with pepsinogen to form the active protein-digesting agent, **pepsin**.

$$HCl + \text{pepsinogen} \rightarrow \text{pepsin}$$

Pepsin causes large protein chain molecules to separate into short chains consisting of 4 to 12 amino acids.

Sometimes a small part of the stomach lining may be damaged or destroyed. The protective mucus secretion (mucin), which coats the walls of the stomach to protect it from its own digestive enzymes, may also be inadequate. When either of these conditions occur, the enzymes and hydrochloric acid which function to digest the meat and the other proteins that we eat, may start to digest the stomach lining itself. A **peptic ulcer** may result. Exactly why the lining breaks down and becomes ulcerated is not known. It is thought that excessive quantities of gastric juice in an empty stomach may irritate the lining and initiate such damage. Hydrochloric acid is usually produced in large quantities during times of stress and strong emotional pressure. For this reason this type of ulcer is sometimes called an "executive" or "stress" ulcer.

The **pyloric sphincter** is located at the lower end of the stomach. This controls the flow of partially digested food out of the stomach. Food at this stage is well-mixed, partly broken down, and in a semiliquid state, known as **chyme**. The time taken for food to pass through the digestive tract varies with the type of food eaten and the amount of fibre contained in the diet. The total time from entry into the body to waste elimination may vary from one to three days.

(kym)

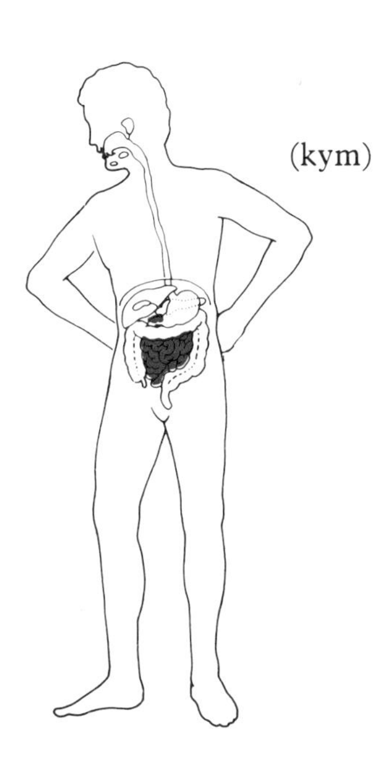

The Small Intestine

Structure of the Small Intestine and Its Secretions

The small intestine is a long, coiled, and looped tube about 2.5 cm in diameter. It may attain 7 m in length and fills most of the lower half of the abdomen. The small intestine appears to be a jumble of tangled loops, but in fact the loops are carefully and orderly attached to the rear wall of the abdomen by a thin membrane called the **mesentery**. This

ALCOHOL AND THE HUMAN BODY

Ethyl alcohol or ethanol (C_2H_5OH) is the active ingredient in beverages such as beer, wine, whiskey, gin, and brandy. Beer and wines contain between 2.5 and 14 percent alcohol while spirits and liqueurs contain between 35 and 50 percent alcohol.

Alcohol is absorbed from the intestine (about 80%) and from the stomach wall (the remaining 20%) and goes into the blood stream. Milk and fatty foods impede its absorption, whereas the addition of mixes such as carbonated "soft drinks" hastens absorption. The alcohol level in blood reaches its maximum within one-half to two hours after consumption. Alcohol enters various organs, including the brain, lungs, and kidneys.

In the liver, an enzyme changes alcohol to another compound that will then be broken down into carbon dioxide and water.

Body mass and the amount of food in the stomach primarily determine the effects of alcohol concentration in the body. After consumption of two bottles of beer or 75 g of whiskey or 150 g of wine, the average-sized male of about 75 kg will have a blood alcohol concentration of 0.05%. Consumption of one more drink could bring the blood alcohol level to the legal limit of 0.08%. The body must chemically change this alcohol for the level to return to zero percent. It takes the body about one hour to reduce the blood alcohol level of 0.025% to zero. For some people a reading of 0.03% could start impairment, but impairment is evident in everyone at a reading of 0.08%. (In the courtroom, this amount is expressed as 80 mg%.)

Alcohol is a depressant of the central nervous system. All kinds of motor performance, for example the control of speech, eye movements, ability to stand, and nerve reflexes are affected. Therefore, the ability to control an automobile is adversely affected by alcohol.

Alcohol affects the gastrointestinal tract. Frequent complaints from people who drink excessively are distension of the

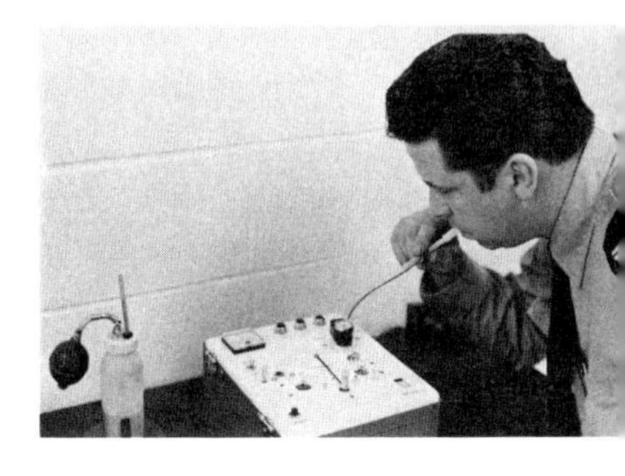

abdomen, belching, and "burning stomach". Excessive consumption of alcohol can lead to erosion of the stomach lining, causing ulcers, and promotes the enlargement of the liver, leading to cirrhosis. The normal liver functions are slowed down greatly. The pancreas can also be affected, leading to obstruction of pancreatic enzyme production.

Alcohol has also been known to affect the unborn. Scientists have found that children born to women who drink excessively or even moderately while pregnant may have a pattern of physical and mental birth defects - called the fetal alcohol syndrome. Most affected youngsters have smaller brains, narrow eyes, and low nasal

membrane serves several functions. It attaches and supports the small intestine preventing the loops from becoming entangled. It also carries blood and lymph vessels which supply oxygen and carry away nutrients absorbed from food in the intestine.

The small intestine is divided into three parts, the duodenum, jejunum, and ileum. The first 25-30 cm of the intestine comprises the **duodenum**.

The next part forms the jejunum, while the last portion is known as the ileum. These three sections are differentiated on the basis of their microscopic structure.

bridges. No one knows how much alcohol is too much. Pregnant women should not drink.

The statistics in Canada relating to alcohol-caused traffic infractions and accidents are staggering. Over 1000 people are killed in alcohol-related traffic accidents in Canada each year. In the province of British Columbia alone, approximately 25 000 charges for impaired driving per year have recently been made. The total cost in terms of necessary law enforcement, hospitalization, and lost productivity in that province is estimated at about $130 million per year.

Police throughout Canada are using the breathalyzer analysis to test suspected drinking drivers. If a person exhales over 0.08% blood alcohol, a charge for impaired driving will be laid under Section 236 of the Criminal Code of Canada. It is a criminal offense to drive with a reading of 80 mg% or 0.08% blood alcohol.

Abuse of alcohol is destructive both to one's mental and physical well-being.

The effects of alcohol on the brain.

PARIETAL LOBE
Sensory control is affected by 0.10–0.30% blood alcohol. Senses are dulled or distorted. Difficulty in writing well, speech may be slurred, technical abilities reduced.

OCCIPITAL LOBE
Visual perceptions are impaired by 0.20–0.30% blood alcohol. Colours may be distorted, double vision may occur. Poor judgment of distance and speed is evident.

CEREBELLUM
The co-ordination of muscles is affected by 0.15–0.35% blood alcohol. Difficult to walk or turn. Balance difficult to maintain.

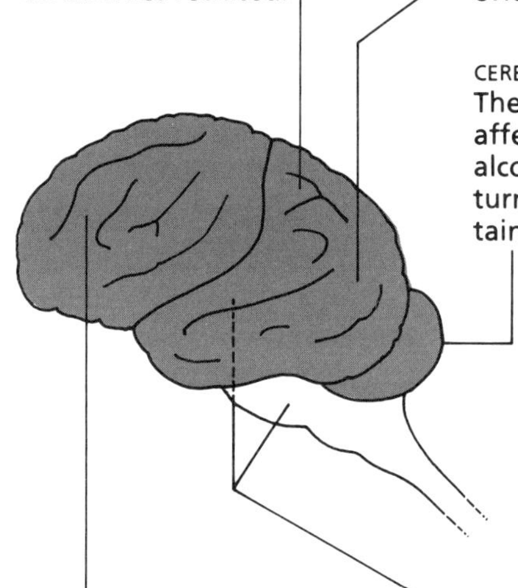

FRONTAL LOBE
Reason and self control are affected by blood alcohol levels of 0.01–0.10%. The alcohol removes the inhibitions we usually have and weakens the self-control. There is usually a feeling of well-being and a false confidence. Judgment becomes impaired, a person usually talks more and listens less.

THALAMUS AND THE MEDULLA
Autonomic nervous system affected by 0.25–0.50% alcohol. Tired, apathetic attitude, respiration and circulation depressed. Eventually a lowering of body temperature, stupor, shock, and possibly death.

The walls of the small intestine are similar in structure to those found in other areas of the digestive tract. The outer coat is associated with the supporting mesenteries. Inside this are two layers of smooth muscle, one arranged longitudinally and the other surrounding the intestine in a circular fashion. These muscles are responsible for the peristaltic contractions of the digestive tube, which move food along inside the gut.

The inner walls of the small intestine are lined with special glandular cells which are found at the bottom of short tubes located between the villi. (For a description of villi, see the

(1) Frontal lobe
0.01–0.1 %
affects reasoning, self control

(2) Parietal lobe
0.1–0.3 %
affects sensory control in writing - speech

(3) Occipital lobe
0.20–0.3 %
affects visual perceptions

(4) Cerebellum
0.15 – 0.35 %
muscle co-ordination.

(5) Thalamus & the Medulla 0.25–0.5%
autonomic nervous system.

section on absorption in the small intestine, further along.) These glands, known as the **crypts of Lieberkuhn**, secrete a digestive juice containing six enzymes. **Erepsin** completes the digestion of proteins, while **maltase**, **sucrase**, and **lactase** cause the final breakdown of complex sugars into single-sugar molecules. **Lipase** completes the digestion of fats, and **enterokinase** activates one of the enzymes from the pancreas after it arrives in the small intestine.

As a result of digestive processes occurring in the small intestine, the three basic food substances (carbohydrates, fats, and proteins) are broken down into molecules small enough to pass through the gut wall and enter capillaries of the circulatory system. The enzymes produced by the crypts of Lieberkuhn aid the process considerably. However, most of the digestion that occurs in the small intestine results from the activity of the liver and the pancreas, which empty their secretions into the duodenum.

The Liver

The liver, which has a mass of about 1.5 kg in an adult, is the largest gland in the body. It is located high in the abdominal cavity, directly beneath the diaphragm on the right side. It partially overlaps the stomach. Two important blood vessels enter the liver. The **hepatic artery** brings blood rich in oxygen. The **portal vein** carries blood to the liver, which is laden with nutrients from the intestines. The blood leaving the liver empties into the inferior vena cava.

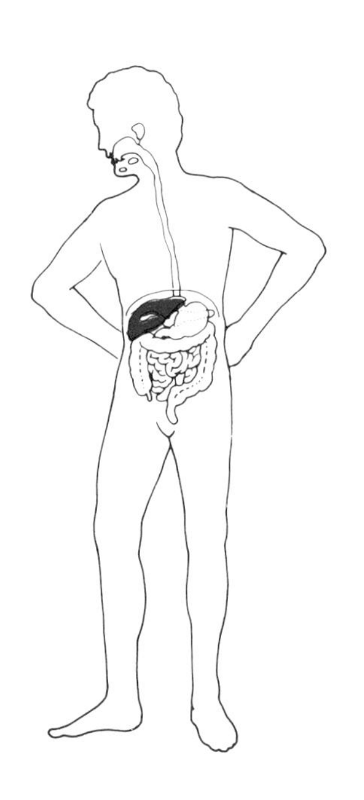

The liver produces **bile salts** that help to emulsify fats. The emulsification process produces results similar to those obtained when using a detergent on greasy dishes: large blobs of fat are broken down into tiny droplets. Bile provides physical breakdown rather than chemical digestion of the fats into fatty acids. Bile is produced continuously and carried by small ducts to the **gall bladder** where it is stored until required. When fats are present and need to be emulsified, the liver can produce as much as 450 mL of bile per day. (See Figure 13.11.)

The liver is one of the most vital organs in the human body, because of the many functions it performs. It acts in part as a factory producing several important chemical compounds, including the protein needed for the clotting of blood. Many of the specialized body proteins, such as gamma globulins, are also produced in the liver, as is Vitamin A.

The liver also serves as a warehouse, storing such valuable

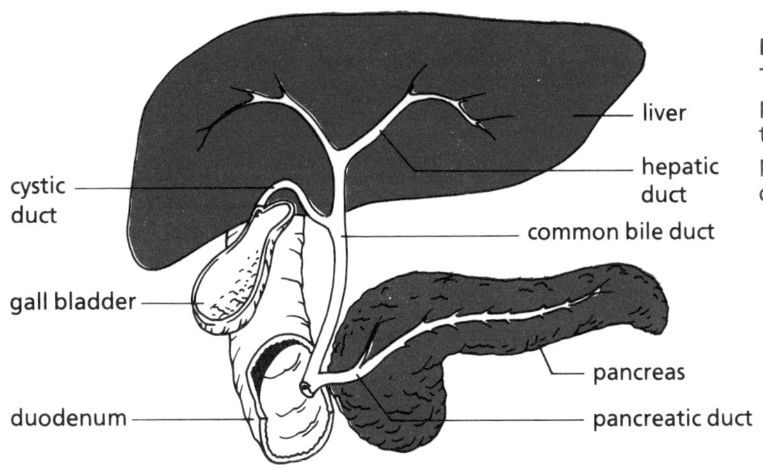

Figure 13.11.
The liver, gall bladder, pancreas, small intestines, and the ducts which deliver bile and pancreatic juices to the duodenum.

supplies as glycogen, iron, and Vitamins A, D, and B_{12}. When the body has an urgent need for glucose, the liver converts its glycogen store into glucose and releases it into the blood stream. The liver can also convert protein and glycerol from fats into glucose when other sources are not available. It is believed that about 60% of the breakdown of fatty acids into sugar takes place in the liver, although this process can also take place in the individual cells of the body.

Additionally, the liver also functions in waste disposal. It extracts and excretes bile pigment, urea, and other toxic substances such as alcohol and certain drugs. Some hormones are deactivated by the liver when they are no longer needed.

HEPATITIS

Hepatitis is a general term that means an inflammation of the liver. Two types of viruses are usually responsible for hepatitis. **Infectious hepatitis** is most common in teenagers and young adults. The first symptoms of this disease are often tiredness, poor appetite, an upset stomach, or pains in the abdomen. A fever may occur and the liver is often enlarged and tender. The most common symptom of hepatitis is **jaundice**. This is not a disease, but a characteristic condition in which the skin and the whites of the eyes become yellow. There is no medicine that will cure hepatitis. Bed rest and a high-protein diet will speed recovery and help to prevent permanent damage to the liver. Infectious hepatitis is spread when people live in crowded conditions with poor sanitation. This occasionally occurs in summer camps or in dormitories.

The other form of hepatitis, **serum hepatitis**, is transmitted by needles entering the arteries or veins. Blood transfusion equipment or the hypodermic needles of drug users can transmit this virus from one person to another. If the needle used to pierce the ears for earrings is not sterilized, it can also transmit the disease.

The best protection from hepatitis is good personal hygiene; washing your hands at appropriate times and never allowing the use of a needle that has not been thoroughly sterilized.

It is fortunate, in light of the many important functions performed by the liver, that this organ possesses an extraordinary ability to regenerate. If a part of the liver is removed, the remaining cells will gradually replace those that have been lost.

The Pancreas (pan-cree-us)

The pancreas is a leaf-shaped gland found close to the stomach in the curve of the duodenum. It measures about 20 cm in length. The pancreas has many important functions, including an involvement in digestion. Its primary role in that regard is to produce **pancreatic juice**, which contains about 28 known digestive enzymes. The juice is alkaline and therefore neutralizes the strong acid mixture which arrives in

Note: The names of the enzymes usually end in -ase. The first part of the enzyme name gives a clue to the food substance that it acts upon, e.g., sucr*ase* acts on sucrose and lip*ase* breaks down lipids (fats).

Table 13.1. Summary of the major digestive agents and their end products.

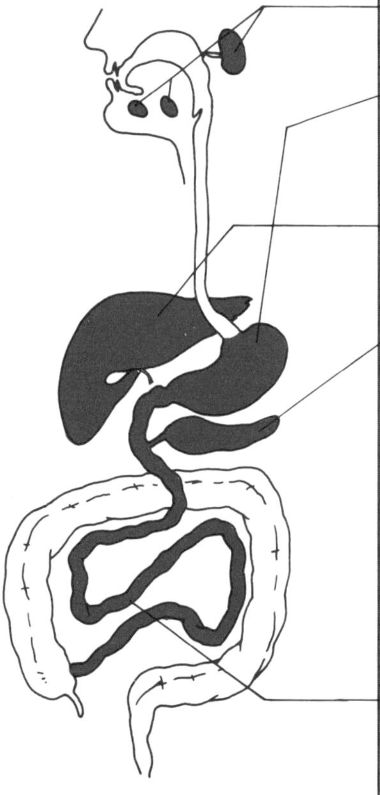

ORGAN	DIGESTIVE SECRETION	ACTIVE DIGESTIVE AGENT	ACTION ON FOOD
Salivary gland	Saliva	Amylase	Reduces starch to maltose.
Stomach	Gastric juice	Hydrochloric acid Pepsin Rennin	Reduces a protein to short peptide chains. Clots milk.
Liver	Bile	Bile salts	Emulsifies fats. Neutralizes acids.
Pancreas	Pancreatic juice Insulin (this is an endocrine function of this gland)	Sodium bicarbonate Lipase Amylase Trypsin, Peptidase	Neutralizes acids. Reduces fats to fatty acids and glycerol. Breaks down starch to maltose. Continues protein breakdown to amino acids.
Intestine	Intestinal juice	Maltase, Sucrase, Lactase	Completes digestion of sugars to glucose.

the duodenum from the stomach, changing it from a pH of about 2 to a pH between 7 and 8, which is almost neutral. The enzymes found in this juice act on each of the three food types. Pancreatic amylase continues the breakdown of carbohydrates into sugars. Pancreatic lipase splits fats into fatty acids and glycerol. Trypsin and peptidase split proteins into small, 2-unit, amino-acid groups. These enzymes flow into the duodenum through a short tube called the **pancreatic duct**. (See Figure 13.11.)

Among the other functions of the pancreas is the production of **insulin**. Insulin plays a vital role in carbohydrate metabolism and in maintaining the balance of blood sugar. A detailed discussion of the action of insulin can be found in the chapter on endocrine glands.

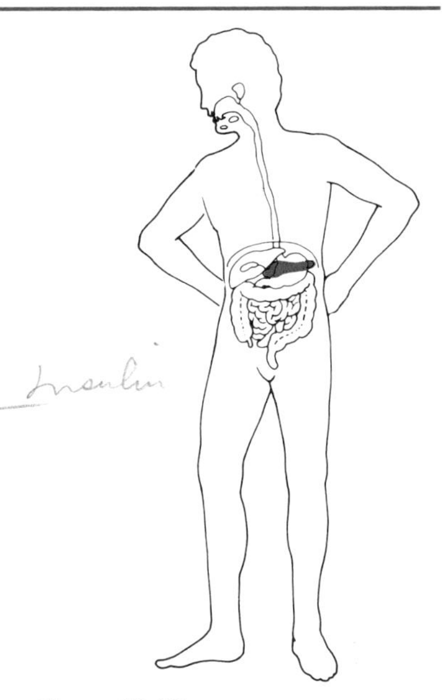

Absorption of Nutrients

The surface lining of the small intestine has small projections which improve its ability to absorb the end products of digestion. These small, fingerlike structures are called **villi**. (See Figure 13.12.) The purpose of the villi is to increase the

Figure 13.12.
The villi in the small intestine. These structures increase the surface area of the small intestine for more efficient absorption

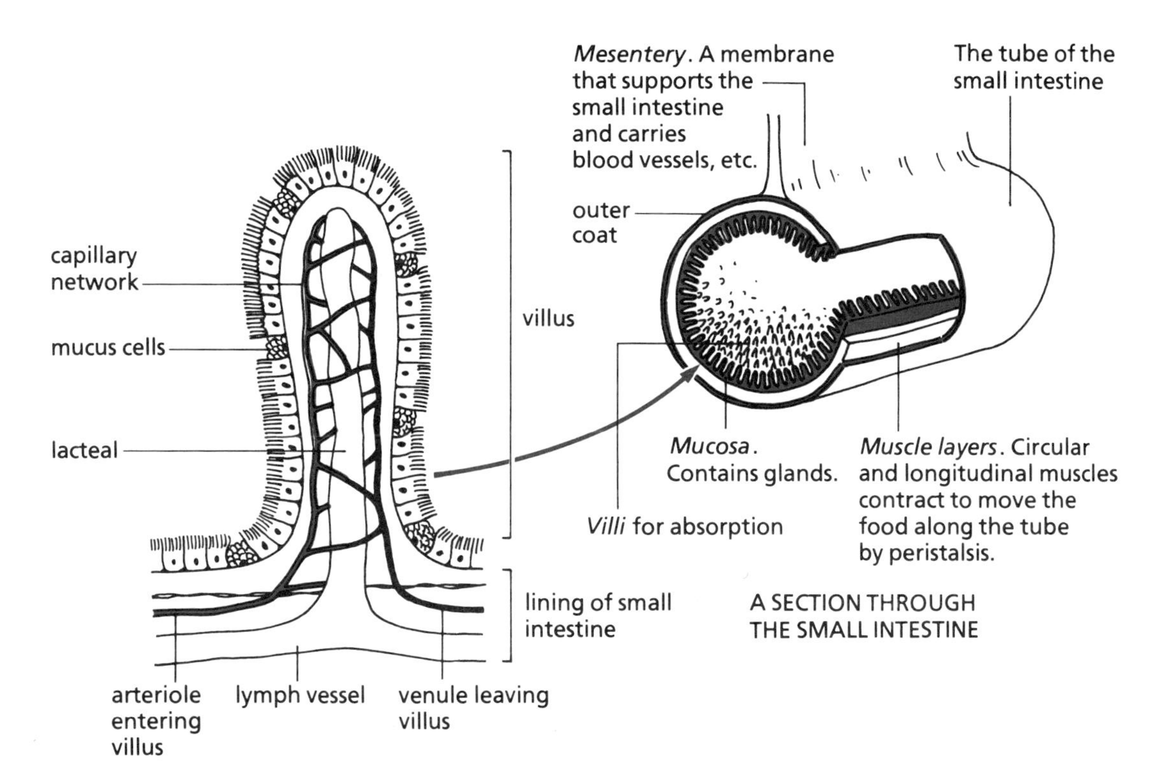

Figure 13.13.
Approximate times taken for food to pass through the digestive tract. The time taken varies greatly, depending on the type of diet (especially the amount of fibre present).

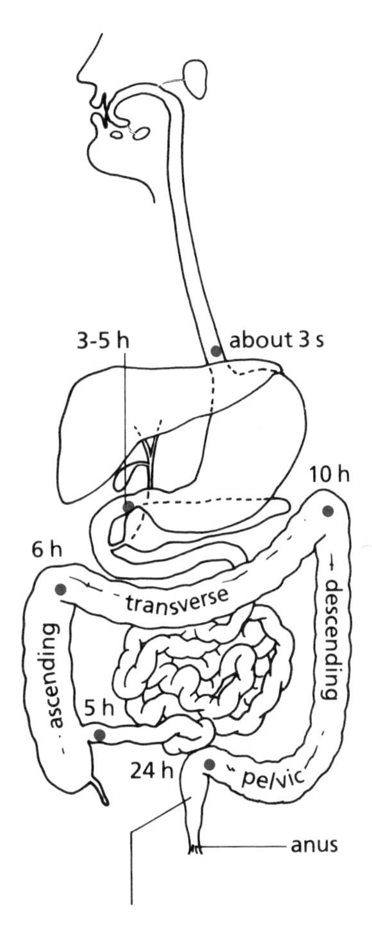

Times taken for food to pass along the digestive tract
The times indicate how long the first part of a meal would take to reach the different parts of the digestive tract.

Food taken in during the last part of a meal may take from 3 to 5 h to leave the stomach.

internal surface area of the small intestine so that food particles can pass through the lining cells and enter the circulatory blood vessels more efficiently and rapidly. Thousands of these tiny villi are found surrounded by the thin soup of digested nutrients along the inside of the small intestine. Each villus contains many blood and lymph vessels. These vessels collect the food molecules and transport them to locations where they are needed. The cells that make up the wall of the villi are themselves equipped with even smaller microvilli on the exposed surface of epithelial cells. These further improve the absorptive abilities of the intestine.

Digestion is an active process: food passes along the digestive tract as a result of peristalsis, which also serves to constantly churn and mix the food. The villi wave about with the movement of food and even this small action contributes to the mixing process. The amount of time required for food to pass through the digestive tract will vary with the type of food eaten and among individuals. Figure 13.13 gives an idea of the average time taken to reach particular points along the route.

The Caecum and Appendix

The junction of the small and large intestines is located in the lower right side of the abdomen. Another sphincter, the ileocaecal valve, occurs at this junction. This valve controls the rate at which the contents of the small intestine pass into the large intestine. Immediately beyond the sphincter is a small pouch called the **caecum**. Extending from this pouch is a blind tube about 6-8 cm long which is called the **appendix**. (See Figure 13.14.)

Although humans cannot break down cellulose, it is still an important dietary component that adds bulk to digested food and thus helps to prevent constipation. In herbivores, the caecum and appendix break down the cellulose. In humans these structures are no longer functional.

Unfortunately, food material can easily get into the blind-ended appendix sac and may lodge there, unable to get out. (Rather like a backwater along a small stream, where the water becomes stagnant because there is no current.) If pathogenic (disease-causing) bacteria lodge in this sac, where moisture, food, and warmth provide excellent growing conditions, the appendix may become infected. This disorder is known as **appendicitis**. Surgical removal of the appendix may then be necessary.

Should an infected appendix burst (which can happen if it does not receive proper medical attention), the contents of the appendix including the bacteria may spill out into the abdominal cavity. This then causes a much more serious condition called **peritonitis**. Prompt action to obtain medical help can avoid such dangers.

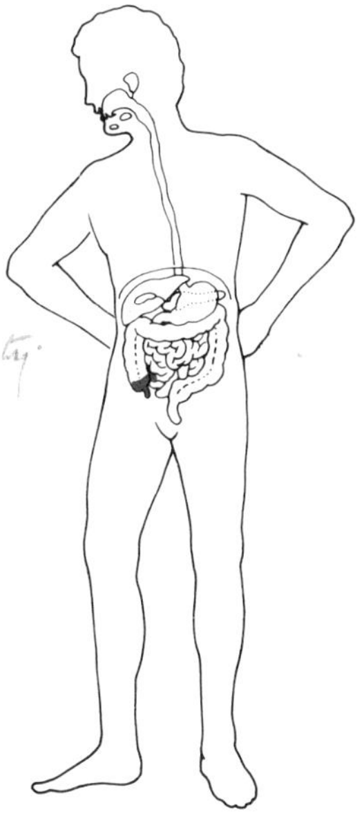

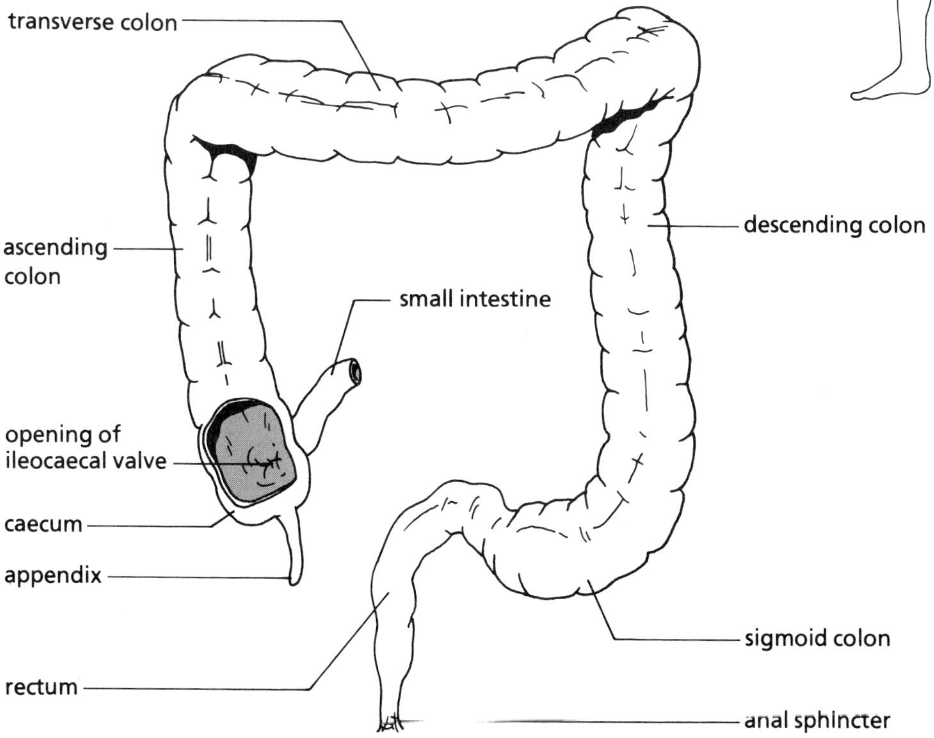

Figure 13.14.
The large intestine and rectum showing the junction with the small intestine.

The Large Intestine

The large intestine is about 1.5 m in length, 7.6 cm in diameter, and makes up almost one-fifth of the total length of the digestive tract. It forms an inverted U-shape that fills the lower abdomen. It is divided into three sections. The first section, the **ascending colon**, rises up the right side of the

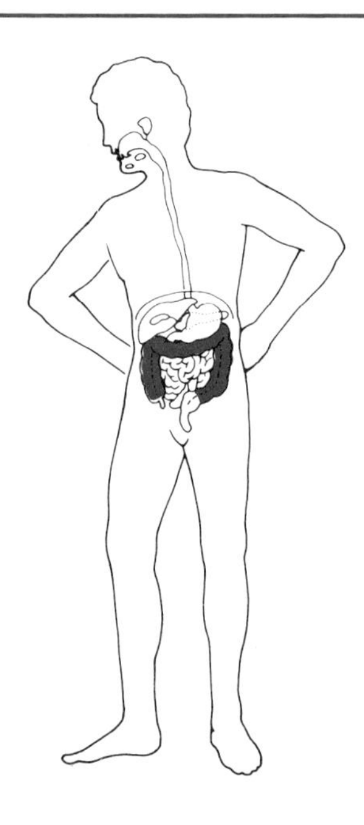

abdominal cavity. The second portion called the **transverse colon**, passes across the middle of the abdomen. The **descending colon** follows down the left side of the cavity and bends towards the centre of the lower body.

The large intestine functions mainly to reabsorb water, which is always in demand for cell metabolism. As this water is reabsorbed, the undigested food residues present in the large intestine are dried into a suitable consistency for defecation. As the colon takes up this water, some dissolved inorganic salts are also absorbed.

The large intestine usually contains cultures of bacteria and other micro-organisms which do us no harm. Some of these organisms operate on the waste material to produce vitamins B and K which are then absorbed for body use.

As the contents of the large intestine are being dried, they are also being mixed by periodic, sluggish contractions of the colon walls. This gradually propels the contents towards the rectum. The waste material, or feces, which are formed in the colon, are made up of about 60% solid material and 40% water. Although the walls of the large intestine possess muscle layers much like those found in other parts of the digestive tract, this organ does not make the same regular, peristaltic contractions of the other organs. Instead, several times a day, there is a strong series of wavelike constrictions, which move the entire contents of the large intestine along a short distance.

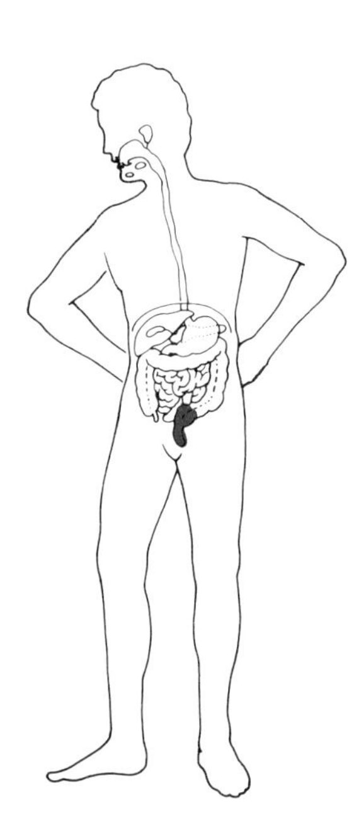

The Rectum

The **rectum** is the last section of the digestive tract. It ends at the **anus** with a ring of sphincter muscle. When the rectum is sufficiently distended and full, nerve endings in the walls send out messages, which produce a mild feeling of discomfort, an indication that the feces are ready for elimination.

The rectal veins are found near the anal opening. These veins sometimes become enlarged and distorted. This increase in size restricts the rectal tube and may make the passage of feces difficult and painful. These enlarged bulges in the rectal veins are known as **hemorrhoids** or piles.

If the diet is lacking in fibre, waste products become drier and more compacted. When sufficient fibre is present, the waste holds more water and is much softer in texture and therefore passes more easily through the rectum.

QUESTIONS FOR REVIEW

SOME WORDS TO KNOW

Match a statement from the column on the left-hand side of the page with a word from the column on the right-hand side of the page. DO NOT WRITE IN THE BOOK.

1. A leaf-shaped gland secreting many enzymes into the duodenum.
2. Substance secreted by the liver.
3. Active digestive agent that breaks down proteins in the stomach.
4. Rhythmic movements of the small intestine that help to move food along the gut.
5. Small fingerlike structures that aid in absorption.
6. A hard material found in teeth that has poor resistance to acids.
7. Rings of muscle that restrict the passage of materials along a tube.
8. The thin sheets of membrane that holds the loops of the small intestine to the back wall of the abdominal cavity.
9. Helps prevent food from entering the trachea.
10. One of the functions of this organ is to reabsorb water.

A. absorption
B. rugae
C. stomach
D. large intestine
E. small intestine
F. pancreas
G. liver
H. bile
I. peristalsis
J. pepsin
K. incisor
L. villi
M. dentine
N. saliva
O. epiglottis
P. sphincter
Q. mesentery
R. mucin

SOME FACTS TO KNOW

1. What is the difference between physical and chemical digestion?
2. Briefly describe the differences in shape and function of the following teeth: incisors, canines, and molars.
3. What is the difference between digestion and absorption?
4. What is the value of fibre in the diet?
5. What is the pH of the stomach and the pH of the duodenum? How is the difference between them produced?
6. What food substances are digested in
   a) the mouth?
   b) the stomach?
   c) the small intestine?
7. What are sphincters? What function do they have?
8. What is an ulcer? How are ulcers caused?
9. What are the functions of the colon?
10. What are the causes and symptoms of hepatitis?

QUESTIONS FOR RESEARCH

1. Select one of the following topics and prepare a report on it:
   - diabetes mellitis
   - Nursing Aid
   - Dental Assistant
   - Periodontist
   - cirrhosis
   - heart burn
   - food poisoning
   - Nursing as a profession
   - Dentist
   - Dental Hygienist
   - peptic ulcers
   - colostomy
   - Orthodontist
   - Food inspector
2. Humans cannot digest cellulose but herbivores, such as cows, sheep, and rabbits, can. Find out what differences there are between the digestive tracts of these animals and that of man.
3. Find out what diet changes are required for a person suffering from an ulcer or diabetes.
4. Read the labels on as many antacid products as you can. Prepare a chart on the contents of the various brands and the cost per tablet. With the help of your teacher design an experiment to determine which tablet is the most effective in neutralizing an acid of specific strength.
5. Suppose that you eat a hamburger. The bun is carbohydrate, the meat contains proteins and fats. Trace the fate of the hamburger from the time it enters your mouth until it leaves the intestine. What enzymes are involved, where are they produced, and what are the end products?

Activity 1: THE SALIVARY DIGESTION OF STARCH

Digestion of starch starts in the mouth with an enzyme present in saliva.

PART I

Materials
25-mL test tubes, test tube rack, 3% starch solution, Benedict's solution, iodine solution, wax marking pencils, water bath

(*Note*: Teachers may wish to substitute diastase for saliva.)

Method
1. Rinse the mouth clean of any particles. Collect about 4 mL of saliva in a test tube and add to this an equal amount of distilled water.
2. Label six test tubes from A to F.
3. Take about one-quarter of the saliva solution, place it in another test tube, and boil it for 6 or 7 min.

4. Add the different solutions to each test tube as shown below.

Tube	Contents	Indicator	Observations	Substance Present
A	3 mL Starch	Iodine		
B	3 mL Starch	Benedict's		
C	3 mL Starch + 1 mL Saliva	Iodine		
D	3 mL Starch + 1 mL Saliva	Benedict's		
E	3 mL Starch + 1 mL Boiled Saliva	Iodine		
F	3 mL Starch + 1 mL Boiled Saliva	Benedict's		

5. Wait 10 to 15 min then –
   To test tubes A, C, and E, add a few drops of iodine
   To test tubes B, D, and F, add about 1 mL of Benedict's solution and heat in a water bath.
6. *Observe any colour changes and record your results.*

Questions

1. *What is the test for starch and the test for sugar?*
2. *What is the purpose of test tubes A and B?*
3. *Explain why each mixture of solutions was tested with both iodine and Benedict's solution.*
4. *What conclusion can you draw about the digestive action of saliva?*
5. *What effect does high temperature have on the action of this enzyme?*

PART II

WHAT EFFECT DOES THE pH OF A SOLUTION HAVE ON THE ACTION OF A DIGESTIVE ENZYME?

Diastase is an enzyme that digests starch.

Materials
HCl solutions of pH 3, 6, 8, and 10. 25-1 mL test tubes, test tube rack, spot plates, iodine solution, 3% starch solution, 1% diastase solution

Method

1. Prepare 4 test tubes with equal amounts of starch solution in each, about 3 mL. *Prepare a chart to record your results.*
2. To test tube A add 3 mL of pH 3 HCl.
   To test tube B add 3 mL of pH 6 HCl.
   To test tube C add 3 mL of pH 8 HCl.
   To test tube D add 3 mL of pH 10 HCl.

3. One student from each bench should watch the time and signal his or her partner at one-minute intervals after the diastase enzyme has been added. To test tube A add 10 drops of diastase solution. Shake gently to mix the contents and start the timing. After one minute, tip a few drops from the test tube into one of the depressions in the spot plate and immediately test with iodine.
4. After every minute, add a few more drops of the starch and enzyme mixture to another depression in the spot plate and test with iodine. Continue to test until you no longer get a positive test for starch. *Record the number of minutes required to digest the starch.*
5. Repeat the experiment using test tubes B, C, and D. If time is short, the teacher may ask different rows of benches to test a different pH effect.

Questions

1. *Which pH gave the best digestion time?*
2. *Relate your results to the digestion of starch in the body. Use your text to review the pH of the duodenum. You may wish to test the pH of the mouth with litmus paper or Hydrion paper. Which of the pH solutions comes closest to the pH of your mouth?*
3. *Draw a graph of your results plotting pH against time.*
4. *Explain why the rates of digestion vary.*

Activity 2: THE ACTION OF LIPASE ON FATS

Lipase, an enzyme produced in the pancreas, acts specifically upon fats to split (hydrolyze) them into fatty acids and glycerol.

Materials

fresh whole milk, pancreatin or lipase, 25-mL test tubes, phenolphthalein solution, detergent, 1 mol sodium carbonate solution, water bath, and hotplate

Method

1. To each of two test tubes add 10 mL of fresh milk and two or three drops of detergent and shake.
2. Add a few drops of phenolphthalein indicator. If this does not produce a deep red colour, add a few drops of sodium carbonate until the red colour is distinct. Mark the tubes A and B.

3. To tube A add a small amount of pancreatin or lipase from the tip of a spatula.
4. Warm both test tubes in a 40°C water bath. *Note the time that you place the tubes in the bath and the time that it takes for the indicator to become colourless.*

Questions

1. *What was the purpose of using the detergent?*
2. *What was the pH of the initial samples?*
3. *What was the purpose of adding the sodium carbonate?*
4. *What does the colour change indicate concerning the actions of lipase on the fats present in the milk?*
5. *Is there any significance to the 40°C temperature of the water bath and the time required for the colour change to take place?*
6. *What are the final products of fat digestion?*

Activity 3: PROTEIN DIGESTION

From your text or class notes, review the following questions.

- *What substances are required for digestion in the stomach?*
- *What substances are required for digestion in the duodenum or small intestine?*
- *What pH is found in the stomach and the duodenum?*

TO INVESTIGATE THE EFFECT OF pH ON ENZYMES AND THE DIGESTION OF PROTEINS

Materials

25-mL test tubes, boiled egg white in small cubes, pepsin solution, pancreatin solution, dilute NaOH (pH 8), dilute HCl (pH 2), incubating oven. (*Note*: small strips of meat may be used instead of egg white.)

Method

1. Add small pieces of egg white to six clean test tubes.
2. Add 3-mL amounts of the solutions listed below.
3. *Prepare a table in your notebook in which to record your results.*

Test tube 1. Egg white + pepsin.
Test tube 2. Egg white + pepsin + HCl.
Test tube 3. Egg white + pepsin + NaOH.
Test tube 4. Egg white + pancreatin.
Test tube 5. Egg white + pancreatin + HCl.
Test tube 6. Egg white + pancreatin + NaOH.

4. Incubate the tubes for about 24 h at 37°C.

Table

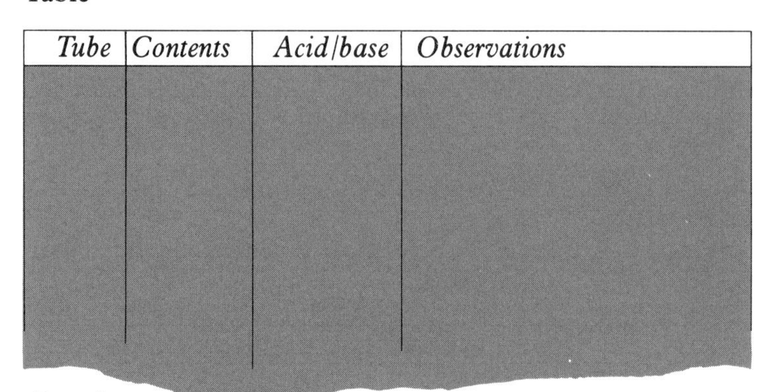

Tube	*Contents*	*Acid/base*	*Observations*

Questions

1. *Why was egg white selected as the test material?*
2. *Under what pH conditions does pepsin act most efficiently?*
3. *What conditions are most effective for the action of pancreatin?*
4. *Compare and comment on your results and the actual pH conditions in the stomach and small intestine.*
5. *What controls were used in this experiment? What other controls could be used to improve the validity of this experiment?*

Chapter 14

Nutrition

- What Determines a Diet
- The Major Food Substances

What Determines a Diet

We often hear people talk about going on a diet. They sometimes refer to some special program that they have read or heard about. In fact, we are all on a diet, all the time. The amount and type of food that we eat from day to day define our diet. Some diets provide enough food but may be poor in quality, while others may be inadequate in both size and quality. Fortunately, many diets are satisfactory in both of these respects.

There are many factors that influence the choice of food we buy and eat. For instance, the food purchased generally reflects the likes and dislikes of the family. Food costs are another important consideration. As one item becomes expensive, it may be replaced in the shopping by a cheaper substitute. Advertising and display merchandising probably influence the purchase of foods.

Fortunately, there has recently been a return to the practice of selecting foods for their nutritional value. Nevertheless, all the influences we have mentioned, and many others, still have some impact on our selection of foods.

For the reasons given above, each of us has to take responsibility eventually for planning his or her own diet. To do this well, the basic needs of the human body and simple principles of nutrition should be understood.

The Major Food Substances

At the beginning of this text, we discussed briefly the major components of food that supply all chemical elements needed by the body. The body requires these elements, in various combinations, to provide energy for the growth and repair of tissues and the general maintainence of the body systems. The food substances that provide these are of six different types:

- Carbohydrates
- Fats
- Proteins
- Water
- Vitamins
- Minerals

Carbohydrates

Carbohydrates are the major source of energy. As their name suggests, carbohydrates are formed from carbon and hydrates. The hydrate portion is composed of hydrogen and oxygen in the same proportions as they are found in water (two parts hydrogen to one part oxygen).

Table 14.1. The simplest carbohydrates – sugars.

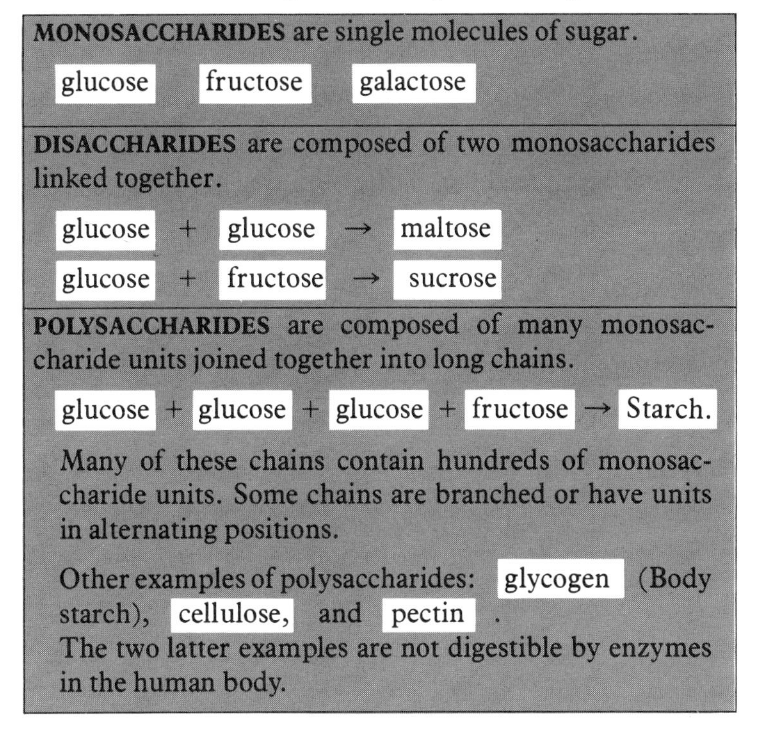

MONOSACCHARIDES are single molecules of sugar.

glucose fructose galactose

DISACCHARIDES are composed of two monosaccharides linked together.

glucose + glucose → maltose
glucose + fructose → sucrose

POLYSACCHARIDES are composed of many monosaccharide units joined together into long chains.

glucose + glucose + glucose + fructose → Starch.

Many of these chains contain hundreds of monosaccharide units. Some chains are branched or have units in alternating positions.

Other examples of polysaccharides: glycogen (Body starch), cellulose, and pectin .
The two latter examples are not digestible by enzymes in the human body.

The simplest carbohydrates are called **monosaccharides** (*mono* means one, *saccharide* means sweet) or single sugars. (See Table 14.1.) Glucose is one of the most important monosaccharides. Others occur naturally in many fruits and vegetables. More commonly, however, they are found bound together in pairs called **disaccharides** or double sugars. For example, table sugar (sucrose) is made up of two monosaccharides joined together.

(mon-oe-sac-ah-ride)

Monosaccharide + monosaccharide → disaccharide
glucose + fructose → sucrose (and water)
glucose + glucose → maltose (and water)

$$C_6H_{12}O_6 + C_6H_{12}O_6 \rightarrow C_{12}H_{22}O_{11} + H_2O$$

When many monosaccharide or disaccharide units are joined together into long chains, they are called **polysaccharides** (*poly* means many). The starch and cellulose found in plants are examples of these complex molecules. Since a large part of our diet is plant material, these substances form a large portion of our daily food intake.

Starch is easily broken down into simple sugars by the action of enzymes during digestion. Cellulose however is not digested in the human body and forms part of the fibre that passes, undigested, out of the body. Another important polysaccharide is **glycogen**. Although this substance is rarely found in the food we eat, it is the major form in which carbohydrate is stored in the body. Glycogen is composed of chains of glucose molecules which become linked together for more efficient transport and storage in the liver and muscles. These chains can easily be broken into glucose units, as demands for energy are made by various parts of the body.

Glucose and Blood Sugar

The glucose molecule is of great importance to the human body. A special body system monitors its presence in the blood at all times. The amount present is referred to as the Blood Sugar Level. After a heavy meal, large quantities of sugar may be present in the small intestine. These molecules are absorbed, carried into the blood stream, and delivered to the liver. In the liver, extra glucose, which is not immediately needed by the body, is converted to glycogen and stored until it is required. The amount of sugar in the blood is influenced by **insulin**, a hormone produced by the pancreas.

Muscles are often called upon to react quickly in an emergency, without special warning. For this reason, glycogen is also stored in the muscle tissues and it is thus available there for instant conversion and use. A more detailed account of these processes can be found in Chapter 17.

Carbohydrates and Kilojoules

Carbohydrates are often condemned as being the major contributors of body fat in overweight people. Excess carbohydrates are, in fact, converted to fat and stored, but so is any excess food that is eaten, whether in the form of carbohydrates, protein, or fats.

Each North American eats an average of over 45 kg of sugar per year. Much of this sugar comes in the form of soft drinks, cakes, candies, syrups, and confections with high sugar content. It is important to realize that energy obtained in this way has little to offer in the way of nutrient value. These sweet foods are eaten usually to satisfy hunger instead of better and sometimes cheaper foods, which do provide essential vitamins, minerals, and proteins.

WHAT IS A KILOJOULE?

The unit by which we measure food and body energy is the kilojoule. Energy is usually defined as the ability to do work or produce heat. The joule is the unit of energy and, in order to avoid very large numbers, the kilojoule, which is equal to 1000 J, is used.

Until recently the generally accepted unit for energy was the calorie. One calorie is equal to 4.2 kJ (approximately).

The following examples will help you "think" in joules. It requires about 400 kJ of energy to meet the needs of an average adult for one hour.

- 100 g of carbohydrate provide about 1680 kJ

Average Kilojoule Requirements per Day

SEX	AGE (years)	WEIGHT (kg)	HEIGHT (cm)	KILOJOULES (kJ)	PROTEIN (g)
Male	15-18	60	175	13 400	54
	19-22	66	175	13 400	54
	23-50	70	175	11 300	56
Female	15-18	54	165	8 800	48
	19-22	58	165	8 800	46
	23-50	58	165	7 900	46

Adapted from Food and Nutrition Board. RDA revised 1979. (NRC)

- 100 g of protein provide about 1680 kJ
- 100 g of fat provide about 3780 kJ
- A chocolate milkshake (350 mL) contains about 2200 kJ
- A piece of fudge (3 cm^3) contains about 480 kJ
- 20 french fries provide about 1300 kJ
- 1 glass of milk (small 250 mL) contain about 800 kJ

Fats

Fats, like carbohydrates, provide energy for body activities. Fats are present in the foods we eat as the stored energy supplies of the plant or animal product that we are eating. In plants, such as corn or cotton, fat in the form of oil is usually stored in the plant seeds. In animals, the body fat represents the animal's surplus energy store.

Sources of Fat

Meat, eggs, and dairy products are important sources of fat, although the fat in some of these sources, such as eggs, may not be readily visible to the eye. When the body turns surplus carbohydrates into fats for storage, these fats are stored in the adipose tissues of the body. They exist in a liquid state inside the cells, since the temperature of the body is considerably higher than that at which fats liquefy. Most fats are liquid (oils) at high temperatures.

Natural fats and oils are usually composed of two main parts: an alcohol, called glycerol, and three fatty acids. The fatty acid molecules consist of long carbon chains with a special combination of atoms, which makes the molecule an acid, tacked onto one end. When three of these fatty acid molecules bond with a glycerol molecule, they are called **triglycerides**. (See Figure 14.1.)

(try-gliss-er-ide)

Fig. 14.1
The composition of fats

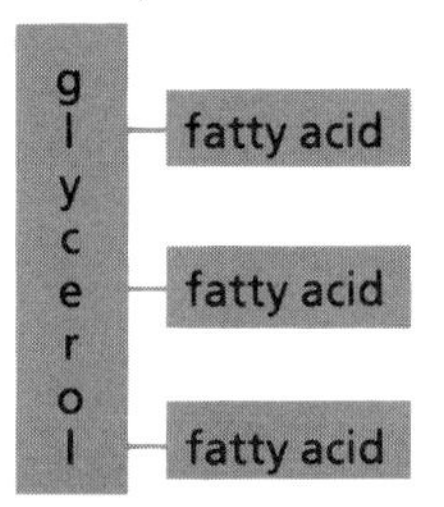

Diagrammatic representation of a triglyceride molecule with three fatty acids.

a saturated fatty acid

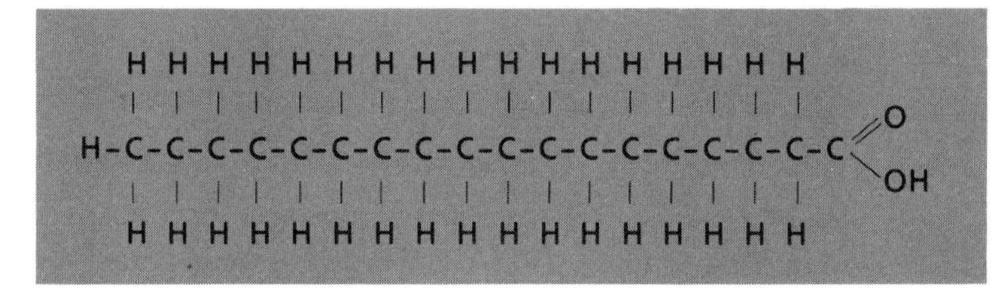

Stearic acid melts at 69.6°C

an unsaturated fatty acid

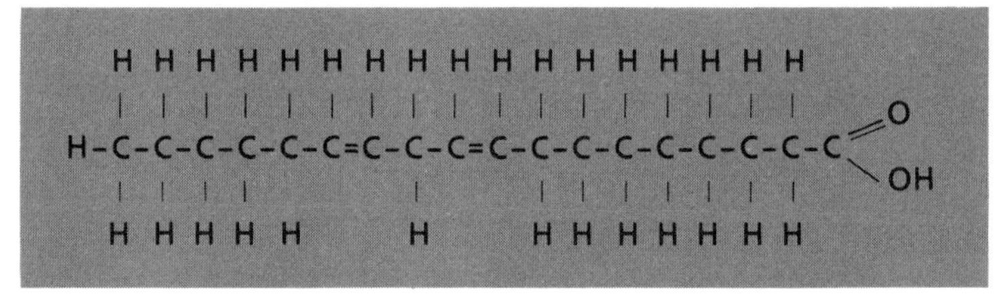

Linoleic acid melts at −5°C

An unsaturated fatty acid contains some double bonds and a smaller number of hydrogen atoms than a saturated fatty acid. It also has a lower melting point.

Saturated and Unsaturated Fats

All the available bonding sites on the carbon molecules of some fatty acids are filled by hydrogen atoms. In others, some of the carbon atoms have double bonds between them; these have fewer hydrogen atoms. When all of the sites are filled and no double bonds are present, the fatty acids are called **saturated**. Those with some free spaces, and double bonds, are called **unsaturated** fatty acids. The type of fatty acid involved determines the type of fat formed. Fats are thus labelled accordingly as either saturated or unsaturated. (See Figure 14.1.)

Saturated fats are solid at room temperatures. Animal fats tend to be made up of a higher proportion of saturated fatty acids, while vegetable fats and oils consist more of unsaturated fatty acids. Vegetable fats are usually liquid at room temperature.

Cholesterol

Cholesterol is believed by many to influence the occurrence of coronary heart disease. In some types of heart and circulatory disorders, fatty subtances cling to the walls of large blood vessels, thus reducing their diameter and restricting the flow of blood. Cholesterol is a major component of this fatty deposit. The gradual reduction in the diameter of these important blood vessels is a leading cause of heart attacks.

(ko-less-ter-all)

Some cholesterol is obtained directly from the foods we eat, such as egg yolks, meat, and butter. Additional cholesterol is made by the liver. If the amount of cholesterol in the diet is cut down, it would seem reasonable that the amount present in the blood stream would be reduced. Tests show, however, that when there is little cholesterol in the diet, it is synthesized in the body from fatty acids, glucose, and ethyl alcohol.

The types of fatty acids in the diet do, however, affect cholesterol concentration. Polyunsaturated fatty acids (fats containing more than one unsaturated fatty acid) tend to reduce the amount of cholesterol, while monounsaturated (one unsaturated fatty acid only) fatty acids seem to have little effect. Saturated fatty acids significantly increase the cholesterol levels. It seems that the most vital dietary factor determining the blood cholesterol level is the total amount of *saturated* fatty acids eaten.

Fats in the Body

Fat may be both good and bad for a person. Internal fat protects and cushions the body organs. Subcutaneous layers of fat help to insulate the body against heat loss through its surface. If too much fat is deposited, however, an individual may become overweight, thus placing extra strain on the heart, muscles, and bones, as well as causing some emotional distress.

Proteins

Proteins are essential for the building, repair, and maintenance of body tissues. They may also provide energy.

Proteins form a major part of the muscles, the internal organs, the brain, nerves, skin, hair, and nails. The vastly

Figure 14.2.
The sequence of amino acids in a molecule of insulin.

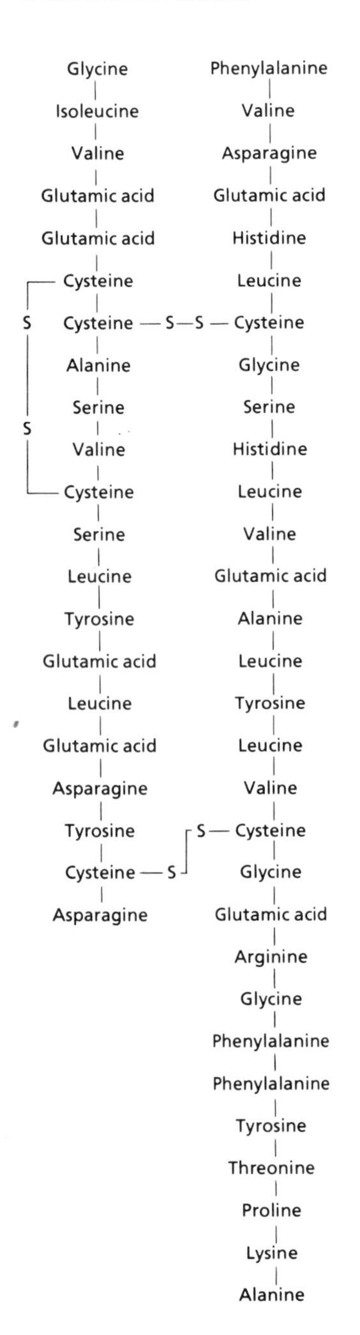

different nature of these body structures is evidence that there is a wide variety of proteins. Although there are great differences in the physical characteristics of proteins the chemical composition of the different types is very similar.

Proteins are giant molecules made up of hundreds, sometimes thousands, of units called **amino acids**. These are composed of nitrogen, carbon, hydrogen, and oxygen atoms. There are only 22 different kinds of amino acids in the human body. The order, or sequence and number of these amino acids determine the type of protein. (See Figure 14.2.)

When proteins enter the body in the food we eat, they pass into the digestive tract There, enyzmes disassemble the long complex chains into separate amino acid units. These are small enough to pass through the pores of cell membranes. After travelling through the blood they are then reassembled into new chains inside various body cells.

Essential Amino Acids

The human body can make some of the 22 amino acids it requires, but is unable to make others. These must be supplied in the diet. Amino acids that the body cannot make for itself are called **essential amino acids**. Lack of any one of these essential amino acids blocks the production of proteins that contain this unit. (See Table 14.2.)

Table 14.2. The essential amino acids.

methionine	threonine
tryptophan	isoleucine
leucine	lysine
valine	phenylalanine

The Nutritional Value of Proteins

The quality of a protein, in terms of its food value to the human body, is determined both by the number of essential amino acids it contains and by the current needs of the body for particular types or amounts of amino acids. The value of a protein depends on how important it is for tissue growth and maintenance.

Animal and Plant Proteins

The sources of food supplying essential amino acids may be conveniently divided into two groups: animal and plant proteins.

Animal proteins, which include those found in meat, fish, eggs, and other animal products, contain rich amounts of the essential amino acids. Plant or vegetable foods, such as flour, rice, cereals, peas, beans, and nuts, also contain protein. However, the proteins are usually of poorer quality, either entirely lacking or containing only small amounts of essential amino acids.

Fortunately, our desire for variety in diet helps to ensure that we get our quota of essential amino acids. Sources low in one essential amino acid must be coupled with another source which contains that missing amino acid. Foods must be eaten together or within a short period of time, because the body cannot stockpile some amino acids while waiting for others to arrive.

Body Protein Requirements

Protein is the basic material of which cells are made. It is needed, not only to build new cells, but to repair old and damaged ones.

During adult life, proteins are required primarily for the purpose of repair and maintenance of body tissues. While the body is growing to maturity, its demand for protein is much greater because new cells must be produced as well as old ones repaired. The more rapid the growth rate, the more protein is needed. It is especially important that the body has sufficient protein during the growth spurts that occur in the first two years of life and again in adolescence. Pregnant women and nursing mothers also have higher protein requirements because of the rapid cell production taking place in the unborn child and because of the need to supply the nutrients necessary in mother's milk.

Storage of Excess Nutrients

When fats and carbohydrates are eaten in excess of body needs, they may be stored in the tissues for future use. Proteins cannot be stored in the same way. If the body demands proteins which are not available from the food intake (during an illness, for example), tissue proteins are broken down for use in the emergency. Any surplus of amino acids not used for energy purposes are lost, because amino acids cannot be stored. Should the amount of available amino acids exceed the body's requirements, their molecules are converted first to glucose and then by further reactions to fat for storage. During this process, the nitrogen component of the amino acids is excreted. The body cannot therefore reconvert these fats to proteins. To use proteins for energy is thus an inefficient use of a valuable food component.

Water

Water is essential for life. It is almost as important as oxygen and more important than food. Although it is possible to live without food for several weeks, without water we would die in a few days.

The Need for Water

We need water to dilute and help dispose of the body wastes and toxins in our systems. Each body cell must be bathed in fluid, to enable dissolved nutrients and wastes to pass in and out of the cells. Water is needed to move the nutrients, wastes, and blood cells along in the blood stream. If water is not available for these functions, we must take it in. Our body signals its need for water by the sensation of thirst.

In hot weather, when we lose water by sweating to keep cool, there is an increase in our desire to drink. Fevers, vomiting, diarrhea, or the use of antihistamines to control allergies, all cause water to be lost from the body and thus increase our thirst.

Water Loss

Loss of water occurs in several ways. The largest amount passes out of the body as urine, heavily charged with dissolved wastes. Some is also lost when solid wastes are excreted. Water is lost by evaporation from the skin to assist in cooling the body, and from the lungs in expired air. You have seen the water vapour in this expired air condense as your breath cools in the air on a cold morning.

Water is replaced by fluids in the diet. Some foods, such as fruit and vegetables, are composed primarily of water. Watermelon is about 90% water, and lettuce has an even higher percentage of water content.

Compare the food values of the beverages given below. You might also find out the prices of each 340-g drink. It is not difficult to see that, apart from milk, most drinks contain only water and sugar and at fairly expensive costs. (See Table 14.3.)

Table 14.3. Comparative nutrient values for some common beverages

	Energy	Protein	Fat	Carbohydrate	Calcium	Iron	Vitamin A	Thiamin	Riboflavin	Niacin	Vitamin C	Amount
	kJ	g	g	g	mg	g	IU*	mg	mg	mg	mg	g
Milk 3.5%	1008	14	14	18	432	0.2	525	0.11	0.62	0.3	3	345
Milk 2%	773	14	8	18	451	0.2	258	0.15	0.66	0.3	3	340
Apple Juice	756	tr.	-	45	22	2.0	-	0.03	0.06	0.3	120	343
Orange Juice	756	3	1.5	39	40	0.7	700	0.35	0.12	1.5	186	343
Tea	8											343
Coffee	13											340
Cola	609			37								340
Ginger Ale	483			29								340
Beer	630	1		14	18				0.11	0.12		340
Spirit	441											42

*IU = International Unit (based on the activity of the vitamin)

Vitamins

More than two hundred years ago, in the days of wooden sailing vessels, it was not unusual for two-thirds of a ship's crew to die during a voyage as the result of a mysterious disease called scurvy.

It was a very long time before the chemical substance in citrus fruits, important for the prevention of scurvy, was identified as ascorbic acid, or what we now commonly call vitamin C. We now know that many other similar substances exist and are vital to health. These compounds are know as **vitamins**.

Vitamins have no food value themselves, but are chemicals usually obtained from the food. There are a few vitamins which the body itself can produce. Vitamin D is produced when the skin is exposed to sunlight. It is also found in milk, since cows produce it while feeding in sunny pastures. Ultraviolet light is the important agent in this process (although ultraviolet rays can also have a deleterious effect upon the skin). Some of the types of beneficial bacteria living in our large intestine produce vitamin K and some B vitamins. These are then absorbed through the lining of the colon for use by the body.

Obtaining Your Vitamin Supplies

A well-balanced diet, should provide all the required vitamins without the need to resort to vitamin supplement pills. There are several reasons, however, why even a balanced diet may be short of some vitamins.

Garden soils are sometimes lacking in essential elements. Vegetables grown in such soils may lack the vitamins that would ordinarily be present. Fruits and vegetables transported long distances, or fruits harvested before they are ripe, can also be low in vitamin content. Freezing and storing foods depletes vitamins, as do cooking and canning processes. It is not uncommon to find that the water in which vegetables are boiled contains more vitamins than the food which is eaten. Vegetables should be cooked lightly and served quickly. In some vegetables and fruits, the best vitamin supply lies in the skin and is thrown away if they are peeled. (See Table 14.4.)

Table 14.4. Factors affecting vitamin retention during food manufacture.

Vitamin	Soluble in:	Lose effective properties when exposed to:				
		Acids	Bases	Light	Heat	Oxygen
A	fat					•
D	fat					
E	fat		•	•		
K	fat	•	•	•		•
Thiamin	water		•		•	•
Riboflavin	water		•	•		
Niacin	water					
B_6	water		•	•		•
B_{12}	water		•			
C	water		•		•	•

If you decide to take a vitamin supplement, read the label. High price or claims of "Natural" or "Organic" ingredients are no indication of value. "Natural" is a much abused term.

Vitamins are needed in very small amounts, but they have very important functions. Although there are many attitudes about nutrition, experts are agreed that many vitamins are interdependent - if one type of vitamin is in short supply, it affects the efficiency of many of the others. Conversely, an overdose of some vitamins may have serious effects. Water-soluble vitamins, if present in excess, are simply excreted into the urine; fat-soluble vitamins, if present in excess, may cause poisoning.

Types of Vitamins

As has been indicated, there are two types of vitamins: **water-soluble vitamins** and **fat-soluble vitamins**. Vitamin C and the B Complex vitamins (which include riboflavin, niacin, and vitamins B_6 and B_{12}), are all soluble in water. Vitamins A, D, E, and K are soluble in fats. If fats are not present in the digestive tract when the vitamins are taken in, they cannot be absorbed. Vitamin supplements should, therefore, be taken at meals, rather than with a glass of water on an empty stomach. Fat-soluble vitamins present in foods are already in oil solutions, since they are found in fatty foods, such as butter. Only vitamins A and D are stored in the body. The others, if not absorbed, are excreted.

Table 14.5 lists the major vitamins, their value or function, and the foods in which they are found.

Minerals

Minerals are inorganic elements needed in small amounts to assist in a variety of essential body functions. Proteins, fats, carbohydrates, and vitamins are all carbon-containing compounds (organic compounds). Minerals do not contain carbon molecules and are readily absorbed in solution without digestion.

The minerals found, both in the body and in our food, in the largest amounts are calcium, iron, phosphorus, potassium, sodium, iodine, fluoride, and chloride. In addition, there are a number of elements found in very small or trace amounts. Even heavy metals like gold, silver, mercury, zinc, and magnesium are found in the body. We do not yet know, however, if the presence of all of these is important for good health. In most cases, heavy metals (such as lead) are toxic to the body.

The best food sources of minerals are fruits, vegetables, meats, milk, eggs, cereals, and water. Milk is a very important contributor of calcium, and liver provides an excellent supply of iron. Despite the small amounts in which most of them are found, minerals are a very important part of our diet and are crucial to good health. Many minerals are components of important molecules, such as hemoglobin, vita-

Table 14.5. The Vitamins.

	VITAMIN	FUNCTION	DEFICIENCY SYMPTOMS	DAILY REQUIREMENTS (16-18 years of age)	FOOD SOURCE
FAT-SOLUBLE VITAMINS	A	The beauty vitamin. Needed to maintain healthy skin, hair, eyes, etc. Improves resistance to infections. Helps break down fats.	Rough, dry skin, low resistance to infections, night blindness.	girls 4000 IU boys 5000 IU	Whole milk, liver, butter, carrots, eggs, green and yellow vegetables.
	D	Needed for calcium and phosphorus absorption to produce good bones and teeth.	Rickets, softening of bones in adults, and poor teeth.	400 IU	Fish liver oils. Sardines, salmon, liver. Some made by the skin in sunlight.
	E	Helps in the formation of red blood cells, muscle, and other tissues. Prevents abnormal breakdown of fat.	Circulation problems, loss of sexual and body vigor, muscular and heart problems.		Vegetable oils, whole grain cereals.
	K	Aids in blood clotting.	Rare, generalized bleeding.		Green vegetables.
WATER-SOLUBLE VITAMINS	Thiamine (B_1)	Needed for oxidation of carbohydrates. Insures proper use of sugars.	Loss of energy, depression, poor appetite, skin problems.	girls 1.1 mg boys 1.5 mg	Whole grain cereals, dry yeast, pork, fish, lean meat.
	Riboflavin (B_2)	Needed for energy metabolism in cells. Helps synthesize fats.	Tissue damage, eye strain, fatigue, itching, sensitivity to light.	girls 1.4 mg boys 1.8 mg	Liver, milk, cheese, green leafy vegetables, beans.
	Niacin (B_3)	Involved in energy reactions in cells.	Lack of concentration, headaches, insomnia, backache, poor memory.	girls 14 mg boys 20 mg	Meat, poultry, fish, whole wheat and enriched grains.

Vitamin B_{12}	Builds genetic molecules. Essential for proper functioning of the nervous system.	Anemia, bowel disorders, poor appetite, and poor growth.		Liver, kidney, fish.
C	Helps to maintain normal development of bones, teeth, gums, and cartilage.	Scurvy, bleeding gums, easy bruising, low resistance to infections.	45 mg	Citrus fruits, green vegetables, potatoes.

WATER-SOLUBLE VITAMINS

Table 14.6. The Minerals.

MINERAL	FUNCTION	DEFICIENCIES	SOURCES
Calcium	Forms bones and teeth, aids blood clotting, and nerve impulse connection.	Poor calcification of teeth and bones.	Common in many foods: milk, cheese, cereals, beans, and hard water.
Phosphorus	Also required for teeth and bones, some cell reactions.	Poor development in teeth and bones.	Meats, fish, dairy products, grains; common in many foods.
Iron	Needed for hemoglobin. Helps cells obtain energy from foods.	Anemia, lack of energy, especially needed in young children, girls, and women.	Liver, heart, meats, green leaf vegetables, whole wheat bread, cereals, and nuts.
Iodine	Formation of hormones in thyroid gland.	Goitre, swollen thyroid gland.	Seafood, iodized table salt.
Sodium	Regulates water between cells and blood.	Dehydration.	Very common: table salt, bread, canned meats, and vegetables.
Potassium	Needed for synthesis of proteins in cells.	Weakness in muscles.	Meats, cereals, milk, fruits, green vegetables.
Fluoride	Strengthens teeth, especially during development.	More rapid tooth decay.	Drinking water by natural or artificial addition.

NOTE:
This table lists the major minerals only. There are a number of other trace minerals that are very important for body functions and health.

mins, hormones, and enzymes. Some minerals are required for the normal functioning of nerves and muscles.

Table 14.6 shows the function of the most important minerals, and the foods in which they are found. The major roles of minerals in the body can be grouped into categories:

1) They control water balance;
2) They regulate the acid-base balance of body fluids;
3) They form part of many complex molecules, such as enzymes and hormones;
4) They form a structural part of bone and cartilage;
5) They help in many enzymatic activities in the body.

Fibre in the Diet

We can absorb most carbohydrates when they are digested and reduced to a small molecular size. **Fibres** cannot, however, be digested. We lack the specific enzymes that break down fibre substances, even many of them composed of sub-units similar to those in starch (which we *can* break down). Fibre remains in the digestive tract and is moved along into the large intestine for eventual elimination. Fibre contains cellulose, lignin, and certain polysaccharides and pectins.

Functions of Fibre in the Diet

Fibre helps to hold water in the large intestine by combining with the materials present. There, it functions to conserve water and thus prevent the fecal matter from drying out. In this way, it acts to prevent constipation. Fibre also has functions which relate to cholesterol and bile salt metabolism. Diets low in fibre are also believed to increase the potential for gallstone formation and coronary heart disease.

North America has a much higher incidence of bowel cancer than do many developing countries. It may be that the high-protein, low-fat diet in this part of the world has contributed to these problems.

Problems Known to Result from Low-Fibre Diets

Several common complaints are linked to low-fibre diets. For instance, when the feces are small and dry, the walls of the colon must provide greater muscle force to move them along. They are more difficult to transport and thus move only sluggishly. This may cause the formation of pockets (diverticula) in the bowel, causing irritation and discomfort. When diets are high in fibre, waste materials flow freely in and out of the appendix, whereas when low amounts of fibre are consumed, food tends to remain in the appendix. It then decomposes and becomes a potential source of bacterial infection. Statistics show that appendicitis is more likely to occur in individuals with a low-fibre diet. Constipation is considered an almost universal complaint in Western countries. In North America, over 250 million dollars is spent annually on laxatives! Hemorrhoids have also been shown to be caused by fibre-deficient diets.

The problems of a low-fibre diet can be reduced by simply changing food habits. Fibre can be added to the diet by eating such foods as whole grain breads, or by adding bran to morning cereals.

Sources of Fibre

About 2 g/d of crude fibre or 7 g/d of dietary fibre have been shown to produce considerable change in colon (lower intestine) functioning. The foods that contain the richest sources of dietary fibre for their mass are cereals which have not been highly processed. Green vegetables and fruits provide the least amount of fibre. Fruits and vegetables do supply fibre but must be eaten in large amounts to satisfy the body's fibre requirements.

Some comparisons:
30 g of bran contains 7 g of dietary fibre.

To get this same amount you would have to eat:
5 slices of whole meal bread, or
1½ heads of lettuce, or
3 large carrots, or
5 apples.

The process by which white flour is produced involves taking out and throwing away the valuable bran components of grain. Whole wheat bread is made from flour that has not been refined as much and, therefore, has not lost its fibre content. When breads are described as being "enriched" or "restored" this usually means that some of the vitamins or minerals, which were lost during processing, have been replaced. It does not mean that any of the fibre or certain amino acids, which were also lost, have been replaced.

QUESTIONS FOR REVIEW

SOME WORDS TO KNOW

Match a statement found in the left-hand column with a suitable term in the right-hand column. DO NOT WRITE IN THIS BOOK.

1. Material found in the large intestine that helps to retain water.
2. An example of a monosaccharide.
3. Major form of carbohydrate stored in the human body.
4. Hormone involved in the control of blood sugar.
5. Part of the fatty deposit that may clog blood vessels in some persons.
6. Major food substance used in building cells and repairing tissues.
7. A vitamin produced in the large intestine by the action of bacteria.
8. An example of a mineral.
9. A problem substance in "junk" foods.
10. Iron is an important part of this molecule.

A. carbohydrate
B. protein
C. fats
D. minerals
E. vitamins
F. glucose
G. fibre
H. monosaccharide
I. cholesterol
J. glycogen
K. insulin
L. vitamin C
M. vitamin K
N. calcium
O. hemoglobin
P. sugar

SOME FACTS TO KNOW

1. What is the primary value to the body of:
   a) carbohydrates?
   b) fats?
   c) proteins?
2. Why is such a large part of our diet made up of carbohydrates?
3. Explain how excessive amounts of cholesterol become a problem in the body.
4. What is the importance of the essential amino acids?
5. Why does the body need so much water?
6. Give four ways by which vitamins, that may be present in the food, are sometimes lost before they are eaten.
7. List two fat-soluble and two water-soluble vitamins. Give two examples of food that contains these vitamins, which you have eaten this week.
8. What are the best sources of the following minerals?
   a) iron,
   b) calcium,
   c) iodine.
9. Why is brown whole wheat bread more nutritious than white bread?
10. Of what value is fibre in the diet?

QUESTIONS FOR RESEARCH

1. Select one of the following topics and prepare a report on it:
   - cystic fibrosis
   - obesity
   - health foods
   - food additives
   - how much salt is enough?
   - celiac
   - anorexia nervosa
   - vitamin supplements, costs and contents
   - sugar and disease
2. Find out what special diets are required for persons suffering from ulcers, gall bladder problems, or high cholesterol.
3. Visit a supermarket and prepare a list of the marketing techniques that are used to tempt people to buy food products. Two examples to get you started: cereals packaged with plastic toys, special displays to catch your eye while you wait at the cashier's desk.
4. Discuss the nutrients in food offered by fast food outlets. Why are these establishments so popular?

Activity 1: SUGARS

Sugars form a subgroup of carbohydrates. Monosaccharides are the basic units of which all carbohydrates are composed.

THE IDENTIFICATION OF SUGARS

Materials

Benedict's solution, 25-mL test tubes, solutions of glucose, corn syrup, brown sugar, sucrose, and starch, hot plate, water bath, dilute hydrochloric acid

Method

1. Place about 3 mL of each of the solutions (water, glucose, corn syrup, brown sugar, sucrose, and starch) in different test tubes. Mark the test tubes for identification.
2. Add about 2 mL of Benedict's solution to each test tube.
3. Place the test tubes in a hot water bath and heat gently. *Observe and record any colour changes that take place and the sequence in which they occur.*

Sample	*Original colour*	*Changes*	*Final colour*

4. Take another sample of sucrose, add a few drops of dilute HCl, and boil for a few minutes. Now test with Benedict's solution.
5. Repeat Step 4 using the starch solution. Test with Benedict's solution. *Record all your observations and results.*

Note: Benedict's solution is a specific test for distinguishing between reducing sugars and non-reducing sugars.

Questions

1. *State the general test for sugars.*
2. *What significance might there be to the sequence of colour change from the original blue to the last colour change?*
3. *Devise an experiment that will test for sugars in a quantitive way.*
4. *Explain the results that you obtained after boiling the starch and sucrose in HCl.*
5. *Explain the differences between starches and sugars.*
6. *Research: List two sugars that are reducing and one that is non-reducing. What does the term "reducing" mean?*

Activity 2: CARBOHYDRATES

PART 1. EXAMINING FOODS FOR THE PRESENCE OF STARCH

Materials

spot plates, iodine solution, starch solution, potato, apple, carrot, sucrose solution

Method

1. On a spot plate, place the following substances in a marked sequence. Water, starch solution, potato, apple, carrot, and sucrose solutions. [Potato or carrot may be cut into convenient portions by using a cork borer to make a cylinder of the sample, then cutting it into small slices.] Place a piece of white paper under the spot plate so that the results may be seen more easily.
2. Add a few drops of iodine solution to each sample. Wait a few minutes and then examine the samples for any colour changes and the presence of granules.
3. *Record your results in a suitable table in your notebook.*

Substance	*Colour change*	*Is starch present?*

Questions

1. *State the recognized test for starch.*
2. *Is starch soluble? What evidence is there for your answer?*
3. *Name 8 foods that contain starch. Are they all solids?*

Activity 3: THE IDENTIFICATION OF FATS AND OILS

Fats and oils form an important source of energy in our food supply.

Materials
25-mL test tubes, Sudan IV brown paper, Bromthymol blue indicator, samples of olive oil, vegetable oil, margarine, cream, etc.

Method
1. Pour about 2 mL of any oil into a test tube and add a few drops of Sudan IV. Shake gently.
2. Pour about two mL of water into another test tube and add a few drops of bromthymol blue indicator. Shake the tube gently.
3. Tip the contents of the first tube into the second tube and shake them together. Observe what occurs, then wait a few minutes and note any changes that you see. *Record your observations regarding the mixing of oil and water and the mixing of the colour indicators with oil or water.*
4. Add a few drops of soap solution or detergent and shake the test tube. *Record your observations.*
5. *Look up the word "emulsification". How does it apply in this experiment?*
6. Rub a drop or two of each sample onto a piece of brown paper. Mark each spot with the name of the sample. Hold the paper up to the light. *Record what you observe and state a simple test for fats. What word is used to describe the condition when light passes through a substance but no image can be seen?*
7 Add small quantities of the samples given to different test tubes and add about 2 mL of water to each. Add two drops of Sudan IV to each tube. Shake gently. Observe the tubes after a few minutes. Any oil drops present will take up the red stain.
Record which of the substances contain fats or oils.

Activity 4: IDENTIFICATION OF PROTEINS

Protein molecules are very large and to be absorbed they must be broken down into smaller units called amino acids. Most tests for proteins indicate, not the presence of the whole protein but, the presence of a specific amino acid in the protein chain.

Materials

egg white mixed with water about 50-50, milk, eggshell, gelatin, meat emulsion, Biuret reagent, 25-mL test tube, water bath, and hotplate

Method

1. Add 2 mL of each substance listed to a different test tube and mark them so that you can identify what is in them.
2. Add 6 drops of Biuret reagent to each tube.
3. Place the tubes in a warm water bath or gently heat for a few minutes.
4. *In your notebook, make a suitable table in which to record your results.*
5. *After a few minutes record any colour changes or observations in your notebook.*

Questions

1. *State the test and colour change for Biuret reagents.*
2. *What control could have been used for this experiment?*
3. *Nitric acid can be used to test proteins. Devise a test using dilute nitric acid. Try testing some different known protein substances, such as fingernail clippings and hair.*

Chapter 15

Diet – Knowing What and How Much to Eat

- A Daily Food Guide
- Assessment of Body Size
- Control of Mass

A Daily Food Guide

An important idea emphasized in the last chapter was that individuals must take responsiblity for their own diets and not rely on others to tell them when and what to eat. Basic rules of good nutrition can be used as a guide.

A simple system that has been adopted by The Canadian Department of Health and Welfare divides all the foods we eat into four basic groups (Figure 15.1.):

A. Meats
B. Fruits and Vegetables
C. Milk and Milk Products
D. Bread and Cereals.

Figure 15.1.
The four food groups.

Every day you need...

MILK

3 or more glasses
to drink and in foods like these—

MEAT AND EGGS

2 or more servings

or some of these alternates

Have one dark green
or yellow vegetable.

VEGETABLES and FRUITS

4 or more servings

BREAD and CEREALS

4 or more servings

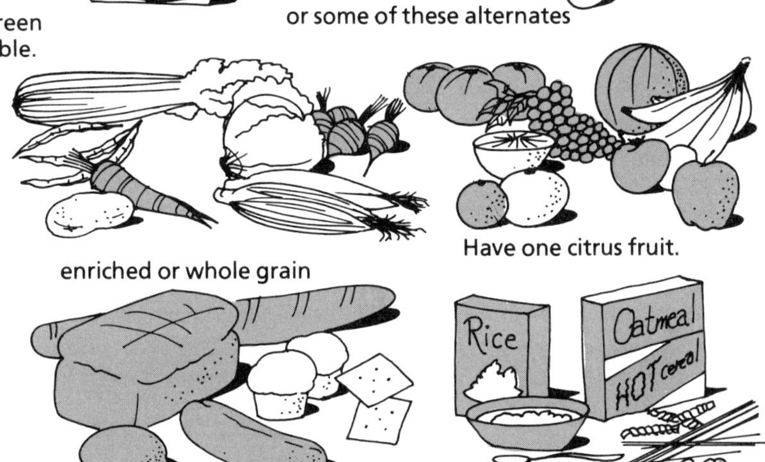

A certain amount of food from each of these groups should be eaten each day – an amount that will provide all the nutrients we require. No single food or group of foods contains all of the nutrients that we need, not even milk or eggs.

What Quantity of Each Group Does A Person Need?

The Milk Group

Milk and milk products provide the calcium needed to build bones and teeth. Unless an adequate amount of milk is consumed, the daily calcium requirement is difficult to meet; no other source provides this mineral in sufficient quantities. (See Figure 15.2.)

Milk also provides proteins and some of the B vitamins, particularly riboflavin. Vitamins A and D are also added to most commercially available milk, because these are lost during processing of the milk to remove fats.

Unfortunately, many people who are anxious about their weight avoid milk, believing that it contains too much fat. Since 2% or fat-free milk is available, there is no reason to avoid this important source of nutrients.

Figure 15.2.
Some foods in the milk group.

What is a Serving?
Here are some examples:

- Milk - 250 mL (approximately 1 cup)
- Cheese - 1 slice Swiss cheese
- Yogurt - 250 mL
- Ice Cream - 250 mL (equivalent to 125 mL of milk)
- Cottage Cheese - 250 mL (equivalent to about 170 mL or ⅔ cup of milk)

Recommended amounts:
- Adults - 2 servings
- Teenagers - 4 or more servings
- Children - 3 servings

Teenagers need a large amount of calcium each day. Yogurt is one food of the milk group that contains many nutrients. It is a substitute selected by many who do not wish to drink two large glasses of milk each day.

The Meat Group

This group contains meat, poultry, fish, and sea foods, eggs, dried beans, peas, and nuts. These foods are the major suppliers of the proteins required for growth and tissue

Figure 15.3.
Some foods in the meat group.

repair. They also supply iron and vitamins of the B complex (thiamin, riboflavin, niacin, B_6, and B_{12}). (See Figure 15.3.)

All foods, including meat, contain stored energy. Most of these protein sources also supply fats as well, even lean meat contains 5-10% fat. A person who eats a large steak (about 350 g) will gain 6048 kJ of energy (more than half the daily total need). The fat contained in this steak would include about 320 mg of cholesterol (300 mg of cholesterol is considered the maximum advisable intake per day).

What is a Serving?

Here are some samples:

- Beef, lamb, ham - 170 g: about 2 slices, 10 × 6 cm.
- Chicken - chicken leg or ½ chicken breast.
- Eggs - 2
- Cheese - 70 g: 125 mL (or half a cup) of cottage cheese, or 2 slices of cheese.
- Fish - 6 sardines, 1 fresh frozen fillet, 4 fish fingers.
- Dried peas/beans - 250 mL
- Peanut butter - 60 mL, half a cup of peanuts.
- Nuts - 125 mL, 60 mL peanut butter.

At least two servings per day in any combination are required. It could be two eggs for breakfast and half a chicken breast for lunch, or two slices of cheese and fish for supper.

The Fruits and Vegetables Group

Figure 15.4.
Some foods in the fruits and vegetables group.

Fruits and vegetables give a variety of flavours, textures, and colours to our diet. Many come ready packaged in a skin - easy to carry for a snack. (See Figure 15.4)

Fruits and vegetables are the most valuable source of vitamins and minerals. They provide 90% of our vitamin C and about 60% of our vitamin A intake. Citrus fruits provide most of the vitamin C that we use, while dark green and deep yellow vegetables are the best suppliers of vitamin A. Vitamin C is the vitamin most often lacking in the average diet.

This food group also provides some minerals (folic acid among them) and fibre. Like all other foods, fruits and vegetables contain some carbohydrates.

What is a Serving?

Here are some examples:

Fresh salads - enough to fill a cup.

Cooked fruits and vegetables - 125 mL (or half a cup).
Raw fruits - half a banana or grapefuit, or 1 orange, or 1 apple.
Fruit juice - (unsweetened) small glass

At least four servings per day are required. A fresh salad might equal two servings, with an apple for a snack and one serving of a vegetable. Another day's intake might include a whole banana and two vegetables at suppertime.

The Bread And Cereals Group

Breads and cereals form a much misunderstood food group. They are among the first foods to be blamed for being fattening. All foods, however, provide energy and bread and cereals are valuable foods, providing carbohydrates, proteins, thiamin, niacin, and iron. (See Figure 15.5.)

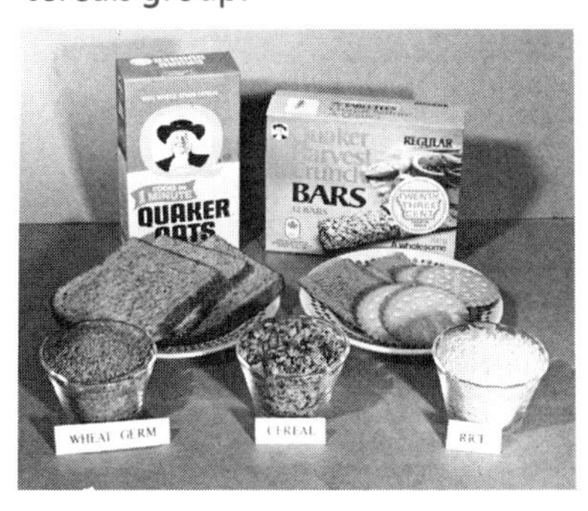

Figure 15.5.
Some foods in the bread and cereals group.

Commercial breadmakers have bleached and conditioned flour until it has lost much of its original nutritional value. Much white bread is now "enriched" by having the vitamins and minerals that were lost during processing replaced so that it should be as nutritious as whole wheat bread.

What is a Serving?

Here are some samples:

- Bread - 1 slice.
- Cooked cereal - 125-200 mL (half to three quarters of a cup).
- Muffins, rolls - 1.
- Macaroni, spaghetti, noodles - 125-200 mL.
- Ready-to-eat cereals - 125 mL.

The recommended minimum amount for adults and teenagers is four servings per day. Two sandwiches (4 slices of bread) at lunchtime often account for the whole allowance. Whole wheat brown breads are highly recommended. Granola or whole grain cereals at breakfast with toast would account for half the allowance, and spaghetti for supper would fill the quota.

Although foods from each of these groups may appear at every meal, this is not necessary. It is important, however, to ensure that the total number of recommended servings from each group is eaten sometime during the day. The number of needed servings varies with the age of the individual as shown in Table 15.1.

Table 15.1. The number of servings in each of the food groups required by people of different ages.

	Children up to 11 years	Adolescents	Adults
Milk and milk products	2 to 3	3 to 4	2
Bread and cereals	3 to 5	3 to 5	3 to 5
Fruits and vegetables	4 to 5	4 to 5	4 to 5
Meat and alternates	2	2	2

NOTE: The first number indicates the minimum recommended number of servings; the second figure is the optional number of servings which may be taken. (Canada's Food Guide)

Other Foods

Obviously some foods have not been mentioned in this guide-items such as butter or margarine, oils, sugars, and desserts. These are not emphasized because they are common in any diet. Some of these foods also contain vitamins, oils for instance, or furnish fatty acids. Primarily, however, they contribute energy.

Understanding the Daily Food Guide

If you develop the habit of including variety in the food you eat, it is easy to cover the range of nutrients needed daily.

If students are involved in heavy work, such as a summer job, they will require a lot of energy and will need to eat more to supply their bodies with energy and the muscle-building nutrients they need. If someone has an office job and reads in the evenings, his or her energy needs are less and, as a result, he or she should eat less. If office workers eat as much as laborers they will soon gain extra body mass. If laborers eat as little as office workers they may lose too much body mass.

Adequate servings of foods from the four groups discussed earlier can supply us with all the nutrients that we need for daily living. If you have an active lifestyle, you will need to increase the number or size of the servings. If your body mass starts to increase, you should cut back on the size of the servings, for you are eating in excess of your needs. Cut back, but don't cut out-reduce the size of portions, but keep your diet balanced at all times.

Breakfasts and Evening Meals

Research has indicated that people who eat breakfast are more alert and productive in the morning than are those who miss breakfast. Students do better in tests and absorb information more effectively when they have had breakfast. They are less likely to tire during the day. The body goes without food all night; it therefore needs to be recharged with food in the morning to provide energy and nutrients for the day's activities.

Heavy meals in the evening are not necessary. When large supplies of food enter the body and are absorbed into the blood stream, the energy contained within the food soon becomes available for use. Common evening activities such as studying, reading, or watching television do not demand much energy.

The Assessment of Body Size

Should I Lose or Gain Mass?

This is a question that many teenagers ask themselves, often quite without reason. Sometimes they are not satisfied with their own answers and start to ask their friends for advice. They may decide on a course of action simply because a favourite item of clothing no longer fits well, ignoring the fact that the body changes are a result of normal growth.

Although an honest appraisal of ourselves will usually give us all the information we need, it is sometimes necessary to check with a physician for a more objective standard by which to judge the amount of body fat present.

Height/Mass Tables

Height/Mass tables rely on data produced by insurance companies for making judgments about people who apply for life insurance. Those with excess mass, for example, are considered a poor risk in comparison to people of average mass. The tables are based upon the average masses and heights of a large number of individuals. They enable a person to compare his or her mass against that of other people of the same sex, height, and age. When using these tables, remember that it is an average figure that is given, so that approximately half the people are above the figure provided and half are below. Usually these tables give a range of masses. If you are a little above or below the average it does not mean that you have too little or too much mass. (See Table 15.7.)

The major problem with these tables is that people of the same mass vary in **frame size**. Some people have larger or more dense bones than others.

Most people tend to gain mass as they get older. The more reliable height/mass tables thus usually include age as a factor.

Skinfold Calipers or the Skinfold Test

This method gives a more accurate picture of where the body mass is located, because it measures the actual fat in your body, rather than comparing you to others or to an average figure. (See Figure 15.6.)

About half of the total fat in the body is subcutaneous fat, deposited just under the skin. In many parts of the body this sheet of fat is only loosely attached to the tissues below, and can be pulled up between the thumb and forefinger, into a fold. Researchers have found that there is a direct relationship between the amount of fat deposited beneath the skin and the amount laid down around the organs of the body, where we cannot easily reach to measure. Therefore, if we measure the extent of subcutaneous fat and apply this to the tables produced by research, we can make estimates about the total fat in the body.

Although the total mass of skin tissue is approximately the same in men and women, the amount of subcutaneous fat varies greatly. Men's bodies have about 11% fat and women's bodies about 24%.

How Much Fat is Enough?

Your body mass should include sufficient fat to help you feel your best and supply your body needs. Specifically, you should have enough to provide adequate reserves of energy, sufficient subcutaneous fat to help insulate the body from adverse temperatures, and enough to protect the organs from injury.

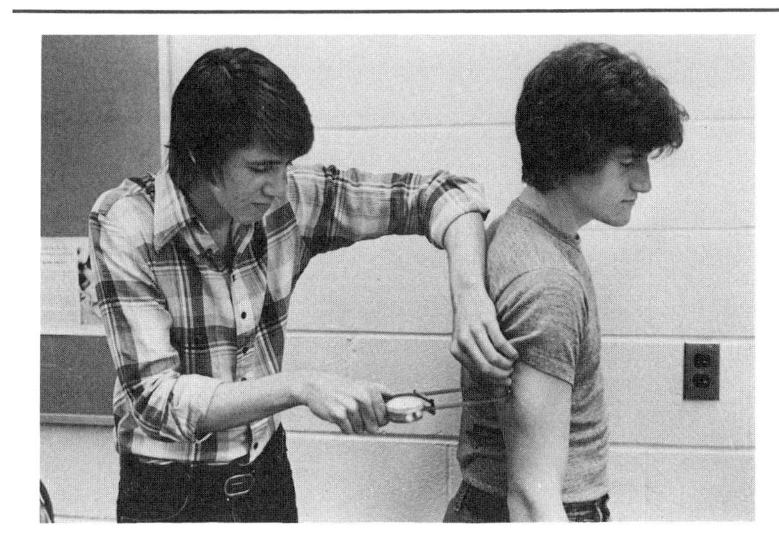

Figure 15.6.
Using the skin calipers.

Control of Mass

The Problems Associated With Being Overweight

Excess mass appears to contribute to some very serious disorders, such as heart disease, high blood pressure, and other circulatory disorders. It adversely affects bones and muscles and crowds the internal organs. It also can have a negative effect on the individual's outlook on life. The extra mass that must be carried causes shortness of breath and constant weariness. The extra effort that must be made for every activity often results in withdrawal from activities that the individual previously enjoyed. As one becomes less active and more sedentary, the problem increases.

People who were once overweight and have lost the excess have great difficulty in maintaining their reduced mass. One of the reasons for this is that in such a person, many cells - adipose tissue cells - have been produced by the body to store the excess food supplies available. Once produced, these cells do not disappear for a long time. When mass is lost these cells start to empty, but the cells are not immediately destroyed - they are still there waiting to be filled up again.

Fortunately, many of us are only slightly overweight. This is a good time to stop the pattern of increasing mass before it gets too far along. So long as we carry only a slight amount of extra mass, we can simply cut back on foods and do some exercise to work off the excess.

In the previous pages, we have discussed the foods that we need in our daily diet and have considered the size of the portion sufficient to supply these needs. To control mass successfully, we need to balance the two sides of the equation, kilojoules of energy consumed and kilojoules expended. (See Figure 15.7.)

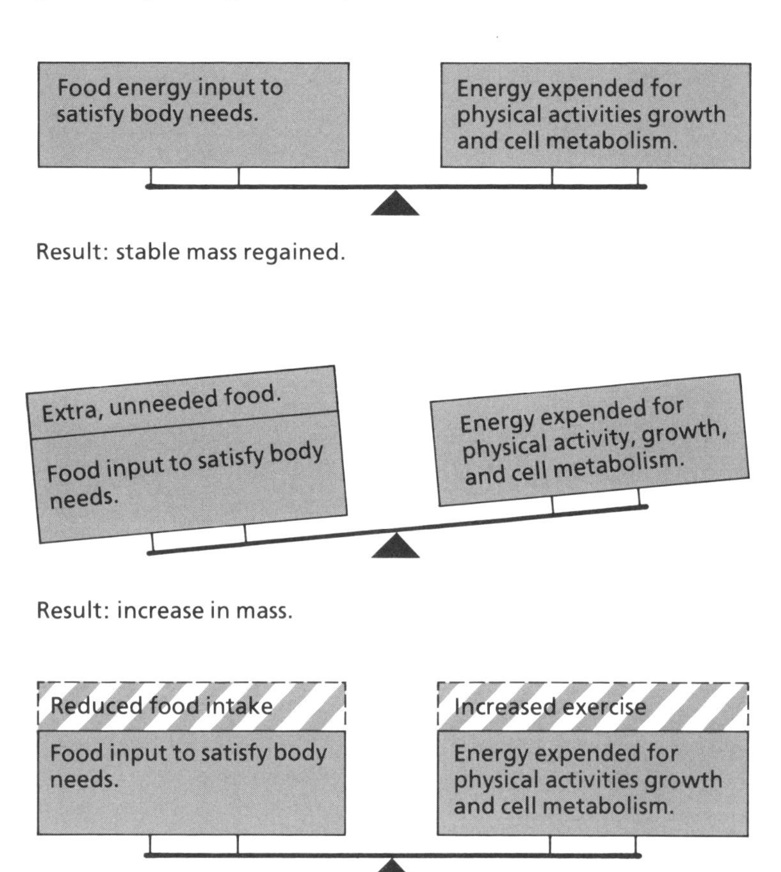

Figure 15.7.
The balance required between the intake of food energy and the expenditure of energy on exercise, body activities, etc., to maintain a stable weight.

Working Off the Excess – by Exercise

Many people have the idea that exercise must be a strenuous, exhausting activity. It does not need to be. (See Table 15.2.) Substituting a half-hour walk to work or school for a four-minute car ride can result in a loss of 6-10 kg a year! The key to continued exercising is to find something that you like

Table 15.2. Food and exercise energy equivalents.

FOOD	MASS (g)	kJ	Time in minutes to burn off kilojoules: WALK	BICYCLE	SWIM	JOG
Coke (227 mL)	240	483	20	16	12	11
White bread (1 slice)	23	252	12	9	7	6
Mars Bar	40	890	40	32	25	21
Peanut brittle	25	462	21	16	13	11
Popcorn (buttered) 250 mL	18	344	16	12	10	8
Choc. chip cookie (1)	11	210	10	8	6	5
Oreo creme cookie (1)	12	168	8	6	5	4
Ice cream sandwich	75	873	40	32	25	21
Ice cream cone	72	672	31	24	19	16
Banana split	300	2494	114	89	71	59
Doughnut (Jelly)	65	949	44	34	27	23
Apple	150	365	17	13	10	9
Banana	150	533	24	19	15	13
Cheeseburger	180	1940	89	69	55	46
French fries (20)	100	1150	53	41	33	27

NOTE:
Walking: At 3.5-6.5 km/h, burns off about 21.8 kJ/min.
Bicycling: At about 11 km/h, burns off 27.3 kJ/min.
Swimming: At an average rate of 30 m/min, burns off 35.7 kJ/min.
Jogging: At alternating 5-min walking–jogging, burns off 42.0 kJ/min.

doing. It is no good deciding to jog if you hate jogging. You won't keep it up. It is better to find an activity that you enjoy, preferably that you do with friends. Most towns have facilities for excellent programs in swimming, basketball, squash, and other sports, as well as fitness classes. You can join an amateur team to play almost any sport. There are also many non-competitive sports–cross-country skiing, hiking, riding, and canoeing. It is important to set aside a regular time for exercise. Regular exercise usually becomes something that you will look forward to and enjoy.

Table 15.8. Snack foods.

Table 15.3. Some exercises that you can do during a short break.

EXERCISE	PURPOSE	WHAT TO DO	COMMENTS
Sit ups	Firms stomach muscles. Maintains good lumbar posture.	Start by hooking your toes under a heavy chair, knees bent. Later try to do the exercise without the toes being held down. After raising the trunk into a sitting position, lower the body slowly; do not fall back.	The exercise can be varied by changing pace (fast-slow), or by adding a body twist at the end of the sit up. Benefit is greater when the chin is on the chest and the spine is "rolled up".
Push ups	Strengthens arms, chest, and abdominal muscles.	Keep the body straight, either from the knees (girls) or feet (boys). Raise and lower the chest keeping hands a comfortable distance apart. Lower the body until the chin or chest just touches the floor.	The exercise can be varied by changing the hand positions, feet positions, or by dipping the chin with each lift.
Leg kicks	Strengthens and firms thigh muscles.	Sit on a chair, stretch your legs straight out, toes pointed. Flutter kick up and down. For variation cross the legs over and under on each kick.	Keep the legs straight. Continue for 30-60 s. Sit on the edge of the chair, holding the seat with your hands.
Isometrics	Improves the arm and chest muscles.	Curl your fingers together, arms across your chest and try to pull your fingers apart. Change to pushing your palms together. There are many variations. Stand facing the wall, about 1 m away. Place hands on wall. Press in towards the wall.	Hold each position for about 6 s and relax for 10 s. Repeat each exercise several times. The possibilities are endless. Use your imagination to think up variations. Be sure to hold each position and then relax, before repeating.
Neck and shoulder flexibilities	Improves neck and shoulder muscles. Helps relax neck and shoulder tensions.	Slowly roll the head around in a circle. Feel the pressure on the neck and produce a stretching sensation in the muscles. Shrug the shoulders moving them around in a circular motion. Place hands behind the neck and press back.	Press for about 6 s and then relax for about 10 s. Repeat 8 times.

For the Very Overweight

So far, we have discussed how to lose just a few kilograms. If you carry really excessive mass, you must, of course, see a doctor to be sure that there is no special medical reason for your condition.

The best advice for reducing mass can be provided by a specializing nutritionist. If the problem is a serious one, a doctor will refer you to such an expert for consultation. Do not attempt major mass reduction without good medical advice.

We have considered two important parts to control of mass: changes in diet and exercise. For continued control, two other factors are important: (1) don't step back into "old" habits, and (2) get support.

On Being Underweight

Few people in North America suffer from this problem. Food supplies are generally adequate and few people are seriously undernourished. The problem of excess body mass has been stressed because of the many health problems associated with it. However, the following case study shows that some people are, indeed, suffering from a psychologically initiated condition of insufficient body mass. Others have too little body mass in proportion to their size because of poor metabolism or other physical conditions.

CASE STUDY. Jean, 16 years of age.

Jean was looking at herself in the mirror, turning from side to side to view her body from all angles. Jean was not happy. Her figure seemed to her to be far too thick and she wanted to be slim. Her lack of appreciation for her appearance would have amazed her friends who thought that she was already extremely thin and often urged her to eat with them at lunchtime. Jean always responded to such invitations with an excuse that she had an appointment to keep, or that she was not hungry, or that she would eat later.

Jean was suffering from **anorexia nervosa**, a disorder that usually affects teenage girls. Initially, such girls may be mildly overweight and start a diet program. Sometimes, when a suitable mass loss brings their body size back to an average mass, they continue to diet, convincing themselves that they still are obese.

Jean constantly devised new ways to avoid eating and also to dodge the pressures others placed upon her to eat more.

Jean would try to wear bulky knit sweaters to prevent others from commenting on how thin she was becoming and whenever she could she exercised to lose still more mass.

Visits to the doctor were unhappy events as the evidence in the doctor's records showed her ever-decreasing mass. The doctor's tests showed low blood pressure and vitamin deficiencies. Jean no longer menstruated, she often became hysterical, suffered from constipation, and felt unwell. The →

doctor showed her some mass tables and compared her mass to the mass of other girls, but Jean could not be convinced that she was underweight or that her symptoms were caused by her very inadequate diet.

The doctor knew from previous experiences, that anorexia nervosa is a **psychosomatic** illness, a physical illness caused by the mind. After many failures to persuade Jean to eat normally, the doctor sent her to a hospital for treatment.

The hospital provided a special diet for Jean and did everything they could to encourage and gently persuade her to eat. Being in hospital limited Jean's freedom and made it possible to keep a more accurate check on her food intake and activities.

Jean ate under protest, and then later, when the opportunity arose, she would run up and down the 10 flights of stairs from her room to the hospital basement to try to burn off the food that she had absorbed. After two weeks, Jean had lost a further 2.5 kg, and the staff had to impose severe restrictions. Jean had a very strong and warm relationship with her family and when the hospital refused to allow any visits by her parents or sisters until she had regained the 2.5 kg, Jean finally made up her mind to try to eat at least a small amount of food.

The severe restrictions were successful and after several months Jean was released from hospital.

Now, three years later, Jean is a slim young woman with a good healthy outlook on her problem. She is aware that the desire to avoid food is still there, but she is determined to overcome it. Large numbers of people who suffer from anorexia nervosa do not recover and eventually starve themselves to death.

Questions for research and discussion:

1. Try to find out more about psychosomatic disorders and how doctors try to help people suffering from these conditions.
2. In recent years, several extensively advertised diet programs have proven to be not only ineffective but dangerous, and in some instances fatal. Investigate the side effects of the use of steroids by people trying to build their bodies and the use of liquid protein diets.

QUESTIONS FOR REVIEW

SOME FACTS TO KNOW

1. List the health hazards of being overweight.
2. List the major things that determine what YOU eat. (Appetite, what is available, location, type of work, social life, friends, etc.) Indicate which of these produce a "good" effect on your diet and which have a poor influence on your eating habits.
3. Name four food groups and give the servings recommended for a person of your age.
4. Make a list of the food that you ate at a recent meal. Arrange the items into the four food groups and report on how balanced the meal was. If any group was not represented, state what substitution you could have made to improve the balance.
5. Why should the amount of food eaten be varied with the amount of physical activity in which you are involved?
6. To lose mass successfully, what are three useful approaches other than cutting back on food?

QUESTIONS FOR RESEARCH

1. Select one of the following topics and prepare a report on it:
   - Dietitian
   - obesity
   - "The Weight Watchers"
   - the school cafeteria
   - Diet Counsellor
   - advertising and body mass control
   - food fads
   - vegetarian diets
2. Write for some brochures that advertise slimming programs, slimming "equipment", or reducing clinics. Analyse these materials and find out just what is being offered for your money. Are there any guarantees? If they claim that a person will lose a specific amount of fat, work out the cost per unit of fat lost.
3. Investigate the values of isometric exercising.
4. Investigate the values of anaerobic and aerobic exercises.
3. Keep an ongoing record of your mass and growth. Note any changes that you make in your diet or in your exercise program. Try to draw some conclusions as to what factors affect your mass.
4. Make a collection of news items that relate to the areas of the world where many people are starving. Bring these to class for discussion.

Activity 1: EATING FOR ENERGY

The purpose of this exercise is to examine your own diet and discover if it is adequate and well-balanced for a person of your age, sex, size, and activity. You will try to determine if you are eating too much or too little of any particular kinds of food and discover the source of most of your kilojoules of energy. From this information you can be reassured about your food habits, or be made aware of potential problems. It will help you decide what changes you might wish to make to reduce or increase your mass as may be necessary.

This exercise is only of value if you are honest with yourself. You must record your intake faithfully, without trying to make it "look good" or hiding any energy sources.

FOOD ENERGY INTAKE

1. *Make a record of ALL the food and drink that you take in over a three-day period. Include meals, snacks, pop, candy, etc. One of the three days should be a Saturday or Sunday, as our food and activity patterns on the weekend are often quite different from those on school days. Keep your record on a piece of paper or in a small notebook and jot down each item, every time you take a bite or drink. You cannot accurately remember everything you eat if you try to record it later.* Before you start your record, find out how much each of the juice or milk glasses that you use regularly will hold. Fill each of them with water and pour it into a measuring jug, so that you will know what quantity of juice or milk you normally drink.

Judging the size of portions of vegetables is not easy and rarely very accurate. The best way is to judge it in terms of "cups" - a half-cup of carrots or beans, for instance.

2. *Record both the kind of food and the quantity you take.*

For example, breakfast might be

Toast (Brown)	1 slice
Butter	1 pat
Jam	1 tablespoon
Orange juice	170 mL (1 average-sized glass)
Milk	225 mL (1 average-sized glass)

3. *Prepare a table for your results.*

Food	*Amount*	*Kilojoules*	*Protein (g)*

Using suitable food tables record the kilojoules and the proteins listed for each of your food items. Try to be as accurate as possible. (See Table 15.4.) (The Department of Health and Welfare will supply more detailed tables free of charge.)

Table 15.4 Some popular fast take-out foods

	kJ
Hamburger (bun included)	1050
Hamburger, double	1365
Whopper	2646
Cheeseburger	1280
Cheeseburger deluxe	2520
Big Mac	2340
French Fries	966
Onion Rings	1260
Fried Chicken - 2-piece dinner	2457
Fried Chicken - 3-piece dinner	4135
Chili Dog	1386
2 Fish, Cole Slaw, Chips	3108
Chopped Steak, 112 g	1373
Chopped Steak, 224 g	2742
Baked Potato	1016
Salad Dressing	630
Pizza (average) 1/2 of 25 cm	1953
Pizza (average) 1/2 of 35 cm	3780
Pizza (average) 1/2 of 45 cm	5040
Egg McMuffin	1310

Hot Cakes, Butter	1142
Milk Shake	1428
Dairy Queen Small Cone	462
Dairy Queen Medium Cone	966
Dairy Queen Large Cone	1428
Dipped Small Cone	676
Dipped Medium Cone	1302
Dipped Large Cone	1890
Dairy Queen Sundae (S)	788
Dairy Queen Sundae (M)	1260
Dairy Queen Sundae (L)	1806
Hot Fudge Sundae	2436
Banana Split	2436

(Extracted and adapted from several sources including the *Fast Food Calorie Counter* by H. Jordan, L. Levitz, and G. Kimbrell.)

4. *Total up the number of kilojoules and the grams of protein that you consumed each day.*

ACTIVITIES: ENERGY OUTPUT

Each activity in which we take part uses up energy. Some activities will use a great many kilojoules each minute, such as a game of tennis. Others will use less, such as sitting watching television. We use energy even when we are asleep to keep the heart, respiratory organs, and other body functions working. If our intake of food energy approximately balances with our output of energy, then our body mass will probably stay fairly constant. If the intake exceeds the output, we can expect to gain mass. In this activity you will be using an *energy factor* that has been determined experimentally for each type of activity (see Table 15.5)

5. *For the same days that you recorded your food and drink intake, record your activities and the time you spent on each one. This sounds like a tall order, but you can group some of them together to make the task easier. For example, if you go to bed at 11:00 at night and get up at 7:20 in the morning, you can easily record it as:*

   Sleep 8 h, 20 min.

   If you have 6 forty-minute classes in the day where you are sitting down to study, you can record:

   Studying 6 × 40 min.

A physical education class involves much more activity, so it will be listed separately:

Running/exercise 40 min.

If you walk between classes and have 7 class changes, this might be shown as:

Walking 7 × 3 min.

Some activities that have similar energy factors can also be grouped together. Dressing, washing, brushing teeth, making a bed, etc., might be listed as:

Getting ready for school – 25 min.

6. *You will find that each activity has been assigned an energy factor in Table 15.5. Walking 11.6, dishwashing 8.1. Place the energy factor shown in the table against each of the activities in your record. If the particular activity is not given, try to judge the effort involved in comparison to some activity that is shown in the table. Beware of overstating exercise. Heavy exercise is rarely continuous. You may run hard for ten minutes, but then stand and rest for several minutes. A game is rarely continuous heavy exercise; adjust for these conditions.*

Table 15.5. Activity and energy factors

ACTIVITY	ENERGY FACTORS	
	kJ/h	kJ/min
Sleeping	4.1	0.07
Sitting	5.2	0.09
Writing & Studying	6.0	0.10
Standing relaxed	6.3	0.11
Singing, Sewing, Dressing, Washing	7.1	0.13
Dishwashing	8.1	0.14
Playing cards, Typing	9.0	0.15
Dusting & Sweeping	10.5	0.18
Washing the car, Cooking, Piano-playing	11.2	0.19
Walking (3.2 km/h)	11.6	0.19
Bowling	13.6	0.23
Canoeing (1.5 km/h)	14.2	0.24
Sailing	15.8	0.26
Bicycling (3.0 km/h), Walking (4.8 km/h)	16.2	0.27
Table tennis	18.0	0.30
Laundry, by hand	18.6	0.31

Walking (6.4 km/h)	20.6	0.34
Volleyball, Roller skating, Badminton	21.5	0.36
Dancing, slow	22.6	0.38
Bicycling (15.3 km/h)	25.8	0.43
Hiking (or hunting), Dancing, fast, Shovelling	27.0	0.45
Water-Skiing, Tennis, Downhill skiing	36.2	0.60
Climbing stairs, Running (8.8 km/h), Swimming (breast stroke 36.6 m/min)	37.5	0.62
Bicycling (20.9 km/h)	40.5	0.67
Rowing, Cross-country skiing	42.0	0.70
Ice skating (vigorous)	48.8	0.81
Swimming (Crawl, 45.7 m/min)	49.1	0.81
Handball	49.5	0.82
Running (12.9 km/h)	62.0	1.03
Cross-country skiing (competitive)	73.6	1.26

(Adapted from *Lazy Man's Guide to Fitness* by K.D. Rose and J.D. Martin.)
Energy expenditures per kilogram of body mass.

7. *Record your mass in kilograms. Now calculate how many kilojoules you expend on each activity by using the following formula:* Energy factor × time spent on activity (hours) × your mass (kilograms).

   Example: Sitting studying - 1.5 h.
   E.F. × *Time* × *body mass*
   6.0 × 1.5 h × 70 kg = 630 kJ

   When you have completed the calculations, add up the kilojoules expended each day and compare it with the number of kilojoules you gained from food.

   Analysing the results
   Do not expect the totals to balance exactly. If they come within 400-800 kJ, you will be doing well. The measurement of nervous energy poses a problem in obtaining accurate results which can make a big difference in your results.

Questions and Discussion

1. *Go through the record of foods eaten and place a cross beside any food items that are composed mainly of carbohydrates. Many of these contain no nutrients: eg., Cola or pop, white bread, chips, alcohol, etc. Total these and see what proportion of your total intake is made up of this kind of kilojoules.*
2. *If you wanted to gain or lose a little mass, what items could you include or omit from your diet? If you need to reduce mass, what more nutritious food could you substitute so that you lose a little mass but don't go hungry?*
3. *Is the intake of protein adequate for a person of your age and size? See Table 15.6.*

Table 15.6. Recommended daily nutrient intake.

	Age (years)	13-15		16-18		19-35	
	Sex	Male	Female	Male	Female	Male	Female
	Mass (kg)	51	49	64	54	70	56
	Height (cm)	162	159	172	161	176	161
	Energy (kJ)	11 700	8 200	13 400	8 800	12 600	8 800
	Protein (g)	52	43	54	43	56	41
Water—Soluble Vitamins	Thiamin (mg)	1.4	1.1	1.6	1.1	1.5	1.1
	Niacin (NE)	19	15	21	14	20	14
	Riboflavin (mg)	1.7	1.4	2.0	1.3	1.8	1.3
	Vitamin B_6(mg)	2.0	1.5	2.0	1.5	2.0	1.5
	Folate (μg)	200	200	200	200	200	200
	Vitamin B_{12}(μg)	3.0	3.0	3.0	3.0	3.0	3.0
	Vitamin C (mg)	30	30	30	30	30	30
Fat—Soluble Vitamins	Vitamin A (RE)	1 000	800	1 000	800	1 000	800
	Vitamin D (μg)	2.5	2.5	2.5	2.5	2.5	2.5
	Vitamin E (mg)	9	7	10	6	9	6
Minerals	Calcium (mg)	1 200	800	1 000	700	800	700
	Phosphorus (mg)	1 200	800	1 000	700	800	700
	Magnesium (mg)	250	250	300	250	300	250
	Iodine (μg)	140	110	160	110	150	110
	Iron (mg)	13	14	14	14	10	14
	Zinc (mg)	10	10	12	11	10	9

Shaded lines correspond to components commonly given in food tables such as Nutrient Value of Some Common Foods, Department of Health and Welfare Canada. Revised 1975. Health and Welfare Canada.

mg = millogram = μg = microgram
NE = niacin equivalent = 1 mg niacin
RE = retinol equivalent = 1 μg retinol

4. *Does your intake vary with the output? For example, if you are going to have a strenuous day, then your breakfast should be larger, to provide the energy you will need. If you have a lazy Saturday, sleeping in until noon, your intake should be proportionately smaller. (If you have a cold or feel unwell, both your food and activity totals will probably be down.) Make a few brief statements about the days you recorded, indicating whether or not each day was a "normal" day for you. Did you have a cold, play in a team competition, or were you upset over some event or argument? List anything you think may have affected your food intake or changed your activities from your regular pattern.*
5. *Try to express your conclusions about how appropriate your diet is for you. Do you need extra or less mass? Consider your muscle tone and posture. Do you need to exercise? Try to write down any ways in which you think you could improve your health by changing your diet. Try to make your statements as positive and constructive as you can.*

Many questions (and probably many answers to questions), will occur to you while you are working on your analysis. Try to recognize that diet is a *dynamic* thing. That is, it constantly needs to be changed. What is an adequate diet for you today may not be tomorrow. You need to make adjustments. Your diet now, when you are involved with daily physical education classes, walking to school, and participating in sports, may be too much and may increase your mass next year, if you become less active.

Activity 2: DO I HAVE TOO MUCH MASS OR TOO LITTLE?

Purpose
To investigate methods of determining the proportion of body mass that is fat.

Materials
scales and tape (or other means of measuring height).

Method HEIGHT/MASS TABLES
Measure your height to the nearest centimetre. Use a proper height-measuring device if possible. If not, stand up close against a wall, feet flat on the floor, with the heels close to the wall. Make sure that your feet, buttocks, and shoulders are against the wall and that the measurement of your height is taken with the scale at right-angles to the wall. Use a chalkboard setsquare, for instance. The height must be

taken without shoes, of course. *Record this measurement.*

Using the most accurate scales you can obtain, weigh yourself and *record this in kilograms*. Be sure that the scales are set to zero and that you stand evenly on the scales, supporting yourself equally on both feet.

Use the height/mass tables (See Table 15.7) and find the average mass for someone of your sex and height. *Record this figure in your notebook.*

Table 15.7. Height and mass table

MEN			WOMEN		
	Average mass (kg) in light clothing			Average mass (kg) in light clothing	
Height	Years of Age		Height	Years of Age	
(cm)	(15-16)	(17-18)	(cm)	(15-16)	(17-18)
150	44.5	51.3			
151	45.4	51.9	144	45.4	46.3
152	46.3	52.4	145	45.9	46.8
153	47.1	52.9	146	46.4	47.3
154	48.0	53.4	147	46.0	47.9
155	48.9	54.0	148	47.6	48.5
156	49.8	54.7	149	48.2	49.1
157	50.7	55.4	150	48.8	49.7
158	51.5	56.1	151	49.4	50.3
159	52.4	56.8	152	50.0	50.9
160	53.3	57.5	153	50.6	51.5
161	54.2	58.2	154	51.2	52.1
162	55.1	58.9	155	51.8	52.7
163	55.9	59.7	156	52.4	53.3
164	56.8	60.4	157	53.0	53.9
165	57.7	61.1	158	53.6	54.5
166	58.6	61.8	159	54.2	55.1
167	50.5	62.5	160	54.8	55.8
168	60.3	63.2	161	55.4	56.4
169	61.2	63.9	162	56.0	56.1
170	62.1	64.6	163	56.6	57.7
171	63.0	65.4	164	57.3	58.4
172	63.8	66.1	165	58.0	59.0
173	64.6	66.8	166	58.7	59.7
174	65.4	67.5	167	59.4	60.3
175	66.2	68.2	168	60.1	61.0
176	67.0	68.9	169	60.8	61.7

177	67.8	69.7	170	-	62.3
178	68.6	70.3	171	-	63.0
179	69.4	71.2	172	-	63.7
180	70.2	72.0	173	-	64.3
181	71.0	72.7	174	-	64.9
182	71.8	73.5	175	-	65.5
183	72.6	74.3			
184	73.4	75.1			
185	74.2	75.8			
186	-	76.6			
187	-	77.2			
188	-	78.0			

Questions

1. *Compare your mass with the average for a person of your height. What difference is there between these values?*
2. *How accurate do you think this method is? What are the disadvantages of this technique?*
3. *Why do insurance companies prepare and use these guides?*

Activity 3: TO CALCULATE YOUR IDEAL MASS

Introduction

Turn to the section in this chapter that describes the skinfold technique and be familiar with the principles of the test.

Materials

Harpenden calipers, scales (See Figure 15.9.)

Method

In each of four test sites, you will measure the thickness of a fold of skin. Each measurement is taken in a similar manner as described below.

Technique

1. Locate the test site as in a), b), c), and d) and mark it with a felt pen. Take a fold of skin between thumb and index finger 1 cm above the mark on the skin. Hold the flesh firmly, pulling it gently away from the body.
2. Place the jaws of the calipers at right angles to the site, over the pen mark, and carefully release the handles of the calipers. Read the measurements after the caliper pressure is fully applied and the needle is no longer drifting.

   Record your result to the nearest 0.2 mm. *Repeat this twice more and, if you obtain differences of more than* 1 mm, *take another measurement and average your results.*

3. Make the tests at the following sites:
   a) **Triceps**: The arm should be bent at right angles, and the site marked at the midpoint between the top of the shoulder and the tip of the elbow. The skinfold is taken parallel to the arm and the calipers applied after the subject has lowered the forearm.
   (b) **Biceps**: The site is marked and taken on the front of the upper arm halfway between the shoulder and the tip of the elbow. The arm should hang loosely at the side of the body. Take the skinfold parallel with the axis of the arm.
   (c) **Subscapular**: This site is measured about 1 cm below the lower angle of the right scapula (shoulder blade). The caliper jaws are held at 45 degrees to the spine.
   (d) **Suprailiac**: Just above the crest of the hip bone, with the fold at right angles to the axis of the body, parallel to the crest of the bone.

Figure 15.9.
Using the Harpenden calipers.

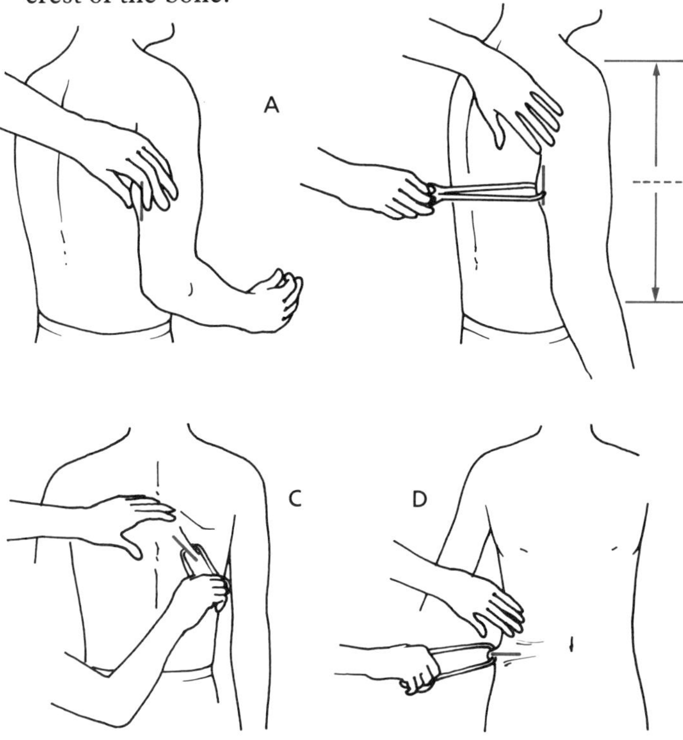

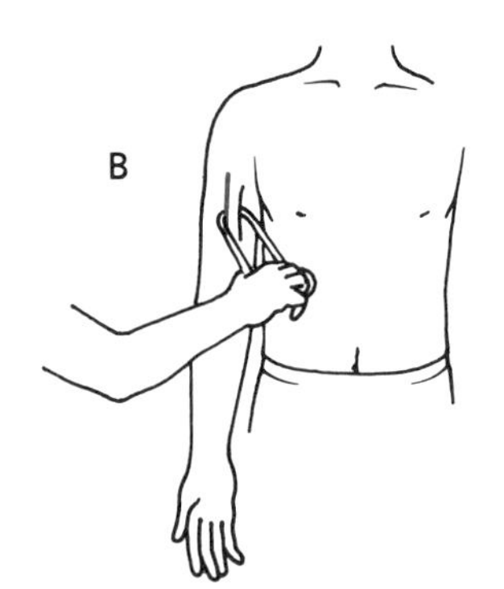

Calculations

To determine the amount of body fat, make a table similar to the format used for the example below (#5):

1. *Add together the four skinfold measurements you have taken.*
2. *Refer to Table 15.8 and find the number nearest to this total in the column under skinfolds that corresponds to your age group and sex. The chart will give you the* ***percentage of body fat.***

Table 15.8. Body fat and skinfold totals

	MALES	FEMALES
Skinfolds (mm)	% Body Fat 17-29	% Body Fat 16-29
15	4.8	10.5
20	8.1	14.1
25	10.5	16.8
30	12.9	19.5
35	14.7	21.5
40	16.4	23.4
45	17.7	25.0
50	19.0	26.5
55	20.1	27.8
60	21.2	29.1
65	22.2	30.2
70	23.1	31.2
75	24.0	32.2
80	24.8	33.1
85	25.5	34.0
90	26.2	34.8
95	26.9	35.6
100	27.6	36.4
105	28.2	37.1
110	28.8	37.8
115	29.4	38.4
120	30.0	39.0
125	30.5	39.6
130	31.0	40.2
135	31.5	40.8
140	32.0	41.3
145	32.5	41.8
150	32.9	42.3
155	33.3	42.8
160	33.7	43.3
165	34.1	43.7
170	34.5	44.1
175	34.9	-
180	35.3	-
185	35.6	-
190	35.9	-
195	-	-
200	-	-
205	-	-
210	-	-

NOTE:
The total percentage of body fat of males (ages 17-29) and females (ages 16-29) has been estimated from measurements of skinfolds. (from standard fitness tests. Ontario)

3. *Record your body mass in kilograms.*
4. *Using the formula, find your Lean Body Mass (LBM). This is the mass of your body if it could be weighed without any fat present at all.*

   LBW = Present mass − (Present mass × % of body fat)
5. *Now determine your ideal mass, which should be your Lean Body Mass plus 15% body fat for males or 20% body fat for females. Calculate your ideal mass by multiplying your Lean Body Mass by 1.18 if you are a male, or 1.25 if you are a female.*

   Example: Female

   Present Mass = 54 kg

   Total of skinfold measurements 55 mm = *27.8% fat.*

   LBM = Present mass − (Present mass × % of body fat)

   = 54 kg − *(54 × 27.8%)*

   = 39 kg

   Ideal Mass = LBM × 1.25

   = *39* × 1.25 kg

   = 48.75 kg *Difference: Actual Mass − Ideal Mass*

   = *54 − 48.75* = 5.25 kg overweight

Questions

1. *Compare your calculated ideal mass with your present mass. Does this test suggest that you need to make any adjustments in your diet or exercise patterns? What do you suggest?*
2. *Do you feel that the test confirms your own subjective evaluation of your mass? Comment on this. (Remember that this test measures* **your** body fat, it is not for making comparisons with others or the average masses of persons in your age group.

Unit VIII

How the Body Removes Wastes from the Blood

Chapter 16

The Excretory System

- The Excretory System
- The Organs of the Excretory System

Excretory System

Every body activity uses energy and generates wastes. If waste products were not removed, they would quickly accumulate in harmful proportions. Some wastes are toxic (poisonous) and pose a serious threat to health if they are not removed promptly. The process of getting rid of metabolic wastes is called **excretion**.

(ex-cree-shun)

Fortunately, atoms and molecules do not wear out, but may be changed or rearranged and used over and over again. Many of the end products of various cell activities can be recycled and used in other processes. As a result, the amount of waste which actually needs to be discharged from the body is very small in relation to the amount of work done.

Excretion Through the Lungs

There are some substances which must be excreted quickly before they have time to build up and become toxic. Carbon dioxide, which is one of these, is excreted through the lungs. The lungs are not often thought of as an excretory organ. Any structure, however, that enables us to eliminate a metabolic waste product plays a part in the excretory process. The lungs can also eliminate alcohol, which, since it is made up of small molecules, can be rapidly passed from the blood stream into the alveoli of the lungs and exhaled. This is why the breathalyzer test is an effective and simple way of determining the amount of alcohol that people have taken into their systems.

Carbon dioxide and water are the major waste products of the energy-forming reactions that are constantly taking place inside the body. When carbohydrates and fats are used to supply fuel to muscles and organs, carbon dioxide and water are released. Although water is only a by-product of the energy forming process, it provides no energy itself; it is not really waste, and the body needs all the water it can get. This "waste" water is thus used in many of the other important processes and activities carried on by the body. Not all of the water can be saved; some water vapour is lost from the lungs and some water must be used to dissolve the wastes we excrete as urine. Water also evaporates from the skin. (See Table 16.1.)

Table 16.1 Water gain and loss by the body.

WATER INTAKE		WATER OUTPUT	
Liquids	1200 mL	Urine	1500 mL
Food	1000 mL	Feces	150 mL
Oxidation of food	300 mL	Lungs	300 mL
		Skin	550 mL
	2500 mL		2500 mL

Excretion Through the Skin and Anus

Salts, as well as some urea and water, are excreted through the pores of the skin. If you taste the sweat on your skin, you will find it has a distinctly salty flavour.

Bile and other substances are secreted into the digestive tract to help in digesting food. These substances will eventually be excreted in the feces, along with the food materials, such as fibre, that we cannot digest. Solid waste excretion (egestion) is discussed at the end of the chapter on digestion. (See Table 16.2.)

Table 16.2 The products excreted by various organs of the body.

EXCRETORY ORGAN	PRODUCTS EXCRETED
Kidneys	Nitrogenous wastes resulting from protein metabolism, such as urea, poisons, water, mineral salts
Lungs	Carbon dioxide, water
Intestine	Waste residues of digestion (mostly cellulose), metabolic wastes, such as bile pigments
Skin	Water, mineral salts

The Organs of the Excretory System

(you-ree-ter)
(you-ree-thrah)

The excretory system consists of two **kidneys** and two **ureters**, which carry the urine produced in the kidneys to the **urinary bladder**. Another tube, the **urethra** carries the urine out of the body through the urogenital organs, either the penis or vagina. (See Figure 16.1.)

Figure 16.1a.
The structures of the urinary system.

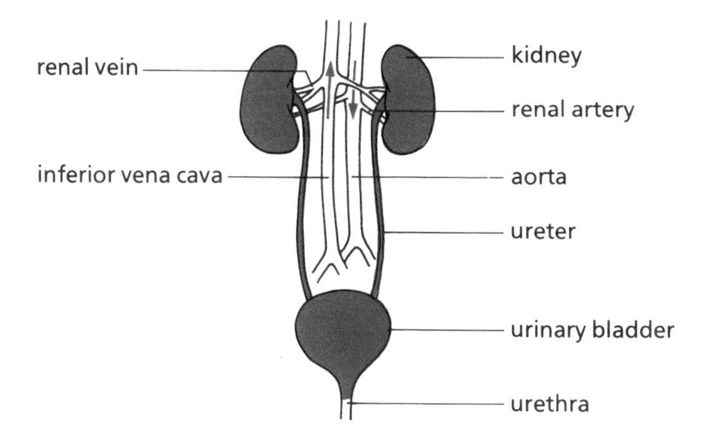

Figure 16.1b.
X-ray of lower abdomen showing the ureters and bladder.

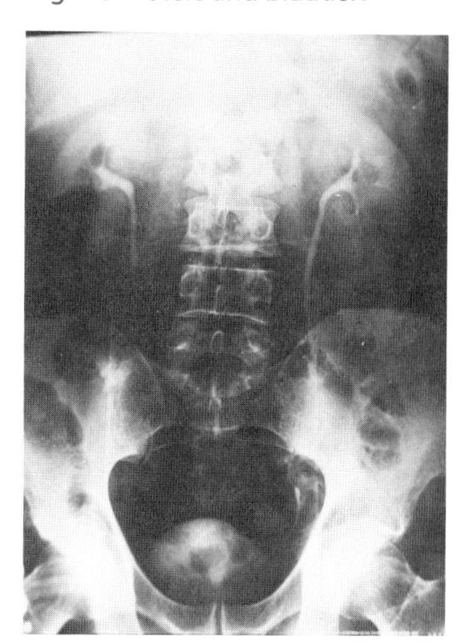

The Kidneys

The kidneys are the major excretory organs of the body. They are found in the posterior wall of the abdominal cavity, one on either side of the spine, just below the level of the ribs. They are snugly secured in a packing of fat and are provided with a rich supply of blood, which comes directly from the aorta. This blood enters the kidney through the **renal arteries**, which carry about 25% of the blood pumped out during each contraction of the heart. The **renal veins** carry blood back from the kidneys to the inferior vena cava.

(ree-nul)

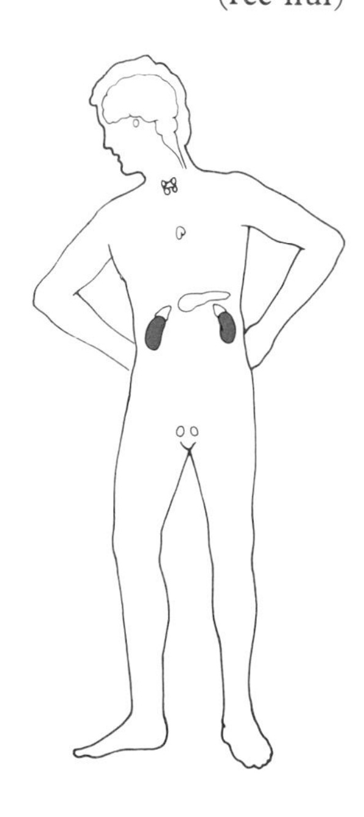

The Functions of the Kidneys

The major function of the kidneys is the elimination of nitrogenous and other dissolved wastes. Wastes released from cells throughout the body are dissolved in water and carried by the circulatory system to the kidneys. The kidneys filter these wastes, making them more concentrated in preparation for excretion and, in the process of doing this, retaining as much water as possible. Thus, the second major function of the kidneys is maintenance of water balance.

Thirdly, the kidneys play an important role in regulating the acid-base balance of the body. Excessive amounts of either acidic or basic (also called "alkaline") materials are constantly being extracted and excreted by the kidneys. Because a normal diet includes more acid-producing than alkaline-producing foods, getting rid of, or neutralizing extra acid is a problem for the body. As well as being able to excrete such unwanted chemicals, the kidneys can also manufacture ammonia, an alkaline substance that acts to neutralize acids, thereby removing their harmful effects.

The Structure of the Kidney

The kidney is shaped like a lima or brown bean and is about the size of a small fist. It has a dark-coloured core which is densely packed with blood vessels. These vessels form a network of capillaries and small tubes known as the **medulla**. The medulla is broken up into a number of triangular-shaped divisions called the **pyramids**. (See Figure 16.2.)

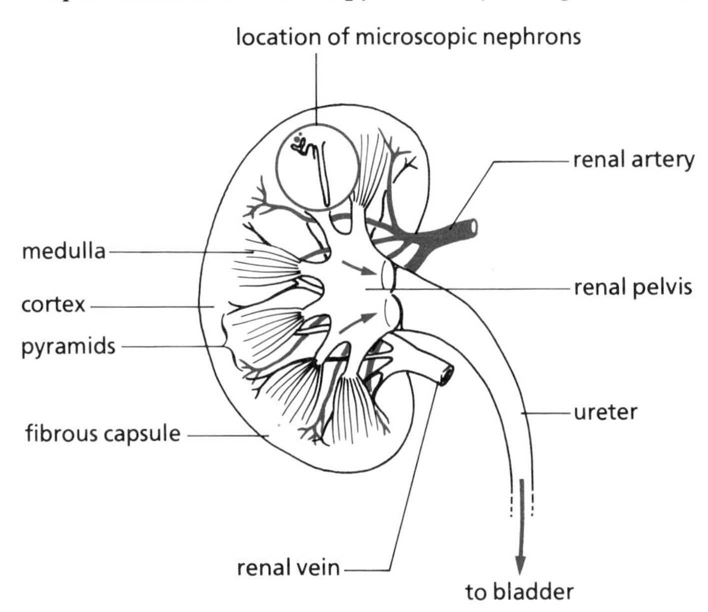

Figure 16.2.
A longitudinal section through the kidney.

(nef-ron)

The filtration of blood takes place in minute structures called **nephrons**, found in the **cortex**, which is the outer layer of the kidney. These structures are very small; there are estimated to be more than one million nephrons in each kidney. In spite of their microscopic size, the nephrons filter a total of more than 180 L of blood every 24 h. Most of this filtered fluid is reclaimed by the body, and only about 1.5 L of urine is passed each day. Sometimes, in addition to the wastes we would expect to find in the urine, some useful substances are present. Physicians use the analysis of a urine sample as a common diagnostic technique. The presence or absence of a particular substance in the urine may direct the physician's attention to a problem or provide information about the malfunction of a particular organ.

The Nephron

Each nephron consists of a ball of capillaries called the **glomerulus**, which lies inside a cup-shaped structure known as the **Bowman's capsule**, and a series of tubules. (See Figure 16.3.)

(glom-air-you-lus)

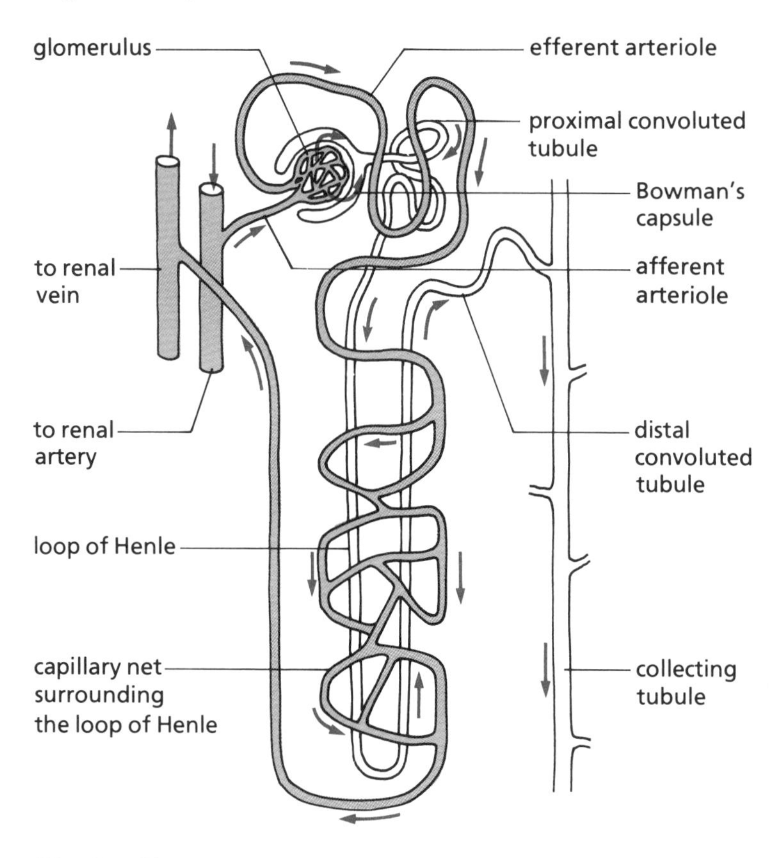

Figure 16.3.
The nephron. The microscopic filtering unit of the kidney. There are more than one million of these units in each kidney.

Blood Delivery to the Nephrons

Blood arrives at the kidney by the renal artery. It then flows into increasingly smaller vessels until it is eventually delivered into an afferent arteriole leading into a glomerulus. After passing through the ball of capillaries in the glomerulus, it leaves through the efferent arteriole and the renal veins.

The blood flowing into the glomerulus is rich in oxygen and contains all of the normal constituents of blood: blood cells, proteins, amino acids, glucose, nutrients of all kinds, salts, and wastes. The nephron's function is to remove the wastes, while retaining the other useful substances. It is rather like trying to clean out a drawer that has been allowed to collect many different items. It is difficult to pick out just the junk items and is easier to tip out the entire contents, replace the things you wish to keep, then throw away what is left. The nephrons work in a similar fashion. Both wanted and unwanted substances cross into Bowman's capsule. Other parts of the nephron then remove the useful materials and pass them back into the blood. The remaining substances, the wastes, are delivered to the bladder for disposal as urine.

The Formation of the Filtrate

As described above, blood supplies to the kidney are carried by the large renal artery. This large vessel rapidly divides into many smaller vessels as soon as it enters the kidney. The result of this sudden decrease in diameter of the tube is to produce an increase in the pressure of the blood flowing through. In much the same way, if you reduced the size of the opening at a drinking fountain you would increase the pressure of the stream, and the water would squirt much farther than usual. In the kidney, the blood pressure in the capillaries is 60-70 mm Hg, whereas elsewhere in the body capillary blood pressure is only 25 mm Hg or less. As a result of this higher pressure in the glomerulus, about 20% of the contents of the blood plasma entering the kidneys is passed rapidly and easily into Bowman's capsule. (See Figure 16.4.)

Figure 16.4.
The production of the filtrate between the glomerulus and Bowman's capsule.

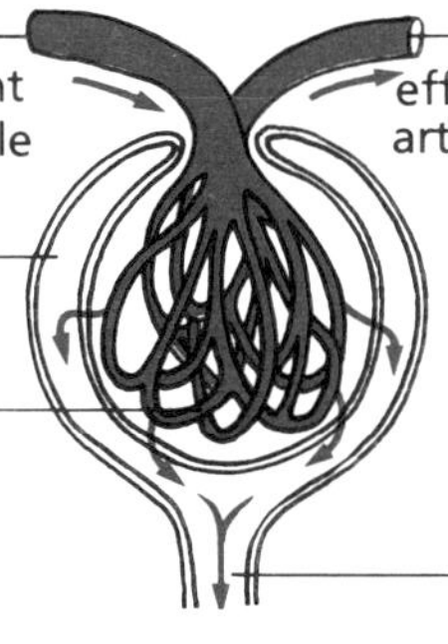

Blood containing cells, plasma, metabolic wastes, salts, glucose, proteins, amino acids, water, etc.
afferent arteriole
efferent arteriole
Blood which has lost about 1/5 of the plasma but proteins, cells, and large molecules remain the same.
Bowman's capsule
glomerulus (ball of capillaries)
Filtrate contains the plasma lost from the blood: water, sugars, wastes, amino acids, salts, etc., but no cells or large molecules.

Those substances that do pass from the glomerulus into the capsule comprise what is called the filtrate. Blood cells and large protein molecules are too large to pass through the tiny pores in the capsule capillaries. Other substances, such as salts, sugars, water, and wastes are made up of smaller molecules. These can pass easily through the pores to enter the capsule. About four-fifths of the plasma component of the blood entering the kidney is left behind and does not become part of the filtrate, but re-enters the blood stream.

The one-fifth that does enter Bowman's capsule contains metabolic waste products and other materials not needed by the body. It also, however, contains salts, water, amino acids, and other useful substances which must be retained and recycled back into the blood.

The Tubules

The filtrate now passes through a series of looped tubes. The short **proximal convoluted tubule** is followed by an extended loop, called the **loop of Henle**. The filtrate then flows into the **distal convoluted tubule**, which finally empties into the **collecting tubules** and the renal pelvis.

(hen-lee)

It is the function of these tubules to sort out the "wanted" from the "unwanted" substances in the filtrate. They then pass the useful substances back into the blood keeping only the wastes which will be excreted as urine. The chemical processes that accomplish this are very complex and beyond the scope of this book. It is important, however, to understand the basic principles involved. (See Figure 16.5.)

Reabsorption in the Tubules

In an earlier chapter diffusion was described as the movement of molecules from an area of high concentration to an area of lower concentration as a result of random molecular motion. For example, if a skunk releases its odourous message, the closely packed molecules of the released chemical gradually spread out, getting farther and farther apart until a few of them reach our noses. As explained, diffusion across a membrane requires no input of energy and, as such, is known as **passive transport**.

Sometimes it is necessary to move molecules in the direction opposite to that taken by a diffusing substance – from a low concentration to a high concentration. Such movement is known as **active transport**. It requires energy and the presence of special carrier molecules, which temporarily associate with the substance to help it pass through a membrane.

As the filtrate passes through the proximal tubule of the nephron,glucose, amino acids, and some salts (also sodium potassium, and calcium) are forced back out of the tubules into the blood stream by active transport. (See Figure 16.6.)

In Figure 16.3 you can see that the efferent arteriole, which leaves the glomerulus, branches out into a network of capillaries which surround the tubules. It is this bed of

Figure 16.5.
How the nephron works.

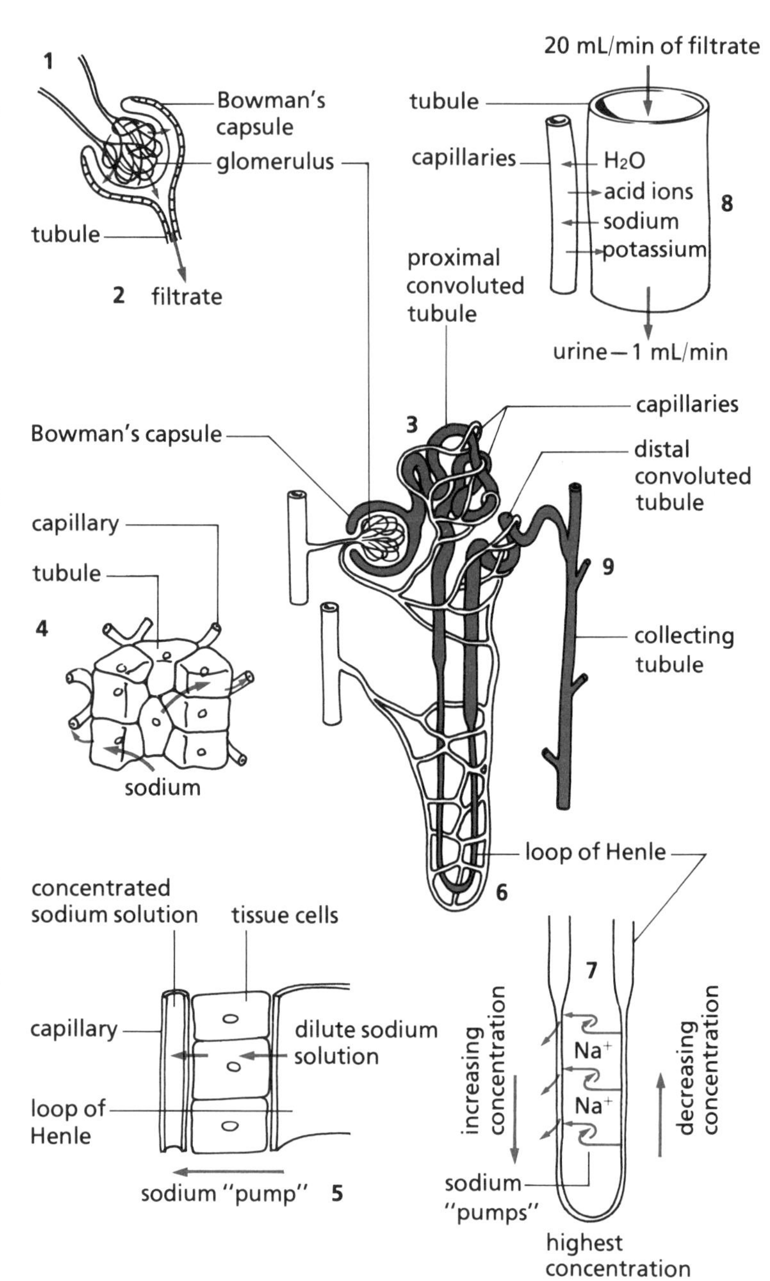

1
Blood pressure forces 120 mL of blood plasma through the capillary walls of the glomerulus into the Bowman's capsule every minute.

2
Filtered plasma enters tubules. Blood cells and large proteins are too large to enter the Bowman's capsule.

3
Capillaries around the proximal tubule reabsorb glucose, amino acids, and potassium ions. Some sodium and water is also reabsorbed.
Wastes – creatine, sulphates, urea, etc., are left in the tubule.

4
Active reabsorption of sodium ions starts. Cells around the tubules take sodium ions from the filtrate and pump them back into the capillaries.

5
Sodium ions must be moved from a dilute solution to a more concentrated one. This requires energy to work the sodium "pumps".
Chloride ions follow the sodium and help form a salty solution that draws water out of the tubules for reabsorption by the capillaries.

6
The highest concentration of sodium develops at the bottom of the loop. A high osmotic pressure is produced in the cells surrounding the loop drawing out still more water for reabsorption.

7
Sodium ions are pumped out of the ascending loop and into the descending loop of Henle, making the concentration of sodium in the tubule more and more concentrated.

8
About 20 mL of the initial 120 mL of filtrate remains in the distal tubule. Here hormones determine how much sodium and water will be reabsorbed to balance the body needs.

9
Acid ions, some combined with ammonia, are exchanged for sodium ions and pass into the tubule. Potassium ions may also be exchanged for sodium and enter the collecting tubule.

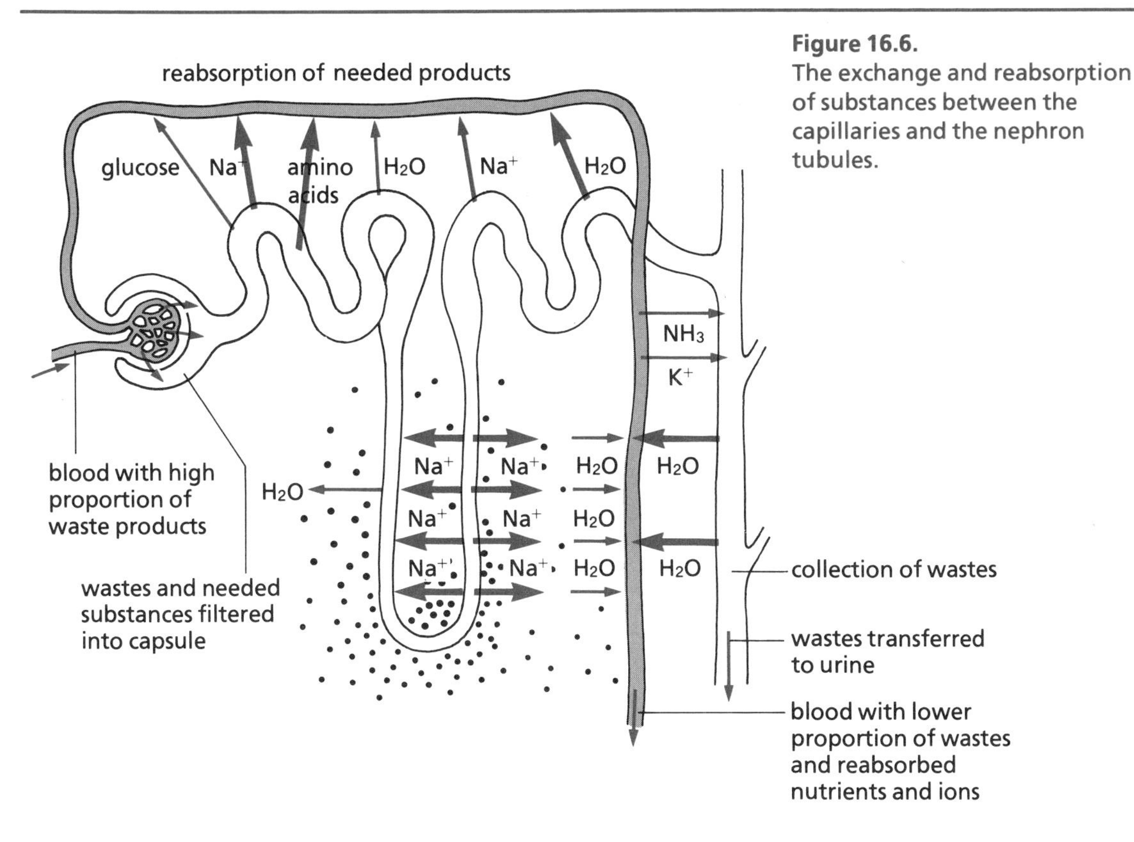

Figure 16.6.
The exchange and reabsorption of substances between the capillaries and the nephron tubules.

capillaries that is constantly reabsorbing the needed products from the filtrate and enabling them to re-enter the blood stream. The waste substances remain in the tubules, becoming gradually more concentrated until they are formed into urine. The cells of the proximal tubule are filled with mitochondria and are even more active and demanding of energy than are the muscle cells.

The cells of the proximal tubules also contain microvilli which increase the efficiency of absorption by increasing the surface area of the tubule. As salts (and ions) are forced out of the tubule the concentration of the filtrate gradually changes. Eventually, the situation reverses itself so that there are more salts outside, and more water inside the tubules. This establishes the condition under which normal diffusion takes place; water then flows out of the tubules by osmosis and eventually re-enters the capillaries.

As mentioned, not all of the dissolved substances leave the proximal tubule to return to the blood stream. Most of the urea remains behind, together with many salts. In the loop of Henle, the concentration of dissolved materials in the filtrate

THE ARTIFICIAL KIDNEY

The dialysis machine. This machine operates as an artificial kidney.

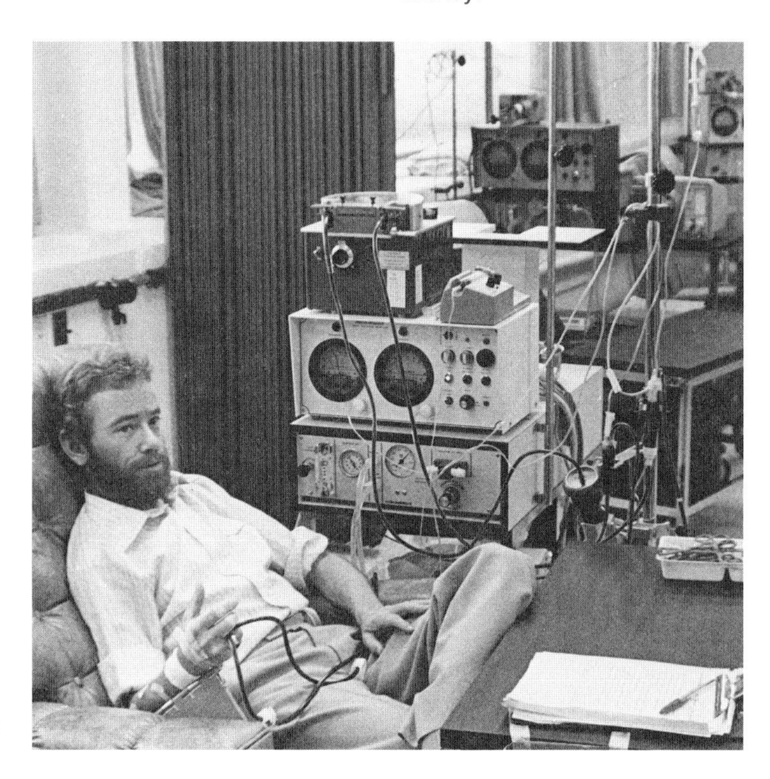

The artificial kidney machine is used to purify blood when the kidneys fail to do their work. This most commonly happens when kidneys become diseased.

In many cases, a permanent plastic "shunt" is implanted between the artery and vein in the arm or leg of the patient (see illustration at the end of this box). During the **dialysis** or blood cleansing, arterial blood is led to the artificial kidney and returned to the vein.

In this machine, a semipermeable membrane such as cellophane tubing is used to do the filtering normally done by the kidney. The entire apparatus is immersed in a tank containing a dialysing fluid, a liquid with a concentration of salts and other substances the same as that of blood. The basic feature of all kidney machines is a semipermeable membrane with tiny pores from 0.0004 to 0.0008 μm in diameter. The membrane pores are of a size that will allow water, urea, uric acid, ammonia, and salts to pass through but will hold back larger molecules such as glucose, fatty acids, amino acids, and proteins. Sodium and potassium will pass freely into the dialysing fluid. These salts are essential to the body so their levels as well as others must be kept constant. It is difficult to prevent the movement of these salts out of the blood into the fluid. That is why the dialysing fluid contains the same concentration of these salts as the blood does. The migration of these salts can go

is approximately equal to that of blood, although this balanced condition changes when extra sodium is pumped out of this part of the nephron by active transport. The walls of the ascending loop of Henle are impermeable to water, however, so that no water follows the sodium by osmosis as it would if the membrane did not block it.

The lining of the distal tubule can become more or less permeable to water depending on the needs of the body. If the body is dehydrated and thus needs water, the cells lining

The Artificial Kidney Machine.

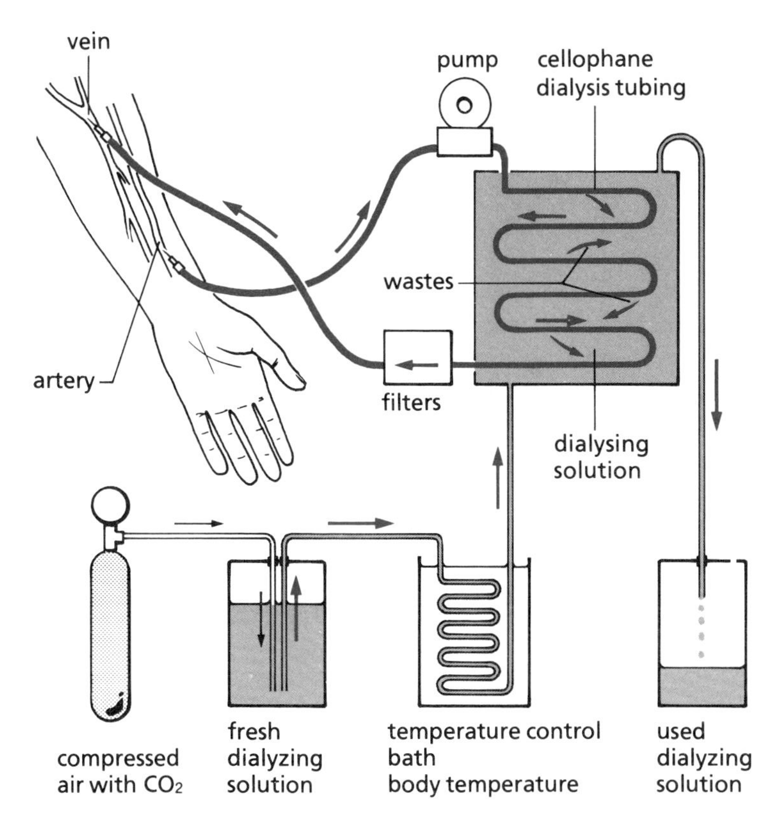

both ways across the membrane.

Provided that the machines are available, patients today can do their own dialysis at home. The hemodialysis takes from three to six hours and is usually done three times a week.

Another kind of dialysis is now possible for patients with kidney failure. It is called continuous ambulatory peritoneal dialysis (CAPD for short). It has made life much easier for patients with kidney failure because it does not require a large machine.

This method uses the body's own abdominal lining called the peritoneum as the semi-permeable membrane for filtering. A fluid with the same salt concentrations as the blood is introduced through a tube (called a catheter) directly into the abdominal wall. A person who uses this method has a catheter implanted into the abdominal cavity. The catheter is simply capped when not used. First, the fluid is inserted through the catheter. Then, the peritoneum acts as a filter so that diffusion and osmosis take place between body fluids and blood supplying the stomach and intestines.

Patients on this peritoneal dialysis have to replace about two litres of the fluid (the dialysate) four times a day and replace it with fresh dialysate. The whole process takes about thirty minutes and can be done almost anywhere.

Research is being carried out by engineers and physicians to test a permanently wearable artificial kidney.

the distal tubule become more permeable and water passes into the blood capillaries to be returned to the body. If there is plenty of water present in the body, these membrane linings become less permeable and water remains in the tubules to dilute the urine. More urine will then be passed and the water balance of the body will be kept approximately constant. A summary of the functions of each part of the nephron appears in Table 16.3.

Table 16.3 Summary of functions within the nephron.

PART OF NEPHRON	FUNCTION	SUBSTANCE RELOCATED
Glomerulus	Filtration of plasma	Dissolved substances and water, no large proteins.
Proximal tubule and the loop of Henle	Reabsorption both by diffusion and active transport. Reabsorption of water by osmosis.	Sodium ions, glucose, and amino acids. Chloride ions, water
Distal and collecting tubules	Reabsorption by active transport and diffusion. Water absorption by osmosis. Secretion by active transport.	Water, sodium ions, ammonia, potassium ions, certain drugs.

The Collection of Urine

(kae-liks)

The distal tubules empty into straight collecting tubules found in the renal pyramids. These collecting tubules, each containing the filtered products of hundreds of nephrons, eventually drain into cuplike **calyxes**. From these structures the urine flows out of the kidney through the ureter.

Although each nephron only produces a minute quantity of urine, the amount produced by two million nephrons together is considerable. About 180 L of filtrate are formed each day (more than twice the individual's body mass!). From 80 to 85% of the water in the filtrate is reabsorbed by the tubules, however. It has been estimated that more than a kilogram of salt, about half a kilogram of sodium, and 180 g of sugar are reabsorbed with the water. An impressive recycling job!

The Antidiuretic Hormone (ADH)

(an-tee-die-yur-e-tik)

Antidiuretic hormone is released by the pituitary gland into the blood stream where it affects the permeability of the collecting tubules. The hormone's function is to match the reabsorbing ability of the tubules to the water needs of the body. At the same time it helps to control the blood-water concentration.

The Composition of Urine

Normal urine is an amber or yellow, transparent fluid, which is usually rather acidic. These characteristics can vary a great deal and still be within the normal range, however, depending on a number of factors. If we eat a standard, mixed diet, acid-producing foods predominate and the urine will be acidic. If we consume a vegetable diet, the urine will be alkaline. The quantity of urine also varies with the temperature and the quantity of liquid ingested. The average amount of urine passed each day is about 1.5 L. About 95% of this urine is water and the rest is made up of about 60 g of dissolved solids. Table 16.4 gives a summary of the characteristics of normal urine.

Table 16.4 The physical characteristics of normal urine.

Amount (per 24 h)	1500 mL. This varies with intake of fluid, temperature, amount of sweating, etc.
Colour	Straw-coloured or amber. If only small amounts are passed the colour is darker. Changes with diet (red from beets).
Clarity	Clear or transparent, but cloudy after it is left standing.
Odour	After standing for a while it has a characteristic "ammonia" odour.
Acid-Base balance	Normal range. pH 4.8-7.5, usually about pH 6. May be less acid if diet is mainly vegetables. Acid on a high protein diet.

The Ureters

Once urine has been formed by the nephrons, it drains into the collecting tubules and then empties into the renal pelvis. (See Figure 16.7.) From the pelvis the urine flows into the ureters (one from each kidney) and is carried down to the urinary bladder.

The ureters, which are 25-30 cm long, become somewhat larger and thicker-walled as they descend. Their inner cell-lining secretes mucus which provides protection from the urine. Layers of muscle in the ureter walls contract in peristaltic fashion to move the urine along.

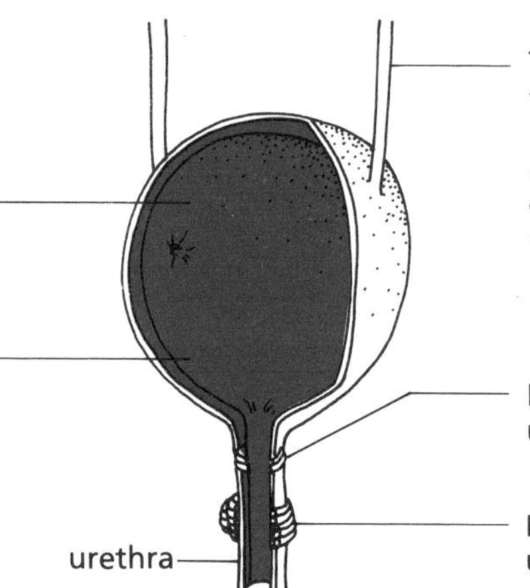

Figure 16.7.
The urinary bladder and its associated tubes.

The Urinary Bladder

The urinary bladder is a hollow organ found in the pelvic cavity, just behind the pubic bone. In men it is directly in front of the rectum and in women it is in front of, and under, the uterus. It is loosely held in place by folds of the abdominal lining (**peritoneum**).

At the base of the bladder is the opening to the **urethra** surrounded by circular muscle fibres forming an internal sphincter. Just below this is another ring of skeletal muscle, the external sphincter. (See Figure 16.7.)

Urination

The quantity of urine contained in a full bladder may vary from 200 to 400 mL. When the bladder is full, stretch receptors in the walls of the bladder send impulses to the spinal cord making an individual conscious of the need to urinate. The brain then sends messages to the external sphincter when an opportunity to empty the bladder occurs (**micturation**). Although emptying the bladder is basically under reflex control, the external sphincter muscles are under voluntary control. The external sphincters over-ride the reflex action and we can control the need to urinate by conscious effort. Infants under two years of age are unable to control the bladder voluntarily without training.

The Urethra

The urethra, a tube that leads from the bladder to the exterior of the body is quite narrow, only about 6 mm in diameter. In females, it lies posterior to the pubic bone and is embedded in the anterior wall of the **vagina**. In males, the urethra passes first through the prostate gland, then through the penis. In the male, the urethra forms a duct common to both the reproductive and excreting systems. It carries both urine and semen through the **penis**.

QUESTIONS FOR REVIEW

SOME WORDS TO KNOW

Match one of the statements from the column on the left with a suitable term from the column on the right. DO NOT WRITE IN THIS BOOK.

1. The tube leading from the kidney to the urinary bladder.
2. Microscopic filtration units in the kidney.
3. The name of the artery carrying blood to the kidney.
4. Minute "tufts" of capillaries in the nephron.
5. The major organ of excretion.
6. The tubule carrying filtrate from Bowman's capsule to the loop of Henle.
7. The tube that carries urine away from the bladder.
8. The fluid that passes from the glomerulus to the Bowman's capsule.
9. A ring of muscles that can close off a tube.
10. A waste product, high in nitrogen, that is found in the urine.

A. Bowman's capsule
B. glomerulus
C. loop of Henle
D. nephron
E. ureter
F. urethra
G. filtrate
H. proximal tubule
I. distal tubule
J. urea
K. aorta
L. sphincter
M. proteins
N. urine
O. kidney
P. micturation

SOME FACTS TO KNOW

1. Draw a sketch and label the major parts of the excretory system.
2. Give the sequence of structures and tubes through which a molecule of water would pass as it is filtered out of the glomerulus, until it enters the urethra.
3. Why are there so many nephrons in each kidney (about one million)?
4. What causes the filtrate to cross from the capillaries of the glomerulus to the Bowman's capsule?
5. Why are large proteins and blood cells not found in the filtrate?
6. What is the function of the antidiuretic hormone?
7. Define "diffusion", "active transport", and "osmosis".
8. What are the normal characteristics of urine?
9. How does the composition of urine differ from that of the filtrate?

QUESTIONS FOR RESEARCH

1. Select one of the following topics and prepare a report on it:
   - cystitis
   - kidney stones
   - the artificial kidney machine
   - urinalysis
   - the signs and symptoms of common urinary disorders
   - gout
   - nephritis
   - urology
2. Sweat is mainly water, but it also contains inorganic and organic substances. Research the substances that are excreted in sweat. Some substances that are harmful to us can also be absorbed through the skin. Try to find out what these substances are.
3. During a physical examination the physician may ask for a urine sample - usually the first morning sample. What kinds of tests does the physician require the laboratory to perform? Find out what each test signifies to the physician.
4. Compare the contents of plasma, filtrate, and urine. Note both differences and similarities.

Activity 1: ANALYSIS OF URINE

Purpose

To Test:

a) the pH of the urine (acidity or alkalinity).

b) the presence of proteins (not normally found in quantity).

c) the presence of glucose (urine sugar is not normally found in quantity).

d) the presence of **ketones** (substances formed in the liver when fat instead of carbohydrate is the main body fuel; not normally found).

e) the presence of **bilirubin** (associated with abnormal bile pigment; not normally found in urine).

f) the presence of hemoglobin (not normally found in urine).

Materials

one urinalysis Combistik for *multiple testing* per student (available from a pharmacy), one comparison colour chart per student, watch or clock with sweep second hand

Method

Follow the procedure as outlined in the directions in the Combistik package.

1. Do not expose the test strip to direct sunlight or heat. Do not touch the test areas of the reagent strip with your fingers.

2. Set out the colour comparison chart, urinalysis report table, watch, and pencil within easy reach.
3. Examine the test strip. The test areas on the strip are in the following order:

   a) pH
   b) protein
   c) glucose
   d) ketones
   e) bilirubin
   f) blood (hemoglobin)

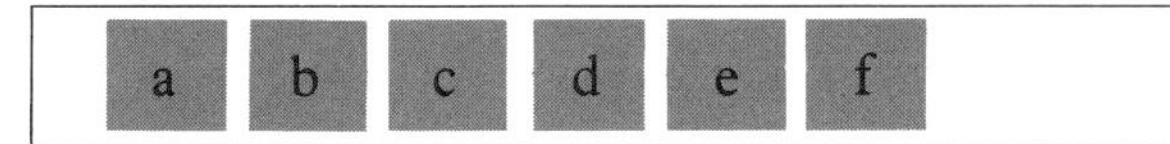

4. Moisten the test strip in the urine stream. Do not use the urine at the beginning of the stream. Do not over-moisten the strip, as some of the chemicals in test areas may be washed out.
5. Compare each of the area colours on the test strip with the comparison colour chart. After the comparison has been made, *record your result by circling the appropriate word or number in the table below.*

URINALYSIS REPORT

Glucose	Negative	Light	Medium	Dark	
Ketones	Negative	Small	Moderate	Large	
Bilirubin	Negative	Small	Moderate	Large	
Blood	Negative	Small	Moderate	Large	
pH	5	6	7	8	9
Protein	Negative	Trace	Light	Medium	Large

Questions

1. *What does an abnormally high sugar level in urine indicate?*
2. *What is a ketone? What should be the normal level?*
3. *What is pH? What are the generally accepted levels in normal urine?*
4. *Where is bilirubin manufactured?*
5. *If urine contains blood or proteins, what portions of the nephron are non-functional?*

Unit IX
Chemical Control of the Body

Chapter 17

The Endocrine System

- The Endocrine System
- The Organs of the Endocrine System

The Endocrine System

It could be said that the endocrine system is what makes heroes! You may have read or heard stories of how, in an emergency, an individual was able to perform some feat of strength or endurance far beyond his or her normal capacity - swimming against an especially strong current to save a child from drowning perhaps, or lifting an unusually heavy object to free a trapped victim. One of the functions of the endocrine system is to mobilize the body's energy reserve, thus giving normal strength, speed, or endurance a boost during moments of special danger and stress. During periods of calm the system acts to conserve energy so that a supply of it is always available for such special occasions. The endocrine system has additional important regulatory functions, which are described in this chapter.

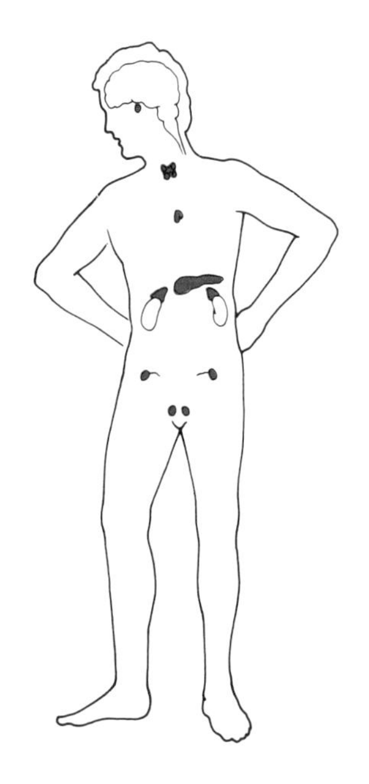

What are the Endocrine Glands?

There are two kinds of glands in the human body. **Exocrine glands** secrete their products into tubes for transportation either onto some surface or into some chamber. For example, the salivary glands secrete saliva into the mouth and the sweat glands secrete solutions onto the body surface. The **endocrine glands**, which together form the **endocrine system**, secrete substances directly into the blood stream (*endo* means within). The endocrine glands are the chemical regulators of the body. They produce substances called **hormones**, chemical messengers sent via the blood stream to target organs which they regulate.

The Role of the Endocrine Glands

In a single-celled organism, one cell carries out all the work of metabolism, respiration, excretion, digestion, and all other body processes. In the human body, cells have become specialized; different types possess special shapes and structures enabling them to carry out special functions. These different cells and functions must then be co-ordinated.

Co-ordination of functions in the body is achieved by two-partner systems that work together. The first is the network of nerves comprising the autonomic nervous system. This system is built for speed, producing a swift reaction to a stimulus. The second system is made up of the endocrine glands. Endocrine hormones generally produce a slower action than do the nerve impulses, although two of the hormones produced by the adrenal glands can give a rapid and vital response in an emergency.

The nervous system enables us to make continual, rapid adjustments to changes that occur in our environment. The endocrines tend to regulate those processes which involve a more sustained adaptation to change, such as growth, sexual maturation, or control of sugar levels in the blood.

Hormones

Each hormone regulates one or more chemical reactions in the body, but different hormones cause very different effects. Hormones are made, like everything else in the body, from materials in the food we eat. Most hormones are proteins and are therefore made up of a combination of amino acids. They are pumped through the blood vessels of the circulatory system and travel rapidly through the body to reach their target organs. As these substances pass through the body, they come into contact with many cells which remain unaffected by the hormones; these cells possess no appropriate receptors to recognize the hormone and absorb it. When a particular hormone does reach its target organ, its presence is detected by special receptors. These receptors, which are sensitive only to it and not to other hormones, combine with the hormone and carry it into the cell where it activates certain responses. Some hormones alter the cell membrane, making it easier (or more difficult) for glucose to pass into the cell. Others control the rate at which certain enzymes carry out reactions, either speeding them up or slowing them down. Hormones may trigger the release of chemicals from other cells, speed up the action of heart muscles, or alter the rate of respiration.

Hormones are produced in only minute amounts. Most of the endocrine organs themselves are small in relation to the large effect they can produce. Hormones are released in response to a specific signal from the nervous system, or from some other gland, or in response to the presence of some special substance in the blood.

The Organs of the Endocrine System

The location and function of each of the organs in the endocrine system are shown in Figure 17.1. As we study the various endocrine glands, we shall find that we are often discussing what happens when the glands are either overworking or not working hard enough. When these glands are functioning normally, we are not conscious of their operation. However, when something upsets their usual activities, the results are often quite obvious.

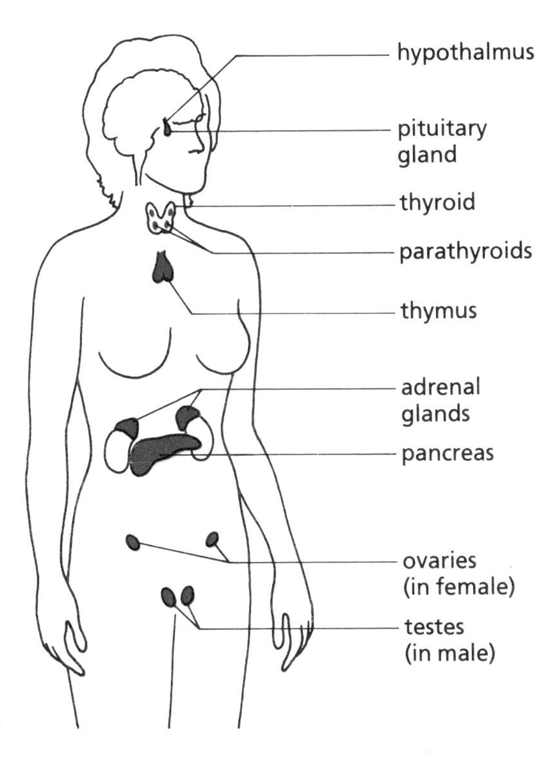

Figure 17.1.
The glands of the endocrine system and their general endocrine functions.

PITUITARY: the master gland, controls or influences all the other endocrine glands.

THYROID GLAND: influences the metabolic rate, decreases blood calcium.

PARATHYROID GLANDS: increase blood calcium.

THYMUS: aids in immunity reactions in younger people.

ADRENAL GLANDS: help to prepare the body for stress.

PANCREAS: endocrine function is to control the blood sugar.

OVARIES: produce the female sex hormones; influence the secondary sex characteristics.

TESTES: produce the male sex hormones; influence the secondary sex characteristics of the male.

The Thyroid Gland

The thyroid gland straddles the "Adam's Apple" (in the neck) and extends slightly below it. It is shaped like the letter H, consisting of a left and right lobe connected by a narrow piece of tissue. It is about 3 or 4 cm high and about 2.5 cm across, and has a very rich blood supply. As well as other functions, the thyroid has an important effect on growth, particularly that of the long bones.

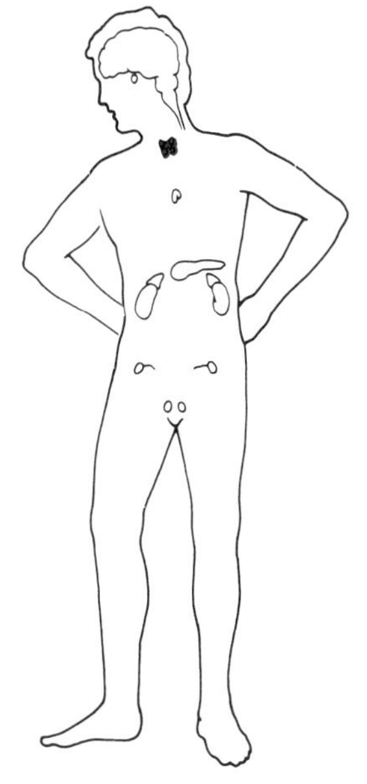

The thyroid gland contains many small hollow sacs that contain a jellylike substance made from proteins. These tiny containers act as a warehouse, holding either the actual hormones or the raw materials, such as iodine, needed to make these hormones. The most important hormone produced by the thyroid is **thyroxin**. When the body demands thyroxin, or other thyroid hormones, it can call upon these reserves, or manufacture more from the supplies on hand. Thyroxin causes an increase in the rate at which some chemical reactions take place in the cells. As a result of its action, more oxygen is taken up and an increased amount of energy becomes available for cell reactions.

Calcitonin is also produced in the thyroid gland. It is responsible for the metabolism of calcium and phosphate in the body. Calcitonin is believed to promote the incorporation of these minerals into bone for storage, thus lowering excess levels of these substances in the blood.

(cal-si-tone-in)

Thyroid Disorders

More than one hundred years ago, doctors noticed that certain disorders were common in certain geographical areas and almost entirely absent in others. The condition known as **goitre** involves an enlargement of the thyroid gland, which causes a very large swelling to appear in the front or side of the neck. (See Figure 17.2a.) This disease was common in mountainous and inland regions, but rarely occurred near the sea. It was discovered that iodine is necessary for the normal functioning of this gland, and, that in mountain regions, this element washes quickly out of the soil. Vegetables grown in soil near the coast, however, contain much higher quantities of iodine and seafoods are a good source of the substance.

(goy-ter)

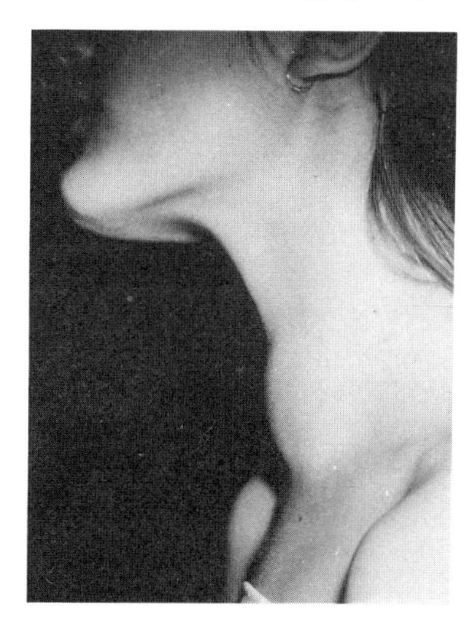

Figure 17.2a.
Simple goitre, possibly caused by lack of iodine in the diet.

Cretinism is a disorder that occurs in young children when underactivity of the thyroid gland results in insufficient amounts of growth and other hormones being produced. Such children are both mentally and physically handicapped. Low levels of iodine in the mother's diet during pregnancy are often the cause of this disorder. Legislation in Canada now requires that small amounts of iodine be added to salt (producing iodized salt). This has almost entirely eradicated cretinism and goitre from the country. (See Figure 17.2b.)

When insufficient thyroid hormone is produced in adults, a condition known as **myxedema** results. (See Figure 17.3.)

(miks-ed-ee-ma)

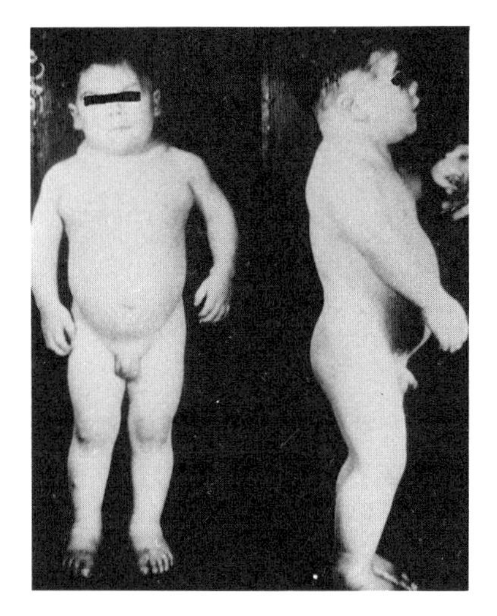

Figure 17.2b.
A 12-year-old boy with reduced mental and physical development caused by cretinism.

Some of the symptoms of this disorder are thickening and puffiness of the skin, increased mass, coarse and brittle hair, and very much slower mental and physical reactions. In general, there is a slowing down of body functions – heart rate, respiration, thought processes – combined with general apathy and tiredness. Both of these disorders, cretinism and myxedema, whether in children or adults, can be relieved with thyroid extracts. Increased amounts of iodine in the diet can also help to relieve cretinism.

Excess amounts of thyroid hormones result in faster oxidation of nutrients than normal, causing an individual to become excitable, irritable, and nervous. The extra energy being burned also causes the body temperature to rise. Appetite increases but mass decreases as the body reserves are burnt off. In some cases, removal of part of the thyroid gland will reduce hormone levels to normal and control the disorder. Even after surgery, however, complex rebalancing of hormones must be achieved.

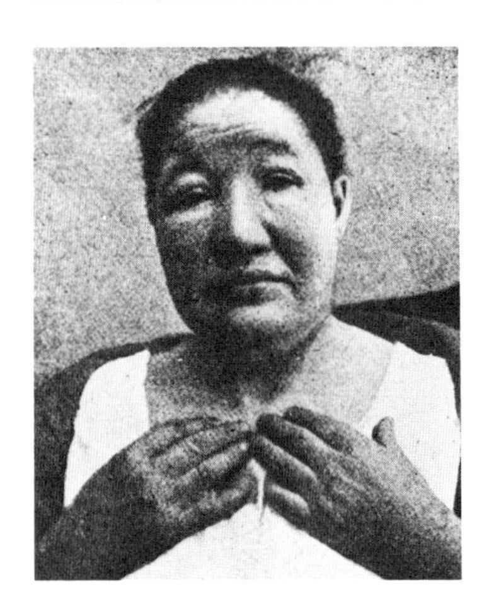

Figure 17.3.
Myxedema. (A) Before and (B) after treatment. Note the puffy facial features and general appearance of tiredness.

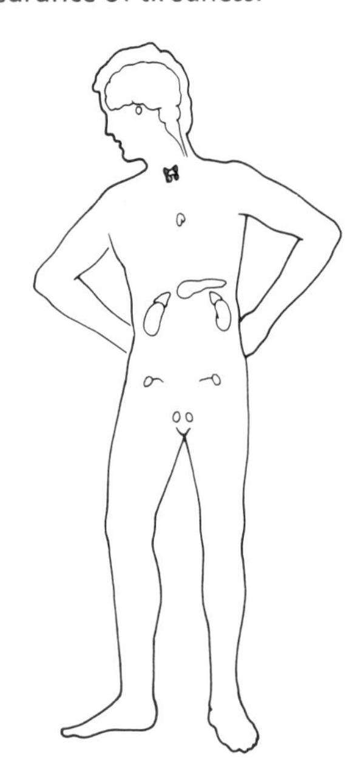

Parathyroid Glands

These tiny glands, the smallest in the body, are found partially buried in the tissues of the thyroid gland. In size and shape, each resembles a split pea. Although the number varies from individual to individual, there are usually four, two in each lobe of the thyroid gland.

The parathyroids secrete a hormone called **parathormone**, or PTH. Its function is to control the amounts of calcium and phosphorus in the blood stream. These minerals play a role in bone formation and in the proper functioning of the nervous and muscular systems. The action of the parathyroid gland is, in turn, controlled by the amount of calcium and the balance of calcium and phosphate ions in the blood.

If there is insufficient production of PTH, or if the parathyroids are removed, the muscles will tend to contract causing painful cramps. The contractions become continuous and the muscles are unable to relax (are in a state of tetany). Nerves also are affected and the body may go into convulsions. Death may be the end result. Early attempts to control excessive activity of the thyroid glands involved their surgical removal. Death inevitably followed such an operation because the parathyroid glands were also removed during this process and their importance had not been realized.

The size of the parathyroids is related to the amount of calcium present in the diet. If calcium levels are low, the size of the glands increases.

Low levels of parathormone cause bone formation to cease and has detrimental effects on the nervous system and in the clotting of blood. Such low levels can be controlled with injections of calcium compounds or by swallowing tablets of calcium lactate. The production of too much hormone causes the calcium level in the blood to increase. This extra calcium may be obtained from storage reserves held in the bones, which then become weak and brittle.

The Pancreas

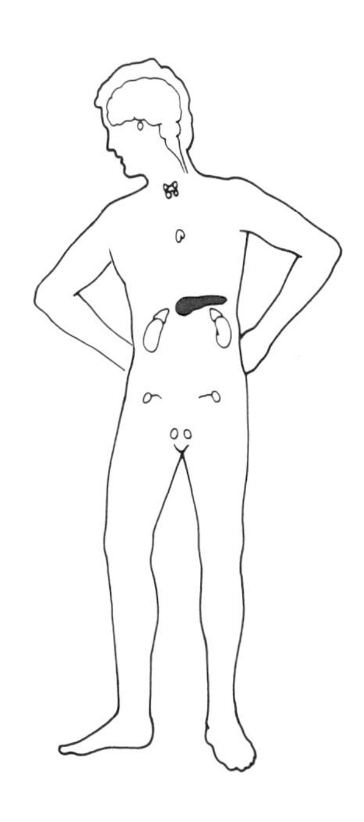

The pancreas is a comparatively large gland which lies beneath the stomach in the loop of the duodenum. It is made of two major segments. One of these parts is composed of a network of blind tubes surrounded by secretory cells. The tubes funnel secretions of the cells into the duodenum through the pancreatic duct. This portion of the pancreas thus functions as an exocrine gland. Pancreatic juice, as the secretion is called, contains enzymes that break down food substances in the small intestine.

Scattered among tubes and enzyme-forming cells are small patches of specialized cells which release their products into the blood stream rather than into the pancreatic ducts. The pancreas is, therefore, both an endocrine and an exocrine gland. These small patches of cells look like small islands in a sea of other tissues. They are called the **islets of Langerhans** after the German biologist who discovered them. There are two kinds of cells in the islets: alpha and beta cells. Alpha cells produce **glucagon** and beta cells produce another hormone called **insulin**. These two hormones work together to control the level of glucose in the blood.

Insulin and the Control of Blood Sugar

Insulin and glucagon regulate the amount of glucose that is used by cells. The two substances have opposite effects. Insulin decreases the levels of glucose in the blood, while glucagon increases the amount present. Working together, they keep the sugar levels in the blood constant. As the body uses glucose to provide cellular energy, the amount of glucose in the blood stream falls. Glucagon then signals the liver to release some of its stores of glycogen (a form of glucose sugar) and replace the sugar lost from the blood. The liver can also convert some fats and proteins into sugar to meet the energy demands of the body. An excess of glucose in the blood (after a heavy meal for instance) signals insulin to take effect; it causes the glucose to be withdrawn from the blood and placed in storage.

THE DISCOVERY OF INSULIN

The first indication that the pancreas had functions other than producing digestive enzymes was discovered by accident. Researchers studying the digestive functions of the pancreas had removed the pancreas from several test animals. The researchers noticed that ants were attracted to the urine of these animals after removal of the pancreas. They analysed the urine and found that it had a high sugar content. In addition, the animals appeared to have symptoms similar to those shown by humans suffering from **diabetes mellitus**.

Based on this knowledge about the pancreas, Dr. Frederick Banting, a young Canadian researcher, performed a set of experiments in 1921.

Dr. Frederick Banting had difficulty convincing others at the University of Toronto that his ideas about pancreas function were worth pursuing. By persistent effort, he overcame the objections, was granted use of an old, disused laboratory and some animals for the experiments, and was given eight weeks to complete his experiments.

Dr. Banting was fortunate that Charles Best, who was just completing his studies at the university, volunteered to help him with the chemical analysis that Banting's experiments required. The young researchers tied off the pancreatic ducts of the test animals. Six weeks later they removed the pancreas from each of the test animals and found that most of the organ of each animal had deteriorated; however, small patches of cells (called the islets of Langerhans) were still quite healthy. The two men ground up this tissue and injected their preparation into test animals suffering from diabetes. These animals quickly improved, whereas control animals that were not given the injections died.

Frederick Banting and Charles Best had discovered **insulin**. For this work they shared the Nobel Prize for Medicine in 1923.

Insulin Imbalance

If insulin is not available because of a disorder of the insulin-secreting cells in the islets of Langerhans, sugar levels in the blood may be allowed to rise to dangerously high levels. In the absence of a command to stop producing and start storing sugars, protein and fats begin to be broken down to produce still more glucose, while carbohydrates, which would normally be used, are not metabolized. Fat stores eventually become depleted and muscles are unable to use glucose effectively and become weak. If the disorder is not treated, progressive weakness and drowsiness develop, followed eventually by coma and even death. If too much insulin is produced (**hyperinsulinism**), the blood sugar levels drop. An individual then becomes irritable, sweats heavily, and is constantly hungry. In untreated cases, this too can lead to giddiness, then coma, and death.

Fortunately, insulin is readily available and can be administered, under medical supervision, to those needing help. Insulin is extracted commercially from pork and beef pancreas. About 17 600 kg of these organs are required to produce 1 kg of insulin crystals. People who suffer from this disorder are required to eat at regular and frequent intervals. This helps them to meet their body's heavy demand for glucose. Eating sweet foods can also help to build up low levels of sugar quickly but temporarily.

Adrenal Glands

The two adrenal glands sit on top of the kidneys. Each gland is made up of two parts: an inner **medulla** and the **cortex**, the slightly larger mass that surrounds it. The medulla produces two hormones, **adrenalin** (also called **epinephrine**) and **noradrenalin**. These are released during emergencies when the body is suddenly exposed to shock or injury. They are often called the "flight and fight" hormones. (See Figure 17.4.)

(ep-in-ef-rin)

The Adrenal Medulla

The medulla is closely linked to, and controlled by the nerves of the hypothalamus in the brain as well as the sympathetic nerves. Any sudden fright or special excitement will cause the release of adrenalin (and some noradrenalin) into the blood stream. As these hormones rapidly circulate throughout the body, they "turn on" some activities and "shut down"

CASE STUDY. Bob, 18 years of age.

It was Saturday, the day of the big track meet. The coach had been pushing the team hard and Bob had been required to fit extra training sessions into an already busy schedule of activities.

Bob worked for a pop distributing firm on Saturday mornings and didn't normally finish work until 1:00 in the afternoon, but by starting early that morning, he had completed his work just as it was time for the track meet. Bob drank a couple of bottles of ginger ale and ate a chocolate bar as he rushed to the track meet.

The coach had high hopes for Bob in two or three events at the meet, and was annoyed by Bob's slightly late arrival and his obviously heavy morning work schedule. He offered Bob some **glucose** tablets to give him some extra energy, and Bob gulped them down and started his warm-up routine.

The first race went off as expected, with Bob achieving an easy second place. He accepted the congratulations of his team mates and went off to rest before entering his second event.

Lying on the grass, Bob began to feel a little uneasy. He didn't feel like himself at all. He started to shake - his limbs were trembling and he felt faint.

A team mate, resting beside him, noticed that Bob was shaking. Thinking that he was worried about the next race, the team mate tried to distract him by talking about a party planned for that evening. Bob interrupted him and asked him if he thought there was a doctor on the field because he thought he was really sick. His friend, after a better look at him, became alarmed and ran off to get help.

The parent of one of the competitors was a physician and came over to see Bob. The doctor saw the sweat on Bob's forehead and noted that he was trembling. Bob told the doctor that he thought he had blacked out for a moment or two. When asked how he felt, Bob groaned and said "hungry". The doctor asked someone standing by to go for food - any food - but to be quick about it. A hamburger and a carton of milk were soon brought and Bob ate ravenously. About fifteen minutes later, Bob said he felt fine and turned to the doctor to ask what caused the attack. "Hypoglycemia," the doctor replied. Bob paused: the term was familiar. Then with a shock, he realized where he had heard the word. A school mate was in hospital, seriously hurt after he had a momentary blackout while driving. The car, without someone to control it, had plunged off the road and hit a tree.

In Bob's case, his exercise demanded a rapid supply of energy, and he had recently consumed a large supply of sweet things - pop, chocolate, and glucose tablets - so it might be expected that there would be plenty of sugar available. However, the large sugar intake had caused the release of insulin, which controls the amount of sugar in the blood, diverting the extra glucose for storage. When Bob's body made a large demand for sugar energy, the insulin supply could not be turned off quickly enough, so the body was busy putting sugar into storage instead of making it available to supply his muscles for running.

The resulting imbalance produced very low levels of sugar in the blood - hypoglycemia. "Hypo" means too little, "glyc" means sugar, and "emia" tells us the problem relates to the blood. Sometimes a better term to use than hypoglycemia, is "impaired glucose tolerance" or "high glucose sensitivity", because in some persons the automatic release of insulin occurs too rapidly or too soon, diverting the sugar before the needs of the body are satisfied.

Sugar regulation is not an instant effect, it requires time. If Bob had eaten a combination of foods, proteins, fats, and carbohydrates, which require digestion, the energy would have been made available more slowly and the rapid changes avoided. Insulin would then have been released in a more moderate and controlled manner.

Questions for research and discussion:

1. Why is eating a good breakfast important when a day of strenuous activity is planned?
2. What types of food provide a gradual release of energy?
3. Many people find that their ability to concentrate is considerably reduced at about 11:00 in the morning and about 3:00 in the afternoon. Try to explain why this occurs.

Increased blood supply to skeletal muscles, heart, lungs, and brain. This ensures maximum delivery of oxygen and sugar energy.

Dilates the pupil of the eye to admit more light.

Relaxes the smooth muscles of the bronchiole walls to provide a better supply of air to the alveoli of the lungs. Stimulates respiration.

Increases the heart rate and the amount of blood pumped out by the heart. Raises the blood pressure.

Contracts muscles in the skin so that hairs stand on end. Produces "gooseflesh". Reduces the supply of blood to the skin.

Increases the rate at which blood coagulates. Redirects blood to areas where it is most needed during an emergency.

Converts liver and muscle stores of glycogen into glucose. Blood sugar level is quickly raised and delivered to muscles where needed.

Contracts the ureters and sphincter muscles of the bladder.

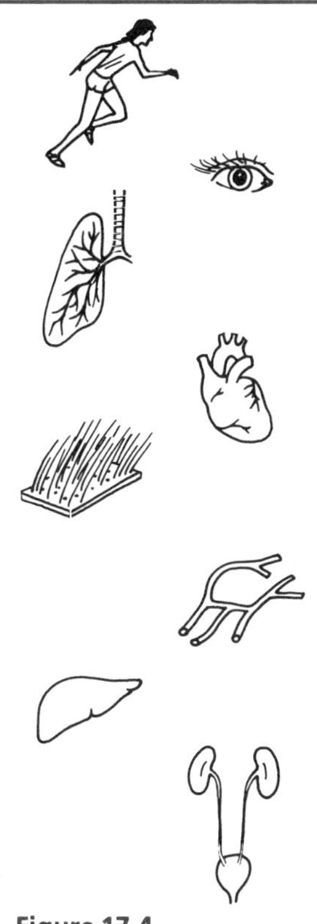

Figure 17.4.
The action of adrenalin. During the moments of special excitement or stress, adrenalin is released into the bloodstream to prepare the body for the emergency. Shown here are the effects this hormone has on various organs in the body.

others, to prepare the body for the emergency. The amount of glucose in the blood is quickly increased to ensure that adequate supplies of energy are available for the muscles. To avoid wasting energy on unnecessary activities during the emergency, blood supplies to the digestive tract and skin are greatly reduced as the heart rate and blood pressure increase. Respiration rates also increase in order to bring in extra supplies of oxygen. In addition, the sugar and oxygen supply to the brain is increased so that rapid responses can be made. If you observe a person reacting to an emergency, you may notice that his or her face goes white as blood supplies to the skin are reduced. The muscles may tense enabling an instant reaction.

The Adrenal Cortex

Unlike those found in the medulla, the cells of the cortex can regenerate if some are removed. The secretions of the cortex also help us adapt to stress and strain. This portion of the adrenal gland is not as closely connected to the nervous system as is the adrenal medulla, however. The cells of the cortex secrete hormones that control many of the body's normal activities. The cortex produces several substances

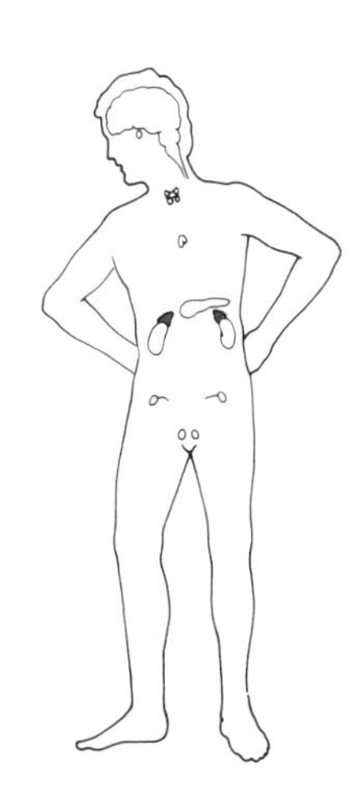

called **corticoids**, which are released in response to the secretion of ACTH (adrenocorticotrophic hormone) by the pituitary gland. The functions of corticoids may be arranged into three major groups:

1. They regulate the amount of mineral salt and the amount of water in the fluids of the body. High corticoid levels will cause water to be retained, making the body puffy.
2. They regulate the rate at which carbohydrates and proteins are used by the body. In particular, they stimulate the formation of sugars from proteins.
3. They affect the sexual characteristics of females. Excessive production of these hormones leads to the development of more typically masculine features, such as greater amounts of hair on the body.

The secretions of the cortex are vital to life. Whereas the medulla responds to immediate danger or emergencies, the cortex helps us adapt to prolonged stress, such as that caused by pregnancy, illness, or everyday changes to which we must adapt.

One of the most important corticoid hormones is **aldosterone**. It helps to maintain water balance and to regulate the amount of sodium and potassium in the body. **Cortisone**, another such hormone, has been widely used to help relieve arthritis. Normally it is involved with the metabolism of carbohydrates, fats, and proteins. It also stimulates the liver to both manufacture glucose and store glycogen.

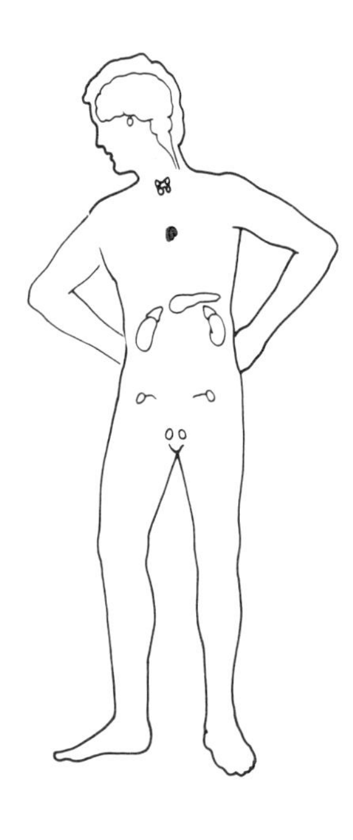

The Thymus Gland

The **thymus gland** is a small mass of lymphatic tissue located beneath the sternum in the thoracic cavity. It is quite large in children, reaching its maximum size at puberty. It then gradually shrinks in size and is replaced by fat. Its main function involves the development of immunity against disease and it is closely related to the lymphatic system. It is believed that the thymus gland secretes a hormone that enables white blood cells to produce antibodies for the defence of the body.

The Gonads: The Ovaries and Testes

Since we shall discuss these organs in detail in a later chapter on the reproductive system, discussion here will be confined to the primary hormones secreted by these organs.

The Male Hormone

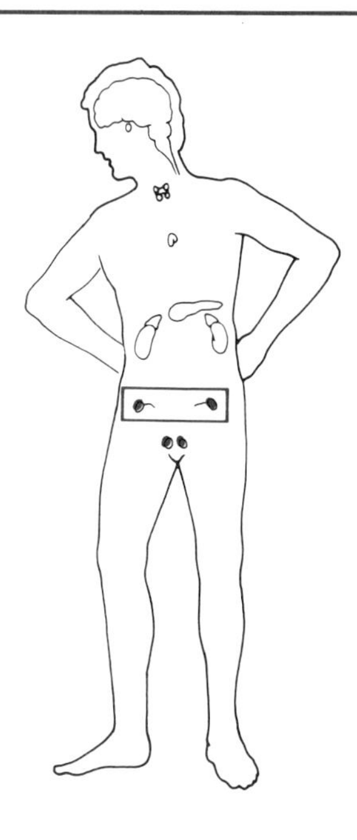

In males the hormone **testosterone** is secreted by cells of the testes. The production of this hormone is stimulated by ICSH (Interstitial Cell Stimulating Hormone) from the pituitary gland. Testosterone is responsible for the development of male secondary sex characteristics and for the proper function of the male reproductive organs. This hormone is responsible for the lower pitch of a male's voice, his extra growth of facial and pubic hair, and for the general shape and strength of the male body.

The Female Hormones

Estrogen is produced by the ovaries. It is the hormone responsible for the growth of the female reproductive system. It stimulates the development of secondary sexual characteristics at puberty and helps to ensure that they are maintained during adult life. Estrogen is involved in the menstrual cycle, producing changes in the wall of the uterus, repairing it after the menstrual period ends, and increasing the thickness of the uterine lining. It affects growth and body shape, the development of pubic hair, and the hair beneath the arms. It also causes the enlargement of the breasts. The secretion of estrogen is controlled by FSH (Follicle Stimulating Hormone) released by the pituitary gland.

Progesterone is produced by a patch of cells in the ovary called the corpus luteum. These cells are activated by another pituitary hormone, LH (Lutenizing Hormone). Progesterone prepares the uterus for the implantation of a fertilized egg cell and keeps the egg provided with nourishment during its growth into an embryo. It works together with estrogen to influence the extra development of the mammary glands in pregnant women.

The Pituitary Gland

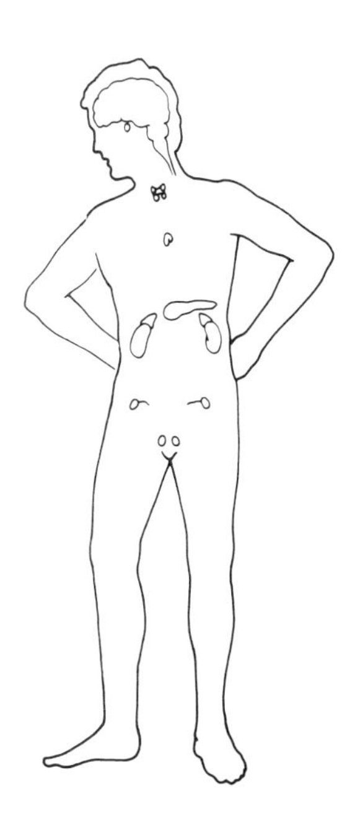

The pituitary is a tiny gland only as large as a pea. It is attached to the hypothalamus at the base of the brain by a thick stalk. The pituitary is really two glands which are quite different, both structurally and functionally. The **anterior lobe** of the pituitary is made up of glandular cells. The **posterior lobe** is composed primarily of nerve cells and has very few glandular cells. (See Figure 17.5.) The nerves of the posterior lobe are directly connected to the hypothalamus in the brain. The hormones of the anterior pituitary can be divided into two main groups: 1) Those that have an effect on the growth and metabolism of the body, and 2) those that influence the adrenal glands and the gonads.

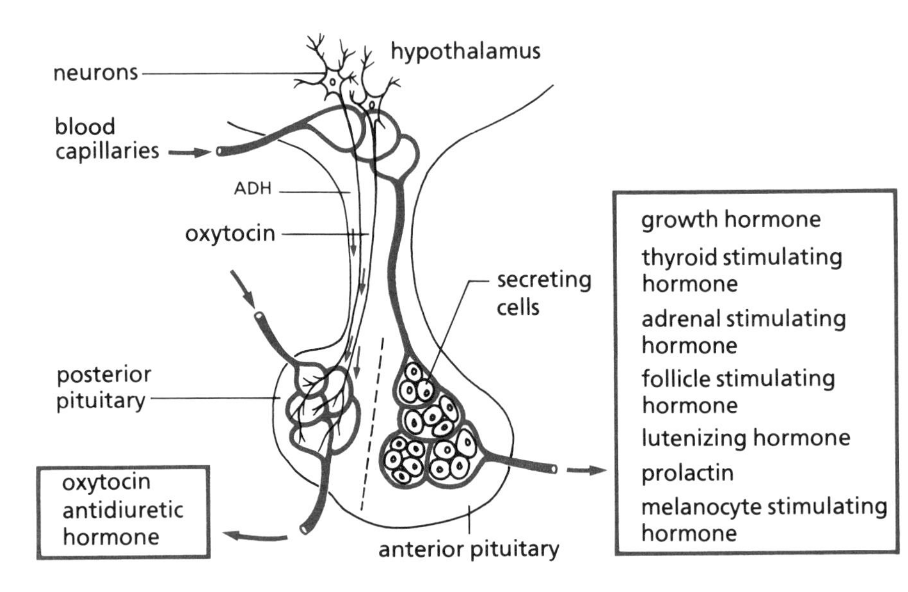

Figure 17.5.
The hormones of the pituitary gland. Hormones of the posterior pituitary are secreted from the hypothalamus, trickle down into the posterior lobe, and enter the blood stream from there to be carried through the body. Anterior pituitary hormones are secreted by cells in the anterior lobe and enter the blood stream for transport throughout the body.

Hormones of the Anterior Pituitary Gland

The Growth and Metabolism Hormones

Growth hormone (GH) is concerned with the growth of bones, muscles, and organs of the body. (See Figure 17.6.) For this growth to occur, raw materials are needed to produce extra tissue. GH aids this process by speeding up protein synthesis and cell division. It also stimulates the conversion of fats to sugars in order to provide energy for this activity.

The **thyroid stimulating hormone** (TSH) stimulates the thyroid gland to secrete thyroid hormone.

The Adrenal Hormones

(ad-reen-oe-cor-tic-oe-troe-fik)

Adrenocorticotrophic hormone (ACTH) is a very long name, but is not nearly as bad as it looks when broken down. "Adreno" tells us that the adrenal gland is the target organ. "Cortico" indicates that it is the adrenal cortex which it affects, and "trophic" means "to seek", so the whole title just says–the hormone that seeks (and stimulates) the adrenal cortex! ACTH controls the amount of corticoid hormones secreted by the adrenal cortex.

(mel-an-oe-site)

Melanocyte stimulating hormone (MSH) is responsible for the production of melanin and thus for colour pigmentation in the skin.

Figure 17.6.
Pituitary hormones, their target organs and effects.

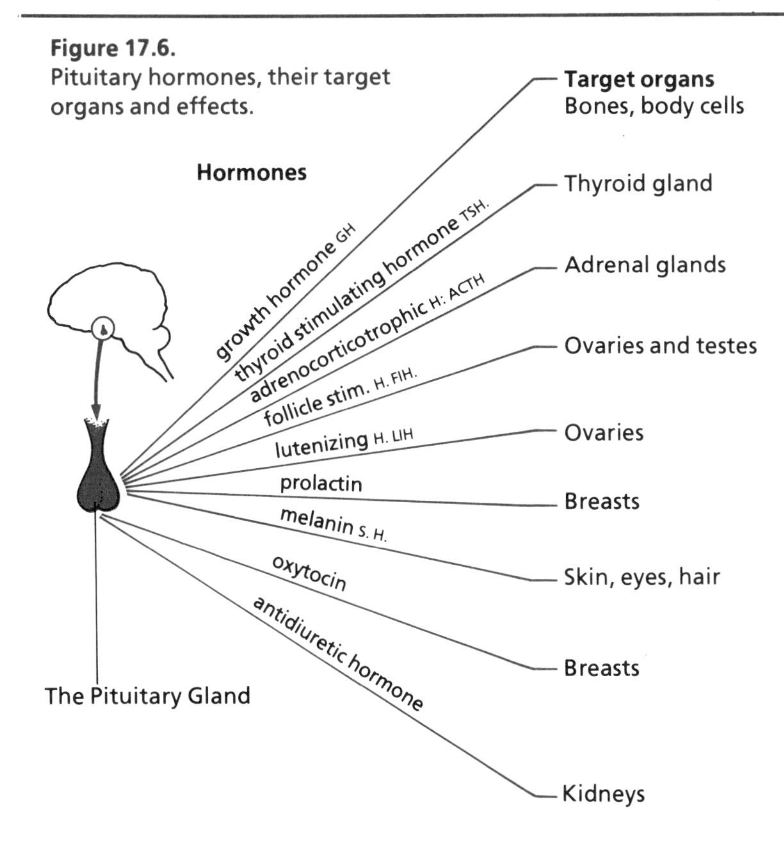

The Gonadotropins

Gonadotropins control the growth, development, and function of the testes and ovaries. There are two major hormones that control these sex organs.

Follicle stimulating hormone (FSH) stimulates follicles present in the **ovary**, causing one of them to begin development. This in turn stimulates the production of estrogen in the ovary. In the male, this same hormone stimulates the development of the seminiferous tubules and the maturing of sperm.

Lutenizing hormone (LH) continues the process started by FSH in the ovary, and causes the ovum to be expelled from the ovary. It also promotes the development of the corpus luteum from those cells left behind when the ovum is released. The corpus luteum, in turn, stimulates production of progesterone.

Interstitial cell stimulating hormone (ICSH) is a male hormone which stimulates the production of testosterone.

Hormones of the Posterior Pituitary Gland

Only two hormones are released by the posterior pituitary and even these appear to be actually produced by the hypothalamus and passed down to the posterior pituitary lobe, to be secreted when needed.

(an-tee-dy-your-et-ic)

The main function of the **antidiuretic hormone** (ADH) (also called vasopressin) is to cause the nephrons in the kidney to release more water from the renal tubules. This concentrates the urine and decreases the amount of water eliminated. ADH also causes contraction of the smooth muscles in the walls of blood vessels, thereby raising the blood pressure.

Oxytocin has two major effects. Firstly, it is believed to cause (during birth) the muscles of the uterus to contract at the end of pregnancy. Secondly, it is involved in causing milk to be released from glands in the breast, so that the mother can feed her baby. Impulses are sent to the hypothalamus upon dilation of the cervix (just before birth) or when the infant sucks the nipple of the breast. The hypothalamus then directs the release of oxytocin, thus affecting the secretion of milk.

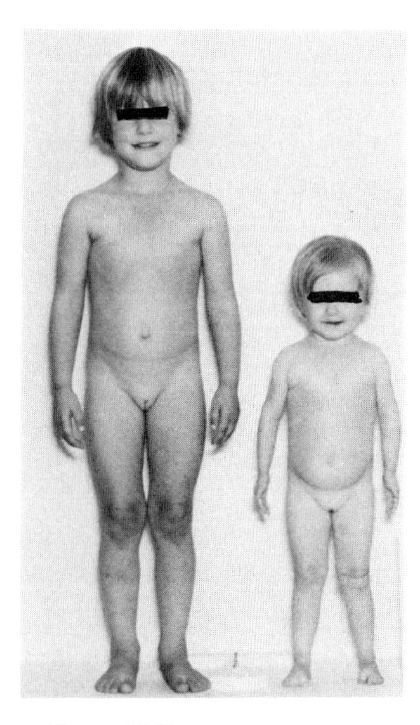

Figure 17.7.
Twins aged three years, eleven months. The reduced growth in one child is due to a lack of growth hormone from the pituitary gland. Note: Body proportions are normal.

Pituitary Disorders

When the pituitary gland fails to function adequately, or when it overworks, it can produce some rather dramatic results. A child with a deficient anterior pituitary lobe produces insufficient amounts of growth hormone and, as a result, fails to grow. This disorder produces a Lorain dwarf. Skeletal and sexual developments are delayed, but the child is alert, intelligent, and quite well-proportioned. Special and very expensive extracts of human growth hormone can be used to correct this deformity. Another type of dwarf (Frohlich's dwarf) results when disease strikes both the anterior and posterior lobes of the pituitary. This type of dwarf is often obese, short in stature, delayed in sexual development, lethargic, and mentally subnormal. (See Figure 17.7.)

If the pituitary is overactive, it may lead to giantism. The long bones grow particularly large and height at maturity may reach 200 to 250 cm. If a similar overproduction occurs in adults, a condition called **acromegaly** results. The skin of these sufferers becomes thick and coarse and the bones of the jaw, hands, and feet thicken. The face enlarges and the features become very coarse. (See Figure 17.8.)

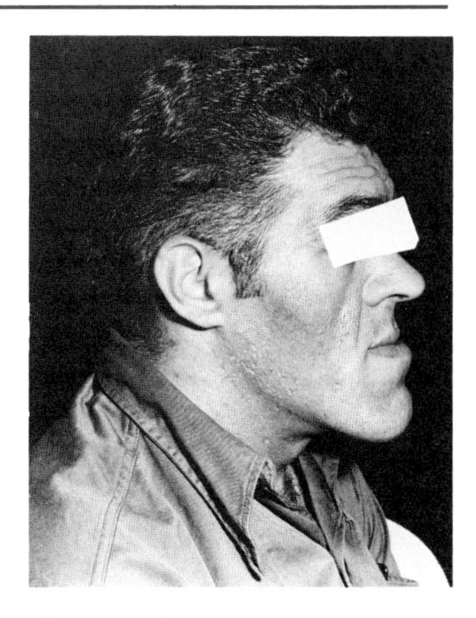

Figure 17.8.
Acromegaly in an adult. Note the coarseness of the facial features and enlarged jaw.

Homeostasis

At many points in this text, **homeostasis** has been mentioned - the maintenance of a steady state in the processes of the body. The body's homeostatic control is rather like a thermostat in the home. As the temperature falls, a message is sent from the sensor (thermostat) to the furnace to boost the heat supply. As soon as the temperature rises sufficiently, another message is sent to turn off the furnace. The temperature may fluctuate up and down a small amount, but a fairly steady balance of heat is maintained.

The endocrine glands have similar but much more complex feedback systems which help to maintain each body function in balance. For example, if body temperature rises, blood vessels in the skin dilate and heat is lost. If the volume of water in the body decreases, ADH is secreted by the pituitary gland so that more water is absorbed and less excreted. When blood calcium goes up, the parathyroid, which monitors this condition, releases a parathyroid hormone to control the situation. When the calcium level falls, the secretion of the hormone is reduced.

The body functions properly only so long as all mechanisms within each system work in a controlled, balanced manner - homeostasis must be maintained.

The Partnership Between the Endocrines and the Autonomic Nervous System

In an earlier chapter, we observed that the sympathetic nervous system reacts during emergencies and that the parasympathetic system acts to ensure that the body returns to a normal state after such an emergency and conserves its resources when possible. The endocrine glands also respond in a similarly balanced fashion. The hypothalamus is the "watch dog" of all physiological events taking place in the body. Its sensors detect changes in temperature, blood chemistry, and blood pressure, while its connection with the higher brain centres keeps it informed of emotional stress and other events. As soon as a particularly frightening or stressful event occurs, an alarm reaction is triggered; the sympathetic nervous system and the adrenal medulla respond rapidly. Nervous impulses and hormones speed through the body establishing immediate alarm reactions. Once this has been accomplished, a resistance reaction takes over; the anterior pituitary is affected and it, in turn, activates the adrenal cortex.

Although this latter system is slower to act, its effects are of longer duration. It prepares the body for a "siege", rationing out body energy reserves and slowing down processes that have least priority. At the same time, it maintains a check for any signs that may indicate that the stress is diminishing so that the body can return to normal functioning as soon as possible.

QUESTIONS FOR REVIEW

SOME WORDS TO KNOW

Match a statement found in the left-hand column with a suitable term from the right-hand column. DO NOT WRITE IN THIS BOOK.

1. A disorder of the thyroid gland.
2. Controls the amount of calcium in the blood.
3. Produces insulin and glucagon.
4. Important hormone in emergencies.
5. Stimulated by ACTH.
6. Produces estrogen and progesterone.
7. Hormone controlling water balance in the kidneys.
8. Glands which secrete their products into the blood.
9. Part of the brain closely associated with the pituitary gland.
10. Chemical messengers in the blood.

A. glucagon
B. insulin
C. adrenalin
D. antidiuretic hormone
E. endocrine
F. exocrine
G. hypothalamus
H. hormones
I. adrenal
J. goitre
K. islets of Langerhans
L. testes
M. ovaries
N. parathyroids
O. thyroxin
P. pituitary
Q. diabetes mellitus
R. endrometrium

SOME FACTS TO KNOW

1. Briefly explain what is meant by homeostasis. Give an example.
2. List six effects resulting from the release of adrenalin into the blood stream during an emergency.
3. Why is the pituitary gland sometimes called the master gland?
4. Name the two main parts of the pituitary gland and list the secretions of each.
5. What is the function of the parathyroid glands? Where are they located? What happens if they are removed?
6. What are the two parts of the adrenal glands and what hormones does each part secrete?
7. What system works together with the endocrine system? State briefly the role of each of these systems.
8. Define the following terms: "hormone", "target organ", "metabolism".
9. Why should diabetics carry some sugar or a sugar-rich food with them?

QUESTIONS FOR RESEARCH

1. Select one of the following topics and prepare a report on it:
   - Endocrinologist
   - Addison's disease
   - Cushing syndrome
   - bio-energetics
   - diabetes mellitis
   - stress syndrome
   - circadian rhythms
2. What is bio-feedback? Discover all you can about this recent development in medical science.
3. Pituitary extract is used in some pituitary disorders. The extract is in very short supply and only severe cases can be treated. Find out how this substance is obtained.
4. Research the means by which organs of the body can be donated to persons who have had organs removed or whose organs (e.g., kidneys) are not functional.
5. Research findings on the hormonal control of maleness and femaleness. Some interesting discoveries have been made that indicate that the levels of progesterone and estrogen in the mother during pregnancy may determine the degree of masculinity and femininity of her offspring when they become adults.
6. Select one of the following endocrine disorders and investigate it fully: cretinism, acromegaly, giantism, goitre, hyperthyroidism.

Unit X

Reproduction and Heredity

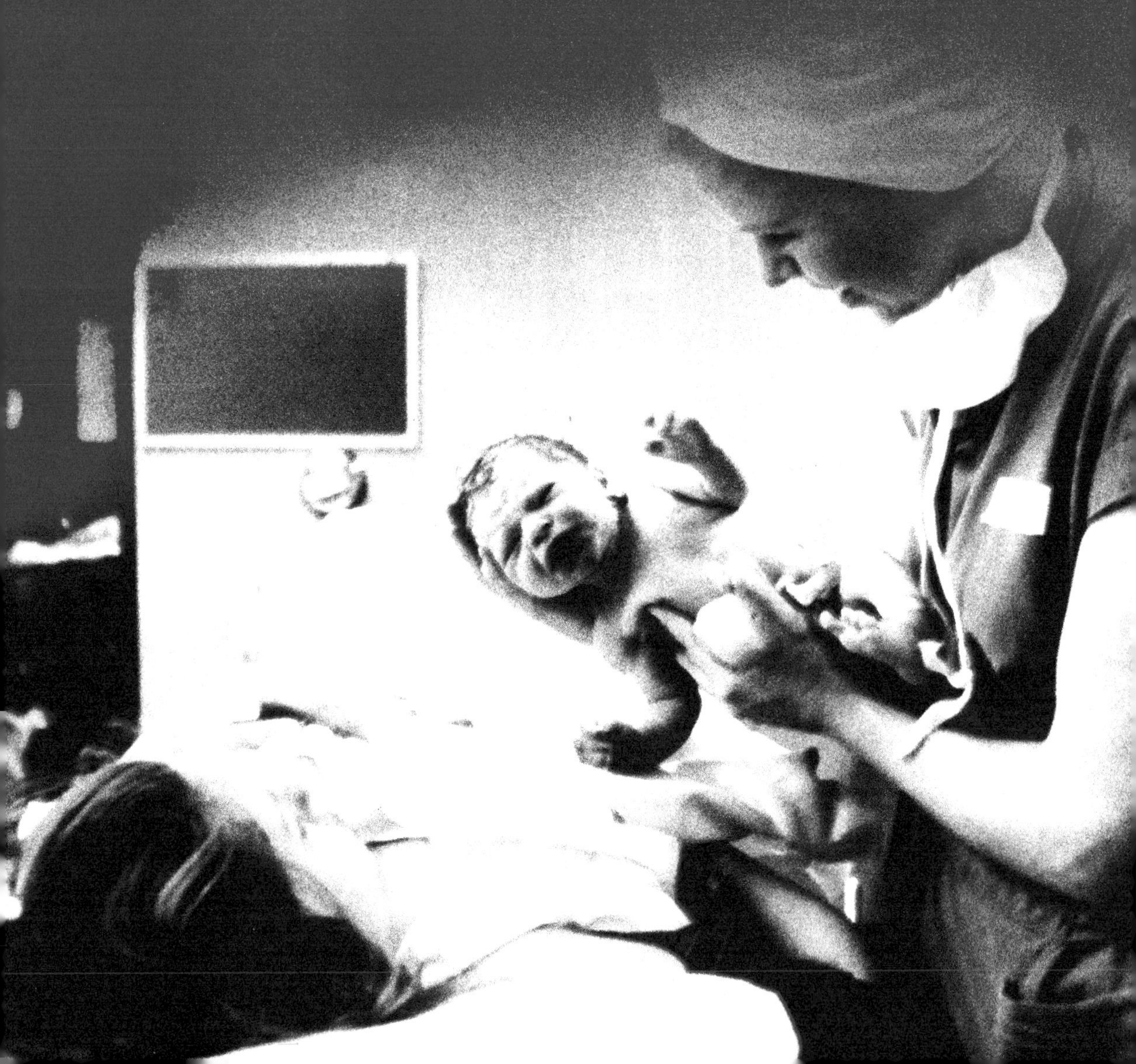

Chapter 18

Changes in the Reproductive System

- The Male Reproductive System
- The Female Reproductive System
- Copulation and Fertilization
- Sexually Transmitted Diseases

Changes in the Reproductive System

Before studying the human reproductive system, it is worth considering some of the changes that occurred during the development of complex animals from more primitive ones. This should help explain why some of the structures of the human reproductive system are present.

Oysters and some other marine organisms produce millions of sperm and millions of eggs and release them into the surrounding water at approximately the same time. Under these conditions the union of the sperm and egg occurs more or less accidentally and the chances of union taking place at all are sometimes remote.

Fish, like oysters, also release their gametes into the water, but often much greater care is taken to see that the union between sperm and egg takes place. For example, the female salmon lays her eggs (roe) in a sandy or gravelly stream bed and the male fish releases his sperm (milt) directly above the eggs.

In frogs and many other amphibians this process is further refined. The male frog mounts the back of the female frog and the sperm are shed over the eggs as they leave the female's body. The eggs are coated with a jellylike substance which protects and nourishes them during development.

Millions of years ago, when animals began to establish themselves on land, the problems of water loss and drying out were considerable and frequently caused death. Such features as leathery skins and body scales eventually evolved enabling animals to avoid dehydration. On land, sperm and eggs could no longer be shed, or they too would quickly dry up and die without a watery medium. This problem was solved by the development of internal fertilization.

The Male Reproductive System

The male reproductive system consists of two testes, an arrangement of excretory ducts and accessory glands, and an organ for sperm transfer to the female. The ducts are called the epididymis, vas deferens, and the ejaculatory ducts. The accessory glands and structures include the prostate gland, seminal vesicle, Cowper's gland, and the penis. (See Figure 18.1.)

Figure 18.1.
The structures of the male reproductive system.

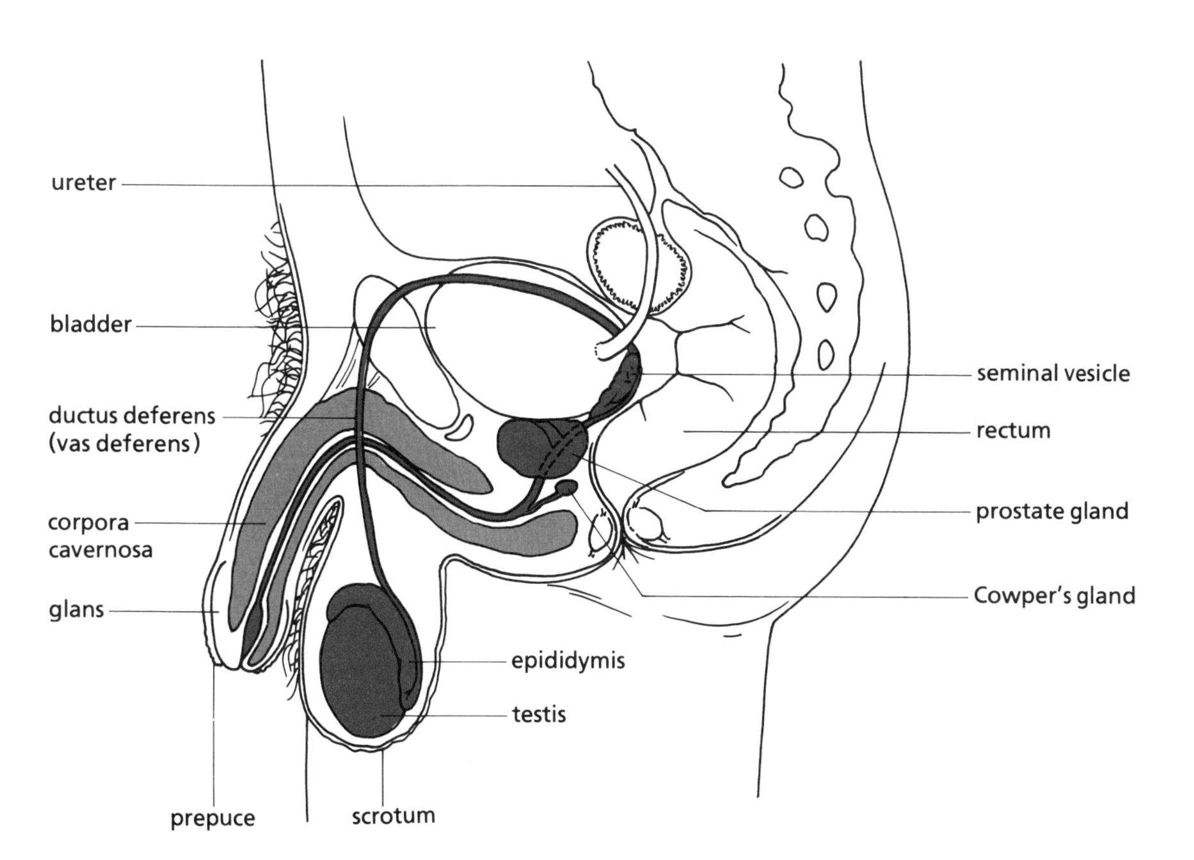

The Testes

The testes begin developing early in the embryonic growth of a male child. They appear initially high in the abdomen, close to the kidneys. They move downward, passing through a small gap between the muscles which form the base of the abdominal wall (the **inguinal ring**). They appear outside the body shortly before birth, in a small sac called the **scrotum**. The external placement of the testes is important for production and development of sperm. Sperm require a lower temperature than that of the normal body for maturation and development. If the testes fail to descend and remain in the abdominal cavity, the ability to produce live sperm is restricted and the male may become sterile.

After the testes descend, the canal through which they passed on their way into the scrotum usually closes up. The area may, however, remain somewhat weakened. If, in later years, an unusual strain forces this weakened area between the muscles to re-open, a loop of the intestine may get pushed into the opening. This is called a **hernia** (rupture) and may need to be corrected by surgery.

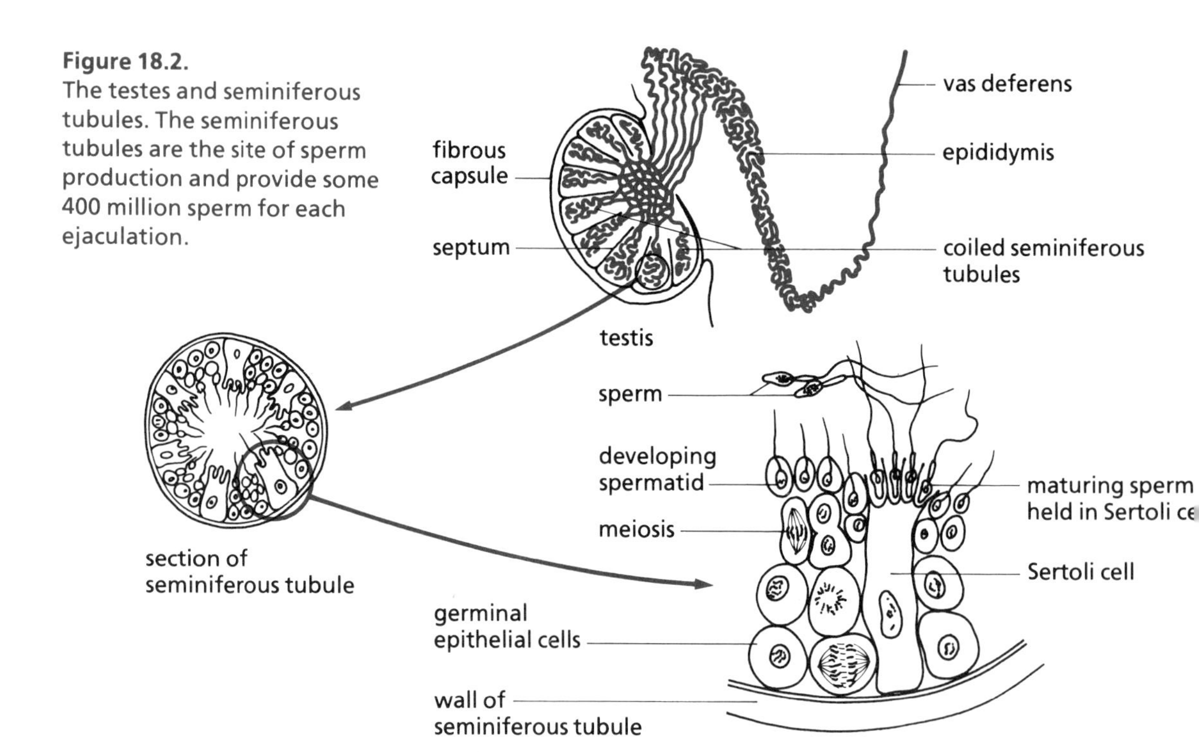

Figure 18.2.
The testes and seminiferous tubules. The seminiferous tubules are the site of sperm production and provide some 400 million sperm for each ejaculation.

The Structure of the Testes

The adult testes are oval bodies about 4 cm in length and 2.5 cm in diameter. They are covered with a thin fibrous coat, inside which they are divided into many compartments or lobules. These lobules contain the tightly packed coils of the **seminiferous tubules**, where millions of sperm are continuously being produced. Although only one sperm is needed to actually fertilize an ovum (egg), the presence of many sperm is necessary to sufficiently affect the coating around the ovum to permit one sperm to penetrate and effect fertilization (See Figure 18.2.)

The seminiferous tubules are lined with germinal epithelial cells. The word germinate means "start to grow". These cells divide by mitosis to form two similar cells. One of these cells remains behind to divide again, while the other divides to form two sperm cells. This second cell goes through the process of meiosis in which each pair of chromosomes is halved. As a result, only one of each chromosome pair is transferred to each tiny sperm cell. Meiosis is discussed in more detail in the chapter on genetics.

The Production of Sperm

The full name of the gamete is **spermatozoa**, which is commonly abbreviated to sperm. There are estimated to be more than 400 seminiferous tubules in each testis, each of which is about half a metre in length. It is here that sperm production takes place. Four hundred million sperm commonly leave the male's body in a single ejaculation. The system of sperm production is highly efficient, and furnishes very large numbers of gametes.

The sperm cells begin their development in the germinal epithelium. As they grow, they gradually mature and take on the appearance of the typical sperm. They then move slowly toward the centre of the seminiferous tubule ready to pass down the hollow of the tube towards the epididymis.

A number of long cells called **Sertoli cells** are located between the developing sperm. They provide nourishment and anchor the growing sperm. Sertoli cells hold sperm back until they are sufficiently mature to continue their journey. Other cells located between the lobules of the testes secrete the male hormone **testosterone**. This substance is essential for proper functioning of the male reproductive system. It stimulates the appearance of secondary sexual characteristics when a boy reaches puberty.

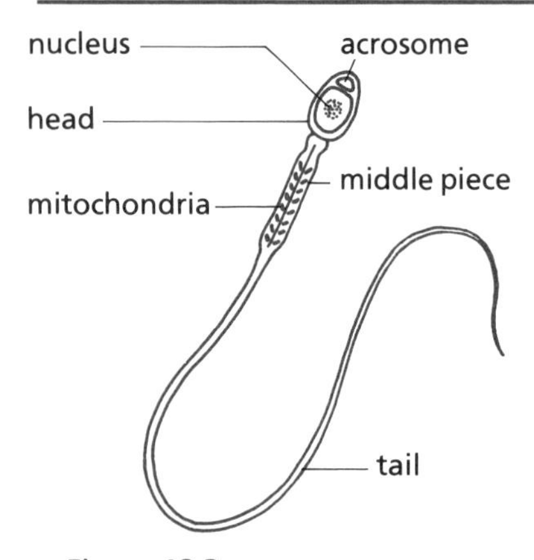

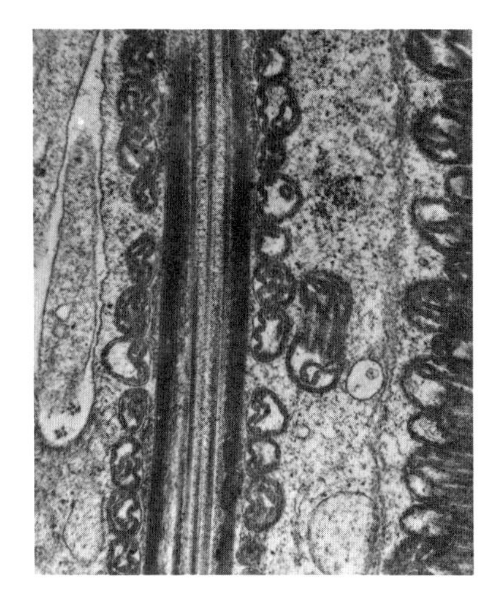

Figure 18.3.
a) The sperm is specially adapted to transport its 23 chromosomes to the ovum and penetrate the protective ovum membrane.
b) Longitudinal section of sperm tailpiece showing the central filaments and mitochondria.

At puberty, FSH (follicle stimulating hormone) supplied by the pituitary gland stimulates the production of sperm in the testes. ICSH (Interstitial Cell Stimulating Hormone) stimulates the production of the male hormone testosterone and thus promotes the development of male organs and secondary sex characteristics. These features, such as a deep voice, extensive facial and body hair, and a muscular frame are the outward signs of sexual maturity.

After leaving the penis, sperm live for only 24 to 72 h. The average length of a sperm is about 0.05 mm, of which the head comprises only 0.003 mm. Sperm consist of three main parts, known simply as the **head**, **middle piece**, and **tail** (See Figure 18.3.) The rounded head portion contains the nucleus and its chromosomes. Also present is a small capsule (the acrosome) containing an enzyme which helps the sperm to penetrate the egg. The middle piece is the "motor", providing the energy to propel the sperm. It contains many mitochondria which provide the ATP needed for this movement. Finally, there is the tail piece, a whiplike flagellum, which moves rapidly back and forth to propel the sperm after it is released.

The Ducts of the Male Reproductive System

The sperm have a long way to travel from the testes to the end of the penis. The first tube that the sperm enter, after leaving the seminiferous tubules, is the **epididymis**. This is a long, narrow structure that sits on the posterior surface of the testes. It consists of a long coiled tube in which sperm continue to mature. The sperm would not have sufficient stored energy to travel through the male tubes on their own. They are, therefore, helped along by muscles in the walls of these tubes and by minute cilia which propel them towards their destination.

The **vas deferens** (or *ductus deferens*) conveys sperm from the epididymis to the ejaculatory duct. (See Figure 18.4.) These tubes form a complete loop over the pubic bone and the bladder. They are forced to take this long route because the testes have descended into the scrotum.

The vas deferens is accompanied by blood vessels and nerves which connect with a tube from the seminal vesicles to form the **ejaculatory duct**. This duct leads through the prostate gland and connects with the **urethra**. The urethra in the male carries both urine from the bladder and semen from the reproductive organs.

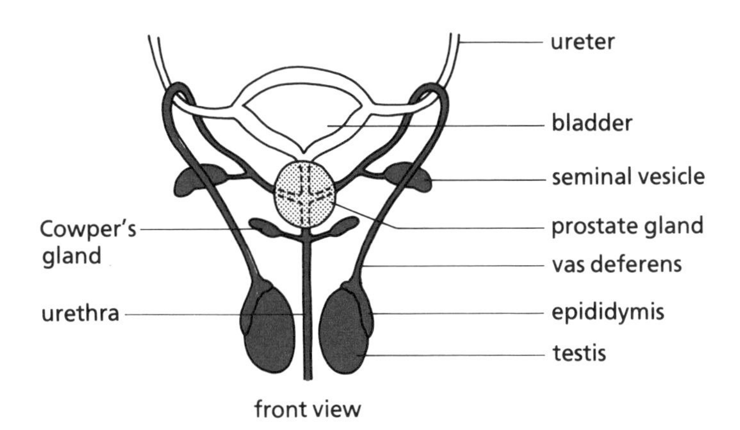

Figure 18.4.
The male reproductive system. Note: with the exception of the prostate gland, the other glands and structures are all paired.

Accessory Structures

The accessory glands produce a number of secretions which surround and nourish the sperm. The **seminal vesicles** secrete a thick, yellow, alkaline substance which contains a considerable amount of fructose (sugar) and citric acid. This flows into the ejaculatory duct, nourishing the sperm and changing the acid-base balance. This improves sperm motility. During sexual intercourse, it helps to neutralize the acidic nature of the vagina, which otherwise tends to retard the sperm.

The Prostate and Cowper's Gland

The **prostate gland** is a firm, muscular organ about 4 cm across. It surrounds the urethra and parts of the ejaculatory ducts. Contraction of smooth muscles in the prostate gland help to push semen out explosively during ejaculation. The secretions of this gland also help to make the semen alkaline, which improves sperm motility and ensures survival of the sperm in the acidic vaginal secretions.

Cowper's gland secretes a substance which affects the walls of the male urethra prior to ejaculation. It neutralizes any acidity left by urine and provides a more suitable environment for the passage of the sperm.

Semen

The semen (seminal fluid) is a mixture of sperm and all the secretions of the accessory glands. About 3-4 mL of semen containing about 400 000 000 sperm are expelled during an ejaculation. As already stated, the urethra carries both semen and urine. During sexual intercourse, the tube leading from the urinary bladder into the excretory tube, is closed off by a ring of sphincter muscles. Semen is a thick fluid, which coagulates within seconds after ejaculation. However, its own enzymes normally maintain the semen in a liquid condition so that the sperm can move freely.

The Penis

Figure 18.5.
A section through the penis showing the spongy corpora cavernosa.

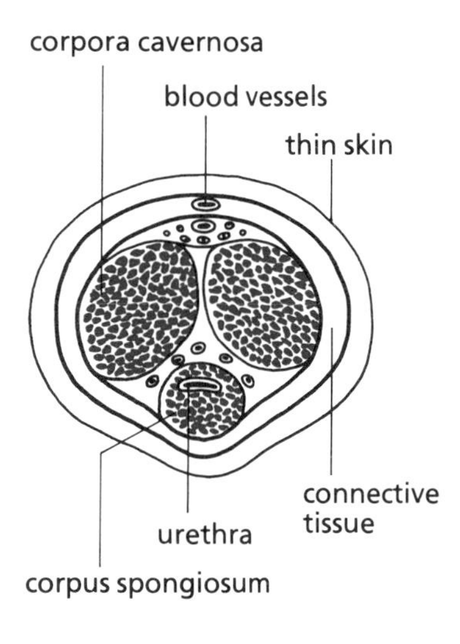

The **penis** is the external organ of the male system. It is made up of three masses of spongy tissue, held together by bands of elastic connective tissue. This spongy tissue, the **corpus cavernosum** forms two cylinders lying side by side. Below these, a third cylinder of spongy tissue carries the urethra through its centre. (See Figure 18.5.) This tissue forms a blunt cone at the end of the penis called the **glans** which is covered with a fold of skin called the **foreskin**. This fold is often removed soon after birth in a process known as circumcision.

Normally the penis is soft and relaxed. At this time the tissues of the spongy corpus cavernosum contain very little blood. During periods of sexual excitement, the blood vessels which supply the penis expand and more blood flows into the spongy chambers producing a higher pressure within the tissues. The tissues then become enlarged compressing the veins that usually carry blood away from the penis and prevent them from doing so. This building of blood pressure makes the penis firm and erect (erection) and capable of transferring sperm into the vagina.

Ejaculation

Ejaculation consists of a sequence of two events. The first occurs when smooth muscles in the reproductive glands and vessels move semen into the urethra, at a point close to the prostate gland. The second, and main event, involves contraction of a small muscle that propels the semen through the urethra and out through the opening in the penis.

The Female Reproductive System

Puberty

Very little development of the primary or secondary sexual features of a female occurs before puberty. At puberty, which normally occurs between ten and fourteen years of age, the secondary sexual characteristics develop. The breasts, uterus, and vagina increase in size and mature; pubic and axillary hair (under the arms) becomes noticeable; and the general contours of the body become more rounded by deposits of fat beneath the skin. Puberty ends with the onset of the first menstrual period, which may occur at anywhere from 10 to 18 years of age. The age at which menstruation begins also varies with different cultures and with such factors as diet. In western cultures the age of females at first menstruation has been getting steadily younger over the last half century. In both sexes, Puberty is due to increased pituitary secretions, although what triggers these secretions is not known.

The Ovaries

The two ovaries are the primary sex organs of the female. The ovaries, which resemble almonds, are similar and are located in the back of the abdominal cavity near the rim of the pelvis. They are held in place by ligaments. During pregnancy, as the uterus and the fetus enlarge, the ovaries are pushed aside. The uterus must, therefore, have some flexibility in its attachment. To allow for these changes during pregnancy, the broad ligaments and the ovarian ligaments hold the ovaries loosely to the uterus, and the suspensory ligaments suspend each ovary from the wall of the pelvis.

The Development of the Follicles

Each ovary is contained within an outer layer of special epithelium (germinal epithelium). Inside the ovary there is a network of connective tissue, which contains small groups of cells called **follicles**. Within each follicle is an egg or **ovum** (*ova* plural). At birth, there are some 400 000 tiny follicles present in each ovary. Each ovum, with its surrounding follicular cells, is called a **primary follicle** and is only partly developed at birth. It will remain in this state until a pituitary hormone re-starts the development of the follicles at puberty. Only about 400 of these primary follicles will ever reach full maturity and be released, one per month, during the reproductive life of the female. The other follicles will gradually degenerate.

After puberty, under the influence of FSH (follicle stimulating hormone), the primary follicles develop into **secondary follicles**. The follicular cells surrounding the ovum multiply and produce fluid which accumulates in these cells. The ovum also changes during the development of the follicle. By the process of meiosis the number of chromosomes present is halved, so that each cell contains one chromosome of each original pair of chromosomes. One cell remains, the other degenerates. (See Chapter 20, on genetics, for more detail.)

Although a number of follicles may start to develop each month, usually only one will reach maturity. The mature follicle is greatly enlarged and causes a bulge in the side of the ovary. Eventually this mature follicle bursts through the ovarian membrane in the process known as **ovulation** and begins its journey along the Fallopian tube. (See Figures 18.6 and 18.7.)

Female Hormones Produced by the Ovaries

As the follicular cells develop, they produce the hormone **estrogen** which is secreted into the blood stream for transfer to other organs. After ovulation, some of the follicular cells

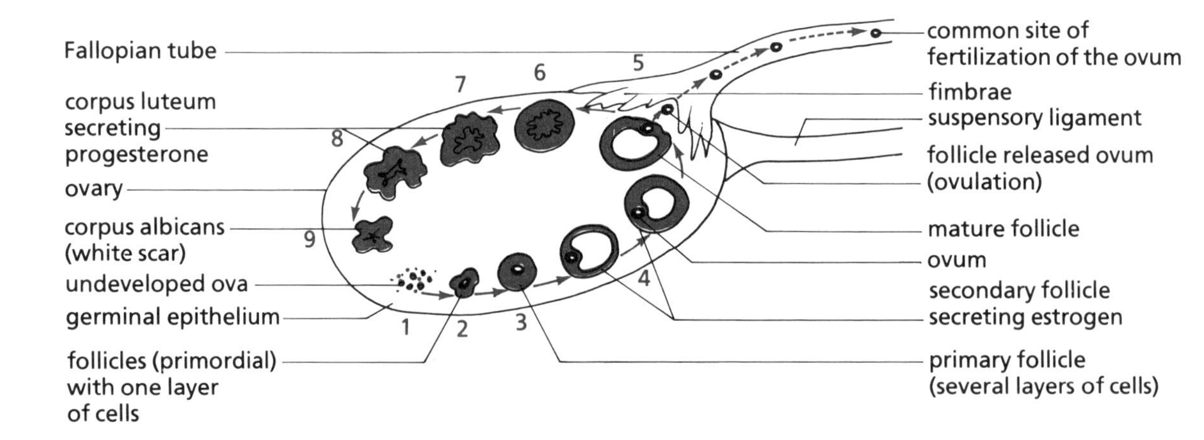

Figure 18.6.
The sequence of follicle development, ovulation, and changes to the corpus luteum and corpus albicans.

are left behind in the ovary. Under the influence of another pituitary hormone, LH (lutenizing hormone), these cells become organized into a yellow mass, called the **corpus luteum**. The corpus luteum produces the female hormone **progesterone** which influences maintenance of the endometrial lining in the uterus.

If fertilization of an ovum occurs, progesterone will continue to be secreted for 2 to 3 months. After 10 days, if the ovum is not fertilized, the corpus luteum shrinks and degenerates until it is just a small white patch of scar tissue, known as the **corpus albicans** (*corpus* means body and *albicans* means white).

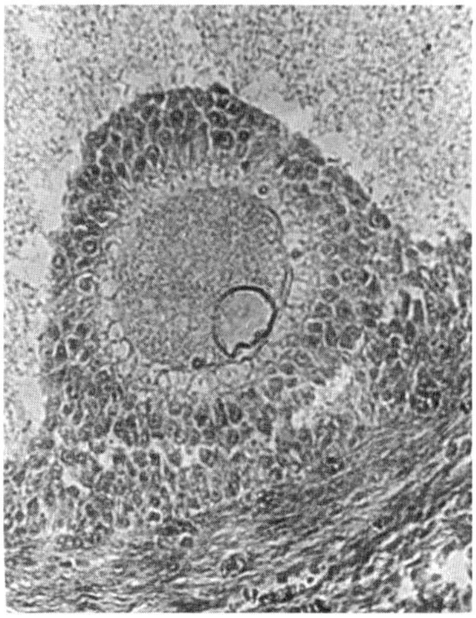

Figure 18.7.
Developing follicle in the ovary.

The Fallopian Tubes

The **Fallopian tubes** (*oviducts*) conduct the ovum from the ovary to the uterus. They are attached to the upper part of the broad ligament and are 10-12 cm in length. The end nearest to the ovary has a funnel-shaped opening surrounded by a fringe of tiny projections called **fimbrae** (Figure 18.8).

Figure 18.8.
The female reproductive organs. Midline section.

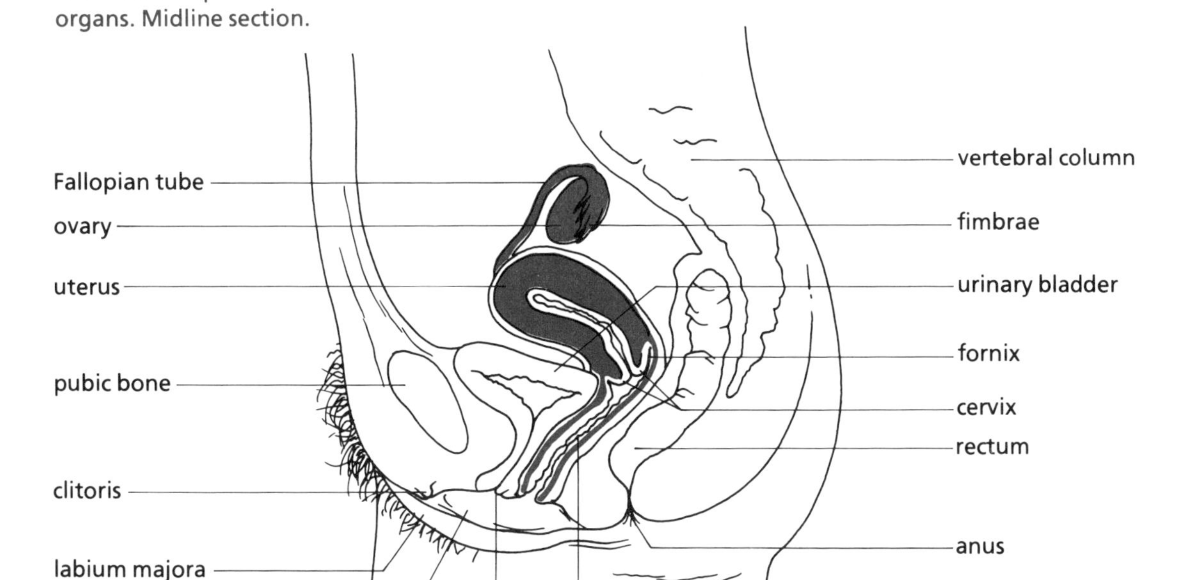

The ovary is not directly attached to the Fallopian tube. The fimbrae almost entirely enclose the ovary and when a mature follicle is releaed, these fimbrae "catch" it and, with wavelike movements, sweep it into the Fallopian tube. Since the ova have no means of propulsion, they are moved along the tube by peristaltic movements of the tube muscles. There is also a current of fluid moved along by the sweeping action of tiny cilia which line the Fallopian tubes. Many secretory cells present in the walls of the tubes help to produce the fluid in which the follicle is carried. The other end of the Fallopian tube leads directly into the uterus.

The Fallopian tube is the usual site of fertilization. (See Figure 18.9.) The follicle containing the ovum will be car-

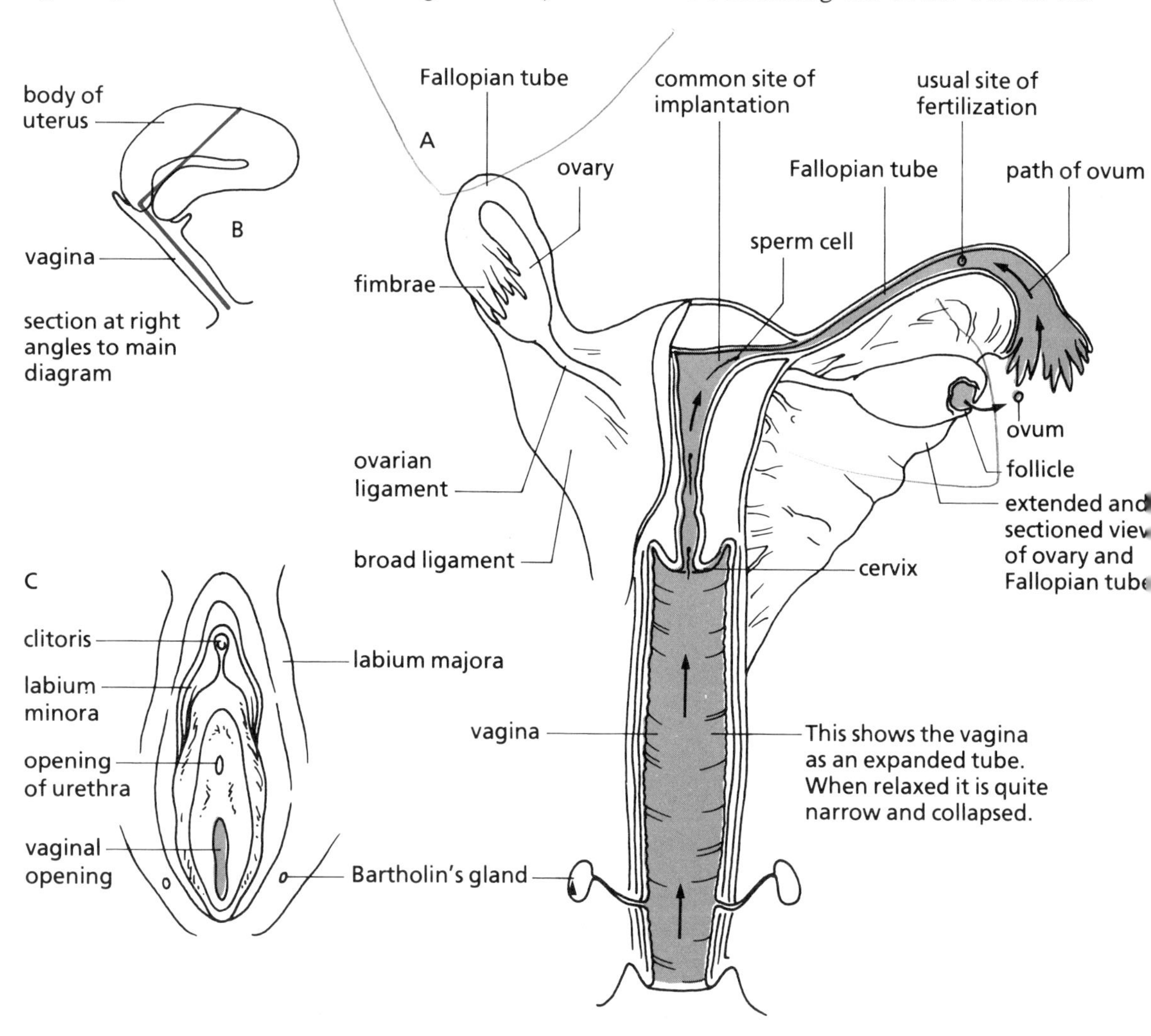

Figure 18.9.
The female reproductive organs and the pathways of sperm and ovum to the normal site of fertilization. Note: from the main diagram (A) it would appear that the vagina and the uterus are in a straight line. Diagram (B) shows that from the side, the two organs are at right angles to each other.

ried along the tube over a period of three to five days. If the ovum is not fertilized it will normally die within 24 h. However, the life span of an ovum varies considerably in different women.

The life span of the ovum is so short that normally the sperm must reach it while it is still travelling along the first third of the Fallopian tube to ensure that fertilization takes place. Furthermore, the fertilized ovum must reach the uterus within a fixed period or it will be rejected, because progesterone levels will have started to decline.

The Uterus

In women who have not borne children, the **uterus** is a hollow, thick-walled, and muscular organ, about the size and shape of a large pear. The upper part of the uterus is free and movable. It rests on the top of the urinary bladder. The bottom part of the uterus is embedded in the floor of the pelvis, between the bladder and the rectum. The uterus is securely attached to the walls of the abdomen. During pregnancy, however, the ligaments that perform this function are able to adjust to the increasing size of the uterus. The ligaments support the uterus firmly so that any unusual or abrupt body movements do not endanger the safety of the developing fetus.

The upper part of the uterus, or body, is much larger than the narrow lower end which forms the **cervix**. The cervix, which is like the neck of a bottle, opens into the vagina. The cavity inside the uterus forms the shape of a wide capital T. At the top of the uterus, between the openings of the two Fallopian tubes, is an area known as the **fundus**. This is a common site for the implantation of a fertilized ovum.

The uterus contains three layers: an outer layer of connective tissue (the **peritoneum**), a middle layer of smooth muscle tissue (**myometrium**) which makes up much of the uterine wall, and an inner layer (**endometrium**).

The Endometrium and the Menstrual or Ovarian Cycle

This important lining is controlled by hormones produced in the ovaries. The endometrium is made up of two layers. The deeper, base layer, produces the replaceable cells of the inner layer, which acts as the lining of the uterus. The deeper layer of cells remains unchanged during menstruation. The inner lining, however, is variable in thickness and changes greatly during the menstrual cycle, being lost completely during the menstrual flow.

The menstrual cycle involves the building up and breaking down of the inner lining of the uterus. This is quite a complex process. Three organs are involved (see Table 18.1):

Table 18.1. The organs and hormones involved in the menstrual cycle.

ORGAN	HORMONES PRODUCED	ORGAN AFFECTED
Pituitary gland	FSH - follicle stimulating hormone	ovary
	LH - Lutenizing hormone	ovary
Ovary	Produces ovum and the hormones estrogen and progesterone	uterus
Uterus	Prepares soft "bed" to receive fertilized ovum. Preparation of endometrium and its shedding, when not needed, is controlled by estrogen and progesterone. Provides feedback to pituitary.	pituitary gland

Normally, the menstrual cycle occurs regularly from puberty to menopause at intervals of from 25 to 35 d, except during pregnancy or while a mother is breastfeeding her baby. The average cycle lasts 28 d, although it varies considerably in different individuals from month to month, due to health, stress, or other factors.

Changes Taking Place in the Uterus

We have already discussed the development of the ovum and its follicular cells. We are now concerned with what takes place in the uterus before and after the arrival of the ovum.

There are three major phases that take place in a continuous process.

1. The **menstrual phase**. This covers the time of menstrual flow, when the endometrial wall is shed.
2. The **follicular phase**. This involves the building up of the endometrium as a result of estrogen being received by the uterus.
3. The **luteal phase**. This occurs after the ovary has released the ovum and progesterone is maintaining the endometrial lining.

Study Figure 18.10.

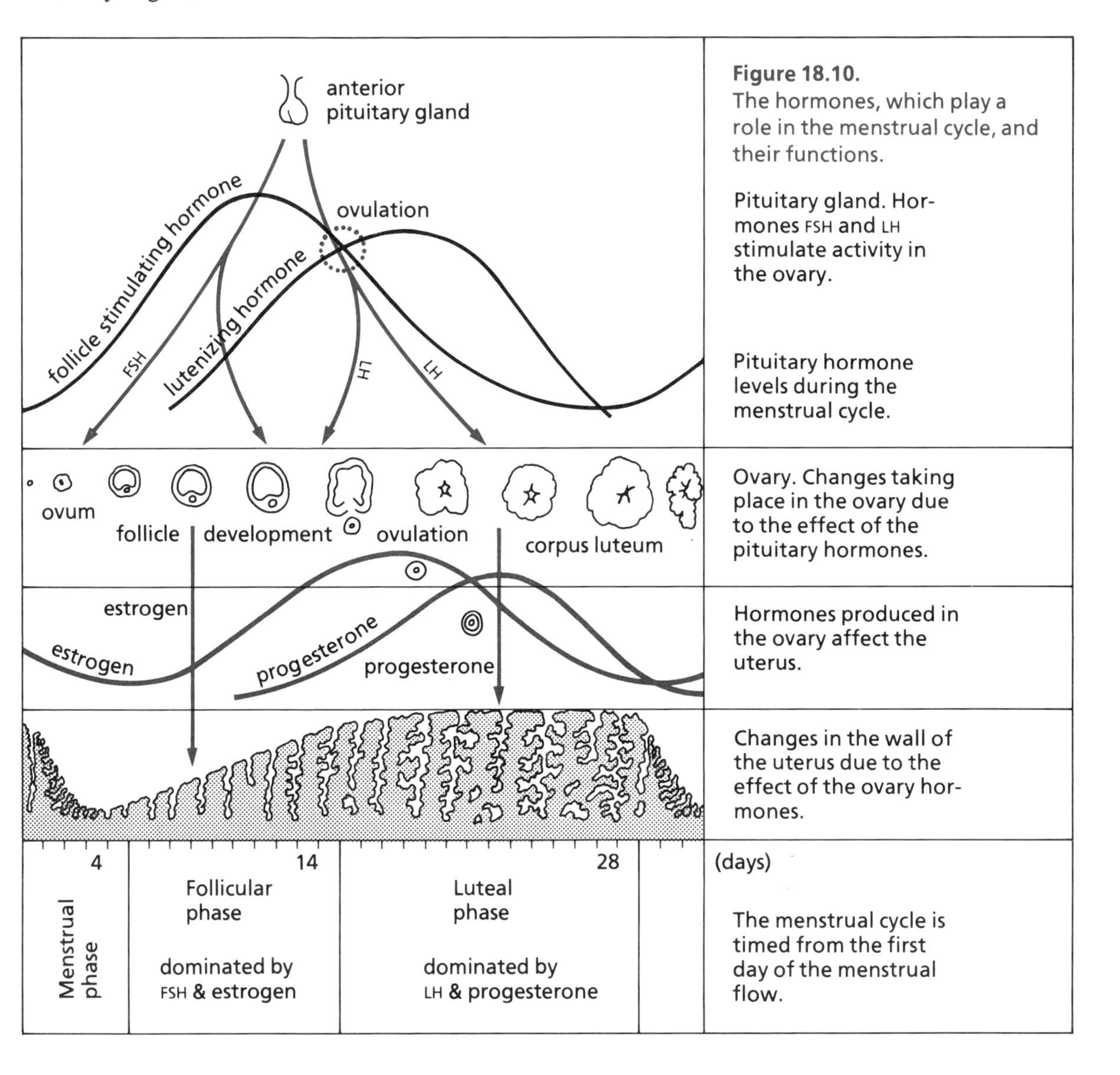

Figure 18.10.
The hormones, which play a role in the menstrual cycle, and their functions.

Pituitary gland. Hormones FSH and LH stimulate activity in the ovary.

Pituitary hormone levels during the menstrual cycle.

Ovary. Changes taking place in the ovary due to the effect of the pituitary hormones.

Hormones produced in the ovary affect the uterus.

Changes in the wall of the uterus due to the effect of the ovary hormones.

The menstrual cycle is timed from the first day of the menstrual flow.

The Endometrium during Menstruation

The purpose of the endometrium, which forms the inner lining of the uterus, is to provide a soft protective "bed" in which a fertilized ovum may become attached and develop. This endometrial lining of the uterus will provide a "home" for the developing ovum and eventually the fetus. It will aid in the transfer of nutrients to the fetus and provide a pathway for the removal of wastes. If the ovum arrives in the uterus unfertilized, the preparations which have been made are not required. The lining of the uterus then breaks down and is released during the menstrual flow.

The three phases which cover the general changes are under the influence of hormones received from the ovary, which is, in turn, governed by the anterior pituitary.

The **menstrual phase** is really the end of the cycle of events, but the onset of the menstrual flow is easy to recognize and, is, therefore, used to mark the first day of the cycle. The flow may vary from one to eight days, with an average of four or five days. During the menstrual phase, the connective and glandular tissues of the outer endometrium start to disintegrate. The blood vessels to the outer layers of tissue are constricted, and without an adequate blood supply these tissues die. Pools of blood develop from these tissues and small patches of cells break away. Eventually, the disintegrated outer layers of the endometrium and the fluids and blood contained in them are discharged and form the menstrual flow.

Even before the flow ceases, the endometrium starts rebuilding. New cells form from the underlying basal endometrium. A new ovum develops within the ovary and a new cycle begins.

The Follicular Phase

The follicular phase begins when the pituitary gland releases **follicle stimulating hormone** (FSH) into the blood stream. FSH stimulates a follicle to develop in the ovary. As the follicle develops, it releases estrogen into the blood stream which in turn, stimulates the lining of the uterus. Estrogen causes cells in the uterine lining to multiply and thicken in preparation for the arrival of the ovum.

During the 9 or 10 d before ovulation, the endometrium builds up to almost its full thickness. Estrogen eventually reaches a certain level in the blood stream, which causes a switching off of FSH secretion by the pituitary gland, and instead stimulates the release of LH. This hormonal change causes the release of the ovum from the ovary.

The Luteal Phase

Ovulation, the release of the ovum, occurs on about the fourteenth day of a 28-d cycle, although this varies greatly. It is important to recognize this variability, especially if birth control measures, such as the rhythm method are being used.

The lutenizing hormone, as its name suggests, stimulates development of the corpus luteum from the follicular cells left behind in the ovary. The corpus luteum produces progesterone which also flows through the blood stream to the uterus.

Hormonal Feedback to the Pituitary

Progesterone is responsible for the final preparation of the endometrium for receiving the fertilized ovum and for the subsequent maintenance of this layer. The endometrium becomes richly supplied with blood vessels, and the glands start to secrete supplies of glycogen (stored energy). If the ovum is not fertilized, the high levels of ovarian hormones (estrogen and progesterone) provide feedback to the pituitary gland indicating that a supply of LH is no longer required and that it is time to begin a new cycle. The corpus luteum then starts to degenerate. As the level of LH falls, the corpus luteum becomes the corpus albicans and the production of estrogen and progesterone are cut off. The corpus albicans is a small white scar of tissue that is no longer functioning.

If a fertilized ovum *is* received and implanted in the uterus, the production of progesterone is maintained for two or three months. The cells developing around the fertilized ovum produce a hormone (chorionic gonadotropin) which is very similar to LH. The action of this hormone enables the corpus luteum to maintain progesterone secretion until the tissues of the embryonic placenta start production of this hormone. Chorionic gonadotropin is carried in the blood stream and some of it gets into the urine. Pregnancy tests for the hormone may detect its presence about 10-14 d after the first missed menstruation or by the end of the first lunar month (an average of 28 d).

The Vagina

The vagina is a muscular, collapsible tube, about 8 cm in length. The upper end of the vagina encloses the cervix, which opens into the uterus where the cervix extends into the vagina; it forms a narrow circular recess around it and between the walls of the vagina. At the open end of the vagina, is a thin folded membrane called the **hymen**. The hymen partially covers the vagina, leaving a small central opening.

The lining of the vagina is composed of smooth muscle and connective tissue. It is covered with a mucous membrane which has many folds. These folds permit the enlargement of the vagina during childbirth, when it serves as the birth canal. The vagina is the organ which receives the penis during sexual intercourse.

The External Genital Organs

These features have a minor role compared to the primary organs already described. The external genital organs consist of the following parts:

The **mons pubis**, a rounded pad of fatty tissue in front of the pelvic bones of the pubic symphysis. It has a thick covering of skin and after puberty is covered with hair.

The **labia majora**. These are the two fatty folds of skin which surround the opening to the vagina. These folds contain many sebaceous sweat glands.

The **labia minora** are two small folds of skin inside the labia majora. At the top, the folds come together and form a small hood (prepuce) that partly covers the clitoris. The folds of these labia enclose two openings, the urethra from the urinary bladder and the opening of the vagina.

The **clitoris** is a small structure made of erectile tissue and is considered to be a vestigial penis. It is richly supplied with blood vessels and nerve endings.

Behind the labia and clitoris is a small chamber (the vestibule) which contains the urethral and vaginal openings. Two small glands, the Bartholin's glands, are found one on each side of the vaginal opening. These secrete a lubricating fluid and are active during periods of sexual excitement and intercourse.

THE PAP SMEAR

The cervix is a common site for cancer in women. Like many diseases, if treatment is started when the disease is in its early stages, considerable success in effecting a cure can be achieved. The Pap smear is a simple and painless process. It involves wiping off a few cells from the surface of the cervix by a physician and examination of them under a microscope. If cancer cells were present, they would be quickly identified. With regular examinations, the disease can be detected while in its very early stages.

The Mammary Glands

The mammary glands or breasts are located on the surface of the pectoral muscles between the second and sixth ribs. Their shape and size is caused by the presence of various amounts of glandular and adipose tissue. The mammary glands are designed to provide nourishment for babies by secreting milk, in a process known as lactation. The size of the breasts does not determine the amount of milk produced.

The mammary glands have a structure similar to that of sweat glands. (See Figure 18.11.) Each is composed of 15 - 20 lobes made up of glandular tissue and fat. Each lobe is connected to the nipple by a **lactiferous duct**. Lying along these ducts are sinuses (small chambers) which store the secreted milk until the baby suckles.

(lak-ti-fer-us)

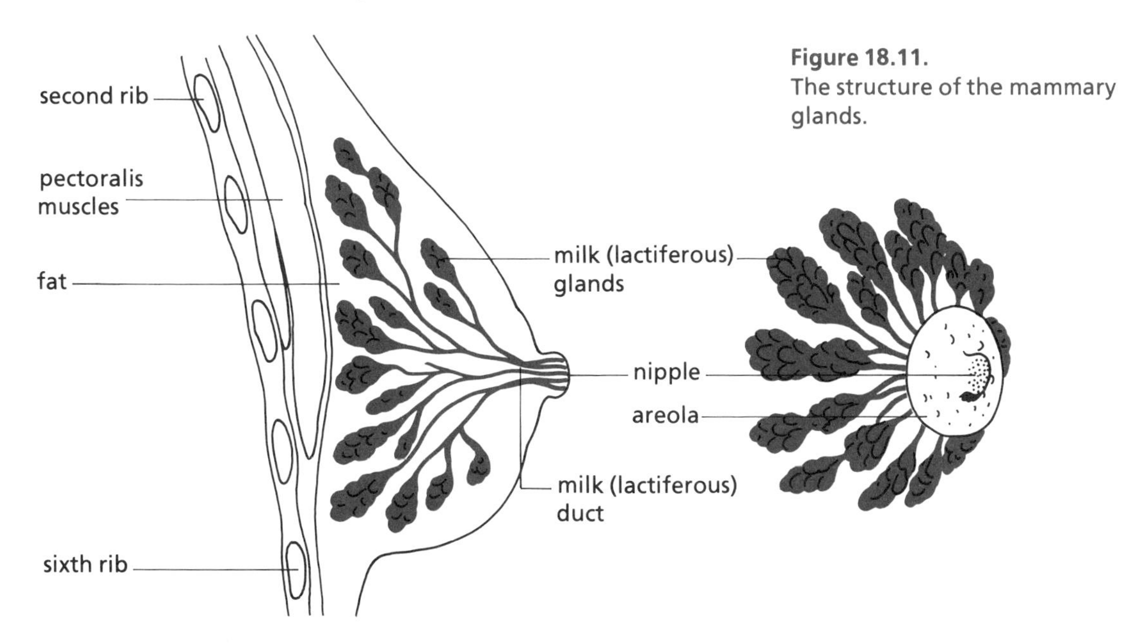

Figure 18.11.
The structure of the mammary glands.

The nipple contains many openings through which the lactiferous ducts empty to the exterior. Around the nipple is a circle of darker pigmented skin, called the **areola**. It has a rough bumpy surface because of the modified sebaceous glands present. At puberty, the female breasts develop, the ducts become larger and considerable amounts of fat are deposited. The size of the nipple and areola ring increases and becomes darker in pigmentation. These changes are influenced mainly by the secretion of estrogen and progesterone by the maturing ovaries.

Copulation and Fertilization

If fertilization is to take place, sperm must be deposited in the vagina close to the time of ovulation. Fertilization is defined as the union of a sperm and ovum. It normally occurs in the Fallopian tube (within 24 h of ovulation) when the ovum is about one-third of the way along the tube. Sperm are transferred to the vagina and then to the Fallopian tube by copulation. When the male becomes sexually excited, blood vessels supplying the penis enlarge and fill the spongy sinuses, thereby causing the penis to increase in size and become erect. When erect, the penis is able to penetrate the vagina.

Back and forth movements of the penis inside the vagina increase sexual tension in the male until ejaculation occurs. The walls of the male ducts contract and push the sperm rapidly through the vas deferens into the urethra. Various fluids from the accessory glands are added to the sperm to provide nourishment and a suitable chemical environment. Together the fluids and sperm form a substance called semen. The semen passes through the male urethra and is projected into the vagina.

Sexual excitement is evident in the female as in the male. Touch and sensory stimulation arouse nerve endings in the female genitalia, especially in the clitoris, which becomes erect. Impulses to the Bartholin's glands and the walls of the vagina cause the secretion of lubricating fluids, which facilitate sexual intercourse. Reflexes in a woman initiate emotional and muscular reactions similar to those occurring in the male. This response is referred to as orgasm (or climax).

Fertilization

After the sperm have been deposited in the vagina at the opening of the cervix, they move forward, partly due to a whiplike motion of their flagella, and partly due to muscular movements of the uterus. The sperm pass from the vagina, through the uterus, and into the Fallopian tubes, where union with the ovum may take place. Sperm must be present in the female genital tract for about 4 to 6 h before the ovum can be fertilized. This amount of time is required for an enzyme (hyaluronidase) contained in the acrosomes of sperm to dissolve part of the membrane that protects the ovum.

It appears that many thousands of sperm must be present to produce enough of this enzyme to dissolve the ovum's protective membrane. Only one sperm, however, will penetrate and achieve fertilization. Immediately after this first sperm has penetrated, a chemical membrane barrier is instantly formed, which prevents the entrance of any other sperm present. (See Figure 18.12.) After entry, the tail of the sperm is lost and the 23 chromosomes which are present in the sperm unite with 23 chromosomes present in the ovum. This results in the normal number of 23 *pairs* of chromosomes being established in this, the first cell of a new individual.

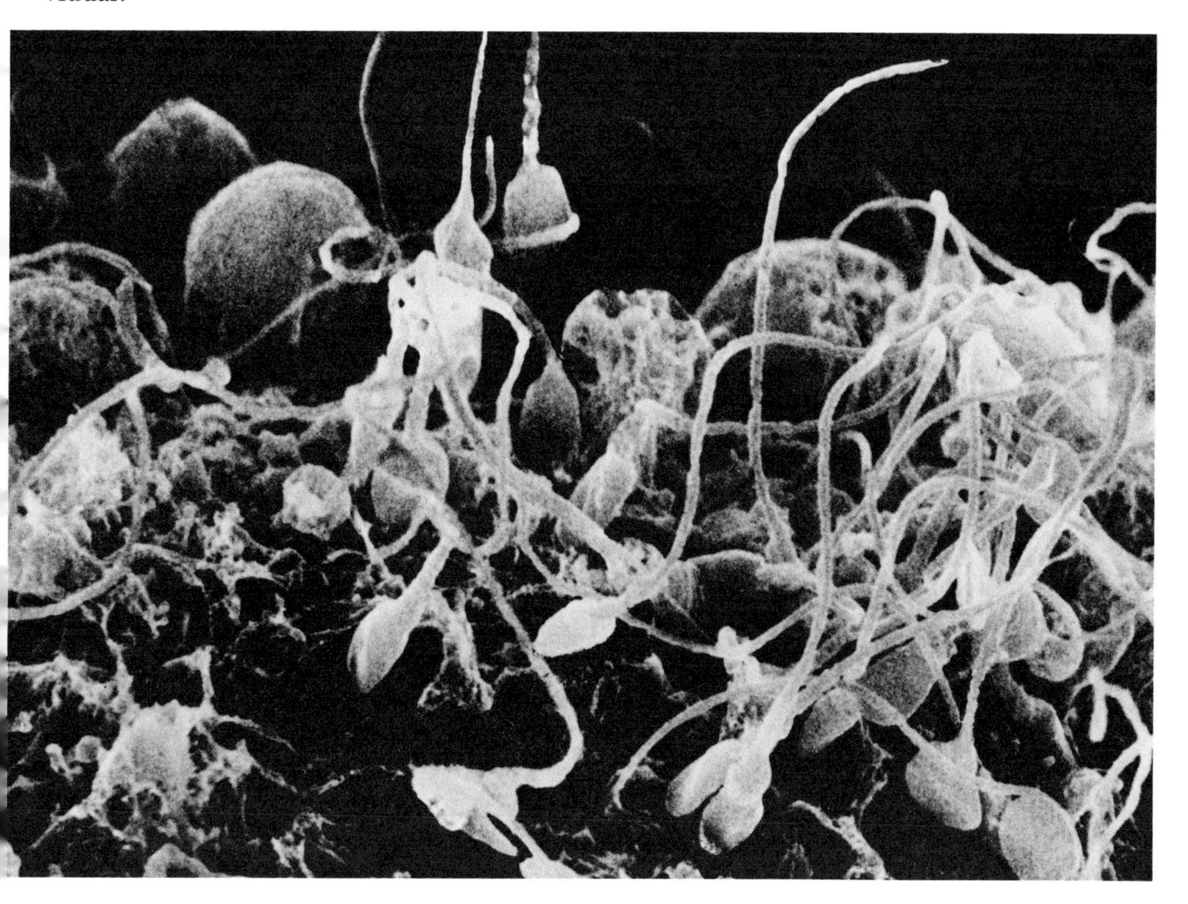

Figure 18.12.
Sperm surround a portion of the ovum in the Fallopian tube. The large cells are follicular cells (mag. 2000x).

This initial cell, with its nucleus and chromosomes, cytoplasm and surrounding membrane, is called a **zygote**. Immediately after fertilization takes place, the zygote undergoes rapid cell division as it passes down the Fallopian tube to the uterus for implantation. (See Figure 18.13.)

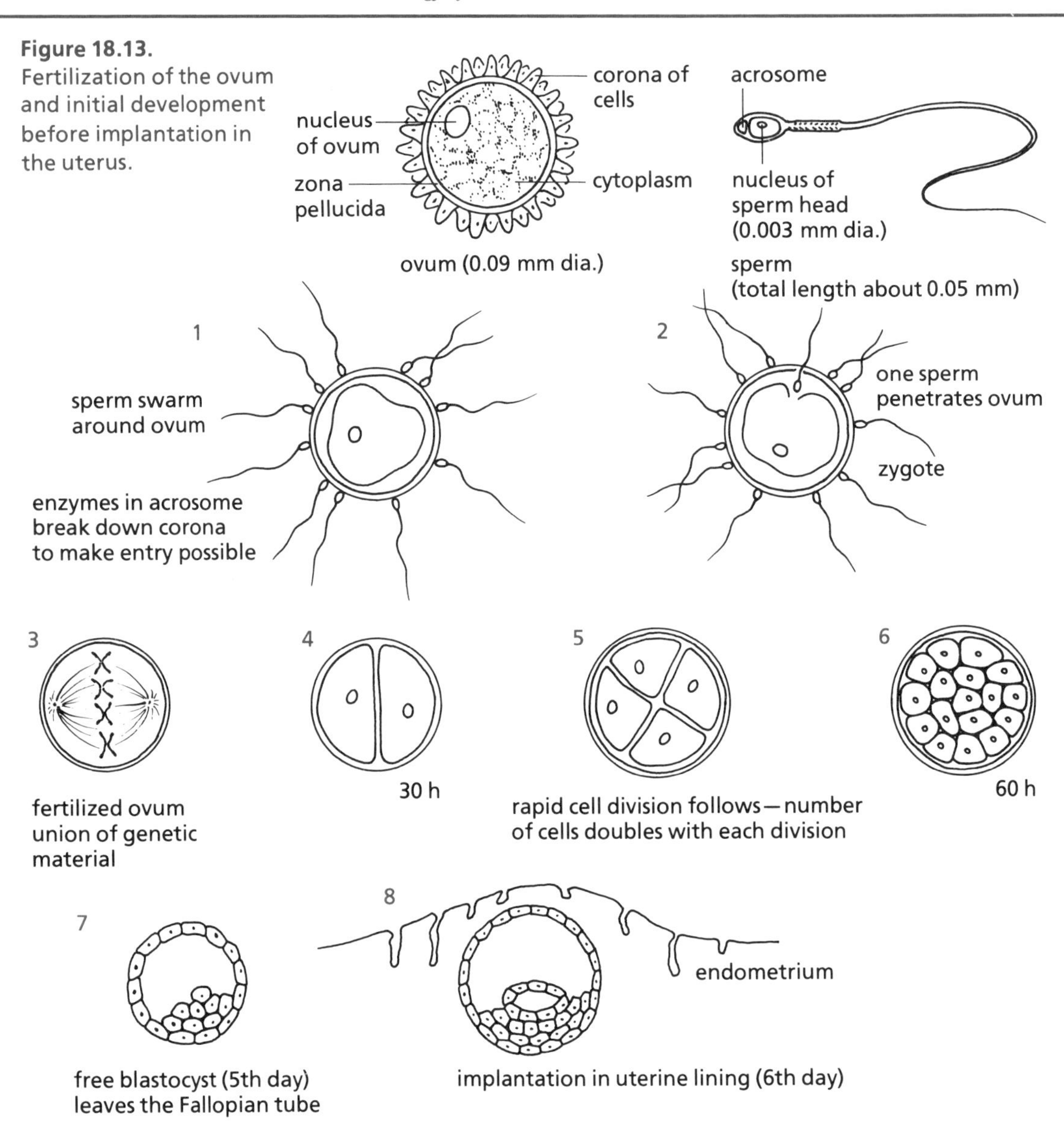

Figure 18.13. Fertilization of the ovum and initial development before implantation in the uterus.

Menopause

When a woman reaches 45 - 50 years of age, internal changes in her body result in the ovaries becoming less sensitive to pituitary hormones. Eventually the ovaries no longer produce ova or secrete ovarian hormones. The menstrual period, with its monthly flow, ceases and there is a gradual change in the internal organs of the uterus and Fallopian tubes. The body sometimes requires several years to adapt to the changes in hormone balance.

Sexually Transmitted Diseases

What is STD?

STD is the abbreviation for "sexually transmitted diseases". Two of these diseases, syphilis and gonorrhea, are commonly called venereal diseases (VD). The term STD is now being used to include VD and other diseases that are transmitted through sexual contact. The spread of some of these diseases has reached epidemic proportions.

Why Worry About STD?

Not all sexually transmitted diseases cause serious illness. But they all cause some discomfort and, if left untreated, some of them can have very serious results. These diseases can be successfully treated in the early stages, and most of them can be prevented if certain precautions are taken. It's important to know the facts.

Some General Facts About STD

For the most part, STD is spread by sexual contact. There is no immunity against any of the sexually transmitted diseases and, as a result, one can have the same disease several times. To date, no vaccines have been developed to prevent STD.

Since each sexually transmitted disease is caused by a different organism, it is possible to have more than one condition at the same time. If a person has an STD, he or she may not be aware of it, because the symptoms are not always obvious. There are many different STDs. Two of the most common are discussed here.

Syphilis

Syphilis is the most serious sexually transmitted disease, and if not treated, can be fatal. It is spread by intimate contact with an infected partner.

The primary stage of syphilis shows up between ten and ninety days after contact, as a sore or chancre, at the place where the germs entered the body.

The chancre is a small painless ulcer which may go unnoticed. However, the chancre is full of syphilis germs, called spirochetes, and the disease is easily spread to other people who come into intimate contact. If the chancre is on or in the mouth, the disease may be spread simply by kissing.

The chancre disappears within a few days to a few weeks without treatment. This does not mean that the disease is

CASE STUDY. Linda, 16 years of age.

Linda was unhappy and worried. She badly needed to talk to someone. She usually shared her personal concerns with her best friend Maria, but this time it wasn't easy to talk to anyone. Linda kept wondering if Maria would keep a confidence this time.

Eventually, Linda called up her friend and suggested that they get together. A request to meet or visit in each other's homes was a common event. The girls went up to Linda's room and put on some records. For some time, they talked of classes, boys, and new records, but Linda couldn't seem to keep her mind on the topic and Maria, becoming impatient, asked her what was bothering her. It took some coaxing to get Linda started, but eventually she blurted out her problem, "I've got VD." Maria was stunned and didn't know what to say at first and then questions tumbled out. "When?", "How do you know?", "Who was it?" Linda was upset, but gradually she told her friend what had happened.

First there had been a phone call from a health nurse who said that she had been given Linda's name as a contact. Linda's name had been produced by someone currently being treated for gonorrhoea.

Linda had immediately asked the nurse who had given her name, but the nurse would not identify the person being treated. She said that Linda wouldn't like it if she gave her name on the phone to someone else. Linda reluctantly agreed. She had then asked the nurse if she had contacted her mother. The nurse explained that no one else knew, because it was entirely a matter between the two of them. The nurse asked Linda the name of her physician and said that she wanted her to see the physician as soon as possible. She told Linda that she would be telephoning the physician, explaining why Linda was coming in. The nurse also said that she would phone the physician again later to see if Linda had kept the appointment.

As Linda talked with her friend, she gradually became more calm, and in response to Maria's questions, told her about the appointment with the doctor. The doctor first asked Linda how long it was since the contact had taken place, and whether she had been sexually intimate with more than one person. The doctor had examined Linda and, using a sterile swab, had taken a sample from her vagina. The sample was sent to a laboratory where a smear could be examined under a microscope and the swab used to start a test culture.

Linda received another call from her doctor a week later. She was asked to make another appointment. When Linda went the second time, the doctor explained that the test was positive and that she did have gonorrhoea.

Linda protested. She had no sores, no pain, no discharge, nothing. She didn't want to believe the results.

The doctor also explained to Linda that, if venereal diseases (or STD–sexually transmitted diseases) went unchecked and untreated, there was a serious risk of a female becoming sterile and unable to have children later. If she had sexual contacts with other persons, she would very probably transmit the disease to them. The doctor said that if everyone with the disease were treated, it would be possible to wipe it out completely. The doctor asked Linda if she was allergic to any antibiotics and then gave her some medication (ampicillin) and some water. She was asked to take it while in the office. She was told that the one treatment would probably clear up the problem, but she was directed to come back in a week's time for a second check, to be sure that she was completely cured.

Maria wanted to know how the nurse knew to call her. Linda explained that, in order to try to stop the spread of the disease, each person was asked to either speak to anyone with whom they had been sexually active or to give the doctor their names and the health nurse would call them, completely confidentially. Linda said that in her case she was sure thatthey had kept their promise and that no one else knew.

Questions for research and discussion:

1. Find out if there is any vaccination for venereal infections.
2. Where can you get information on venereal diseases?
3. Where are VD treatment clinics located in your area?

cured, but indicates that the germs are attacking other parts of the body.

The secondary stage of syphilis usually shows up about the time the chancre disappears, 2.5 to 6 months after initial infection. Signs and symptoms of this stage often resemble those of other diseases and they may include a skin rash. Other symptoms which may appear are sore throat, fever, sores in the mouth, headache, bone pain, and patchy loss of hair. The disease is still highly infectious at this stage.

Like the chancre, these symptoms will disappear without treatment. Unless treatment is obtained, the disease goes "underground".

After the second year, the disease will not spread to sexual partners. This latent stage can last for years and, in about 65% of the cases, there will be no further symptoms.

The other 35% of syphilis cases will experience such serious ailments as heart disease, paralysis, blindness, or insanity. These effects may not occur until fifteen years or more after the first infection.

A pregnant woman can pass the disease to her unborn child. This condition, called congenital syphilis, can result in the baby being born deformed or dead. Every expectant mother should have two blood tests for syphilis, one early and one late in pregnancy. Proper treatment, given in time, will cure infection and increase the woman's chances of having a healthy child.

A physical examination and laboratory tests are needed to diagnose syphilis. Tests may not be positive until three months after infection so that more than one test may be required to confirm the diagnosis. Syphilis can be treated with injections or oral doses of antibiotics.

Gonorrhea

Gonorrhea is spread by sexual intercourse. The disease organism is the **gonococcus**, a sensitive bacterium which cannot survive outside the human body. For this reason, it is impossible to catch the disease from toilet seats, door handles, or other surfaces.

The symptoms of gonorrhea in men are a burning sensation when urinating and a discharge of pus which usually appears two to ten days after intercourse with an infected person. Most men will be aware of these symptoms. However, some men and most women will have no symptoms. Because of the short incubation period and the contagious nature of the disease, people with no symptoms can continue to spread the disease for months.

Symptoms will disappear eventually. Unless treatment is received, however, the disease remains infectious and can cause serious problems. Untreated gonorrhea may result in a skin rash and/or an extremely painful form of arthritis which can cause permanent crippling. It affects the genital organs and can lead to sterility.

An infected pregnant woman can transmit gonorrhea to the baby's eyes as it passes down the birth canal. Treatment must be provided or the child may become blind. Hospitals routinely treat the eyes of all newborn babies as a precaution.

Gonorrhea is diagnosed by laboratory tests. Because the incubation period is so short and many people show no symptoms, it is important for all recent sexual partners of an infected person to receive treatment. Gonorrhea is treated with antibiotics. Follow-up care by a physician is necessary to ensure that there has been a cure, because persons who have not yet been cured may have no symptoms.

Vaginitis

Vaginitis is the most common infection of the female reproductive organs. It refers to inflamation of the vagina. There are a number of forms of vaginitis, the symptoms of which range from an itch in the genital area to a discharge. Vaginitis is not dangerous, but can be very uncomfortable and it should be treated as soon as possible. Men can transmit the organisms that cause vaginitis to female sexual partners.

Herpes Genitalis

This disease is caused by a virus which is very similar to the one that causes cold sores or fever blisters. It is thought that the **herpes virus II** is spread by sexual intercourse.

Groups of small, painful blisters appear on the sexual organs. If the sores become infected, they may discharge pus or blood. The sores usually clear up within twenty days, but some people have recurrent outbreaks.

A herpes virus II infection may make women more susceptible to cervical cancer, although this relationship has not been proven to date. No cure is yet available for herpes genitalis. Methods of treatment are currently being investigated.

Can The Spread of STD Be Stopped?

As the foregoing sections indicate, treatment and cure are available for most of the sexually transmitted diseases. Anyone who has contracted one of these conditions should seek medical treatment immediately.

Because many people, particularly women, develop no symptoms after contracting certain sexually transmitted diseases, it is vital that each infected person being treated provide the means to contact all of his or her sexual partners. These partners may then be examined and, when necessary, treated. In this way, the spread of STD can be arrested.

Self-diagnosis and treatment are unwise and can result in serious complications. Self-treatment may hide signs of infection and a person may think that he or she is cured when actually it is not so. Laboratory tests often are required in order to determine if an infected person has been cured of the disease.

Parts of "Sexually Transmitted Diseases" are based on material published by the Ministry of National Health & Welfare, Ottawa, Ontario.

QUESTIONS FOR REVIEW

SOME WORDS TO KNOW

Match one of the statements found in the column on the left with a suitable term from the column on the right. DO NOT WRITE IN THIS BOOK.

1. Hormone which initiates the development of a new ovum.
2. Hormone produced by the corpus luteum.
3. Organ in which the fetus develops.
4. Tube in the penis through which semen passes.
5. The inner lining of the uterus.
6. Hormone produced by the follicular cells in the ovary.
7. Spongy, blood-filled tissues in the penis.
8. Male accessory gland.
9. Major male hormone.
10. Sperm cells are produced in these tubules.

A. fimbrae
B. menopause
C. puberty
D. testosterone
E. progesterone
F. estrogen
G. FSH
H. LH
I. urethra
J. ureter
K. seminiferous
L. Fallopian
M. corpus cavernosum
N. uterus
O. ovary
P. prostate
Q. testes

SOME FACTS TO KNOW

1. List the ducts and organs (in order) through which a sperm passes after developing in the seminiferous tubules, until it leaves the penis.
2. What are the functions of the male accessory glands?
3. What is the function of the following
   a) Sertoli cells,
   b) prostate gland,
   c) corpus cavernosum,
   d) epididymis.
4. Explain the sequence of stages through which the ovum passes as it develops in the ovary.
5. Make a table as follows:

Hormone	Organ where produced	Organ that hormone affects	General effect of the hormone

   List the following hormones and then complete the table: progesterone, estrogen, FSH, LH.
6. Where exactly are the ovaries located and how are they held in place?
7. What is meant by the following terms?
   a) implantation,
   b) fertilization,
   c) ovulation.
8. Why do sperm and ova only live for a very short time?
9. Describe the structure and function of the endometrium.
10. Briefly describe the three phases of the menstrual cycle.

QUESTIONS FOR RESEARCH

1. Select one of the following topics and prepare a report on it:
   - obstetrician
   - tubal ligation
   - contraception
   - problems of communication about human reproduction
   - vasectomy
   - sexually transmitted diseases
2. The Herpes virus is becoming almost epidemic in some parts of North America. At present there is no cure for this disease. Find out all you can about Herpes, its symptoms, and its present treatment.
3. What is a Pap test? How can this test be used to help avoid cervical cancer?
4. What causes infertility in men and women?
5. How does the contraceptive pill work? Are all contraceptive pill prescriptions the same?

Activities: EXAMINATION OF REPRODUCTIVE TISSUES AND CELLS

Material
prepared slides of rat or human testis, sperm cells, rat or human ovary with Graafian follicle and corpus luteum.

Method 1
Examine the cross-section of a seminiferous tubule of the rat testis under the low power of your microscope. Turn to your high power and examine a section of the tubule wall. Refer to Figure 18.2. Sketch any four specialized cells. Repeat the above procedure with the *human sperm cell*. Sketch and label the acrosome, head, middle piece, and flagellum.

Questions
1. *Which cells are haploid, diploid?*
2. *What happens to the spermatids as they migrate into the centre of the tubule?*
3. *Sertoli cells are difficult to observe. Why are they so much larger than the surrounding cells?*
4. *Observe the slide of the rat/human testis under low power. Observe the cells between the seminiferous tubules. These are interstitial cells. What is their function?*
5. *What is the function of the acrosome?*

Method 2
Examine the slide of the rat or human ovary at 100X magnification. There is no need to use your high power. Sketch a mature follicle with its oocyte (future egg) and label it. Examine the corpus luteum under low power. Refer to Figure 18.6.

Questions
1. *What is the function of the follicle cells?*
2. *When does the corpus luteum appear in the female menstrual cycle and what is its primary function?*
3. *What would be the level of each hormone just prior to ovulation?*

Chapter 19

Pregnancy and Birth

- Growth and Development
- Pregnancy
- Birth
- Growth

Growth and Development

Growth is one of the major characteristics of all living things and it sets them apart from inanimate objects. Animals grow by metabolic processes – by taking in food substances, breaking them down into simple molecules and then reassembling them to form the complex structures of the new body.

Growth is basically an increase in size produced by the multiplication of cells. These cells eventually differentiate to perform different functions. Growth involves an ordered sequence of development stages which produces an irreversible progression from conception to adulthood and eventual death.

The Cell

The adult body contains 60 000 000 000 000 cells, all of which developed from one initial fertilized egg cell. Constant cell division creates identical daughter cells, each of which contains half the mass of the parent cell. Each daughter cell receives from the parent a set of chromosomes which are identical in every way to those of the parent cell.

For cells to multiply, they must be constantly supplied with raw materials. These provide energy to do the work of the cell and, more importantly, supply the protein necessary for building the new cells. Apart from water, proteins are the most vital group of compounds in the body and account for about three-quarters of the dry mass of the human body.

The heart of a newborn baby is only one-sixteenth the size of that in a grown adult but it contains the same number of cells. These cells increase in size as the child grows, taking care of the enormous task of pumping blood throughout the body during every moment of its life. Unlike those of the heart, the cells of most organs and tissues are constantly multiplying, allowing not only for growth of other organs, but also for the replacement and maintenance of the tissues.

Early Development

The first cell formed by the union of a sperm and ovum is known as the **zygote**. Fertilization usually takes place in the first third of the fallopian tube and the zygote is then moved slowly down the tube towards the uterus, a journey that

CLONING

The word "clone" is derived from the Greek "klon", meaning twig or slip, referring to asexual reproduction of plants. Plant cloning, using cuttings of twigs or leaves, has been practised for centuries. Today, single cells of plant tissues can be cultured in the laboratory. For example, a single carrot cell will reproduce, making many copies of itself, and eventually, a replica of the original carrot from which it came.

Animal cloning proved successful several years ago using frogs. The nucleus from an intestinal cell from a frog was transplanted into a frog egg that had been irradiated so that its own nucleus was nonfunctional. The egg cytoplasm then contained only a functional nucleus from the intestinal cell. The "new egg", which contained genes from the frog whose intestinal cell nucleus was used, went on to divide to form a tadpole that developed into a frog which was identical to the donor frog.

More recently, the first mammalian clone was produced. Nuclei from mouse embryos were removed and inserted into mouse eggs whose nuclei had been removed. These eggs containing the transferred nuclei were inserted into the uteruses of mature mice. The mice that were born (the clones) were identical to the mouse whose embryo nuclei were used.

These first successful experiments with mammals bring much closer the time when decisions will have to be made about the possibility of cloning people. This will be an important biological *and* social issue. Will it be possible to clone people? Will we allow identical copies of people to be made? If so, *which* sort of people will be cloned? *How* will we decide? *Who* will decide?

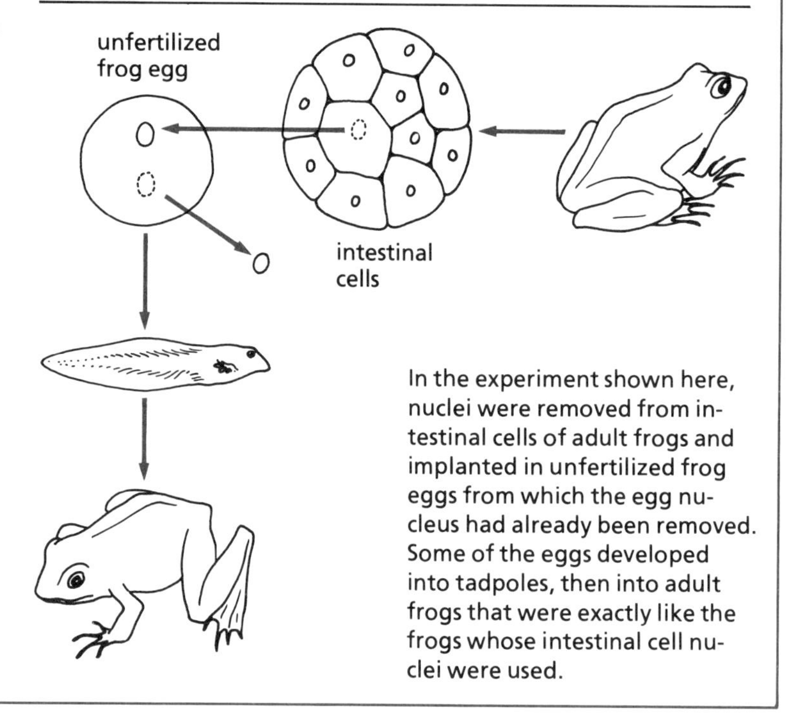

In the experiment shown here, nuclei were removed from intestinal cells of adult frogs and implanted in unfertilized frog eggs from which the egg nucleus had already been removed. Some of the eggs developed into tadpoles, then into adult frogs that were exactly like the frogs whose intestinal cell nuclei were used.

usually takes two or three days. Immediately after fertilization, before it has even reached the uterus, rapid changes begin to take place in the zygote. The cell divides repeatedly (mitosis), doubling the number of cells present with each division, until a hollow ball of cells is formed, known as a **blastocyst**. (See Figure 19.1.)

(blas-toe-sist)

(koe-ry-on)

As the fertilized ovum passes down the Fallopian tube, a membrane, known as the **chorion**, forms around the mass of dividing cells. The membrane is covered with tiny villi secreting enzymes which help to form a path into the tissues of

Figure 19.1.
Development of the embryo.

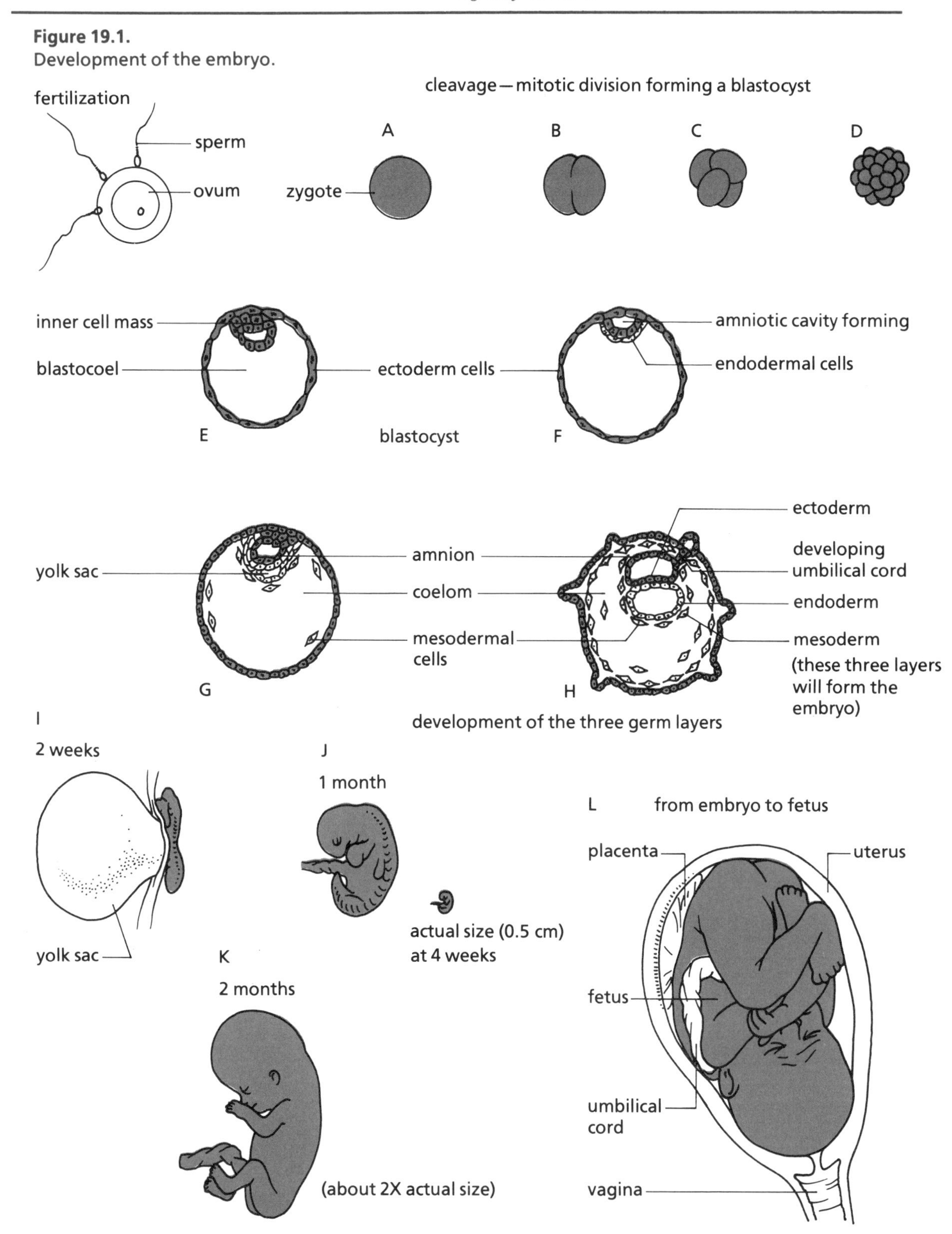

the endometrium so that the blastocyst can nestle close to the maternal blood supply. The blastocyst gradually sinks between the soft cell tissues of the thickened endometrial wall of the uterus. When this process called **implantation** has occurred, a successful pregnancy has been established.

Pregnancy

The implanted blastocyst continues to divide rapidly, producing more and more cells. Eventually two major groups of cells are produced. One group will eventually become the embryo (the developing baby). The second develops into several structures and tissues that support the growth of the embryo while it is in the uterus.

Structures within the Uterus

The **amnion** is a thin sac filled with a watery fluid. The embryo develops inside this sac, supported and well protected from bumps by this fluid. (See Figure 19.2). The **placenta** develops in close contact with the wall of the uterus. It contains blood vessels both from the mother's circulatory system and from the embryo's blood system.

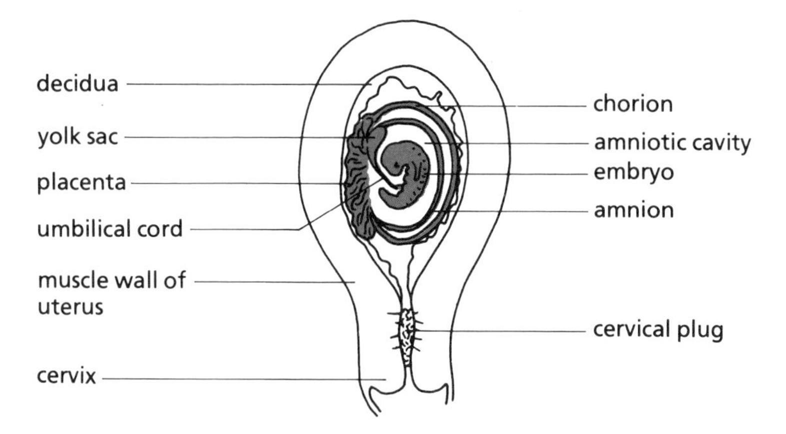

Figure 19.2.
Uterus containing the early embryo and showing the amnion, yolk sac, placenta, and uterine wall.

Since there is no direct connection between these two systems, the blood does not mix. In Chapter 9, on the circulatory system, we discussed problems that can arise when two different blood types are mixed. Commonly, a developing baby has a blood type different from that of its mother; for this reason the systems are kept separate. However, nutrients, oxygen, and other substances in the mother's blood are transferred to the embryo across the placenta in which capillaries of both systems come close to one another. (See Figure 19.3.) Wastes and carbon dioxide resulting from the cell activities of the embryo diffuse out of its blood vessels and pass into the mother's circulatory system for disposal. The placenta then acts as both a barrier and a bridge, it is the exchange site for all materials entering or leaving the embryo. (See Figure 19.4.)

The **yolk sac** functions only temporarily in human development, although it plays a major role in animals that hatch from eggs. In humans it produces blood cells for the embryonic circulatory system before the true blood producing tissues are available.

Figure 19.3.
The placenta forms a barrier between the fetus and the mother which prevents blood cells and large proteins from crossing from one system to the other. The blood vessels of the two systems are so closely associated in the placenta that nutrients and other needed substances can easily pass into the fetal circulation. Waste products can also pass from the fetal blood into the mother's blood for excretion.

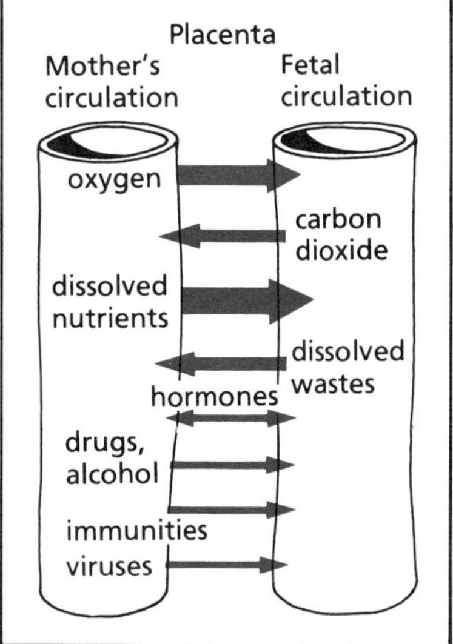

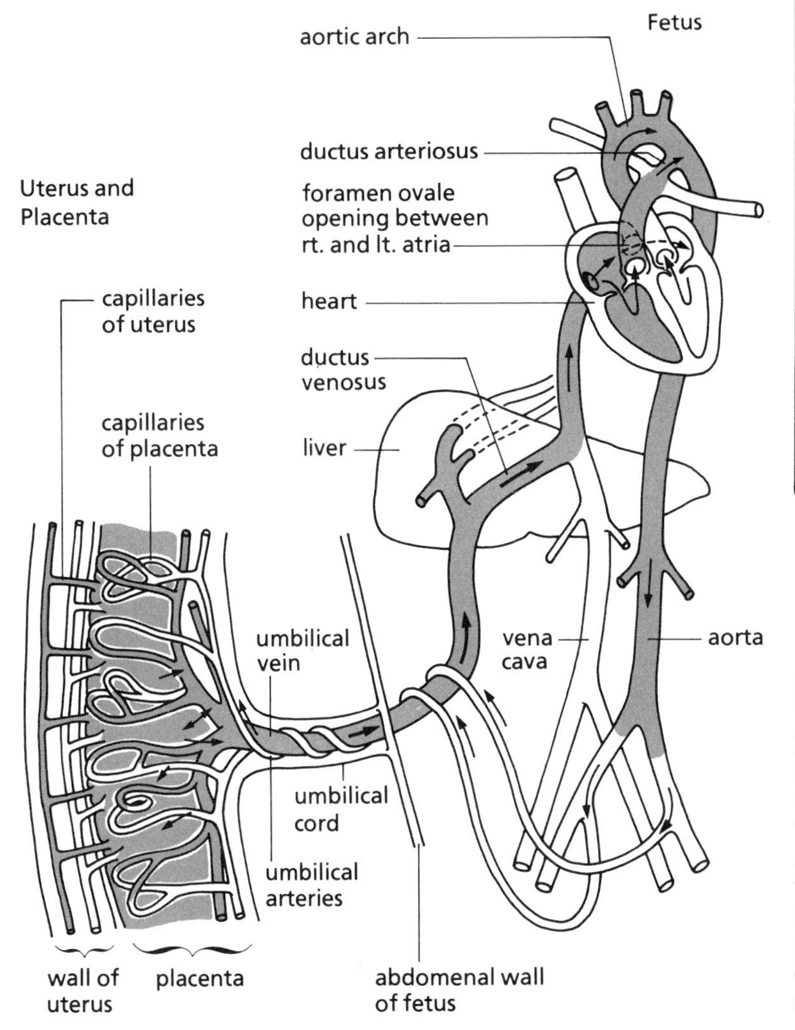

Figure 19.4.
The fetal lungs and liver are not yet being used for their adult purposes. As a result only minimal blood is passed through these organs. Three structures are present in the fetal circulation that are not present in the adult circulatory system. These are the **foramen ovale**, the **ductus venosus**, and the **ductus arteriosus**.

AMNIOCENTESIS

Amniocentesis is the process of inserting a needle into the amniotic sac within the uterus of a pregnant woman to remove a small quantity of the fluid which bathes the developing embryo. This is done to help determine whether or not the fetus is normal. This process is done at the 16th week of pregnancy.

The collected fluid is separated into cellular and non-cellular components. Cells in the fluid have been shed from various parts of the fetus such as the fetal skin, respiratory tract, and umbilical cord. Though many of the cells are dead, a few are alive and can be multiplied by culturing them for about two weeks. The chromosomes from these cells can be stained and examined to determine if there are any abnormalities. Biochemical studies are also possible.

The most common chromosomal abnormality seen in amniotic fluid cells is an extra chromosome (# 21) making a total of 47 instead of the normal 46. If there is an extra # 21 chromosome in the karyotype, the fetus will have a condition known as Down syndrome.

Mothers who are 35 years of age or over have an increased likelihood of giving birth to Down syndrome children. Either the mother or the father may provide the abnormal reproductive cell.

Experience so far indicates that in performing amniocentesis the risk of harming the fetus is very small. To make certain, an ultrasound scan is always performed before the amniocentesis. This is a method using high-frequency sound waves to locate the placenta and determine the size of the fetus and whether or not there is a twin pregnancy. Amniocentesis is usually available to women over 35 years of age, to women who have had a previous child with a chromosomal defect, to women who are known to be carriers of a disease that could affect their offspring, and to women who have had a previous child with a spinal defect known as spinal bifida.

Not all abnormalities can be detected by amniocentesis. For example, most cases of mental retardation cannot be detected, nor can hereditary conditions that have no visible chromosome abnormality or any way of being identified by chemical tests, such as heart defects, etc. However, there have been many successful diagnoses of genetic defects of the embryo through amniocentesis.

The last of the structures associated with the embryo is the **umbilical cord**, which connects the developing baby to the placenta. It carries one vein and two arteries. Remember that arteries are defined as vessels that carry blood *away* from the heart and veins carry blood *to* the heart. In adults, blood travelling to the heart through veins is deoxygenated. However, in the developing baby, the vein in the umbilical cord brings oxygen from the mother. Once this oxygen has been used in the developing baby's system, the blood is carried away from its heart in arteries and transported to the placenta where it picks up more oxygen and discharges its burden of carbon dioxide. The primary reason for this difference in the type of vessel carrying oxygen is, of course, the fact that the lungs of the embryo are not working or filled with air; oxygen is obtained through the placenta using the lungs and blood stream of the mother.

Development of the Embryo and Fetus

When looking at a group of teenagers or adults, it is hard to realize that they were once only 3.5 kg babies. Between birth and about twenty years of age, body mass increases more than 20 times and height about 3.5 times. Even this rate of change, however, is extremely small compared to the changes that take place during the first two months after conception. In the first 8 weeks, the overall length of the embryo increases 240 times and mass increases over one million times as the single cell develops into a miniature baby. The nutrient supplies, vital to this amazing growth, flow through the placenta to reach the embryo, and the mother's diet during pregnancy thus has direct and important effects upon the developing baby.

The group of cells which form the embryo undergo such vast and complex changes that only a brief summary is possible here. After dividing many times and forming the supporting tissues already mentioned, the blastocyst soon separates into three layers, the **ectoderm**, **mesoderm**, and **endoderm**. These layers then differentiate (become different types of cells with different functions) into a variety of tissues. (See Table 19.1.)

Table 19.1 Some of the major tissues and organs that develop from the three germinal layers of the ectoderm, mesoderm, and endoderm.

GERM LAYER	TISSUES AND ORGANS (BY DIFFERENTIATION)
Ectoderm	Nervous system, epidermis, parts of the eye, salivary galnds, pituitary gland, adrenal medulla, skin, hair, and nails
Mesoderm	Connective tissue, bone, muscles, cartilage, blood, blood vessels, lymphatics, spleen, adrenal cortex, parts of the reproductive organs
Endoderm	Epithelium of the digestive tract, linings of the lungs and respiratory passages, liver, pancreas, thyroid, parathyroid, and thymus glands

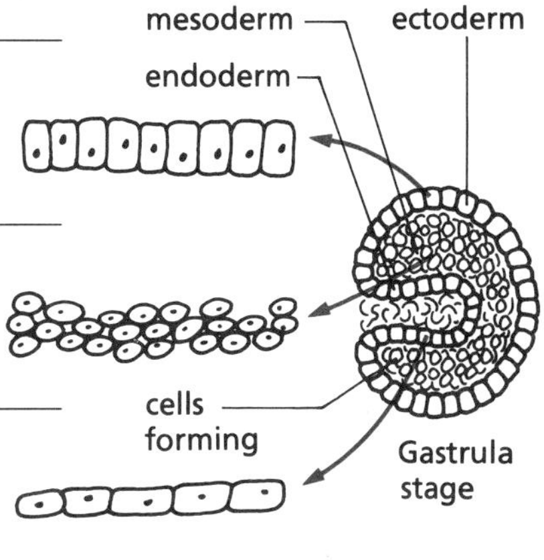

mesoderm
ectoderm
endoderm
cells forming
Gastrula stage

By the end of the first month of pregnancy, the embryo is little more than 3 mm in length. Yet its heart has been pumping blood since the 18th day after conception. It has the beginnings of eyes, a spinal cord and nerves, lungs, stomach, intestines, liver, and kidneys.

In just eight weeks the embryo begins to form the first bone cells. This change marks such a major transition that the term embryo is dropped and the word **fetus** is adopted to describe the developing baby. The total period of time that the embryo and fetus are in the uterus is known as **gestation**. In humans this involves 280 d from the beginning of the last menstrual period. (See Figure 19.5.)

The events which mark the growth and development of the embryo and fetus are outlined in Table 19.2.

Figure 19.5.
Three stages in the development of the human embryo:

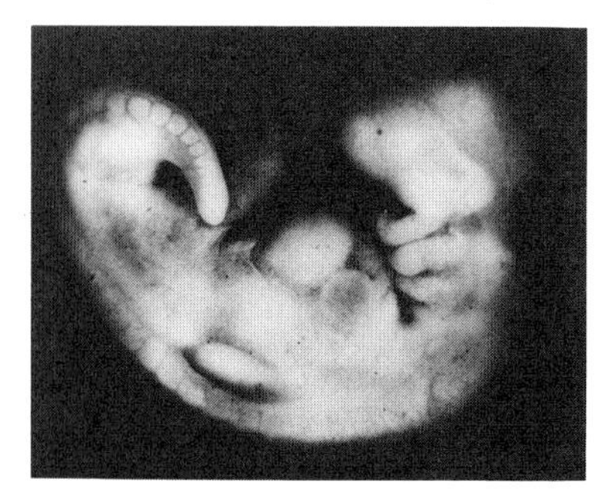

a) 32 d (5 mm). Note the brachial arches, heart, yolk sac, segmentation in the spinal vertebrae, and tail.

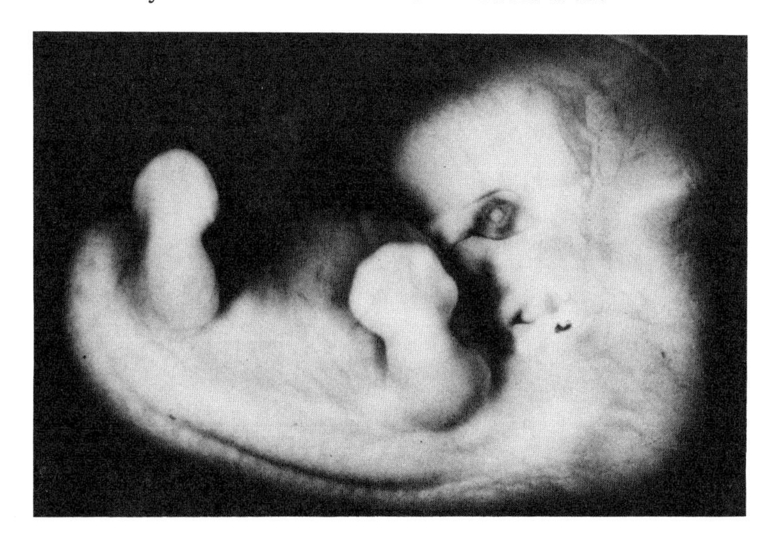

b) 41 d (12 mm) Note the large head, eye, and limb development. The tail is still present and external ear openings are apparent.

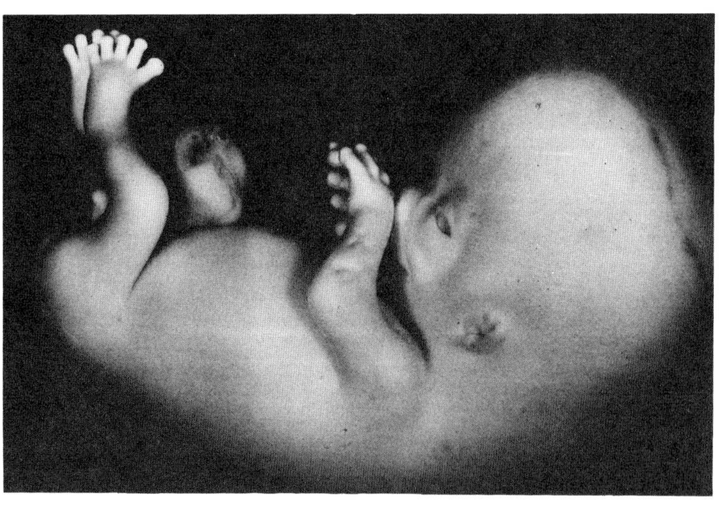

c) 52 d (23 mm) Nose flat, eyes set well apart, formation of external ear, fingers and toes developing. The umbilical cord can also be seen.

Table 19.2 Summary of the major changes that take place in the embryo during pregnancy.

MONTH	SIZE	MAJOR CHANGES IN DEVELOPMENT
1	0.5 cm	The embryo at this point is so small that the mother may not yet know she is pregnant. The head is made up mainly of two large bulges which will form the two halves of the forebrain. The eyes, nose, and ears are not yet formed, but are well started. The beginnings of a spinal cord and nervous system, lungs, stomach, liver, kidney, and intestines are present. Its heart began beating about the 18th day. Small buds, which will become the arms and legs, have appeared.
2	2.5–3.5 cm	During the second month the arms and legs become distinct limbs with tiny fingers and toes. The eyes, far apart, are sealed behind fused lids. A tiny tail, that will later disappear, curves up from the bottom of the embryo's spine. It now possesses all the internal organs of the adult in various stages of development. A tiny yolk sac produces blood cells on a temporary basis until the bones are formed and can take over this process. A tiny mouth is evident, as well as lips and buds for 20 milk teeth. At 8 weeks the term *embryo* is discarded and the term *fetus* is used. The key to this change is the development of bone cells replacing the small moulds of cartilage in the bone sites.
3	6.0–8.0 cm	The eyes continue to mature, external ears are present. The development of bone continues, spreading from the centre of each bone towards the ends. The placenta almost

		surrounds the fetus now, as it floats in the buoyant amniotic fluid. Its respiratory system moves, taking in and giving out the salty fluid. No oxygen is present, but the muscles are now working together. Soon the mother may start to feel the first kick or thrust of an arm, as the fetus flexes its growing muscles in the cramped space of the uterus. The heartbeat can now be detected and the fetus is now easily recognizable as a human baby.
4	12–18 cm	The head, now with distinct human features, may also be showing hair. The skin is pink from the capillaries in the skin. The bones are closing in to form the joints. All systems continue to develop and mature. The fetus has now grown beyond the limits of the placenta and the umbilical cord is stiff with the pressure of circulating fluids. This pressure also helps prevent the cord from tangling and "kinking", which might restrict the flow of blood within it.
5	25–30 cm	The fetus is now very active and constantly flexing its muscles. The body increases in size and the head is now smaller in proportion to the body. All the systems are now developing rapidly. The fetus may have complete vocal cords, although without air in its lungs it is uanble to make sounds. It may suck its thumb, strengthening muscles it will need for feeding.
6	28–34 cm	The eyelids separate and the eyelashes form. The eyes may now open. The skin is pink and wrinkled. From now on there are few major changes; all the systems

		are formed, but require extra time in the uterus to grow in size and strength.
7	36–45 cm	The fetus, at 7 months, is sufficiently developed to survive if born prematurely. It presses against the amnion that drapes it like a veil. The skin has become thicker and is coated with a creamy substance, called the vernix. This substance helps to protect the skin during its long immersion in the amniotic fluid.
8	40–49 cm	Fat is now being deposited, building up reserves for the baby's first days in the outside world. The testes, in a male fetus, descend into the scrotum. The bones of the head are soft and unjoined. The chances of survival for a premature baby are now greatly increased.
9	47–56 cm	Further fat is added, hair and nails have grown considerably, and the fetus is ready to begin life in the outside world.

Birth

When the fetus is finally ready to make its entrance into the outside world (Figure 19.6) a number of important changes take place in the body of the mother. Estrogen which stimulates the changes in the lining of the uterus during the menstrual cycle also causes the muscles of the uterus to contract. Progesterone prevents uterine contractions from taking place. During pregnancy, these hormones are kept balanced. Near the time of birth, the level of progesterone drops below a threshold level and the muscles in the walls of the uterus begin to contract signalling the onset of labour.

The activity that actually initiates the birth process is still not clearly understood, but hormones produced by the fetus greatly influence the levels of progesterone in the mother. Oxytocin from the pituitary gland aids in stimulating the

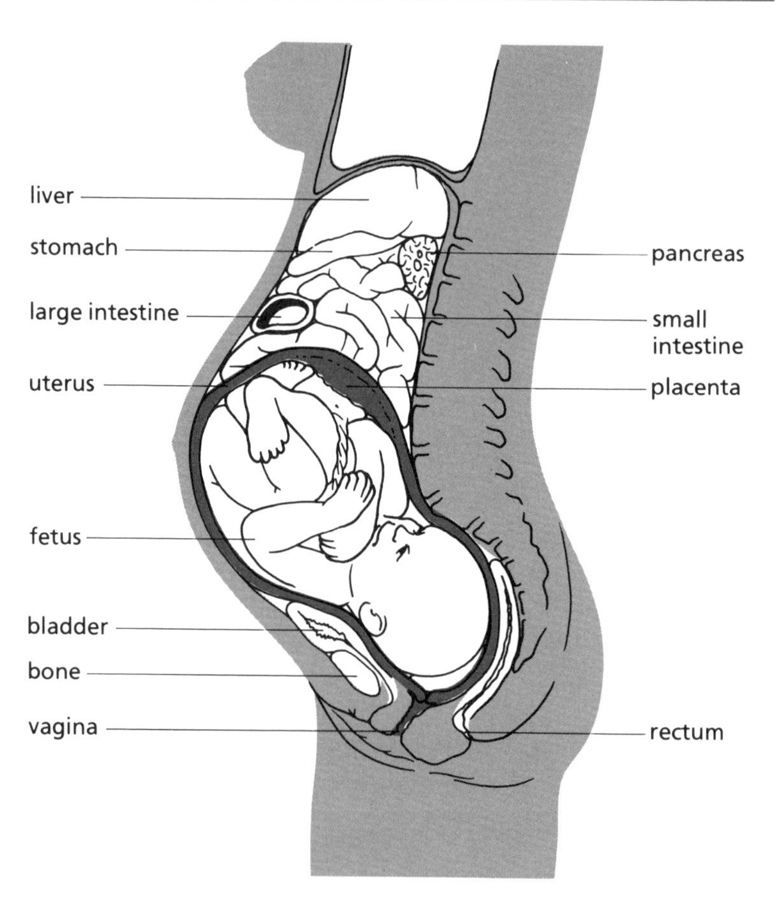

Figure 19.6.
Section through the abdomen. The fetus is shown in position with the head down ready to enter the birth canal at the start of labour.

contractions, which occur in waves, rather like the peristaltic waves that take place in the digestive tract. The contractions are irregular at first, true labour not beginning until the contractions are regularly spaced. As the contractions increase in strength, the amnion which surrounds the fetus breaks and the amniotic fluid is discharged through the vagina.

The flow of amniotic fluid is preceded by the expelling of a mucus plug which has been present in the cervix during pregnancy. (See Figure 19.2.) The passing of this plug, which may be accompanied by some flecks of blood and the waters of the amnion, is usually painless but is soon followed by the onset of true labour.

The muscles of the cervix and the vagina gradually relax and increase the size of the birth canal opening. (Dilation increases the opening of the cervix from about 3 mm to 10 cm.) The ligaments which hold the pubic bones together also relax slightly to allow easier passage of the baby.

This first phase of labour may last for about 16 h in the case of the birth of a first baby, but later births usually occur more rapidly. Strong contractions of the uterus eventually force the fetus down towards the cervix and vagina.

Delivery

Each contraction moves the fetus further down and forces the head into the opening of the cervix. The second stage of labour involves the expulsion of the fetus from the uterus.

The delivery of the baby is followed a short time later by expulsion of the placenta (after-birth). The contractions which separate the placenta from the wall of the uterus and force it out of the body also constrict any blood vessels that may be torn and prevent loss of blood. (See Figure 19.7.)

Figure 19.7. The delivery.

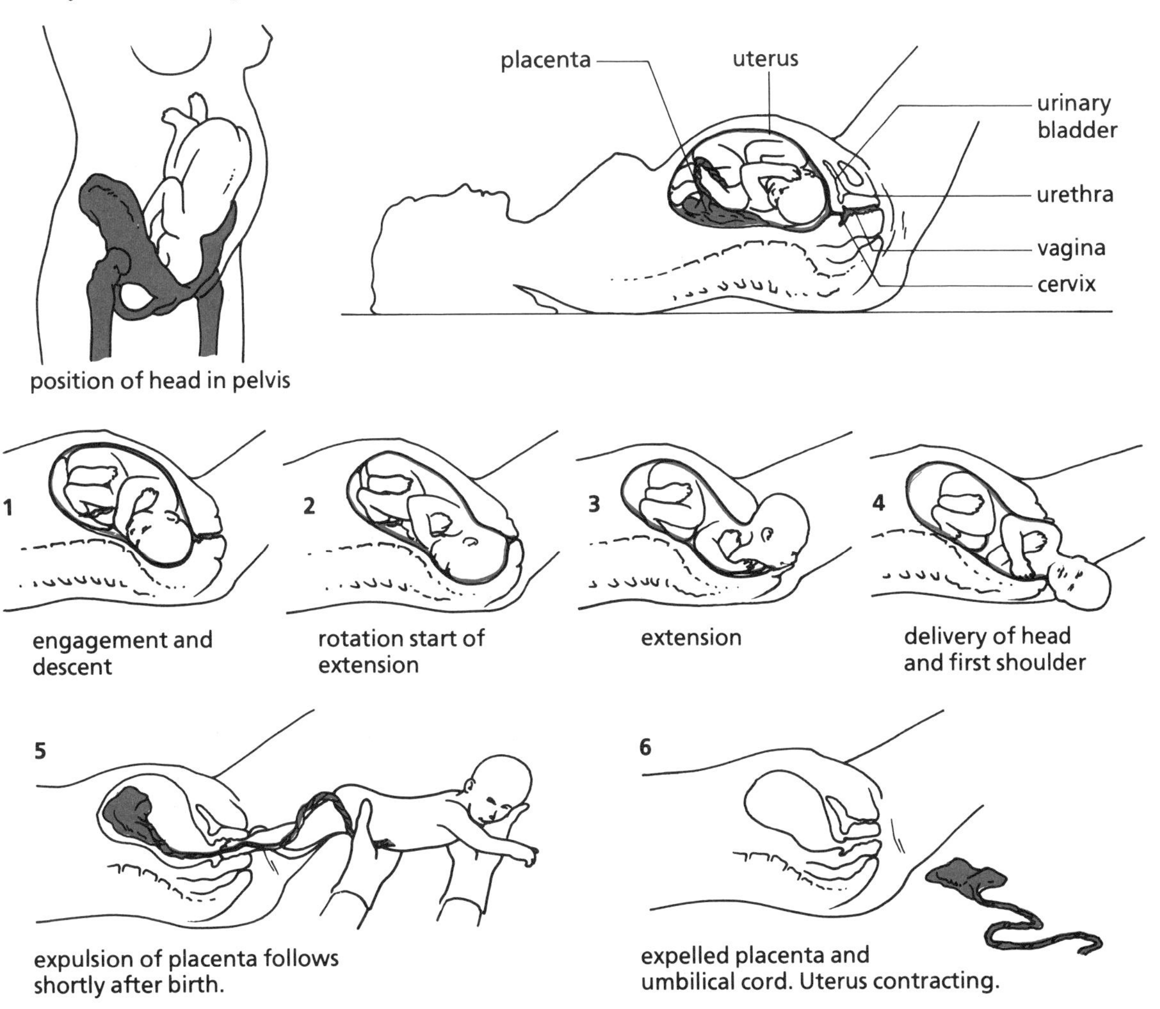

The baby arrives still attached to the umbilical cord and the placenta. The umbilical cord must now be tied and cut on the side away from the baby's abdomen. A jellylike substance present within the cord expands when exposed to air sealing off the arteries and veins, so that there is no loss of blood.

For the baby, delivery is a very disturbing experience. Up until this moment it has been in darkness. Now it is suddenly exposed to light. Its warm, underwater world is suddenly exchanged for a dry one, which is many degrees lower in temperature. Its heart must now send blood through its lungs to pick up its own supply of oxygen. These lungs are wet from long immersion in the amniotic fluid. Sometimes, if the baby does not start to breathe readily, the doctor attending the delivery will give the baby a sharp slap, causing it to cry and expand its lungs. The baby must now set its own systems in action to rid its body of wastes.

Some Other Functions of the Placenta

The placenta does a great deal more than just pass nutrients and wastes between the mother and developing baby. When the human body contains foreign proteins - whether it is a wood sliver, bacteria, or even a transplanted organ, it tries to get rid of it. As described earlier, white blood cells in particular perform this function. The baby inside the mother can also be viewed as "foreign protein"; if it were not for the placenta the mother's immunological protective system would try to get rid of it.

The nutrients in the mother's blood are swiftly passed on to the embryo through the placenta. Unfortunately, this system also works for substances such as alcohol and nicotine. The placenta does not have the ability to selectively prevent passage of some substances carried in the mother's blood; such viruses as German measles and such drugs as thalidomide can pass through the capillary walls and cause deformities in the offspring.

A newborn baby also receives a set of temporary immunities from its mother. If the mother is exposed to certain diseases and becomes immune to them, that immunity, through the placenta, is passed on to the developing baby. The antibodies remain in the blood of the baby for six months or so after birth, which gives it time to become strong enough to withstand many infections. In time, it will build up its own set of protective immunities. A baby has no functional immune system of its own until approximately six months of age. Mother's milk is believed to be an important factor in developing the immune system.

TWINS

There are two kinds of twins, those that arise from one egg and those that result from two eggs.

Sometimes, instead of the usual single ovum being released from an ovary, two are released together. As they are separate cells, they must be fertilized by different sperm. The only "twinning" involved in such a case is that the embryos develop in the uterus at the same time. With respect to being similar, they have no greater chance of sharing characteristics than do any pair of children with the same parents. Twins that result from two different ova are known as **fraternal** twins or dizygotic twins (two zygotes).

Identical twins, or monozygotic twins, develop from a single egg fertilized by a single sperm. The fertilized cell splits into two at a very early stage after fertilization, and each part then develops into separate individuals. These two persons, arising initially from the same ovum and sperm, contain the same genetic information and so will have all their genes in common. Identical twins will always have the same sex, either two boys or two girls. Fraternal twins may be of different sexes.

Usually twins are raised in the same home and experience the same environment throughout their growing years. Such similarity of food, care, medical attention, affection, homelife, and other factors usually results in a considerable resemblance in behaviour as well as physical characteristics. Studies of twins that have been raised separately in different homes and situations show that some differences can be produced. Such results help to distinguish between genetic and environmental influences on the growth of individuals. For example, one set of identical twins may have the genetic potential to reach a certain height. If one is raised in an environment in which good nourishment is available, this maximum height will likely be attained. If the other twin is raised in an environment with an inadequate diet, it is probable that, regardless of the genetic potential, the child will not be as tall.

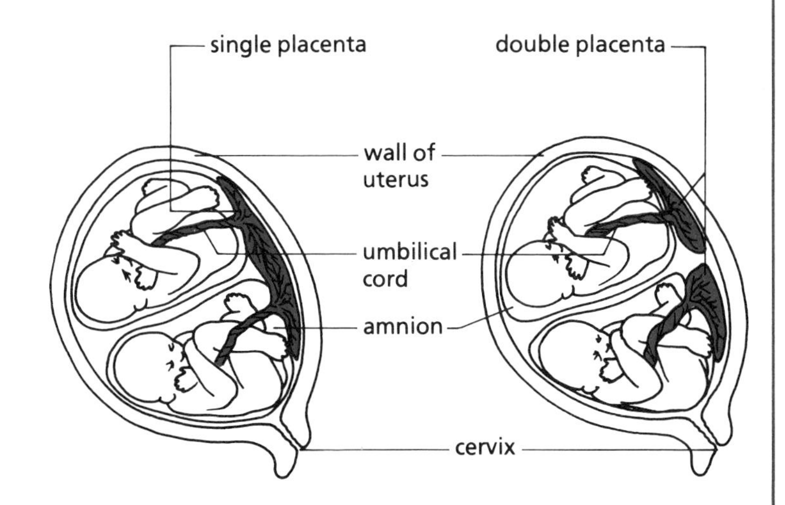

Identical twins develop from a single ovum, fertilized by a single sperm. Identical twins share the same genetic characteristics and are always of the same sex.

Fraternal twins develop from two different ova fertilized by two different sperm. They have no more genetic characteristics in common than ordinary brothers and sisters. Fraternal twins are not necessarily the same sex.

Differences in Growth Between Boys and Girls

There are several differences between the growth rates of boys and girls. Girls, on the average, learn to crawl, sit, and walk earlier than boys. Simple skills are also acquired earlier and the girls' growth spurt that occurs during puberty starts well in advance of that in boys. In general, permanent teeth appear earlier in girls. The difference in height between the sexes becomes much more marked after girls reach puberty. They start their growth spurt at about ten and one-half years of age, while boys, on average, begin at twelve and one-half years of age, during which time the child becomes a young adult, capable of all adult biological functions.

When the boys' teenage growth spurt arrives, it is usually more extensive than that which occurs in girls, and they eventually are taller on average than girls. Boys grow an average of 30 cm and become 20 kg heavier during the growth period which usually peaks at about fourteen years of age. Girls grow an average of 16 cm and increase by 16 kg in mass, with the spurt levelling off at about twelve years of age.

Other body changes occur as well. In girls, these usually include greater fat deposition and an increase in the width of the pelvis. Boys usually become stronger and more muscular than girls. The sexual organs mature in both sexes and secondary sex characteristics, such as enlarged breasts and pubic and axillary hair, become evident. These changes are largely due to the effects of hormones produced by maturing sexual organs.

Nutrition and Growth

The birth mass is influenced by the nutrition of the mother even before conception. The pregnant woman needs a rich store of reserves to ensure an adequate supply for the fetus. Good nutrition during pregnancy, especially an intake of milk, with its high calcium content, provides body-building materials for the growth of the embryo or fetus.

Children especially need a constant supply of nutrients for building body tissues. There is also a heavy demand for energy to support maturation, and, of course, an abundance of good quality protein, minerals, and vitamins is essential during these years of rapid growth.

While development between two and twelve years of age obviously requires the support of good nutrition, it is the accelerated adolescent growth spurt that makes exceptional demands. This period is usually accompanied by changes in food patterns brought about by peer-group influences. Often poor eating habits develop during this period.

If children lack adequate nutrients during these important growth periods, they may still continue to gain in height and mass, but have mineral-poor bones, nitrogen-poor soft tissues, borderline anemia, and a generally decreased rate of metabolism. This is often accompanied by a susceptibility to infections.

Growing teenagers always seem to be hungry. The demands of the body are exhibited by increased mealtime appetite and a constant need for between-meal snacks. Wherever possible, snacks should provide more than just kilojoules. Processed commercial snacking foods provide an expensive and poor dietary substitute for nutritious foods.

Deficits in food nturients may affect the genetic potential of the body, preventing it from reaching its optimum height or frame size, for example. Improper nutrition during early childhood and adolescent growth spurts has the worst effect. Unfortunately, there are no clear signals that indicate nutritional deficiences, not even for doctors.

It has been said that **genetics** is the architect of the body providing the design and **nutrition**, good dietary habits, is the builder in charge of construction. The quality of the materials provided in the diet determines the rate of progress and the final result of the original genetic design.

Approximately 25% of our life is spent reaching emotional and physical maturity, acquiring the skills and knowledge that we need to equip us for an adult life.

By the time we reach twenty years of age, the physical processes cease to improve and we start a slow subtle decline. This approximately matches the beginning of the period in which skills and knowledge that we have acquired are put to practical use.

Eventually, a decline in physical and mental powers occurs, but if the body has not been abused, our powers usually can be maintained at satisfying levels of performance. Both the physical and mental powers must be exercised throughout our lives. Neglect increases the rate of decline.

THE AGING OF HUMAN CELLS

If all deaths resulting from sickness and disease were eliminated, the average life span of the human would still not exceed ninety or a hundred years. This is because of aging. The estimate for the loss of functional capacity of cells after age 30 is about 0.8% per year.

To study how the aging process occurs, scientists look at the way the cells behave, usually using human fibroblast cells (connective tissue cells of the lungs). *In vitro* conditions (in glassware) show that these fibroblast cells divide over a period of months, then slowly stop dividing and die. It has been found that the ability of normal human embryo fibroblast cells to divide is limited to 50 divisions over a period of seven to nine months under *in vitro* conditions. This suggests that the aging process is a built-in property of fibroblast cells.

Other experiments were carried out by freezing fibroblast cells at −179°C in liquid nitrogen. If these cells were frozen when they had divided for the 20th time and later thawed, these cells would continue to divide for 30 more times, then they would die. Several experiments like these were conducted by stopping the division after the 10th time of division and freezing the cells. The thawed cells continued to divide for another 40 times. It seems as though there is a built-in clock that counts the number of cell divisions. Other experiments have confirmed that there is a "biological clock" existing within cells.

The location of the clock that controls the aging of the cell is found to be the nucleus. In experiments in which the nuclei of cells were removed, it was found that only those with nuclei completed a total of 50 divisions.

With more cell division, chances of DNA molecules in the nucleus making "errors" is greater. In the laboratory it appears that by 50 divisions enough errors are made that the cells can no longer divide normally. Since almost all cells are constantly dividing in order to keep us functioning normally, our own aging and eventual death will probably likewise be due to an accumulation of DNA errors – if something else doesn't cause our deaths first.

QUESTIONS FOR REVIEW

SOME WORDS TO KNOW

Match a statement from the column on the left with a suitable term from the column on the right. DO NOT WRITE IN THIS BOOK.

1. A hollow ball of cells that develops after the union of sperm and ovum.
2. Thin sac filled with a watery fluid in which the embryo develops.
3. Small temporary structure that provides the developing embryo with blood cells.
4. Name given to baby during the first few weeks in which it is in the uterus.
5. The process in which the blastocyst becomes attached to the endometrial lining.
6. A layer of cells which develops from the blastocyst and differentiates into specialized tissues and organs.
7. The total period of time that the baby is in the uterus.
8. Connects the baby with the placenta.
9. Result of the fertilization of two different ova at the same time.
10. Human-made component placed in the body to substitute for a defective organ.

A. placenta
B. uterus
C. heredity
D. blastocyst
E. zygote
F. implantation
G. amnion
H. ectoderm
I. yolk sac
J. implant
K. transplant
L. fraternal twins
M. identical twins
N. umbilical cord
O. gestation
P. embryo
Q. fetus
R. genetics

SOME FACTS TO KNOW

1. List the functions of the placenta.
2. How does the production of fraternal twins differ from that of identical twins.
3. What are the three germinal layers found in the blastocyst? Give three major systems or tissues which arise from each layer.
4. Briefly list the sequence of events that takes place during the delivery of a baby.
5. What shocks does a baby experience as it enters the world?
6. Explain the meaning or the function of the following:
   a) amnion,
   b) implantation,
   c) yolk sac,
   d) labour.
7. Make a list of all the functions of the placenta. Give a brief explanation of each.
8. Why is nutrition so important for the healthy development of the fetus and growing children?
9. Research any one of the many new developments that have produced the ability to replace damaged organs or tissues in recent years. Some suggestions: pacemakers, heart valves, hair transplants, organ transplants, etc.

QUESTIONS FOR RESEARCH

1. Select one of the following topics and prepare a report on it:
   - Pediatrician
   - care of the newborn
   - Caesarian section
   - blue baby
   - advantages of breast-feeding over bottle-feeding
   - natural childbirth
   - crib deaths
   - the premature baby
   - Candy Stripers or Hospital volunteers
2. What are some of the dangers associated with childbirth?
3. How does the environment in which a child grows up affect the child's development? There are several aspects to this question, you may wish to limit your research to just one aspect of development.
4. What is involved in good pre-natal care?
5. Investigate one of the special agencies that help old people; for example, Meals-on-Wheels or New Horizons. Your area may have different organizations. Your local Health Unit can help you get started.
6. Interview an elderly person who lives alone and find out the chief difficulties that older people have to overcome.
7. Find out all you can about health insurance or life insurance. What benefits are available, what health problems are not covered? Investigate the costs of health care.

Chapter 20

Human Genetics

- Information for Development
- What Are Chromosomes?
- Inheritance
- Mutations
- Birth Defects

Information for Development

We are all aware that we have characteristics in common with other members of our family. It may be eye or hair colour that we share in common with one parent, or perhaps a blood type or tallness that we share with the other. Brothers and sisters may possess several characteristics in common and be so alike that other people can see the family resemblance without previously knowing that the relationship existed.

The "blueprint" for the development and appearance of each individual is contained within the chromosomes of our cells. This information is so efficiently organized and tightly packed together that every detail of our body structures, the chemical processes of each system, even some of the elements of our personalities, are contained within the 23 pairs of chromosomes located inside the cell nucleus. In spite of the enormous amount of information they contain, these chromosomes are so small that it takes a high-powered miroscope to see them.

What Are Chromosomes?

Chromosomes are made of complex molecules called DNA (deoxyribonucleic acid). In spite of the long name, the general structure of this molecule is not difficult to understand. If you can imagine a ladder with flexible, instead of rigid sides that can be twisted around and around into a spiral, then you have the general shape of this molecule. The sides are made up of alternating molecules of deoxyribose sugar and phosphate molecules. The "rungs" of the ladder, stretched between the sugar molecules, are formed by pairs of molecules called **nitrogen bases**. (See Figure 20.1.)

The nitrogen bases are very important molecules, for the sequence in which they occur in the DNA molecule determines the "code" for all the genetic information. There are four nitrogen base molecules-adenine (A), thymine (T), guanine (G), and cytosine (C). These molecules always occur in the same linked pairs in DNA. Adenine pairs with thymine and cytosine pairs with guanine because of the physical size, shape, and the nature of the bond formed between these molecules.

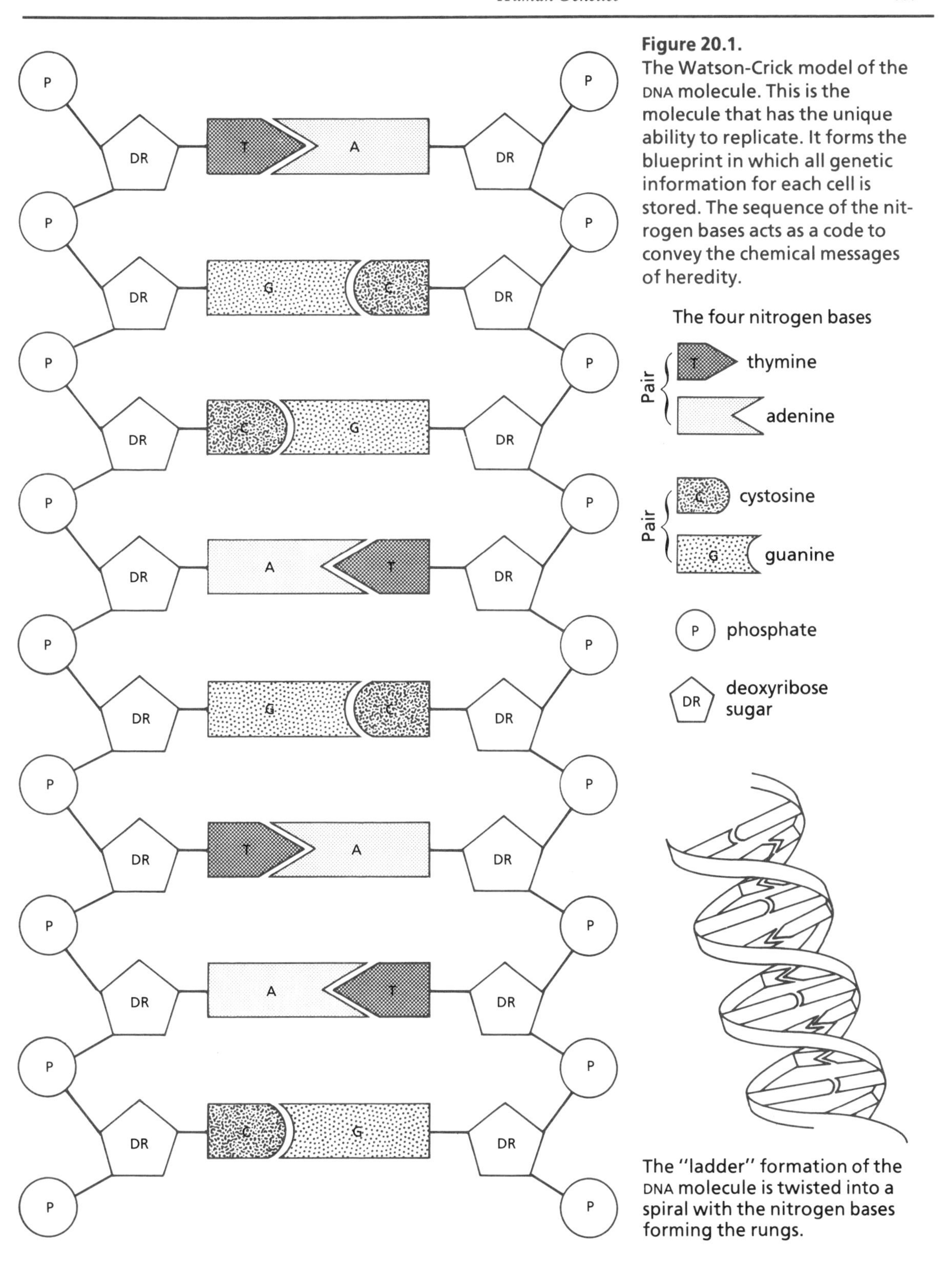

Figure 20.1.
The Watson-Crick model of the DNA molecule. This is the molecule that has the unique ability to replicate. It forms the blueprint in which all genetic information for each cell is stored. The sequence of the nitrogen bases acts as a code to convey the chemical messages of heredity.

The "ladder" formation of the DNA molecule is twisted into a spiral with the nitrogen bases forming the rungs.

The Genetic Code

An example of another code will help explain what is meant by a "genetic code". You probably know that the Morse Code consists of two symbols, a dot and a dash. With this code, any word can be communicated by using these two symbols. For example, the word "Black" can be written

B	L	A	C	K
—...	.—..	.—	—.—.	—.—

The genetic code has four symbols - adenine, thymine, guanine, and cytosine. All the information needed for building all body structures, proteins, hormones, and other molecules is coded into sequences using these four chemical substances. With four symbols, an enormous number of variations is possible in the arrangement of these bases along the DNA chain, for in each position there are four possibilities.

Our knowledge of the chemical nature of the chromosome is relatively recent and there is still much to be discovered. However, the microscopic image of the chromosomes has been quite thoroughly examined. We know that chromosomes are formed from granular material in the nucleus just before cell division and, when stained, appear as short dark threads in the nucleus. They may be counted and recognized as distinctive shapes, similar to each other, yet slightly different from all the other chromosomes. We know that each species of organism has a specific number of chromosomes; there are 23 pairs in humans, fruit flies have 4 pairs, and horses have 33 pairs.

Chemical studies have shown that along the length of the chromosomes are short sections of DNA, with coded information giving instructions for a particular characteristic or characteristics. These short sections are called **genes**. There are estimated to be hundreds of genes along each chromosome. Each gene carries the information for a characteristic, such as hair colour, from one generation to the next.

Each pair of chromosomes contains specific information about many characteristics in the body. One chromosome pair could have information about hair colour, liver structure, body build, etc., while another chromosome pair might have information about entirely different characterics. Two chromosomes with information about the same characteristics make a **homologous pair**.

(hoe-mol-uh-gus)

PROTEIN SYNTHESIS

Most of the proteins of which one person is made are similar in nature to those of which other people are made. Some are different. For example the protein hemoglobin is the same for most people, but some individuals may have an abnormal kind called sickle cell hemoglobin. What causes a person to develop abnormal hemoglobin or to have normal hemoglobin molecules?

To understand how proteins are formed, one has to consider the make-up of the chromosome, the DNA molecule. DNA in the nucleus apparently unravels and each half makes a copy of itself thereby producing single stranded molecules called *messenger ribonucleic acid* (mRNA) and *transfer ribonucleic acid* (tRNA). DNA molecules "read off" the synthesis of mRNA and tRNA. This reading off process, or transcription as it is sometimes called, passes on to mRNA and tRNA specific instructions from the DNA molecule.

mRNA and tRNA then move into the cytoplasm. The job of tRNA is to pick up specific amino acids (the building blocks for proteins) and to bring them along the ribosomes where mRNA has situated itself. Specific bonding occurs between mRNA and tRNA. The multitude of amino acids which the tRNA's have brought to the ribosomes combine with each other in specific ways to form a specific protein molecule. A protein such as hemoglobin is formed in this way in the red bone marrow. If there is an "error" in the DNA make-up then errors in mRNA and tRNA will also occur. One reason for the formation of the abnormal hemoglobin in sickle cell anemia is incorrect information, a single mutation of the DNA. Errors in tRNA and mRNA cause the formation of the wrong protein or an abnormal protein. Fortunately, this process of protein synthesis usually occurs in our cells daily with no errors. Thus, we are able to function normally.

The diagram below summarizes the very important process of protein synthesis.

How DNA provides the plan for the making of a protein

Chromosomes carry the coded information- the "gene" plan for the type of protein to be made. The "plan" is determined by the sequence of nitrogen bases
This master plan remains in the nucleus.

The RNA makes a complementary copy of the DNA plan. This copy is carried into the cytoplasm as a "messenger' to the ribosomes.

There are 22 different amino acids which must be arranged into a particular sequence to form a specific protein.

Energy delivered by ATP activates a bond between the tRNA and the amino acid.

Each amino acid associates with a particular 3-unit group of transfer RNA.

Once activated the tRNA carries the amino acid to the mRNA "plan" and finds a site where the pairs of bases will fit.

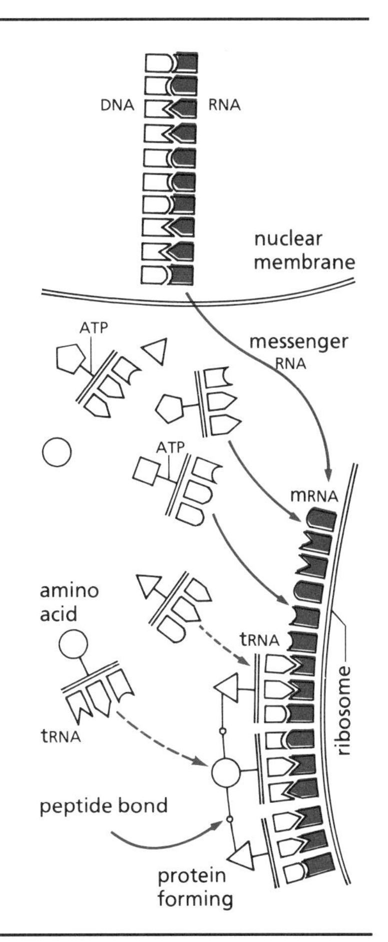

Once the tRNA have found a matching site on the mRNA, bonds (peptide) form between adjacent amino acids and a protein is formed. The bonds between the amino acids and the tRNA are broken and the protein is free to leave. The tRNA also are free to leave and seek new amino acids to fill other sites.
Remember: The sequence of nitrogen bases in the chromosome (DNA) was responsible for determining the final sequence in which the amino acids were joined to form a certain protein.

Meiosis

The division of body cells, called **mitosis**, was discussed in Chapter 2, and the specialized cell division, **meiosis**, that takes place in sex cells was outlined in Chapter 17. It is now necessary to explain in more detail why it is so important that the sex cells divide by meiosis.

Each cell forming the human body has a nucleus containing 23 pairs of **chromosomes** which provide the genetic blueprint for the development of the cell. However, the sex cells (gametes) which are formed in the ovaries or testes have only a half set of chromosomes, one of each pair of chromosomes. If a sperm or ovum had the full complement of 23 pairs (46 chromosomes), when they joined together at fertilization, there would be 92 chromosomes in each of the cells of the embryo. In each succeeding generation, the number would then double. To prevent doubling during the development of the gametes, the number of chromosomes is halved during the process of meiosis. Only one chromosome from each pair is present in each sperm or ovum cell.

Cells that contain 23 pairs or a total of 46 chromosomes are called **diploid** cells (2n). Cells with half the number of chromosomes, one of each kind, are **haploid**, 23 in number (n), (See Fig. 20.2). The only haploid cells in the human body are the egg and sperm.

Sperm Production

The seminiferous tubules of the testes are lined with special germinal cells called **spermatogonia**, which, like all body cells, have the diploid (2n) chromosome number. At puberty, these cells start active division, which continues throughout a male's lifetime. Mature sperm result from this division. First the DNA molecules in the chromosomes of these cells replicate, then the cells divide to form two identical daughter cells. The daughter cells then divide again, *without* the replication of the DNA, and as a result, each of the cells formed has only half the normal number of chromosomes. They are, therefore, haploid (n) cells. These cells become specialized as sperm. (See Figure 20.3.)

Egg Production

A similar process of cell division takes place in the ovary. However, when the special germinal cells called **oogonia** divide, only one cell matures and becomes the egg (ovum).

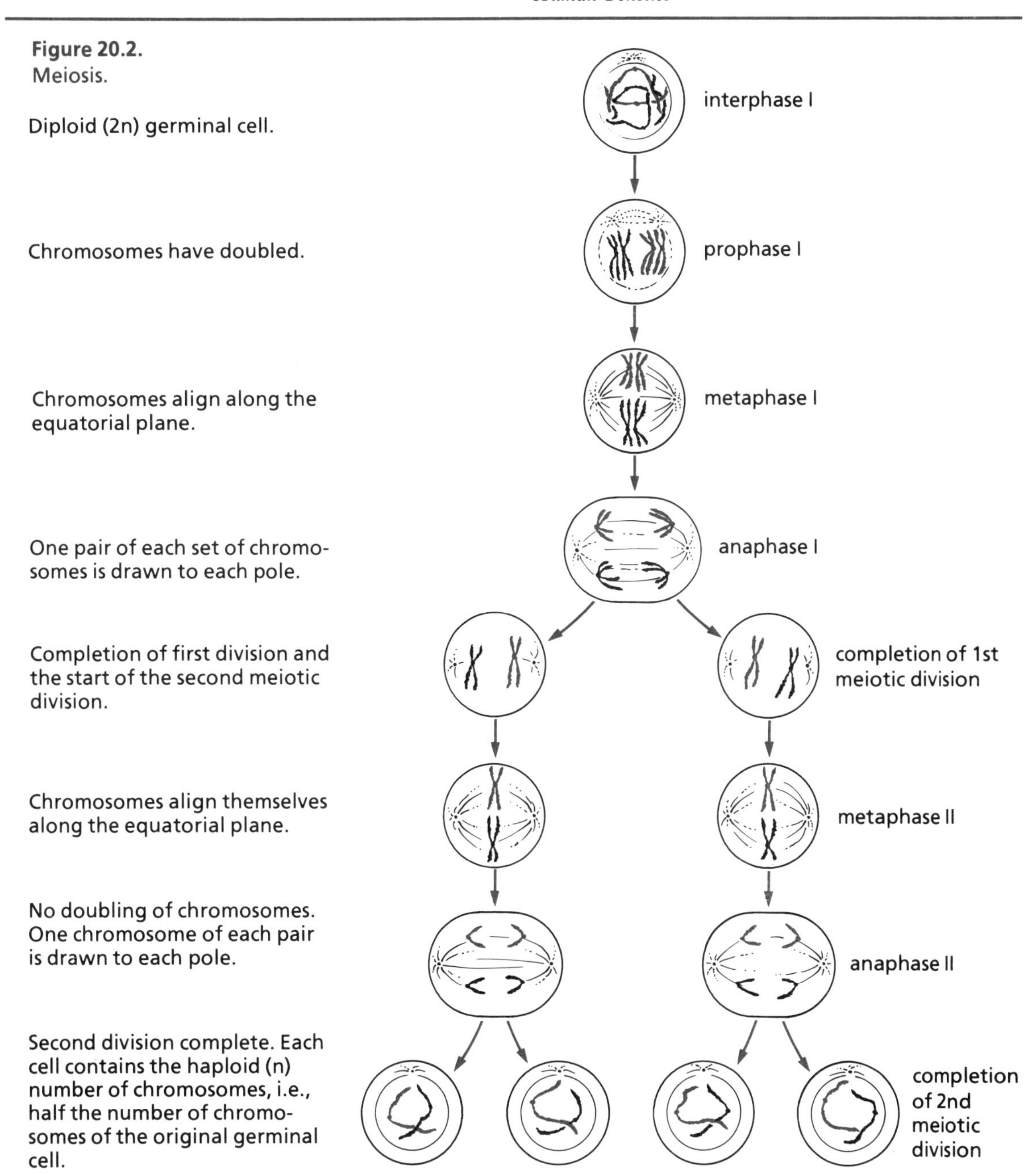

Figure 20.2.
Meiosis.

The other three smaller cells that are produced break down and distintegrate. See Figure 20.3 to compare ovum formation and sperm formation. Unlike sperm production, which occurs throughout the male's lifetime after puberty, eggs are all produced before a female is born.

Figure 20.3.
The development and meiotic divisions that take place in the gametes.

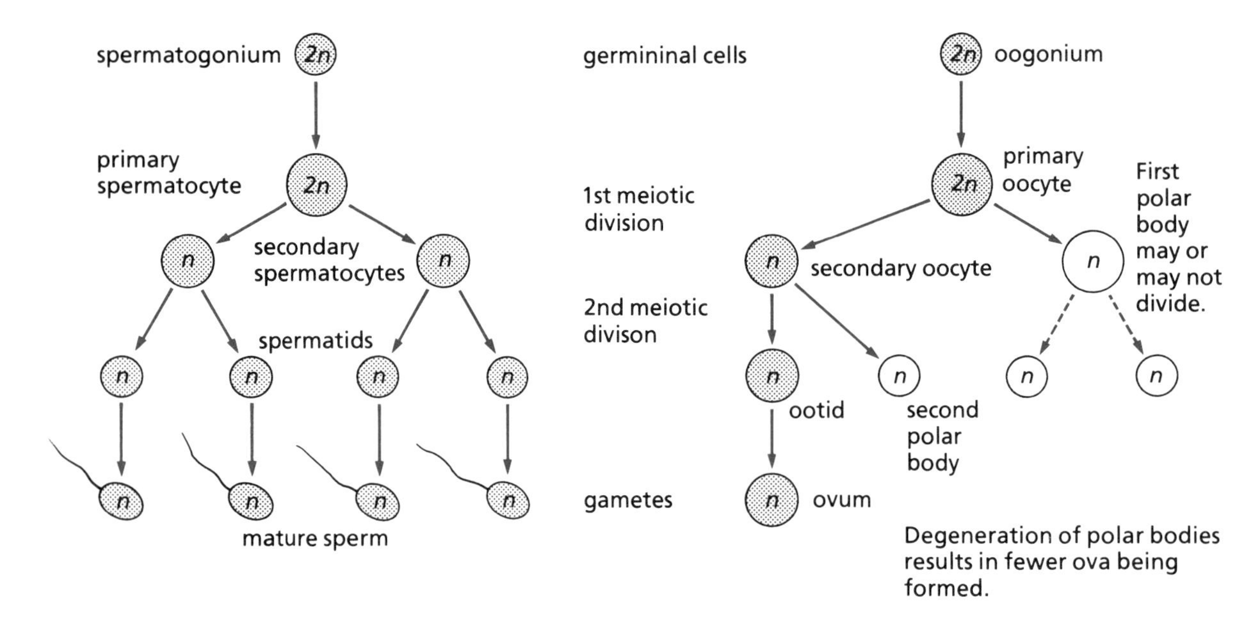

Inheritance

Inheritance is the transfer of characteristics or traits from one generation to another. As the pairs of chromosomes are pulled apart during meiosis, chance alone will determine which member of each pair ends up in the same cell. The following example will show the importance of this "chance" event. To keep the possible combinations simple, we will use only three pairs of chromosomes. (See Figure 20.4.)

Eight combinations (2^3) in a sperm or egg are possible with 3 pairs of chromosomes. However, humans have not just 3, but 23 pairs of chromosomes. Therefore, the possible combinations are enormous. It is calculated that 2^{23} or over 8 000 000 different combinations are possible in each sperm or ovum.

Figure 20.4.

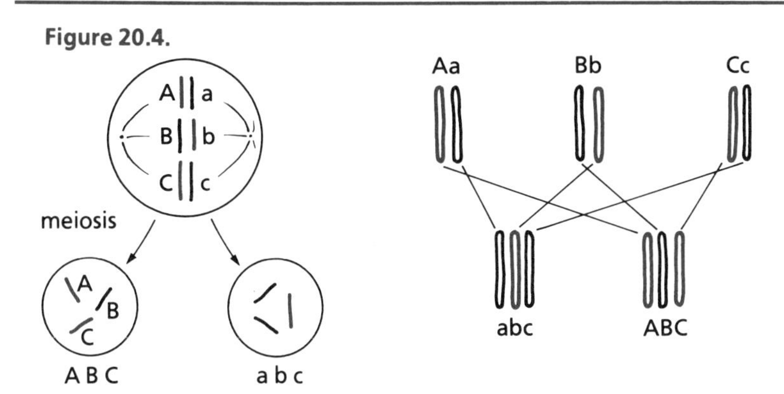

Pairs of chromosomes at the second metaphase ready to pull apart. One of each pair moves to each end of the cell.

Two random chance selections are illustrated. Another six are possible. Can you work out what they will be?

This is only part of the variation that is possible when an egg is fertilized, for now any one of those sperm can unite with an ovum with equally many possible combinations. The combinations possible in the fertilized egg (zygote) are greater than 70 000 000 000 000! No wonder it is impossible to find an exact double for any one human being.

How Genes Act

To illustrate the transmission of genes from one generation to the next, let us consider how a child might inherit either straight or curly hair. Although there are usually many genes involved in the transmission of a characteristic, we will assume that this particular trait is due to the presence of one gene and that only two expressions, *curly* or *straight*, are possible. If the effects of a gene show in the offspring, we refer to it as a **dominant** gene. The so-called "hidden trait" is due to the **recessive** gene. In our example, curly hair is dominant, straight hair is recessive.

In a person's cells, the pair of chromosomes carrying this trait could be found in one of three possible combinations:

Fig. 20.5

	A		B		C	
	Chromosome	Gene				
Information contained in the gene	curly hair	curly hair	curly hair	straight hair	straight hair	straight hair
Person will have:	**curly hair**		**curly hair**		**straight hair**	

In the examples A and C, both genes are the same, therefore the characteristic would be expressed. When chromosomes have the same genes as in A and C they are known as **homozygous** chromosomes. If the genes are not the same, as shown in B where both straight and curly hair genes are present, the chromosomes are called **heterozygous**.

As hair is not both straight and curly, one of these hair types must be dominant over the other; this one will show up in the child. The recessive gene is present in the child but is "unseen".

By tradition, capital letters are used to represent dominant genes and lower case letters are used for the recessive genes. Curly hair has been found to be dominant over straight hair in humans. If we now assign letters to the homologous pairs, we get the following:

Fig. 20.6

	C = curly hair	c = straight hair	
	C C	C c	c c
Phenotype of the person:	curly hair	curly hair	straight hair
Genotype of the person:	Homozygous CC	Heterozygous Cc	Homozygous cc

(fee-noe-type)

(jen-oe-type)

The term **phenotype** is used to describe the visible characteristics of an individual that are produed by the genes. The gene combination that produces the characteristics is the **genotype**. Figure 20.6 shows the phenotypes and genotypes that are possible for curly or straight hair. What's *your* phenotype?

The Transmission of Genes

To show how genes are transmitted from one generation to another, let us assume that, in a family, the father is homozygous for curly hair (CC). The mother is homozygous for straight hair (cc). We can now determine precisely what type of hair will be inherited by their children.

As the same information is present in each chromosome of the homologous pair in the father, each sperm can carry only a gene for curly hair. Since the mother has homologous genes for straight hair, every ovum will carry only a gene for straight hair.

Table 20.1. Some human traits that are due to dominant or recessive genes.

DOMINANT GENES	RECESSIVE GENES
Brown Eyes	Blue or grey eyes
Hazel or Green eyes	Blue or Grey eyes
Congenital Cataract	Normal
Farsighted Vision	Normal Vision
Shortsighted Vision	Normal Vision
Astigamatism	Normal Vision
Dark Hair	Blond Hair
Non-red Hair	Red Hair
Curly Hair	Straight Hair
Long Eyelashes	Short Eyelashes
Normal Hearing	Congenital Deafness
Blood Groups A, B, and AB	Blood Group O
Normal Blood Clotting	Hemophilia (sex-linked)
Normal Red Cells & Hemoglobin	Sickle Cell Anemia
Rh Blood Factor, Positive	Rh Blood Factor, Negative
Ability to Curl Tongue	Cannot Roll Tongue
Free Ear Lobes	Attached Ear Lobes
Ability to taste PTC	Unable to taste PTC

When a sperm and ovum from this couple unite, the fertilized cell that results will have a curly hair gene (C) and a straight (c) gene. It will be heterozygous (Cc).

Since the curly gene (C) is dominant over the straight hair gene (c), each of the children will have curly hair (Cc), the heterozygous condition. This means that the straight gene will be present, but its effects will not be seen in the children.

C = curly hair c = straight hair

Father Mother
CC x cc

Gametes		C	C
Mother ♀	c	Cc	Cc
	c	Cc	Cc

Father ♂

Genotype: All offspring are heterozygous.

Phenotype: All have curly hair.

If one of these children, when an adult, mates with another heterozygous (Cc) curly-haired individual, the children of such a union could have the following possible combinations:

Father Mother

Cc x Cc

		C	c
		Father ♂	
	C	CC	Cc
Mother ♀	c	Cc	cc

Genotypes	Phenotypes
25% CC	Curly hair
50% Cc	Curly hair
25% cc	Straight hair

When both parents are heterozygous (Cc), the chances of having a child with curly hair are three times as great as having a child with straight hair.

Although one child would be homozygous curly and the other two heterozygous curly, you would not be able to tell the difference by looking at them. Only by examining the parents and their future children, would it be possible to determine whether the genes present are heterozygous or homozygous. However, if a child has straight hair, the child must be homozygous for the recessive gene. For recessive genes to be expressed, both genes must be recessive.

The Inheritance of Two Characteristics

When chromosomes are pulled apart during cell division and meiosis, it is a matter of chance which chromosome of each pair is drawn to a particular pole.

If two characteristics controlled by genes on two different chromosomes are considered, you will see how different combinations of characteristics are possible. We will use, as an example, *brown eyes*, which are, in general, dominant over *blue eyes* and a *straight* little finger which is dominant over a *bent little finger*.

Example: Consider the union of a brown-eyed man with a straight little finger and a woman with blue eyes and a bent little finger.

Brown eyes (E) Blue eyes (e)
Straight little finger (S) Bent little finger (s)

If the farther is homozygous for brown eyes and straight little fingers and the mother is homozygous for blue eyes and bent little finger, we can predict that the children will all be heterozygous and will show the dominant traits of brown eyes and straight little fingers.

		Father (EESS)	
		ES	ES
Mother	es	EeSs	EeSs
(eess)	es	EeSs	EeSs

If two heterozygous individuals (EeSs) produce children, then the predicted ratios of the offspring would be as follows:

		(EeSs)	Possible Male Gametes		
		ES	Es	eS	es
Possible	ES	EESS	EESs	EeSS	EeSs
Female Gametes	Es	EESs	EEss	EeSs	Eess
(EeSs)	eS	EeSS	EeSs	eeSS	eeSs
	es	EeSs	Eess	eeSs	eess

Phenotypes:
9 brown eyes, straight little finger
3 brown eyes, bent little finger
3 blue eyes, straight little finger
1 blue eyes, bent little finger

Incomplete Dominance and Co-Dominance

In plants there are examples of incomplete dominance, cases where the cross between a white flower and a red flower result in some intermediate colour such as pink. There are very few examples of this in humans. Where intermediate ranges of a trait are present, such as in skin colour, it is usually caused by the involvement of several gene pairs.

Sometimes two traits are present at the same time and neither appears to be dominant. This occurs in a disease known as **sickle cell anemia**. This disorder affects the oxygen-carrying molecule of hemoglobin that is found in the red blood cells. Homozygous normal genes produce normal hemoglobin and normal red blood cells. Homozygous sickle cell genes cause a severe anemia that is often fatal. The hemoglobin produced is not normal and the red blood cells are distorted into a curved sickle shape. If the heterozygous condition is present, both normal and sickle cell hemoglobin is produced and the person suffers only a mild form of anemia.

Pedigrees

A very useful way of showing how genes are inherited and how they may be traced back through several generations is to use a **pedigree** diagram. (See Figure 20.7.)

Fig. 20.7

Symbols

- ○ female
- □ male
- ○–□ marriage
- children
- twins
- displays trait
- carrier
- ○ □ free from trait

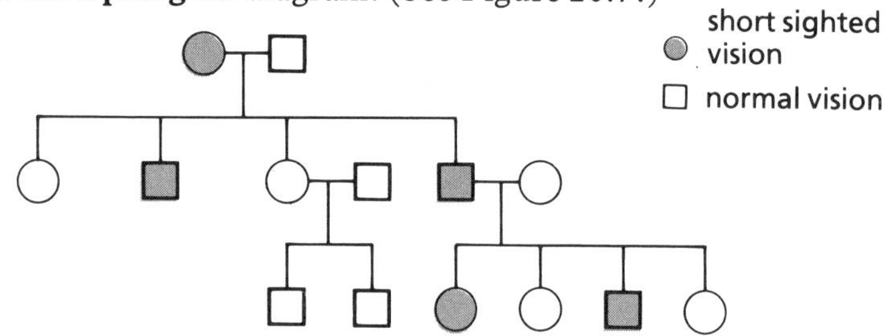

If we examine this pedigree and know that the gene for short-sighted vision is dominant over the normal vision gene, then, where a person has normal vision, they must have two recessive genes. Let S = short-sighted, s = normal vision. We can easily fit the genotypes of all the individuals who have normal vision beneath the symbols. (Step I)

Fig. 20.8

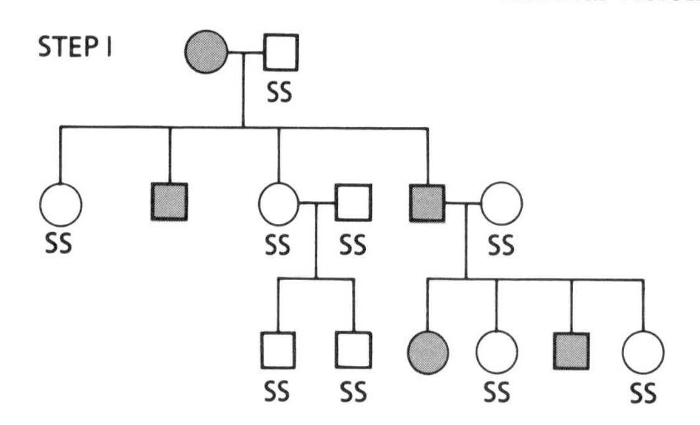

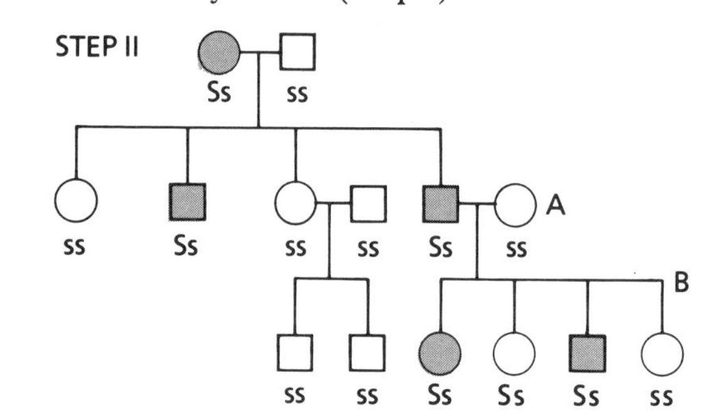

Step II. We know that if the dominant characteristic "short-sighted" vision is expressed, then at least one of the genes for this trait is present. Thus we can now add S to all the shaded symbols. If we now look at the parents who are short-sighted and their offspring, we can determine whether the genotype is heterozygous or homozygous in each case. For instance, at A the mother has double recessive genes, so each ovum must carry a recessive gene and the children must, therefore, all carry at least one s. Thus, the child at B, who is short-sighted, must be heterozygous (Ss).

In the case of inheritance of a recessive trait, a whole generation or several may occur without the characteristic being expressed. Both of the parents must carry the gene for the characteristic to be expressed.

Multiple Genes

In Chapter 9 we discussed human blood groups; these are A, B, AB, and O. Blood Group A carries the A **antigen**, group B the B antigen, AB carries both A and B antigens, while O blood carries neither antigen. In this case, A and B have equal dominance to each other; both are dominant over Type O.

The inheritance of a particular blood type is determined by a gene that controls the production of a specific antigen:

Type A blood individuals have a gene that produces the A antigen.
Type B blood individuals have a gene for the B antigen.
Type AB blood has genes that produce both antigens.
Type O blood produces neither antigen.

Table 20.2 illustrates this complex inheritance of blood types and shows the possible blood types produced by parents of differing genotypes. If we know whether the parents are heterozygous or homozygous, we can determine the possible blood types of their children.

Table 20.2. The inheritance of blood types.

PHENOTYPES (Blood groups)	A BLOOD	B BLOOD	AB BLOOD	O BLOOD
Possible genotypes	A A or A O	B B or B O	A B	O O

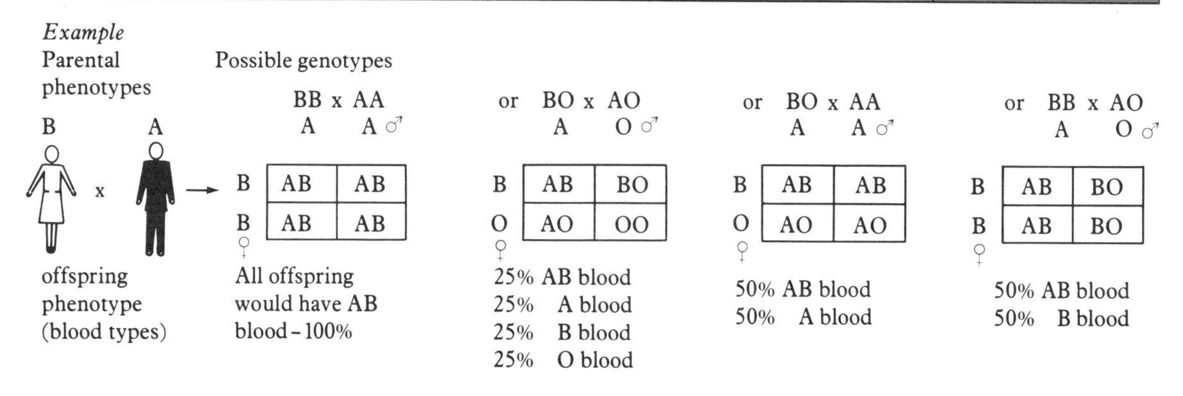

Consider a father with type O blood and a mother with type A blood. The possible number of blood types in the children is then quite small.

Possible Genotypes of the Parents	Possible Blood Types in the Offspring
Mother AA or AO Father OO	A and O

Sometimes an understanding of blood type inheritance is used in courts of law. In a paternity suit, a woman may allege that a particular man is the father of her child. The woman's and man's blood can be tested and may help to substantiate or disprove the allegation. Using the last example, if the child produced has A or O type blood, the father cited in the case may indeed be the father. However, if the child's blood type was B or AB, this man could not be the father.

It is important to note that if the blood type matches, it does not prove that the man was the father, only that he could be the father. Any other male with the same blood type could be the father. Many other factors that are found in blood are also inherited. The Rh factor is one. These and other factors can also be taken into consideration in such cases. Height and skin colour are other examples of multiple genes affecting particular characteristics.

The **gene pool** refers to all the genes present in a population at a particular time. The composition of the gene pool will vary greatly from one population to another. In Scandinavian countries the genes for blond hair predominate, whereas in India genes for black hair are more common. Among the pygmies of the Kalahari desert, genes for tallness are lacking and the population is very small in height. Interaction between populations (gene pools) such as occurred through immigration to North America results in a gene pool which leads to a greater variety in the population.

Sex Determination

If we examine the chromosomes of a cell under a microscope, we see that there are 22 pairs which match up well by their physical appearance. There is also a 23rd pair, which in males is made up of two different chromosomes but in females is made up of two well-matched chromosomes. In females, these are two cross-shaped chromosomes, which are called *X chromosomes*. In males, there is one X, just like those

THE CHROMOSOMES

Prior to 1950, biologists were in dispute as to the number of chromosomes in a human cell. It was believed that each human cell contained 48 chromosomes. In the mid-1950's special techniques were used to culture cells. The cultured cells were treated so that the chromosomes would separate and spread out in the nucleus rather than remain bunched together. It was thus easier to count the number of chromosomes. The number of chromosomes was found to be 46, and not 48 as had earlier been thought.

Today, human white blood cells and, for fetal tests, cells from the amniotic fluid are generally used when chromosome analysis is needed. Cells of these types are cultured and allowed to divide. When the stage of division called metaphase is reached, a chemical (colchicine) is added. Colchicine will stop the cell from dividing any further. Special stains are then used to prepare the material for microscopic examination. The prepared slides are photographed and enlarged. Each chromosome pair on the photographic print can be identified according to its size, structure, and staining pattern. Geneticists can now study all the matching pairs of chromosomes. This arrangement of chromosomes is called a **karyotype**.

From the karyotype, an extra chromosome 21 indicates that the fetus or individual has Down syndrome. An extra or missing sex chromosome may also be discovered by karyotyping.

The karyotype, then, is a very important tool in determining chromosomal disorders and identifying carriers of certain chromosomal disorders.

Over the past few years a special staining technique has been discovered which shows a characteristic "banding" of individual chromosomes. These bands are caused by the uptake of stains and represent gene sites.

In humans, chromosome banding and other new techniques have helped in the identification of gene positions controlling many functions. All human chromosomes have now been partially "mapped".

Male and Female karyotypes.

a) Chromosomes from amniotic fluid for karyotyping. Normal human male karyotype.

b) Chromosome banding produced by a special staining technique helps the geneticist to determine presence of genes on the chromosomes. Note the last chromosome is the Y, which does not have a homologous partner.

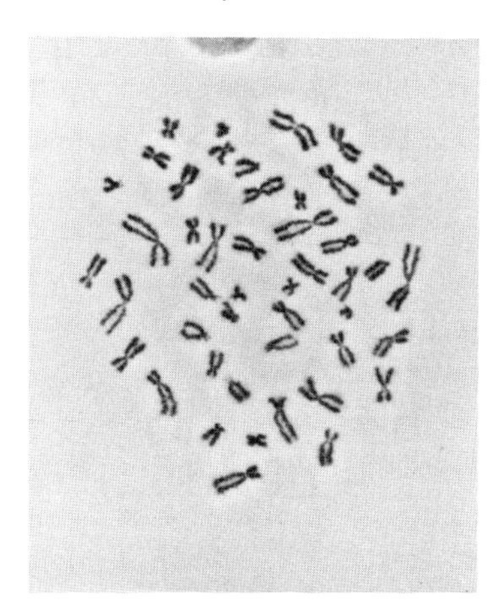

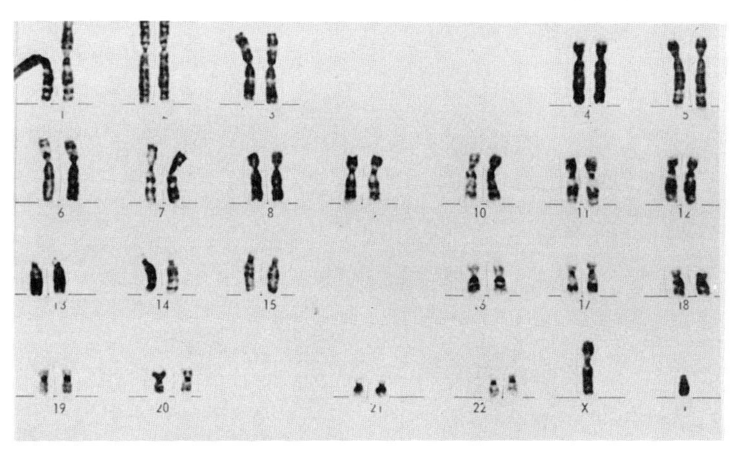

in females, and its mate is a short, hooked chromosome, known as a *Y chromosome*. These are known as **sex chromosomes**. The other 22 pairs of chromosomes are called **autosomes**.

During the development of sperm and ova, when the cells undergo meiosis, women who have two X chromosomes produce ova with a single X chromosome in each. Men, with one X and one Y chromosome, will produce sperm, 50% with an X chromosome and 50% with a Y chromosome present. It is, therefore, the male that determines the sex of a child depending upon whether the sperm that fertilizes the ovum carries an X or a Y chromosome. (See Figure 20.9.)

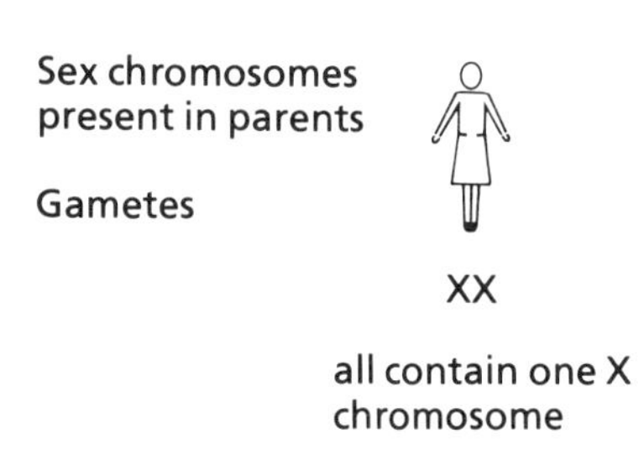

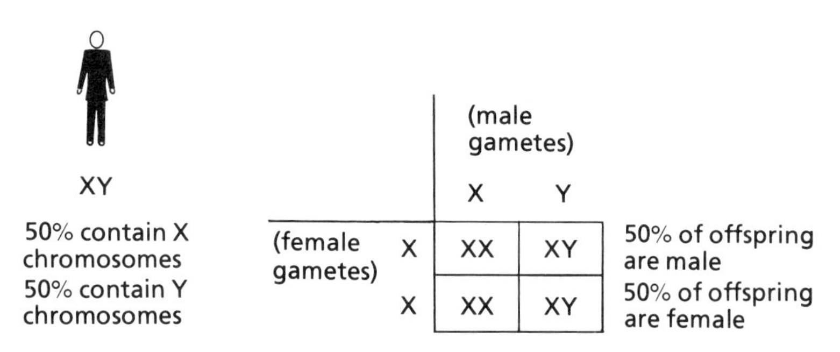

Figure 20.9.
Sex determination. The male gamete determines the sex of the offspring depending on whether an X or a Y chromosome is present in the sperm that fertilizes the ovum. The ovum always carries an X chromosome.

Sex-Linkage

Some disorders that are inherited occur much more commonly in males than in females; such disorders are usually due to **sex-linked genes**. Red-green colour blindness and **hemophilia** are two examples of sex-linked traits that are transmitted to the offspring by genes that are on the sex chromosomes.

Since the X chromosome is larger than the Y, it contains a number of genes that are not present on the shorter Y chromosome (See Figure 20.10.) The following explanation will show why this is so important in the transmittal of sex-linked traits.

The most common of the traits that are "sex-linked" are recessive. For example, red-green colour blindness is recessive (c), normal colour vision is dominant (C). Thus, if a female has one normal X chromosome (C) and one X chromosome with the gene for red-green colour blindness (c), she will simply "carry" the trait (Cc). The dominant normal "C" gene is expressed. However, if a male has an X chromosome with the gene for red-green colour blindness, he will be colour blind. Why? Because his other chromosome, the Y chromosome, due to its shorter length, has *no* gene, normal or otherwise, to pair with the defective recessive gene. With no normal colour-vision gene to dominate

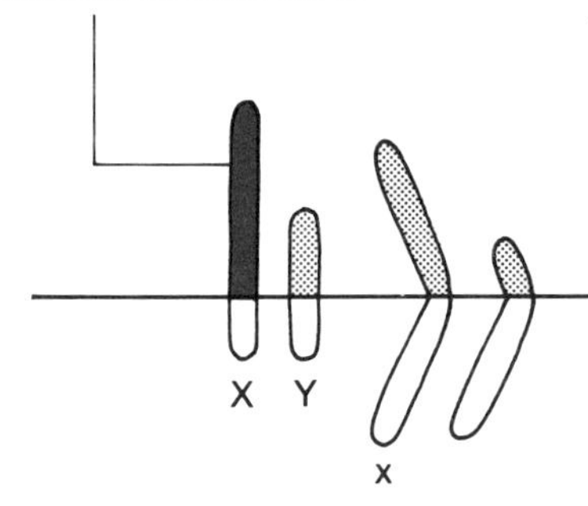

Figure 20.10. Diagrammatic representation of the X and Y chromosomes showing that homologous genes are present in only some portions of the two chromosomes. Sex-linked genes are located in the non-homologous portions of the chromosomes.

the recessive defective gene, he is colour blind. Figure 20.11 shows how colour-blindness could be inherited by boys in a family.

Figure 20.11

Mother carries red-green colour blindness trait — Father has normal vision

X^CX^c —— x —— X^CY

	X^C	X^c
X^C	X^CX^C	X^CX^c
Y	X^CY	X^cY

One-half of the females produced would be normal (X^CX^C) and one-half would be carriers (X^CX^c).
One-half of the males would be red-green colour blind (X^cY).
One-half would be normal (X^CY).

Can females be colour blind? Yes, they can, but only if their father is! If they get a "c" gene from their mother (a "carrier") and a "c" gene from their father (who would be colour blind), they would be homozygous recessive "cc" and, thus, colour-blind. (See Figure 20.12.)

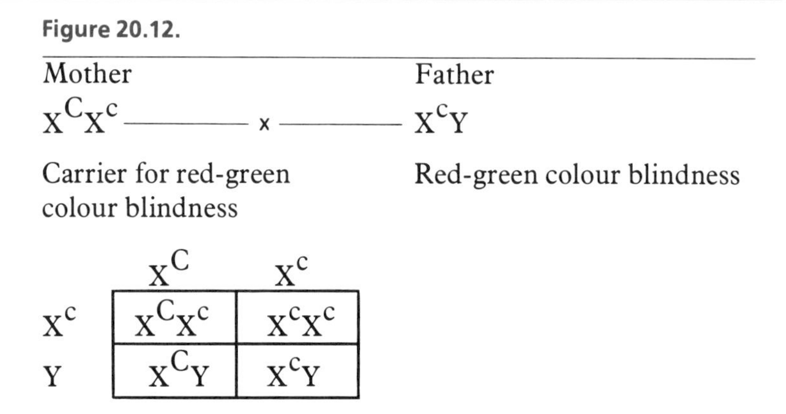

The expected proportion of affected offspring would be one-half of the females and one-half of the males with the red-green colour blindness trait.

Mutations

Not every chromosomal disorder is inherited. Sometimes there is an alteration in the DNA structure in a chromosome, and a variation appears which has not been evident in previous generations. Such changes are known as **mutations**.

Most mutations are harmful, and some cause such drastic changes that offspring cannot survive. However, some mutations are useful and can help to improve an organism's ability to survive. A mutation that produces a fur colour that allows the animal to blend in with its environment is an advantage. White fur on a snowshoe hare is an advantage. However, a white rabbit in a green field would be at a disadvantage because it would soon be caught; therefore, it would not survive long enough to produce more offspring with white fur. If the mutant with a new coat colour has better camouflage, it has a better chance of survival and a greater chance of transmitting its new gene to the next generation.

Mutagenic Agents

We know that there are a number of agents that can upset the normal passage of genetic information from one generation to another. **Radiation** of several types – X-rays, gamma rays, and ultraviolet light – have been shown to release subatomic particles that can penetrate cells and cause chromosomal changes. Mutations are also induced by several chemicals. Mustard gas used in World War I and a number of medical and non-medical drugs have been found to produce chromosomal changes.

Here are the major points about mutations:

1. Mutations are unusual, even rare events.
2. Mutations are usually harmful.
3. Mutations happen randomly (by chance).
4. Mutation defects cannot at present be cured.
5. Mutations are transmitted to the next generation.
6. There are many known causes of mutations, including radiation, chemical, and physical changes in the chromosomes.
7. Mutations in bacteria and other organisms may be experimentally manipulated by scientists in a process known as gene splicing.

Birth Defects

The birth of a baby is a very special event. For parents, grandparents, brothers, and sisters who have anticipated the event for many months, the final arrival of a baby is a moment of delight and excitement. Sometimes, however, the event becomes a time of great concern or disappointment if the baby is not "normal". An understanding of genetics and genetic counselling can sometimes avert such problems.

Some abnormalities are so slight that they are only minor inconveniences. Colour blindness, in the majority of cases, means only that the individual cannot distinguish between red and green. At the other extreme, a baby born with only a rudimentary brain or severe defects of its internal organs, may die almost at once.

In North America, it is estimated that about one in 14 children are born with a serious mental or physical defect; that is, in about 7 percent of all live births. In addition, there are thousands of spontaneous abortions, stillbirths, and miscarriages each year. These problems are sometimes caused by severe abnormality that interrupts the fetal development. About 20% of all human birth defects can be traced to genetic factors. Another 20% are caused by environmental factors

IMPLICATIONS OF GENETIC ENGINEERING

There are now companies throughout the world with such biological names as "Genetech" and "Biologicals." What are these companies manufacturing and selling? Drugs. The drugs help arrest cancer, diabetes, heart attacks, and certain viral diseases, to name a few. These companies are using a new technology to produce these drugs. By splicing the genes from human cells into the plasmids (small pieces of DNA) of bacteria, the bacteria will produce specific proteins and enzymes (some are life-saving drugs).

The first successful gene splicing was done by Dr. Har Ghobind Khorana, while he was head of organic chemistry at the British Columbia Research Council.

The potential benefits from gene splicing are far reaching. Bacteria can now be made to produce interferon, insulin, urokinase (used to dissolve blood clots), endorphin (for pain killing), and the hormone thymosin (which shows promise as a treatment for brain and lung cancer).

There are fears that this research could lead to production of a "superbug" which might get out of control. For example, some *E. coli* bacteria (normally found in the intestine), when infected with certain viruses in these experiments, have been known to cause tumours in mice and have caused human cells in test tubes to become cancerous. The future of gene splicing must be conducted carefully to prevent disastrous results.

that affect the baby while it is developing in the uterus. Still other defects are caused by the interaction of both hereditary and environmental factors.

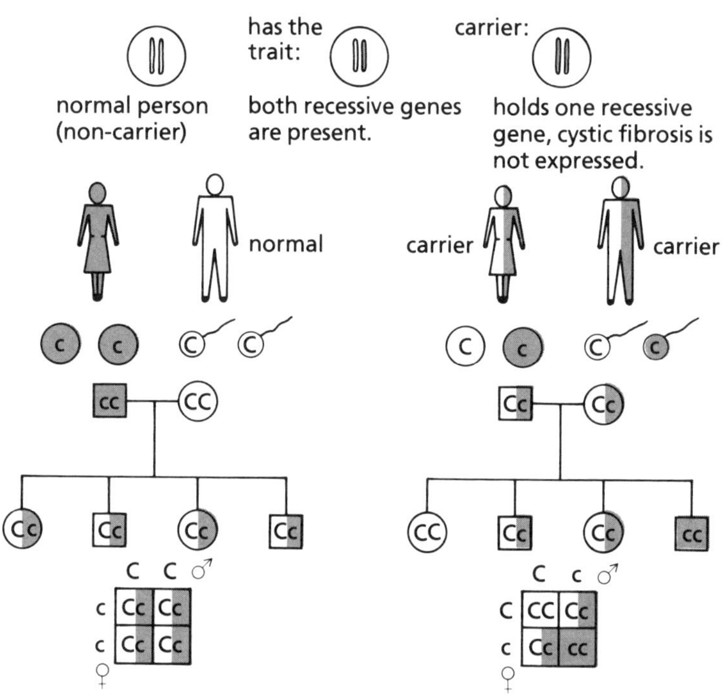

Figure 20.13.
The inheritance of diseases and defects. When a disease is transmitted from one generation to another it is known as a hereditary disease. Cystic fibrosis is the most common fatal disease of childhood. The disease affects the pancreas and the bronchioles of the lungs. Cystic fibrosis is a recessive trait represented here by the letter c. The gene for a normal pancreas and bronchioles is represented by C.

Most of the DNA of *E. coli*, a bacterium common in the human intestine, is found in its large circular chromosome. However, other DNA is in small circles called **plasmids**.

The technique for cleaving DNA from a bacterium and splicing into it, a segment of DNA from another unrelated organism (for example, a segment from human DNA responsible for production of insulin). The rapid reproduction of bacteria ensures a rapid production of insulin.

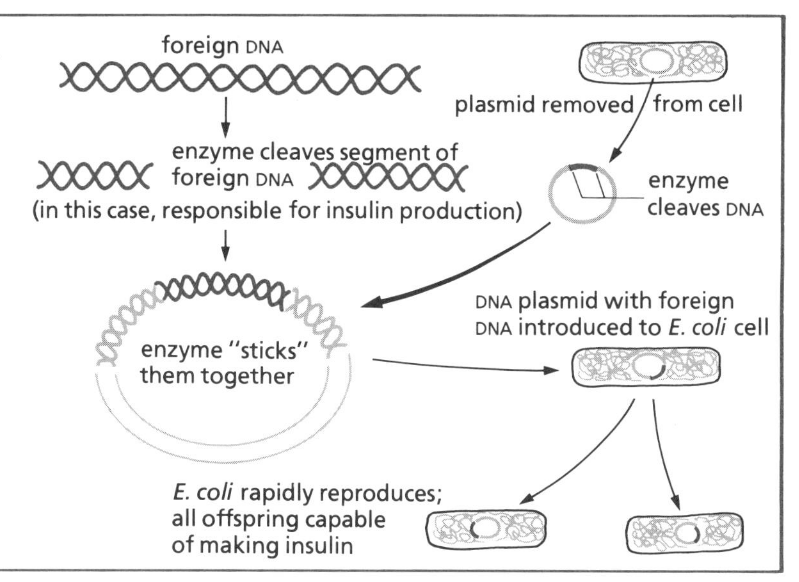

CASE STUDY. Barbara and Jim Wadsworth, 28 and 27 years of age.

Barbara and Jim had been married for six years and had a three-year-old son John. Their son suffered from a disorder that produced a number of physical and mental defects. John had small folds of skin in the inner corner of each eye and a large tongue which protruded slightly from his mouth. His hands were small and his fingers stubby. He had a heart defect and he was mentally retarded as a result of the malformation of his brain. These characteristics are typical for people with the disorder known as Down syndrome.

Barbara and Jim had no illusions as to the future of their son. This was not a condition from which he would recover.

Barbara and Jim wanted to have another child but were concerned that their next child might also have Down syndrome. After many months of indecision, they visited their physician and explained their concern. The physician said that they were fortunate to be able, in their case, to benefit from known information about the disorder. The physician explained that genetic counselling was available for cases such as theirs.

In advance of their first visit to the counsellor, a form was sent to Barbara and Jim which requested information about their own medical history and that of members of their families.

The visit to the genetic counsellor was a long, informative one. The counsellor, Mary Willet, was a knowledgeable and sympathetic person to whom Barbara and Jim were quickly able to relate. Mary told Barbara and Jim that the presence of John's extra chromosome number 21 was probably not caused by a gene that either of them carried. She referred to the medical histories that they had provided and noted that there were no other cases of the disorder in either family. This suggested that it was very unlikely that Jim or Barbara carried a defective gene. It was more likely that their son's Down syndrome was caused by a spontaneous event in egg or sperm cell before fertilization. In fact, in more than 90% of Down syndrome cases, the parents do not carry a defective gene. The "mistake" arises unexpectedly and for an unknown reason.

Women over forty years of age are much more likely to have children with Down syndrome than younger women. As Barbara was still below the age at which incidence of Down syndrome becomes more common, the counsellor said that, in Barbara's and Jim's case, the chance of a second child suffer- →

ing from the same disorder was as small as that for other parents of their ages.

Barbara and Jim were delighted and very relieved. Mary Willet explained that if Barbara became pregnant, amniocentesis could be carried out early in the pregnancy and, in the unlikely event that the baby did not have the normal number of chromosomes, the pregnancy could be terminated if she and Jim so wished.

Before they left the office, Mary gave them information on various care centres and assistance programs that were available in their area for children with Down syndrome. Barbara and Jim went home, feeling that they had finally learned what they needed to know. They were excited about the prospect of having another baby and began planning for the future.

Almost a year later, Barbara gave birth to a normal, healthy baby.

Questions for research and discussion:

1. Why do genetic counsellors ask so many questions about grandparents and other relatives?
2. Why is it necessary to take blood samples from parents and their children when investigating some hereditary disorders?
3. Other genetic abnormalities you may wish to investigate are: cystic fibrosis, Tay-Sachs disease, spina bifida.

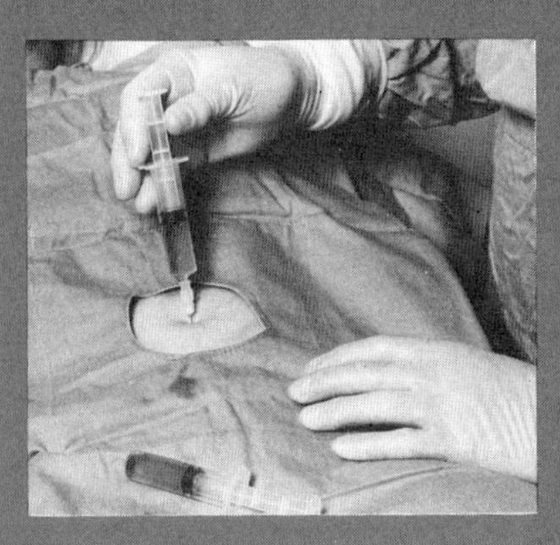

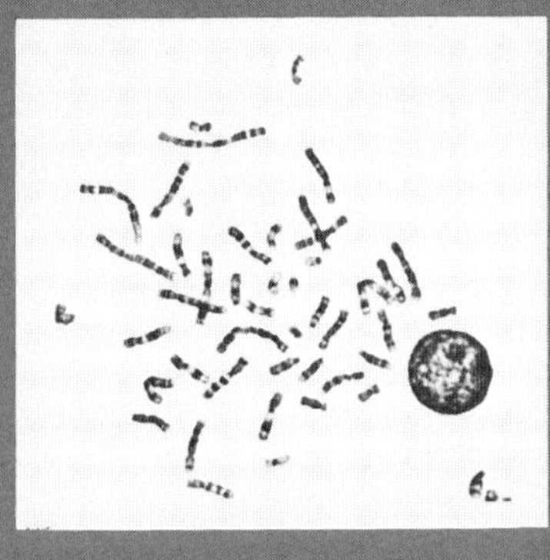

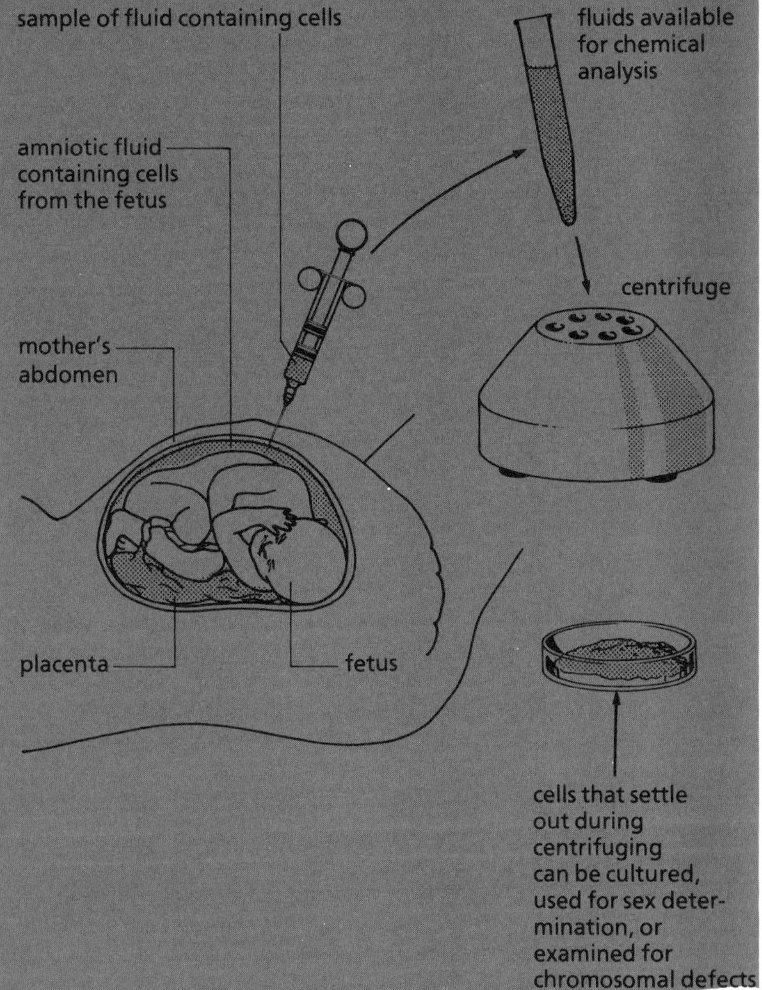

Genetic Factors

We have seen that the information that determines many of our characteristics is coded in the chromosomes that we receive from our parents. If we receive a gene that is responsible for some abnormality, it may be passed on to our children. Whether the defect is expressed or not will depend on such factors as whether the trait is caused by a dominant or recessive gene.

Some genetic defects arise during the process of cell division when an extra chromosome or too few chromosomes might be produced. Chromosomes can also be broken and perhaps reattach in another place, deleting or duplicating the information they carry. Normally, chromosomes are present in the nucleus of each cell in pairs, but in some birth defects, the nuclei contain an extra chromosome in addition to the normal pair. This condition known as **trisomy** will be discussed in greater detail.

Trisomy

Trisomy Involving Sex Chromosomes

The extra chromosome (or chromosomes in some cases) in the cells of the embryo are the result of an error that occurs during cell division of the egg or sperm from which the embryo is formed. The most common cause of trisomy is the failure of a pair of chromosomes to separate during the production of sperm or ovum cells. This lack of separation is known as **meiotic nondisjunction**. Normally a protein filament pulls each chromosome of a pair away from its partner, drawing them to opposite ends of the cell. If one filament is broken, both chromosomes will be drawn to the same end. As a result, a sperm (or ovum) will be formed with two identical chromosomes present. If an embryo results from the union of this sperm with an ovum, that embryo will have trisomy. Another sperm from this male will lack this particular chromosome entirely.

If nondisjunction occurs in the sex chromosomes, the following gamete combinations are possible:

abnormal sperm, XX, O, YY or XY.

Abnormal ova may contain XX or O combinations.

If these abnormal gametes fertilize with a normal gamete, the following combinations are possible:

XO, YO, XXX, XXY, XYY.

People with an XO combination of chromosomes are females. They are very short in height, do not menstruate, and are sterile. The disorder is known as **Turner syndrome**. When XXX rather than the normal XX occurs, some women may appear quite normal while others may be much more severely affected and closely resemble persons with Turner syndrome. When non-separation of chromosomes results in an XXY combination, it is known as **Kleinfelter syndrome**. Such persons are male and appear normal but are sterile with small, poorly developed testes. Often these men are tall and thin, with high-pitched voices, and are usually mentally retarded.

When trisomy occurs in chromosome 21, it results in a condition known as **Down syndrome**. People with this defect are mentally retarded. The face is typically rather broad and flat, the eyes are slanted, and the tongue appears larger than normal. (See Figure 20.14.)

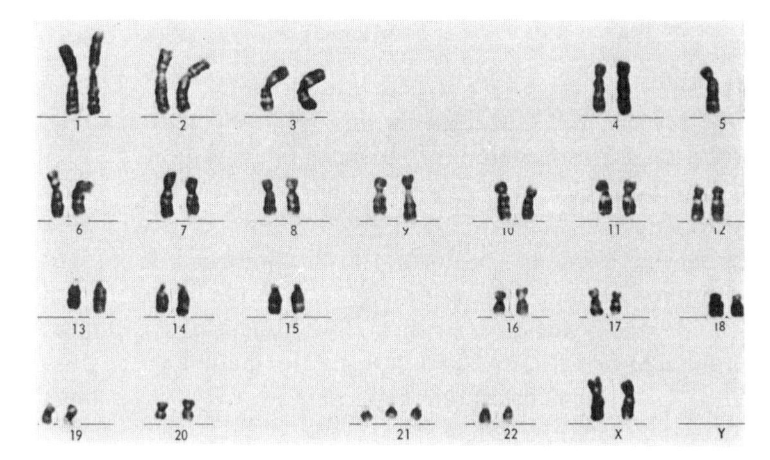

Figure 20.14.
Trisomy. Karyotype of a person with Down syndrome, trisomy 21. The XX chromosomes show this is a female karyotype.

Sickle Cell Anemia

A very small chemical change such as the substitution of a single base in the DNA chain, can produce very large effects in the body. For example, the formation of an abnormal kind of hemoglobin results from a single base substitution and causes the disorder known as sickle cell anemia. In some anemic people the decrease in the number of red blood cells is accompanied by red cells with a distorted sickle shape. When the families of these patients were investigated, it was found that this sickle cell anemia is inherited as a recessive gene. Further examination has shown that individuals, homozygous for the trait, are affected, but heterozygous persons having some sickle cells do not develop the disease.

Environmental Factors

Sometimes something happens during pregnancy that has little physical effect on the mother, but has a drastic effect on the baby she carries.

Viruses

German measles (**rubella**) in a pregnant woman can cause considerable damage to the fetus, although the mother may not suffer seriously from the disease. If a mother contracts German measles in the early months of pregnancy, her baby may be born deaf or with **cataracts**. The child may also be mentally retarded or have a heart defect. Other viruses are also suspected of causing birth defects.

Drugs

A number of medical and non-medical drugs have been found to be harmful to the fetus. For example, the drug **thalidomide**, which was taken by mothers to combat morning sickness and other unpleasant symptoms of pregnancy, resulted in thousands of deformed babies.

Excessive doses of otherwise beneficial vitamins have been associated with mental retardation. The **hallucinogenic** drug, LSD, and some other non medical drugs have been shown to damage chromosomes.

Radiation

The radition from X-rays, which enables physicians to make many diagnostic decisions, can, if excessive, be harmful. If they are used during the early months of pregnancy when rapid cell division is taking place, X-rays may cause changes in the chromosomes. Whenever X-rays pictures are taken, non-target areas of the body should be shielded, especially the germ cells of the ovaries or testes.

Other Factors

In many instances, it is difficult, or impossible, to determine the cause of a birth defect. Diet and the general health of the mother are important. The age at which a mother becomes pregnant also contributes to defects. Very young mothers and mothers of over 40 years of age, are known to bear higher numbers of babies with birth defects than are other mothers.

Timing and the Effects of Drugs

The brief discussion on birth defects may seem very alarming. It is important to know that not all mothers who have German measles deliver babies with birth defects. Not all the mothers who used thalidomide or had X-rays taken during early pregnancy produced defective offspring. Many of them were quite normal.

There are stages during the development of the embryo when a critical development is taking place. Interruption or interference with a process at a vital stage may make a very significant difference. If the mother takes a drug that causes a certain birth defect, a few weeks earlier or later, it may have little or no effect on the fetus. For example, if a defect-causing agent reaches an embryo just as its eyebuds are forming, blindness may result. If this stage of development is complete when the drug is taken, then the likelihood of the agent affecting that specific part may be reduced.

Prevention of Some Birth Defects

If we know what causes a particular problem, we are, in many cases, closer to being able to solve it. Some birth defects can thus be avoided or the risk considerably reduced.

Some Preventive Guidelines

1. Good medical care, normal body mass, and a high level of fitness, resulting in good health of the mother is especially important for producing healthy, normal children.
2. People with family histories of certain hereditary abnormalities can seek genetic counselling. Expert advice can free some potential parents from the concern that their child might have some deformity when there is no need to worry. **Genetic counsellors**, after a close study of each family pedigree, can often give accurate assessments of the predicted ratios for normal and defective children in a particular family. This will give them a good idea of what the chances are that they will produce a child with a particular condition.
3. After the third month of pregnancy, a sample of the cells from the amniotic fluid surrounding the fetus in the uterus can be withdrawn and analysed by a process called **amniocentesis**. The potential for the development of

many abnormalities can be recognized at this early stage by examining the cells in this fluid. Abnormalities caused by the presence of the wrong number of chromosomes or by their abnormal appearance can be detected by amniocentesis. Many enzyme-deficiency diseases can also be recognized by chemical analysis.

4. Taking preventive measures, such as use of the German measles vaccine and determination of the mother's sensitivity to Rh incompatibility, can prevent some abnormalities.
5. Pre-natal care is most important. As soon as a woman thinks she is pregnant, she should see a physician for confirmation and for advice concerning her welfare and that of the baby. The doctor will undoubtedly advise her never to take any drug unless it is prescribed by a physician who knows that she is pregnant. The doctor will avoid the use of X-rays except in an emergency, advise the mother to cut down on smoking or better still to quit, and to avoid contact with infectious diseases. The doctor will watch for symptoms of problems, and prescribe treatment to ensure the mother's own good health as well as that of her baby.

Non-Inherited Diseases

Some diseases of newborn children are not inherited. **Congenital diseases** are caused by an infection in the mother or perhaps by a drug that affected the development of the fetus in the uterus. Some congential disorders are caused by deficiencies in diet of the mother during pregnancy.

If a mother has a syphilis infection, the syphilis bacteria can pass through the placenta and cause damage to the fetus as it develops. When pregnant women have blood tests to check for various potential problems, a routine test is done to make certain that no syphilis bacteria are present in her blood which might cause problems for the baby. Infectious diseases are not inherited, they are transmitted directly from those who already have the disease, from intermediate hosts, or from contaminated sources, such as water, soil, or the air. Another serious infection in pregnant women can cause **conjunctivitis** in their newborn babies. Such infections may be acquired during delivery, as the baby passes through the birth canal. Since this condition can lead to blindness, the eyes of every newborn baby must be treated.

QUESTIONS FOR REVIEW

SOME WORDS TO KNOW

Match the descriptions given in the left-hand column with a word in the right-hand column. DO NOT WRITE IN THIS BOOK.

1. Two different genes or traits on two homologous chromosomes.
2. The appearance or presence of a trait in an organism.
3. A unit of heredity, a location on a DNA molecule responsible for a specific characteristic.
4. Rod-shaped bodies in the cell nucleus.
5. Two similar chromosomes carrying information about the same characteristics.
6. A change in a chromosome producing a new inheritable characteristic.
7. The presence of an extra chromosome in a cell.
8. Two successive cell divisions with only one duplication of chromosomes producing haploid cells.
9. Process of removing some amniotic fluid for culture and analysis.
10. A single set of chromosomes such as are found in gametes.

A. diploid
B. genotype
C. phenotype
D. gene
E. amniocentesis
F. dominant
G. recessive
H. trisomy
I. karyotype
J. haploid
K. chromosome
L. homologous
M. mutation
N. meiosis
O. mitosis
P. homozygous
Q. heterozygous

SOME FACTS TO KNOW

1. Briefly outline the structure and the functions of a chromosome.
2. a) What is the purpose of meiosis?
   b) How many chromosomes are there in a normal human body cell?
   c) How many chromosomes are there in a sperm or ovum?
3. Explain the following terms:
   a) homozygous and heterozygous,
   b) homologous chromosomes,
   c) phenotype and genotype.
4. Explain how it is that there are so many variations of skin colour.
5. Explain how sex is determined genetically.
6. What is a mutation? What causes mutations?
7. a) List some of the causes of birth defects.
   b) How can the risk of birth defects be reduced?
8. What is the value of amniocentesis?
9. What is the difference between an inherited disease and a congenital disease?
10. Produce a pedigree of your family. Cover as many generations as you can by asking questions at home. Select one trait, such as attached or unattached ear lobe, ability to roll your tongue; or choose a trait from the list of dominant and recessive human traits in Table 20.1. Fill in the members of your family that possess the trait on your pedigree.

QUESTIONS FOR RESEARCH

1. Select one of the following topics and prepare a report for presentation to the class:

   - genetic counselling
   - Turner syndrome
   - diabetes mellitus
   - karyotypes
   - genetics and evolution
   - Down syndrome
   - genetic engineeering
   - Huntington's chorea
   - mapping chromosomes
   - hybridization: producing new varieties of plants and animals

2. Select one of the traits listed in the table of dominant and recessive characteristics. Trace as many members of your family as possible who possess the trait and produce a pedigree of your findings. If your family has a trait that is not listed in the short table in this text, you can probably find a good reference book at the library. There are books that list many hundreds of known traits.

PROBLEMS IN GENETICS

1. A woman has cataracts (this is a dominant trait and she is homozygous). The father is homozygous and normal. What will be the predicted phenotypes and genotypes of the children. Use C for cataract, c for normal.

2. In humans, curly hair is dominant over straight hair. If a straight-haired man marries a curly-haired woman, what are the *possible* genotypes and phenotypes of their children? Give the predicted ratios in each case.

3. A pedigree for farsightedness is given below. Shaded symbols indicate the presence of the trait.

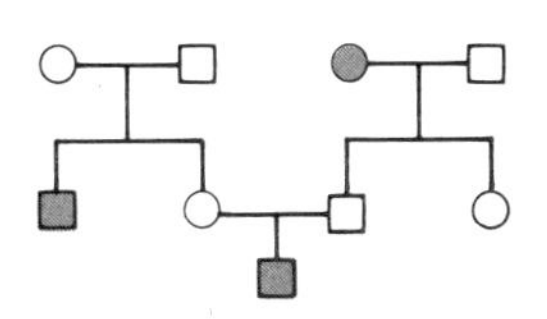

   a) Is the trait dominant or recessive? Explain how you can tell.
   b) Give the genotypes of as many of the individuals in the pedigree as you can.

4. The pedigree shown indicates the occurrence of **phenylketonuria**, a disease that leads to mental retardation.
   a) What evidence is there that this disease is a recessive trait?
   b) Note the cousin marriages. Why might cousin marriages produce an increased risk?

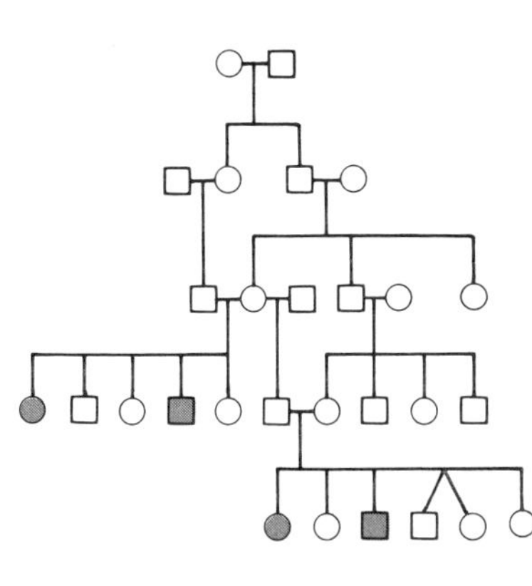

5. If a man with type O blood, has parents who both have type B blood, marries a woman with type AB blood, what will be the theoretical percentage of their children with type B blood?

6. In a paternity law suit a woman with type AB blood sues a man with type O blood. The child has the same blood type as the mother. Could this male be the father? Explain your answer.

7. Hemophilia in humans is controlled by a recessive sex-linked gene.
   a) Could both a father and son be hemophiliac?
   b) Explain why hemophiliac mothers always have hemophiliac sons.

Activity 1: THE ABILITY TO TASTE PTC

Materials

PTC paper

Method

PTC (phenylthiocarbamide) is a harmless chemical. The ability to taste PTC is determined by a single gene. When placed on the tongue the paper may have a different taste for different people; some may experience no taste at all. People who have two recessive genes for the ability to taste PTC will detect no taste. *Test as many members of your family as you can and record the results in a pedigree chart. Let T = ability to taste and t = the lack of ability to taste PTC. If one of your parents cannot taste PTC they will be tt. If they can taste it they will be either TT or Tt. Find out the genotypes of as many relatives as you can.*

Activity 2: IS THE ABILITY TO ROLL THE TONGUE INHERITED?

Most people can turn up the side of their tongue so that, near the tip, the sides nearly touch.

Questions

1. *How many in the class are tongue rollers? Find out if your parents, brothers or sisters are tongue rollers and prepare a pedigree for this trait.*
2. *Based on your class results, do you think the ability to roll the tongue dominant or recessive?*
3. *What are the genotypes of non-rollers and rollers?*
4. *Could two parents who do have the ability to roll their tongues have a child who could not roll the tongue? Explain.*

Glossary

A

abdomen. Area of the body between the diaphragm and pelvis containing the internal organs; belly.

abduction. Sideways movement away from the body's midline; act of turning outward.

absorption. Taking up of substances by the capillaries, skin, or other tissues.

accommodation. The adjustment of the eye for various distances by changes in the shape of the lens.

acetylcholine. A substance released from certain nerve endings; used for chemical transmission of nerve impulses at synapses and end plates. Easily destroyed by the enzyme, cholinesterase.

Achilles tendon. Heel tendon of the calf muscles of the leg. Achilles was the mythical Greek hero who could not be wounded except in the heel.

acid. Any substance donating hydrogen (H^+) ions. An acid has a pH below 7.00, a base has a pH above 7.00.

acid-base balance. The state of equilibrium between acids and bases such that the hydrogen ion (H^+) concentration of the blood is maintained at pH 7.30.

acne. Skin disorder of the sebaceous glands, common in adolescents.

acromegaly. Enlargement of the hands, feet, and jaw because of oversecretion of growth hormone from the anterior pituitary.

actin. A thin-muscle protein which, with the protein myosin, is responsible for contraction.

active transport. A process in which molecules are moved across a membrane *against* the concentration gradient. The process requires energy and may involve carrier molecules or special enzymes.

adduction. Movement toward the midline of the body.

adenoids. A mass of lymph cells; part of the tonsils.

adenosine triphosphate (ATP). The major source of cellular energy; found in all cells.

adipose. Related to fat-containing cells; fatty.

adrenal. An endocrine gland located above the kidney which secretes hormones that prepare the body for stress in emergencies.

adrenalin. The main hormone of the adrenal medulla. (See *epinephrine*.)

aerobic. Requiring oxygen for life or function.

allergy. Sensitivity to a substance. A condition characterized by reaction of body tissues to specific antigens without production of immunity (e.g., hay fever).

alveolus. An air sac of the lungs.

amino acid. Basic unit of structure in proteins. Two groups: essential, not synthesized in body; non-essential, synthesized in body.

amnion. The inner fetal membrane that holds the fetus suspended in amniotic fluid (a clear, watery fluid which protects the fetus).

amniocentesis. Withdrawal of amniotic fluid for tests and analysis. Performed by inserting a needle through the abdominal wall into the uterus.

anaerobic. Not requiring oxygen.

anaphase. The stage of cell division (mitosis) in which chromosomes move toward the poles of the dividing cell from the centre of the cell.

androgen. A substance which stimulates the development and maintenance of sex characteristics (for example, testosterone).

anemia. A condition in which there is reduction of circulating red blood cells or of hemoglobin.

anesthesia. Partial or complete loss of sensation with or without loss of consciousness.

anorexia nervosa. Psychologically induced loss of appetite.

antagonistic. That which counteracts or has the opposite action of something else.

anterior. In front; ventral.

antibiotic. A substance produced by bacteria, moulds, and other fungi, that has power to inhibit growth of or destroy other organisms.

antibody. A protein, formed by the body, that acts against a specific substance (antigen).

antidiuretic. An agent that causes reduction of urine formation.

antigen. A substance (foreign body, e.g., virus) causing production of antibodies.

anus. The lower opening of the digestive tract.

anvil (incus). Middle of the three bones of the middle ear, between hammer and stirrup.

aorta. The main artery of the body arising from the left ventricle of the heart.

appendage. Any part attached to a larger part, such as a limb.

appendix. An appendage. Small structure adjacent to the caecum. It is a blind sac, of little importance to human digestion.

aqueous. Aqueous humor; watery nature.

arachnoid. The central protective membrane covering the brain and spinal cord.

arteriole. A small artery; leads into a capillary or capillary network.

arteriosclerosis. General term referring to many conditions where there is a thickening, hardening, and loss of elasticity of the walls of blood vessels.

artery. Any vessel carrying blood away from the heart to the tissues.

arthritis. Inflammation of a joint, usually accompanied by pain.

articulate. To join together as in the junction of two bones.

asthma. Intermittent difficulty of breathing accompanied by wheezing and coughing, which is caused by an allergy.

astigmatism. Unequal curvature of the cornea and lens causing a distorted image on the retina.

atlas. Top bone of the vertebral column on which the skull rests.

atrophy. Wasting or decrease in size of a tissue or organ.

auricle. The outer ear flap on the side of the head; also called the pinna.

autosome. Any of the paired non-sex chromosomes.

axis. A line running through the centre of a body. The second cervical vertebra.

axon. Part of a nerve cell (neuron) which conducts impulses away from the cell body.

B

barbiturate. A sedative.

basophil. Large, nucleated white blood cell containing large granules in the cytoplasm. Stains with basic dyes.

benign. Not malignant; not dangerous.

biceps. A muscle with two heads or points of origin. Muscle of upper arm.

blastocyst. The hollow, fluid-filled ball of cells formed from the fertilized egg in early pregnancy.

blind spot. Area in the retina in which there are no rods nor cones; point where optic nerve enters the eye.

boil. An abscess or infection in the skin caused by *Staphylococcus aureus* bacteria.

bolus. A soft mass of food ready for swallowing.

bronchi. The first two divisions of the trachea that penetrate the lungs and terminate in the bronchioles.

bronchitis. Inflammation of the mucus membrane of the bronchi.

bronchiole. A smaller division of the bronchi that leads to the alveolar ducts and the air sacs.

bursa. A sac of fluid between the bones at a joint.

C

caecum. The blind pouch that forms the first portion of the large intestine.

calyx. Cuplike division in the kidney that collects urine.

cancer. A malignant tumour or abnormal growth of cells.

carbohydrate. Compound composed of carbon, hydrogen, and oxygen found in foods such as sugars and starches.

cardiac cycle. The period from the beginning of one heartbeat to the next beat.

cartilage. A flexible, rubbery supporting tissue that is found in the nose, ear, between the vertebrae, and between many joints.

cell. The basic unit of structure and function in all living things.

centriole. A cell organelle that is involved in cell division; forms spindle fibres.

centromere. The structure which attaches the spindle fibres to chromosomes.

cerebellum. The dorsal portion of the brain that is involved in co-ordinating muscle activities.

cerebrum. The upper and largest part of the brain; contains motor and sensory areas and association areas that are concerned with mental activities.

chemoreceptor. A sense organ or sensory nerve ending that is stimulated by a chemical substance or change, as in the taste buds.

cholesterol. A fatty compound found in animal fat, bile, blood tissues, etc.

cholinesterase. Any enzyme that catalyzes the breakdown of acetylcholine.

chorion. Outer membrane around the embryo from which villi connect with the uterine lining to give rise to the placenta.

choroid. The middle layer of the eyeball that contains blood vessels.

chromatid. One of the double structures at prophase of cell division that later forms the chromosomes.

chromatin. Readily stained substance within the nucleus, mainly composed of DNA.

chromosome. Small rod-shaped bodies in the cell nucleus. Contains the genes for hereditary characteristics. 46 present in humans (23 pairs).

chyme. The mixture of partially digested foods and digestive juices found in the stomach and small intestine.

cilia. Small hairlike projections from epithelial cells, trachea, or bronchi, etc.

cirrhosis. A chronic liver disease characterized by degenerative changes in the liver structure.

clone. A group of cells descended from a single cell; exact duplicate.

clot. A semi-solid mass of fibrin threads, trapped blood cells, and platelets; to coagulate.

coccyx. Three or four small fused, rudimentary vertebrae at the base of the spine.

cochlea. A coiled tube that forms the portion of the inner ear that contains the organ of Corti, the receptor for hearing.

colon. The portion of the large intestine from the caecum to the rectum.

conception. The union of a sperm with an ovum to initiate pregnancy.

conduction. The transmission of a nerve impulse; transfer of heat or sound waves through a conducting medium.

congenital. Present from birth. Occurs during fetal development; not inherited.

conjunctiva. Mucous membrane covering the inside of the eyelid in front of the eyeball.

constriction. A narrowing. When the diameter of a pupil of the eye or a blood vessel gets smaller.

contagious. Diseases transmitted by direct or indirect contact.

contraception. The prevention of conception.

contraction. The shortening or tightening of a muscle.

convection. Transfer of heat in liquids or gases by means of currents.

coronary. The blood vessels of the heart. Pertaining to the heart.

corpus luteum. Body of cells that fills the ovarian follicle following ovulation. Produces progesterone.

cortex. An outer portion or layer of an organ or structure.

cutaneous. Refers to the skin.

cytoplasm. The cellular substance between the cell wall and nucleus of a cell.

D

death. Permanent cessation of all vital functions.

decibel. The unit used to measure the intensity of sound.

defecation. Emptying solid waste or fecal matter from the large bowel or lower colon.

dendrite. Thin extension of the nerve cell that carries impulses toward the cell body.

dentine. Hard substance, part of the teeth; it surrounds the tooth pulp.

deoxyribonucleic acid. (DNA). Composed of nitrogenous bases, deoxyribose sugar, and phosphate groups.

depression. A mental state of dejection, lack of hope, or absence of cheerfulness; also a physical condition.

dermis. The true skin. Skin tissue lying below the epidermis.

dialysis. Process by which solutes move across a semipermeable membrane.

diaphragm. The muscle and connective tissue partition between thoracic cavity and abdominal cavity (muscle of respiration).

diarrhea. Passage of frequent watery stools; a symptom of gastrointestinal disturbance.

diastole. The period of relaxation for the heart muscles when the chambers of the heart fill with blood.

diffusion. Movement of molecules through a medium from a high concentration to a low concentration.

digest. To break down foods using enzymes or mechanical means.

digitalis. The dried leaves of the purple foxglove plant, used in powdered form as a heart stimulant.

dilate. To become larger in size or diameter.

dilute. To thin down or weaken, by addition of water.

diploid. Cells having pairs of chromosomes present after fertilization.

distal. Further away from the body or the point of reference. The opposite of proximal.

dorsal. The back of an animal. Side of the body where the backbone is present.

dura. Hard or tough; the tough outer membrane around the brain and spinal cord.

dystrophy. Degeneration of an organ resulting from poor nutrition, abnormal development, infection, or unknown causes.

E

electrocardiogram. A chart record of the electrical currents produced by the heart muscle; produced by an electrocardiograph machine.

electroencephalograph. An instrument that records the electrical currents produced in the brain.

embryo. The developmental stages from the fertilization of the egg to the end of the eighth week.

emphysema. Respiratory disease. Condition resulting from rupture or expansion of alveoli of the lungs.

emulsify. To prepare two liquids, which do not mix, so that one, in the form of small globules, is dispersed throughout the other.

endocrine. Producing secretions that pass directly into the blood stream.

endolymph. Fluid contained in the inner ear.
endometrium. The lining of the inner surface of the uterus.
endoplasmic reticulum. The series of tubular structures and membranes found in the cytoplasm of cells; transports substances within cells.
endorphin. Special substance, normally found in the brain, producing the effects of opiates, which dull pain.
enzyme. An organic protein causing alterations in the rate of chemical reactions without being consumed in the reactions.
eosinophil. A granular white blood cell which stains with an acid dye.
epidermis. Outer layer of skin cells; contains no blood vessels.
epididymis. A long convoluted tubule in the testis; conveys sperm to the vas deferens.
epiglottis. A leaf-shaped cartilage which covers the larynx during swallowing. Helps deflect food into the esophagus.
epilepsy. A nervous disorder that may cause convulsions, loss of consciousness, and sensory disorientation.
epinephrine (adrenalin). Hormone produced by the adrenal medulla; it stimulates the sympathetic nervous system; produces cardiac stimulation.
epithelial. The covering or lining tissues of the body; lines tubes and ducts.
estrogen. A female hormone producing or stimulating sexual characteristics and development.
excrete. To separate and expel useless substances from the body.
exocrine. External secretion by a gland, through ducts, e.g., sweat glands.
extension. The act of straightening a limb or muscle.

F

fatigue. A feeling of tiredness.
feces. The waste matter expelled from the bowel through the anus.
fertilization. The union of an ovum (egg) with a sperm.
fetus. The developmental stage from about the ninth week after conception to birth.
fibrin. A white to yellowish insoluble fibrous protein formed when blood clots.
fibrinogen. A blood protein that is acted upon to produce fibrin when blood clots.
filtrate. The name given to the fluid that has passed through a filter or from the glomerulus to the Bowman's capsule in the kidney.
filtration. The process of forming a filtrate by passing a fluid, under pressure, through a filter or selective membrane.
fimbrae. Fringed, funnel-shaped opening of the Fallopian tubes.
flexion. The act of bending a limb at a joint.
follicle. A small, hollow structure containing cells or a secretion. A layer of cells surrounding the ovum which produce hormones.
foramen. A natural opening through a bone or a membrane.
fulcrum. The point about which a lever turns or pivots.
fungus. Simple organisms which lack chlorophyll; some may cause diseases in humans.

G

gamete. A male or female sex cell; a sperm or egg (ovum).
ganglion. A mass of nerve cells, especially outside of the brain or spinal cord.
gene. The unit of heredity responsible for transmission of a characteristic to the offspring; part of the DNA molecule.
genotype. The hereditary makeup of an individual as determined by the genes.
glomerulus. A small group of capillaries in the kidney.
glottis. The opening between the vocal cords in the larynx (voice box).
glycogen. A complex carbohydrate, "animal starch"; stored in liver and muscle.
goblet cell. A one-celled, mucus-secreting gland found among the lining cells of the

respiratory and digestive tract.

goitre. An enlargement of the thyroid gland caused by lack of iodine, in the diet.

Golgi bodies. Small structures in the cytoplasm which produce enzymes and enclose them in membrane.

gonads. Sex glands; the testis or the ovaries.

H

(H^+). Symbol for hydrogen ion concentration.

haploid. Having one-half the normal number of chromosomes; sperm and eggs are haploid cells.

hemoglobin. Protein molecule with iron in red blood cells that combines with oxygen (O_2) or carbon dioxide (CO_2) and acts as a buffer.

hemolysis. Destruction or bursting of red blood cells with release of hemoglobin into the plasma.

hemophilia. Hereditary blood disorder caused by lack of a blood-clotting factor. Excessive bleeding from wounds. Sex-linked recessive trait.

hemorrhoids. Swollen veins in the rectal area.

hepatitis. An inflammation of the liver caused by an infection.

heterozygous. Having one or more pairs of dissimilar genes or chromosomes for an inherited characteristic.

histamine. Substance released during an antigen-antibody reaction, may cause bronchial constriction, arteriole dilation, increased gastric secretion, and a fall in blood pressure, etc.

homeostasis. The state of constancy of body composition and function.

homologous. Similar in origin and structure.

hormone. The chemical produced by an endocrine gland and passed into the blood stream; causes a response in a target organ.

hymen. A membranous fold which partly or completely closes the vaginal opening.

hypertension. High blood pressure.

hypothermia. Abnormally low body temperature caused by exposure. Usually ranges from 32-35°C for the body core temperature.

I

immune. Protected against the harmful effects of a disease.

impermeable. Not allowing passage.

implantation. Embedding of the blastocyst (fertilized and developing ovum) in the uterine lining.

incus (anvil). Middle of the three bones of the middle ear.

ingest. To take foods into the body.

inhibit. To repress or slow down.

insertion. The act of implanting. The attachment of a muscle to the bone that is moved by the muscle contracting.

inspiration. Taking air into the lungs.

insulin. Hormone produced in the pancreas that regulates sugar metabolism.

interferon. A protein produced by the body cells in response to invasion by viruses. Effective in combatting certain infections.

interphase. A 'resting' stage in mitosis, when a cell is not dividing.

interstitial. Lying between, as interstitial fluid lies between vessels or cells.

involuntary. Occurring without an act of will.

isometric. Refers to no change in length; muscle contraction without movement.

isotonic. A solution having the same solute concentration of blood plasma or a reference solution. A muscular contraction in which shortening is allowed and movement does take place.

J

jaundice. Yellow pigmentation in the skin when bile pigment (bilirubin) in the blood is high.

K

karyotype. A mapping or arrangement of the 46 human chromosomes based on the size of the individual chromosome.

keratin. A protein found in the epidermal structures such as hair and nails.

L

lacrimal. Relating to tears.

leukemia. Cancer of the blood-forming organs.

leukocyte. White blood cell. May be granular or non-granular in blood.

ligament. A strong band of fibrous tissue used to connect bones to each other.

lipid. Fat or fatlike substance that is not soluble in water, but is soluble in alcohol.

litre. Metric measurement of fluid volume. Equals 1000 mL, or 1000 cm^3.

lumen. The cavity of any hollow organ, as in the intestine and blood vessels.

lymph. A clear fluid containing some leukocytes, mainly lymphocytes. Fluid that bathes tissues.

lymphocyte. A white blood cell formed in lymph nodes. Used to combat bacteria and various antigens.

M

malignant. Harmful, dangerous, likely to cause death.

malleus (hammer). First small bone in the middle ear; articulates with the anvil or incus.

mandible (jawbone). The horseshoe-shaped bone of the lower jaw which articulates with the skull. It holds the lower teeth.

marijuana. A tall hemp plant; smoked or mixed with food to induce feeling of well-being.

maxilla. Upper jaw holding the upper teeth.

medulla. Central portion of an organ, e.g., the medulla oblongata of the hindbrain.

meiosis. A form of cell division that reduces chromosome number by one-half. Occurs in the ovaries and testes during the production of eggs and sperm.

melanin. Black or brown pigment found in skin, hair, and retina.

membrane. A thin sheet of tissue lining a tube or cavity, or covering a surface. Also a thin sheet of material surrounding cells and cell organelles.

meninges. The protective membranes covering the brain and spinal cord: dura mater, arachnoid, pia mater.

menstrual. Refer to the sloughing of the lining (endometrium) of the uterus. Relating to the monthly flow of blood from the female genital tract.

mesentery. A double-layered fold of the peritoneum attaching various organs to the wall of the body cavity.

mesoderm. The middle germ layer of the embryo; gives rise to connective tissue, muscle, blood, bone.

metaphase. A stage of mitosis in which chromosomes line up on the equator of the dividing cell.

microvillus. Tiny fingerlike extension of a cell surface used to increase surface area and enable more efficient absorption to take place.

millimetre. One one-thousandth of a metre.

mitochondrion. A small structure in the cytoplasm of a cell which transforms energy into a form usable in cell activities.

mitosis. A type of cell division that results in production of daughter cells identical to the parent cell.

monocyte. A white blood cell with a kidney-shaped nucleus; it is phagocytic.

mononucleosis. Acute infection believed to be carried by a type of herpes virus.

motor. Producing movement; denoting nerves that carry impulses from nerve centres to muscles.

motor end plate. The ending of a motor nerve on a skeletal muscle fibre.

mucus. The thick, sticky fluid secreted by a mucus cell or gland.

mutation. A permanent change in the gene (DNA molecule); a variation in an inheritable characteristic whereby the offspring express a trait not present in the parents.

myelin. A fatty substance forming a sheath or covering around many nerve fibres; speeds impulse conduction.

myofibril. Divided into sarcomeres which are the basic units of contraction in muscles.

myopia. Near-sightedness.

myosin. A muscle protein acting as an enzyme to aid in initiating muscle contraction.

N

nephron. The functional unit of the kidney that forms urine, regulates blood composition, and filters wastes.

neurilemma. A thin, living membrane around some nerve axons.

neuron. A nerve cell; basic unit of the nervous system. There are around 28 000 000 000 neurons in the human body.

neutrophil. A white blood cell having a three- to five-lobed nucleus; makes up to 70 percent of white cells; is phagocytic, i.e., engulfs and destroys foreign proteins.

nucleic acids. The type of organic acids found in chromatin.

nucleolus. Small spherical structure inside the nucleus that is involved in cell division.

nucleus. Small spherical structure within the cell. Centre where cell activities are controlled.

O

obese. Excessively fat, overweight.

occipital. Relating to the back of the head.

oral. Refers to the mouth or the mouth cavity.

organ. Two or more tissues organized to do a particular job.

organ of Corti. Structure in the inner ear; contains the cells responsible for hearing.

organelle. Submicroscopic structure within the cytoplasm of a cell that carries out particular functions, e.g., mitochondria or vacuole.

origin. The beginning of a nerve; the fixed attachment of a muscle.

osmosis. The passage of a liquid such as water through a semipermeable membrane to mix with another liquid.

osteocyte. A bone-forming cell; is contained in a space (lacuna) in the solid material of the bone.

osteomyelitis. Infection of bone caused mainly by a staphylococcus.

P

palate. The roof of the mouth; composed of hard and soft portions.

papilla. A small nipplelike protrusion associated with taste, touch, or smell.

paralysis. Temporary or permanent loss of function of some part of the body.

parasympathetic. Refers to the portion of the autonomic nervous system that controls normal body functions.

pathogen. Any organism capable of causing disease.

pedigree. Line of descent, in genetics. A chart showing the patterns of inheritance.

pelvis. A basinlike skeletal structure that supports the spinal column and rests on the lower limbs.

penicillin. A group of antibiotic compounds obtained from cultures of the mould *Penicillium notatum*; inhibits bacterial wall synthesis.

periosteum. A thick fibrous covering of bones; the connective tissue carrying blood and nerves.

peristalsis. The alternate contraction and relaxation of the walls of a tubular structure moving contents onwards.

peritoneum. A layer of tissue that lines the walls of the abdominal and pelvic cavities, enclosing the organs.

permeable. Allowing passage of solute and solvents in solutions.

pH. A symbol used to express the acidity or alkalinity of a solution.

phagocyte. Any cell which ingests bacteria and foreign bodies.

phagocytosis. Engulfing of particles by cells.

phenotype. The visible appearance of an individual.

pinocytosis. Cells take in fluids through the cell membrane by engulfing them.

pituitary gland. Small gland at the base of the brain; secretes hormones that control other glands in the body, growth, and development. Often called the "master gland".

placenta. The structure attached to the inner uterine lining through which the fetus gets its nourishment and excretes its wastes.

plasma. The liquid portion of the blood.

platelets. Tiny bodies found in blood which produce thromboplastin that is used in the clotting of blood.

pleura. Refers to the membrane(s) lining the cavities of the thorax, or covering the lung.

pons. Part of the brain between the cerebrum and the medulla.

potential. Implies a measurable electric current flow or state between two areas of different electrical strength.

pregnancy. The condition of carrying a developing offspring in the uterus.

prognosis. Prediction of the course and outcome of a disease or abnormal process.

prophase. A stage in mitosis characterized by the formation of visible chromosomes.

protein. A large molecule composed of many amino acids.

prothrombin. A substance converted to the active enzyme thrombin during the clotting of blood.

protoplasm. The essential substance which makes up all living cells.

proximal. The part of an appendage that is closer to the main part of the body.

psychosomatic. Involving the relationship between mind and body.

puberty. The time of life when both sexes become functionally capable of reproduction.

pulse. Rhythmic increase in pressure within a blood vessel with each contraction of the heart.

pupil. The opening of the centre of the iris of the eye.

pus. A thick, viscous fluid consisting of white blood cells and bacteria.

R

radial. Relating to a bone in the forearm (radius). Moving in various directions from a central point.

radiation. The emission and projection of energy through space.

radiology. The use of radiation and X-rays for diagnosis and treatment of disease or injuries.

receptor. A sense organ responding to a particular type of stimulus.

recessive. A characteristic that does not usually express itself in offspring because of suppression by a dominant gene.

recipient. One who receives blood or a graft.

reflex arc. A series of neurons serving a reflex (receptor, sensory nerve, centre, motor nerve, effector). Produces involuntary movements or actions.

refractory. Resistant to stimulation.

renal. Refers to the kidneys.

resuscitation. Bringing back to life or consciousness.

retina. The third and innermost part of the eye wall; contains visual receptors (rods and cones).

rhodopsin. A visual pigment found in rod cells; "visual purple".

Rickettsia. A genus of parasitic bacteria transmitted to humans through the bites of infected lice, ticks, or fleas.

RNA. Ribonucleic acid, composed of nitrogenous bases, ribose sugar, and phosphate.

ruga (pl. rugae). A fold or wrinkle such as found in the lining of the stomach and vagina.

S

saccharide. A sugar; a carbohydrate containing one or more simple sugar units.

sarcolemma. Cell membrane of a muscle fibre.

sarcomere. The portion of a myofibril lying between two Z lines.

saturated. Holding all it can. A saturated fat has all the hydrogen it can hold on its chemical bonds.

sebaceous. Refers to sebum, a fatty secretion of the sebaceous (oil) glands.

semicircular canals. Three bony canals at right angles to each other in the inner ear; responsible for balance.

seminiferous. Refers to production and transport of sperm.

sensory. Refers to nerve fibres that carry a stimulus from a receptor to the central nervous system.

sickle cell. A red blood cell that is crescent-shaped because it contains an abnormal hemoglobin. Found in a special type of anemia.

sinus. A cavity in a bone, or a large channel for venous blood to flow in.

spasm. A sudden, involuntary, often violent contraction of a muscle.

sperm. The male sex cells produced in the testes; also called spermatozoa.

sphincter. A band of circularly arranged muscle that narrows an opening when it contracts.

sphygmomanometer. Instrument used to measure blood pressure.

spirometer. An apparatus that measures the rate and volume of breathing.

stapes (stirrup). Innermost of the three small bones of the middle ear; is attached to the oval window and articulates with the incus (anvil).

suture. A stitch used in surgery. An immovable joint uniting the bones of the skull.

sympathetic. Refers to the portion of the autonomic nervous system that controls response to stressful situations.

synapse. A small gap at the junction between two neurons. The gap is bridged by the release of a chemical such as acetylcholine to allow impulses to pass.

synovial. Fluid in the space between the bones of a freely movable joint.

synthesize. To form new or more complex substances from simple substances.

syphilis. An infectious, chronic venereal disease transmitted by physical contact with an infected person or by sexual intercourse.

systole. Contraction of the muscle of a heart chamber.

T

telophase. The last stage of mitosis characterized by cytoplasmic division, and positioning of chromosomes at poles of the cell.

temporal. Relating to the side of the head or temple.

tendon. A fibrous tissue that attaches a muscle to a bone.

tetanus. A sustained contraction of a muscle; a disease caused by a bacterium, characterized by sustained contraction of jaw muscles ("lockjaw").

threshold. The lowest strength stimulus that results in a detectable response or reaction.

thromboplastin. Blood protein used to form thrombin in blood-clotting mechanism.

thyroxin. An active iodine-containing hormone produced by the thyroid gland, important in regulating metabolism.

tidal. Refers to the volume of air normally inspired, as in tidal volume.

tone. A state of slight constant tension or contraction exhibited by muscular tissue.

toxic. Poisonous.

toxin. A poisonous substance; harmful.

transplant. Organ or tissue removed from a donor and surgically transferred to a recipient.

tumour. A swelling or enlargement; may be malignant or benign.

twitch. A single involuntary or sporadic muscular contraction in response to a single stimulus.

tympanum Ear drum.

U

ulcer. An open sore or lesion in the skin or in a mucous membrane-lined organ, e.g., stomach.

umbilical cord. The structure connecting the fetus to the placenta.

urea. A metabolic waste formed in the liver from protein breakdown.

ureter. Duct taking urine from the kidney to the bladder.

urethra. Duct carrying urine from the bladder to the exterior of the body.

uric acid. An end product of metabolism found in urine.

urine. The waste fluid formed by the kidney and eliminated from the body via the urethra.

uvula. Fleshy tissue suspended from the soft palate above the back of the tongue.

V

vaccine. Substance used to promote antibody formation. Killed or weakened pathogens prepared in a suspension for inoculation.

vacuole. Small fluid-filled sac within the cytoplasm.

vector. An organism which transmits pathogenic organisms from one host to another.

venereal. Relating to or resulting from sexual intercourse.

ventricle. One of the two lower chambers of the heart.

venule. A small vein; gathers blood from capillary networks.

vestibular. Refers to the equilibrium structures of the inner ear and their nerves.

villus (pl. *villi*). A small finger-like projection on the wall of the small intestine that increases the surface area of the wall for more efficient absorption.

virus. Submicroscopic infectious organism that attaches to living cells to survive and reproduce.

viscera. The internal body organs, especially those of the abdominal cavity.

vitamin. An essential organic substance required for metabolic processes; it works with enzymes to control body functions.

voluntary. Under willful control; intentional.

W

wart. A small horny outgrowth on the skin, usually of viral origin.

waste. Useless end products of body activity, e.g., feces.

X

XX. Sex chromosomes giving rise to a female individual.

XY. Sex chromosomes giving rise to a male individual.

Z

zygote. A cell produced by the union of an egg and sperm; a fertilized ovum (egg).

Index

Note:
Heavy type indicates a major reference (**134**)
Italic type indicates a reference in a box, diagram, or margin (*344*).

A

abdominal cavity 52
abduction **127**
absorption of nutrients **301**
accessory glands (male) 421-422
accommodation and vision **185**, of lens *188*
acetylcholine and nerve fibres **141**, 163
Achilles tendon 105
acne 63, **72**
acromegaly 410
actin filaments **119**
active transport **28**, 29, 383
acuity, visual **188**
Adam's apple (see *larynx*)
adaptation, eye **185**
adduction **127**
adenine 466-468
adenoids 277
adenosine diphosphate (ADP) 30-32
adenosine triphosphate (ATP) 23, 30-32, 125
adipose tissue **45-46**, 49
ADP (see *adenosine diphosphate*) 30-32
adrenal cortex 404
adrenal glands *398*, **403-406**
adrenalin **403**, *404*, and heart-beat 239
adrenocorticotrophic hormone (ACTH) **408**
affector **156**
afferent nerve fibres 156
afterbirth (placenta) 448, 457
after-image **186**
air sacs (alveoli) **275**, **285**
alcohol and reaction time 140, 162, **163**, **164**, effects on body **312-313**
aldosterone 406
all-or-none response **140**
allergen 264
allergies **264-265**
alveoli **275**, 278, 283-285, ducts 278
amino acids **334**, *469*, essential, **334**, and protein **334**
amniocentesis 450, *488*
amnion 448, fluid 456
amphetamines **163-166**
anaphase, of mitosis *34*, **35**
anatomical position 53, *54*
anatomy, defined 8
anemia 215
anorexia nervosa *361-362*
antagonistic muscles **122-123**
anterior horn, spinal cord 155
anterior lobe, pituitary **407**, hormones, 408
antibiotics **266**
antibodies **219**, **261-262**
antidiuretic hormone (ADH) *388-389*, **410**
antigens **219**, **261-262**
antihistamine 265
anus 320
anvil (incus) 198
aorta *232*, 233
appendicular skeleton 88
appendix **318-319**
aqueous humor 180
arachnoid 154
arches (foot) *101*
areola 433
arm **103-104**
arrector muscle *64*
arteries and circulatory system **227-228**, 230, *236*
arterioles **227**
arthritis **110**
articular cartilage *86*, **105**
articulation of bones 104-106

artificial respiration **228-289**
ascending colon **319**, 320
ascorbic acid (Vitamin C) 341, 352
association fibres **151**, neuron, **156**
asters and mitosis 34
astigmatism **189**
athlete's foot 72
atlas bone *94*, 95
ATP (see *adenosine triphosphate*)
atrioventricular node **237**
atrioventricular valves (see *bicuspid* and *tricuspid*)
atrium **231-232**
auditory canal **196**
auricle (see *pinna*) 196
autonomic nervous system 136, 145, **158-161**, divisions of **159**, functions of **158-159**, **411-412**
autosomes 481
axial skeleton 88
axis bone *94*, 95
axon 46, **137**

B

bacilli 257
bacteria **255-257**
bacteriophage 256
balance, organs of **202-203**
balanced diet (see *Food Guide*) 328, **350**
baldness 64
ball-and-socket joint **106-107**
Banting, Dr. Frederick *402*
barbiturates **164, 168**
basophils 216
bel *201*
Bell, Alexander Graham, and sound *201*
benzedrine **163**, *166*
Best, Charles *402*
beverages, nutrient values *337*
biceps muscle 93, 94, 122, *123*
bicuspid valve 233
Bigelow, Dr. Wilfred *240*
bile 314-315
bile salts **314**, 315
binocular vision **186-187**
birth defects **485-486**, *486*, 492
birthmarks 71
birth process **455-458**
blackheads 63
blastocyst 446
bleeding, stoppage of **217-218**
blind spot **184**
blood 46-48, characteristics of **212**, circulation **227**, composition **212**, donors **221**, functions of **212-214**, sugar **330-331**, **402**, volume 240
blood clots 218
blood plasma 15, **212-213**
blood pressure 228, **241-242**, measurement of **242-243**
blood transfusion 219
blood types **218**, typing 219, genetics **479-480**
body defences *254*, **260-265**
body fluids *15*
body size **355-356**
body temperature, regulation of **68-70**
boils 72
bone(s), articulation of 104, formation of **86-88**, *87*, fractures of **108-109**, as lever **124**, structure of **85-86**
bone tissue **45**, 48
bony labyrinth of inner ear 198-199
Bowman's capsule **381-382**
brain, human, described **141-146**, comparative development 142, protection **153-154**, maps *148-149*
bread group **353**
breasts 433
breathing, control of **287**, mechanics of **281-282**, rates 287, (see also *respiration*)
breakfast 355
broken bones, repair **109-110**
bronchi **278-280**
bronchial tree 278-280
bronchioles **278**, 279
bundle of His **237**
bursa 123
Burton, D. E. 21

C

caecum **318-319**
caffein 163
calcium (in blood clotting) 218
calcitonin **399**
callus **109**
calorie (see *kilojoules*)
cannabis *167* (see *marijuana*)
canaliculi 87
canine tooth 302
capillaries **227-230**
capillary sphincters *260*
carbohydrates 13, 14, **328-330**, energy value of **330-331**
carbon 13
carbon dioxide, transport by blood **234-235**
carbon monoxide 286
carbonic acid 285
cardiac muscle **43-44**, 48, **115**, **231**
cardiac output **240**
carotid artery 236
carpal bones 104
cartilage **45**, *96*, **105**
cataracts 491
cavities, body **52-53**

cell(s) 16-24, growth of **445**, nucleolus 20, 22, nucleus 20, 22, respiration **30**, *31*, size 16, structure of **16**, 16-24
cell body of neuron **137**
cell division **33-35** (see *mitosis*)
cell membrane 19, 20
cellulose **329**
cementum 303, *304*
central fissure 145, 150
central nervous system 136, 140, 151
centrioles 20, 24, and mitosis 34
centromere 34
cereals 353
cerebellum **142-144**
cerebral cortex 141, 145, 148, hemispheres 145
cerebral hemorrhage *244*
cerebrospinal fluid **154**
cerebrum **145-150**, *146*
cerumen glands 196
cervical vertebrae *94*, **95**
cervix (of uterus) 427, 432
chemical composition of body **12**, summary 14
chest cavity (see *thoracic cavity*) 52
cholesterol **333**
cholinesterase 141
chorion 446
chorionic gonadotropin 431
choroid layer **182**
chromatin 20, 22
chromosomes 22, 466, **481**, and meiosis *471*
chyme **311**
cilia *19*
ciliated epithelium **42**, 47
circulatory system **227-247**
citric acid cycle 31
clavicle 103
clitoris 432
cloning *446*
coagulation, blood 217, 218
cocaine *169*
cocci 257
coccyx 96
cochlea **199-200**, duct 199-200
codeine *169*, 193
collagen, connective tissue 49
collapsed lung 280
colon **319**
colour blindness **189**, **482-483**
colour vision **182-183**
columnar epithelium **42**, 47
commissural fibres 151
compact bone 86
compounds, chemical 14
compound fracture **108**
concave lens *188*
conditioned reflex 157
conduction deafness 202
cones, retina **182-184**
congenital diseases 493
conjunctiva **179**
conjunctivitis 179
connective tissue **44-46**
conserving body heat 69-70
contagious diseases **258**
convection, and body heat loss 69
convergence (action of eyes) 186
convex lens *188*
co-ordination 150
copulation 434
cornea **181**
corpus albicans 425, 431
corpus callosum 145
corpus cavernosa 422
corpus luteum 425, 431
corpus spongiosum 422
cortex, adrenal **403-404**, kidney 380
Corti (see *organ of*)
corticoids 406
cortisone 406
Cowper's gland 421
cranial facial complex 90-93
cranial nerves 152, *153*
cranium **91**, **92**
cretinism **399-400**
Crick, F.H.C. and DNA *467*
cristae *23*
crossing over *147*
crown (of tooth) 302
crypts of Lieberkuhn **314**
cuboidal epithelium **41**, 47
curves of spine 98
cuspid teeth 302
cytoplasm 20, **21**
cytosine 466-467

D

dandruff 73
dark, adaptation of vision 185
deafness **201-202**
decibel **201**
deciduous teeth **302**, 303
defecation (elimination) 320
defences, body **260-265**
delivery **455-458**
dendrites 46, **137**, 140
dentine 302-304
deoxyribonucleic acid (see DNA), in cell nucleus 22, structure of *23* (see also *chromosomes*)
depressant drugs **163**, *168*
depth perception *186*
dermis **60-62**
descending colon **320**
development 466
dexadrine 166
diabetes mellitus *402*
dialysis 28, dialysis machine *386*
diaphragm 52, in breathing **281**

diastole **237-238**
diastolic pressure **242**
diet, balanced 328, **350**
diffusion **25-26**
digestion 301, chart of enzymes *316*, in mouth 301, **302-306**, *308*, time to *318*
dihybrid cross (two traits) 476
diploid number **470**
disaccharides **329**, 330
disease, pathogens 255, transmission of **258-259**, types of *266-267*
dislocation of bone **108**
distal end of limb 53
distance, judgment of 186
DNA (see *deoxyribonucleic acid*) **466-468**, *469*
dominance, laws of **473-476**
dominant characteristic **473**
dorsal (posterior) portion of body 53
dorsal root 155
Down syndrome 490
Drugs, for infections **266**, medical/non medical **163-169**, pregnancy 492, reaction time 140
duodenum **312-313**
dura mater **153-154**
dwarfism 410

E

ear **195-204**, external **196**, diagram of *196*
ear drum **196**
ectoderm 451
efferent nerve fibres (neuron) 156
egg production (ovum) **470-472**
ejaculation 422
ejaculatory duct 420
elastic cartilage 45, 49
electrical activity and nerve tissue **139**
electrocardiogram **241**
electron microscope 18, *21*
element(s) 12
embryo, development of *447*, **451-452**, *453-454*, formation of **451**
enamel, tooth 302, *304*
endocrine glands **396-408**, role **396** (see also individual glands)
endocrine system **396-410**, organs **398-408**
endoderm 451
endolymph *196*, 202
endometrium 427-429, in menstruation 430
endoplasmic reticulum 19, 24, 25
endorphins *161*
end plate **120**
energy factors *366-367*
energy needs of muscle **125**
energy value, of food *364-365*, balance **358-359**
enterokinase 314
enzyme(s) *33*, action of **32-33**
eosinophils 216
epidermis 60, **61**
epididymis 420
epiglottis **277**, **306**
epilepsy *151-152*
epinephrine **403**
epithelial tissue **40-42**
erectile tissue 422
erepsin 314
erythrocytes **213-215** (see *red blood cells*)
esophagus 278, 306, **307**
estrogenic hormones **407**, 455
ethmoid bone 93
eustachian tube 198
excretory system **377-391**
exercise **358-359**, **360**
exocrine glands 305, 396
expiration *281*
expiratory reserve volume 282
extension, muscle 118-127
external ear **196**
extracellular fluid *15*
extrinsic eye muscles **177-178**
eye, defects in structure **187-189**, muscles of **177-178**, orbital cavity 176, reflexes **184-186**, protection **178-179**, structure of **176-184**, vision **186-187** (see also individual parts of eye)
eyeball *180*
eyebrows **178**
eyelashes 179
eyelids 178

F

facial bones **93**
Fallopian tubes **425-426**, 446
false (floating) ribs 94
far-sightedness **187-188**
fatigue, muscle 125-126
fats 13, 14, 331, composition **332**, in body 333, sources 331-333
fatty acid 332
feedback (nerve) 145, (hormonal) 431
female reproductive organs **423-432**
femur 100
fertilization of ovum 427, **434-436**
fetal circulation 448
fetus **451-455**
fibre **342**, in diet 342-343, sources of 343

fibrinogen 218
fibrous connective tissue **44**, 45, 49
fibula 101
filaments, muscle *119*
filtrate (in kidney) 382
fimbrae 425
fingerprints *62*
flexion **127**
fluids, body 15; (see also *excretory system*)
fluoroscopy *92*
follicle stimulating hormone (FSH) 407, **409**, **424-425**, 430
follicles, ovary 423-424
fontanel 91
food(s) 350-359, energy value 358-359, 364, composition of *302*
food guide **350-354**; servings 354 (see also *diet*)
foot 101
foramen magnum 92, 154
forebrain 142-145, **144**
foreskin 422
fovea centralis **184**
Fox, Terry *102*
fractures, bone **108-109**, healing 109
fraternal twins *459*
freckles 71
frontal bone 92
frontal lobe **145**
fruits and vegetables group **352-353**
fundus 427
fungi 255, **258**

G

galactose 329
gall bladder **314**, 315
gametes *472*
ganglia *154-155*, **159**
gas(es), exchange between blood and tissues **283-285**, transport by blood **285-286**
gastric glands **310**, 311
gastric juices **310**, 311
gastric lipase 310
gastric secretion **310**
gene(s) 468, multiple 479, gene pool 480, transmission **474-475**
genetics 462, factors **489**, code **468**, **473**
genetic engineering *486-487*
genotype **474**
germ layers *451*
German measles (rubella) 491
gestation 452
gland(s) 41, sebaceous 63, sweat 63, tissue 41
glans 422
glial cells 146
gliding joint **106**, *107*
glomerulus **381-382**
glucagon 402
glucose **329-330**
glue sniffing 164-165
glycerol 331-332
glycogen 329, **330**
glycolysis 31
goblet cells **42**
goitre **398**
Golgi apparatus *19*, 20, 24, 25
gonads **406-407**
gonadotropins 409
gonorrhea 257, 439
goose flesh 68
gray matter 138
green-stick fracture 108
Gingras, Dr. G. 126
growth hormone 408
growth and development **445**
growth, boys and girls 459-461

H

hair **63**, *64*
hair follicle 61, *64*
hallucinogens **165**, 491
hammer (malleus) 198
hand, bone **103**, 104
haploid number **470**
hard palate 275, *276*
Harpenden calipers *372*
hashish *167*
Haversian canal **87**
head, region of body 88-93
hearing 198-202, disorders **201-202**
heart **230-240**, artificial *234*, valves 233
heart attack *238*
heartbeat, chemical control of 239, factors affecting **239-240**, neural control of 235-236, rate *237*, regulation **240-241**, sounds **237**, 238
height/mass tables 355, 370, boys/girls **461**
helix, double 466-467
hemispheres of cerebrum **145**
hemoglobin **214**, 215, **285**
hemophilia 218, 482
hemorrhoids 320
hepatic artery **314**
hepatitis *315*
hereditary characteristics 472-477
heredity and chromosomes **466-491** (see also *genetics*)
heroin **163**, *169*
herpes genitalis **440**
heterozygote **474-475**
hindbrain **141-143**
hinge joint **106**, *107*
histamine **261-262**, 265
homeostasis 6, 68, 411
homologous pair 468

homozygote **474**
Hooke, Robert 18
hormones 14, **396-397** (see also individual hormones)
human genetics **468-491**
humerus 103
hyaline cartilage 45, 49
hydrochloric acid, in gastric juice **310-311**
hydrogen 13
hymen 432
hypermetropia **188**
hypertension **243-244**
hypoglycemia *405*
hypothalamus 68, 144
hypothermia 69
Hz (vibrations per second) 200-201

I

identical twins *459*
ileum 312
ilium 100
immovable joints 104
immunity **263-264**
immunization schedule *264*
implantation 448
impulses, nerve **138**
incisor tooth **302**, *303*, 304
incomplete dominance 477
incus (anvil) 198
infectious diseases **258**
inferior vena cava **231-232**
inheritance (see *heredity*) **472-473**, two characteristics 476
inner ear **198-199**
inorganic salts 14
insertion, muscle **123**, *124*
inspiration *281*
inspiratory reserve volume 282
insulin 317, **330**, 402, discovery *402*, imbalance **403**
integumentary system 50
intelligence 145
intercostal muscles **281-282**
interphase of mitosis 33, *34*, 35
interstitial fluid 15, **244**
intervertebral disc **97-98**
intestine, absorption of nutrients by **311-318**
intracellular fluid 15
intrinsic eye muscles **177-178**
involuntary muscle 48, **115**
iodine and goitre 399
iris, eye **180-181**
iron-deficiency anemia *215*
ischium 100
islets of Langerhans **401-402**
isometric contraction **121**
isotonic contraction **121**
itch 68

J

Jarvik, Dr. Robert *234*
jejunum 312
joint(s) **104-107**

K

karyotype *481*
keratin 61, 63
kidney(s), function of **379-380**, structure **380-387**
kidney, artificial *386*
kilojoules **331**
Kleinfelter syndrome 490

L

labia **432**
labour pains **456**
lacrimal duct 179
lacrimal gland **179**
lacteal 317-318
lactiferous glands (ducts) 433
large intestine **319-320**
larynx 277-278
Law of Dominance **473-476**
Leeuwenhoek, Anton van 18
left brain, right brain *147*
leg **100-101**
lens(es) of eye **181**, 187, **188**
leukemia **217**
leukocytes **216-217** (see *white blood cells*), 247
ligaments 44, 104, **105**
light *183*
light reflex **184**
linkage, sex **482-483**
lipase **310**, 314
lipids 14 (see *fats*)
liver **314-315**
lobes of cerebrum **145**
long bone, parts of 86
loop of Henle **383-384**
loose connective tissue **44**
LSD *167*
lumbar vertebrae **94**, *95*, 96
lung(s) **280-284**, capacity of **282**, 283, disease 291, excretory functions 377
lutenizing hormone 407, **409**, 425
lymph **245**
lymph nodes **246-247**
lymphatic duct **247**
lymphatic system **245-247**, 264
lymphocytes *214*, 214-216, 262
lysosomes 19, 20, 23, 24

M

macrophages *262*
male reproductive system **417-422**

malleus (hammer) 198
maltase 314
mammary glands 433
mandible **93**
marijuana **165**, *167*
marrow **86**
mast cells 264
matrix, of connective tissue 44
maxilla **93**
meat group 351, 354
medulla, adrenal **403-404**, kidney 380
medulla oblongata **142-143**
meiosis **35**, **470-472**
melanin **61**, hormone 408
membrane, cell **19**
membranous labyrinth 199
memory 150
meninges **153**, 154
menstrual cycle **427-430**, hormones *428-429*
mescaline **165**
mesenteries **311-312**
mesoderm 451
metacarpal bones 104
metaphase of mitosis *34*, **35**
metatarsal bones *101*
micron 18
micro-organisms 254
microscopy, size *18*
microtubules 24
microvilli 19, 20, 42
midbrain **144**
middle ear **197-198**
minerals 13, 14, food material **339**, **341**
mitochondria 19, 23
mitosis **33-35**
molar teeth **302**, *304*
moles 71
monocytes 216
mononucleosis **217**
monosaccharides **329**, 330
mons pubis 432
morphine *169*
motor areas, cortex **148-149**
motor impulses, cerebral **148-150**
motor nerve 120
motor neuron **156**
mouth cavity 276; digestion **301-302**
mouth-to-mouth resuscitation *288-289*
movable joints **104**
muscle(s), names of *116-117*, eye **177**, points of attachment **123-124**, stimulation 120, force **124-125**, structure of **119**
muscle contraction **118-120**, 121
muscle fibres **119**
muscle system **115**, **118**, *116-117*
muscle tissue **43-44**, **115**
muscle tone **126**
muscular dystrophy 128
mutagenic agents 484-485
mutation(s) **484-485**
myelin sheath 46, 137
myofibrils **119**
myometrium 427
myopia (near-sightedness) **187-188**
myosin filaments **119**
myxedema **399**, *400*

N

nails, finger and toe 64-65, nail bed 65
nasal bones 276
nasal cavity **275-276**
nasal pharynx 277
near-sightedness **187-188**
nephrons 380, **381**, 382, 384, blood delivery 381, reabsorption 385, functions *388*
nerve(s) cranial *153*
nerve cell 46, **137**
nerve impulse **138**
nervous impulses and heartbeat 235, 237
nervous system **135-140**, autonomic **136**, chemical effects on **163-169**, transmission **140**
nervous tissue **46**, 50
neurilemma 138
neurons **136**, 137, connections between 141, structure of **137**
neutrophils 216
niacin *340*
nicotine 163, 290-291
night vision 183
nitrogen bases **466**
node of Ranvier **138**
nondisjunction 489
non-inherited diseases **493**
noradrenalin 403
nostrils 276
nucleic acids 14
nucleolus 20, 22
nucleus, cell 20-22
nutrition and growth **460-461**

O

occipital bone 92
occipital condyles 92
occipital lobe **145**
oil-soluble vitamins *340*
olfactory cells **205**, *206*
opiates **163-169**
optic nerve **184**
oral pharynx 277
organelles 16, **22-25**
organ, defined 50
organ of Corti 200

organ systems 50, 51
origin, muscle **123-124**
osmosis **26-27**
ossicles 198
osteocytes 45, **86**
osteomyelitis 110
otoliths *203*
oval window, of ear, 198, **199**
ovary 406, **423-424**
overweight **356-357**, problems **357**, 358, 361
oviducts **425**
ovulation 424
ovum **423-424**, production 470
oxygen, and respiration, **275**, **283-287**, transport by blood 235
oxygen deficit **126**
oxyhemoglobin **214**, 215, 285
oxytocin hormone **410**, 455

P

pacemaker **235-237**
pain receptors 67
palates, hard and soft **305**
pancreas 316-317, **401-402**
pancreatic duct **317**
pancreatic enzymes 317
paralysis 128
parasitic worms 255, **258**
parasympathetic division of nervous system **161**
parathormone **400**
parathyroid glands 398, **400-401**
parietal bones 92
parietal lobe **145**
parotid gland **305**
passive transport **383**
patella 100
pathogens **254-263**, types of **255**, 262
PCP *168*
pectoral girdle **103**
pedigree **478-479**
pelvic cavity 53
pelvis **100**
Penfield, Dr. W. *148*
penis **422**
pepsinogen **310**
peptic ulcer 311
perilymph 199, 202, *203*
periosteum, long bone **86**
peripheral nervous system 136
peristalsis **307**, 308
peritoneum 427
peritonitis 319
permanent teeth 302
perspiration (sweat glands) 63
phagocytes 29, **260-261**
phalanges 101, 104
pharynx **277**, 307
phenotype **474**
physiology, definition 8
pia mater **154**
pinna 196
pinocytosis 21, 29
pituitary gland 144, 398, **407-408**, anterior lobe 407, disorders of 410, hormones *409*, 408-410, posterior lobe 407
pivot joint **106**, *107*
placenta **448-450**, **458**
plasma, blood **212-213**, 228, fluids 245
platelets *212*, **217-218**
pleura 280
pleurisy 280
polar bodies 472
polysaccharides **329**, 330
Pomeranz, Dr. B. *161*
pons **144**
portal vein **314**
posterior horn, spinal cord 155
posterior lobe, pituitary 407, hormones of **410**
posterior (dorsal) portion of body 53
posture **122**, **126**
Prebus, Dr. Albert *21*
premolar teeth **302**, *303*
pressure, and blood flow *228*, **241-242**
primary colours *183*
procallus **109**
progesterone **407**, 425, 455
projection fibres 151
prophase, of mitosis *34*, **35**
prostate gland **421**
protein(s) 13, 14, **333-336**, essential (amino acids) 334, value 334, requirements 335
protein synthesis *469*
prothrombin 218
protozoa 255, **258**
proximal end of limb 53
psychosomatic 362
puberty **423**
pubis 100
pulmonary artery **233**, 235
pulmonary circulation **234-235**
pulmonary veins **233**
pulp, tooth 303
pulse **239**
pupil, eye **181**
Purkinje fibres 237
pyramids, kidney 380

R

radiation and mutations 491
radiography 92
radius, bone 104
rapid eye movements (REM) *185*
reaction time **140**
receptors, skin **66-68**, 156
recessive characteristic **473-476**

recommended daily nutrient intake **368**
rectum **320**
red blood cells (erythrocytes) **212**, **213**, functions of **214**, 215
referred pain 67
reflex act **157**
reflex arc **156**
reflexes 144, acquired 157, eye **184-185**
refractory period **139**
regions of body 52-53
renal artery 379
renal vein 379
rennin 310
reproduction, female organs of **423-433**, fetal growth and birth *453-455*, male organs of **417-419**
rescue breathing **288-289**
residual air (volume) 282-283
respiration, cellular 30 (see also *breathing*)
respiratory centre **287**
respiratory system 275-289
respirometer 283
retina **182-183**
Rh factor **220-221**
rhodopsin and vision **183**
riboflavin 340, 368
ribonucleic acid (RNA) *469* (see also *DNA*)
ribosomes 19, 24
ribs 94, *95*, *96*
rickettsia 255, **257-258**
rods of retina *182*, **183**
Roentgen, Dr. W. *92*
root (tooth) 303, *304*
rotation, joint **127**
round window, inner ear **200**
rubella 491
rugae **310**

S

saccule 198, **202-203**
sacral region, vertebral column 96
saliva 304, **305**
salivary glands **305**
sarcolemma 119
saturated fats **332**
saturated fatty acids 332
scala tympani 199
scala vestibuli 199
scapula 103
Schwann cells 138
sclera **181**
scrotum 418-419
sebaceous (oil) glands **63**
semen **422**
semicircular canals 198, **203-204**
semilunar valves **234**
seminal vesicle 421
seminiferous tubules **418-419**
senses (see *eye*, *ear*, *taste*)
sensitized 221
sensory area, of cortex *148*
sensory impulses, cerebral **150-151**
sensory neuron **156**
serotonin *185*
Sertoli cells 418-419
sex, determination of **480-481**, chromosomes 481, *482*
sex organs, female **423-429**, male **418-422**
sex-linked traits 218, **482-483**
sexually transmitted diseases (STDs) **437-441**
shoulder (pectoral) girdle 103
sickle cell anemia 477, **490**
simple fracture 108
sinoatrial node 235-237
sinus(es) 53
skeletal muscle **115**, *116-117*
skeletal system **85**
skeleton, appendicular 88, axial 88, functions of 85, **88-89**
skin, appendages of **63-65**, care of **73**, glands of 63, functions of *59*, patterns of **62**, structure of 60-68
skinfold measurements *356*, tables *373*
skull, human **88-93**
slipped disc 97
small intestine **311-314**
smell, sense of **205-206**
smoking **290-291**
smooth muscle **43**, *48*, **115**
Snellen eye chart 191
sodium ions 139
sound, how it travels **197**
spasm, muscle 128
speech **150**
sperm cells **419-420**, production **470**
sphenoid bone 92
sphincter 310, 390
sphygmomanometer **243-244**
spinal cord 144, **154-158**, injury to **157-158** (see also *vertebral column*)
spinal nerves 97
spindle, and mitosis 33-35
spine disorders *99*
spirilla 257
spirometer *283*
spleen **247**
spongy bone **86**
sprain **108**
squamous epithelium **41**, 47
stapes (stirrup) 198
starches **329**, 330
sternum 94, **95**, *96*, 103
stethoscope *239*
stimulants **163**, *166*
stimulus, nerve **140**

stirrup (stapes) 198
stomach, movement during digestion 310, secretions **310**, structure of **310-311**
stratified epithelium **41**, 47
stratum corneum **61**
stratum germinativum **61**
striated muscle **119**, *120*, tissue **43**
sublingual gland **305**
submaxillary gland **305**
substrate 32
sucrase 314
sugars 329-330
sulcus 145
superior vena cava **231-232**
suture 88, *90*
swallowing **306**, 307
sweat glands **63**
sympathetic division, nervous system **159-160**
sympathetic ganglia **159**
symptoms defined 255
synapse **140-141**, 145
synovial capsule *106*
synovial fluid 105
synovial membrane 105
syphilis **437-439**
system defined 50
systole **237-238**, **242**
systolic pressure **242**

T

tarsal bones 101
taste bud **204-205**
teeth **301-304**
telophase, of mitosis *34*, **35**
temperature (see *body temperature*) 68-70
temporal bones 92
temporal lobe **145**
tendon 44, **105**, **123-124**
tendonitis 124
testes, hormones of 398, 406
testosterone 407
thalamus 144
thalidomide 491
thoracic cavity 52
thoracic duct 247
thoracic vertebrae 94, *95*, *96*
thrombin 218
thromboplastin 217-218
thymine 466, *467*
thymus gland *398*, **406**
thyroid cartilage 277
thyroid gland **398-399**, disorders 399
thyroid stimulating hormone (TSH) **408**
thyroxin 239, **399**
tibia 100
tidal air (volume) **282**, 283
tissue(s) 40, bone 45, connective 40, 44-45, defined 40, epithelium 40, 41, muscle 40, 43-44, nervous 46
tobacco, effects of **290-291**
tone of muscle **126**
tongue **204-205**, **304-305**
tonsils 277
touch receptors 67
toxins **257**
trachea **278-279**
tranquillizers **163-164**, *168*
transverse colon **320**
triceps muscle **122**, *123*
tricuspid valve 233
triglycerides **332**
trisomy **489**
true ribs *96*
trypsin 317
tubule, kidney 383
turbinate bones **276**
Turner syndrome **490**
tympanic membrane **196-197**

U

ulcers, stomach *309*, **310-311**
ulna 104
umbilical cord **449-450**, 458
universal donor, blood 219
universal recipient blood 220
unsaturated fats **332-333**
unsaturated fatty acids 332-333
ureters 378, **389-390**
urethra 378, **390-391**, 421
urinary bladder 378, *379*, **390**
urination **390**
urine 380, 387, **388-389**, 390
uterus 427-430, 446, *448*, **448-449**
utricle 198, **202-203**
uvula 305

V

vaccines **263**
vacuoles 20, **24**
vagina 391, **432**
vaginitis **440**
vagus nerves *153*
varicose veins **230**
vas deferens **420**
vectors, disease 259
vegetables group, fruit and **352-353**
veins **227-230**, *236*
vena cava 231-232
venereal disease 437-441, *438*
ventral (anterior) portion of body *53*
ventral root 155
ventricle **232-233**
venules **227-230**
vermiform appendix 318-319
vertebrae **93-96**
vertebral column **93-94**, 95, 154 (see also *spinal cord*)

vestibule (bony labyrinth) 199
villi, intestinal **317-318**
viruses **255**, 256, 491
vision **176-186**, defects **187-189**
visual acuity **188**
vital capacity **282-283**
vitamin(s) 13-14, **337-339**, chart *340-341*, 352, oil-soluble 339, water-soluble 339, factors affecting retention *338*
vitamin A and rhodopsin **183**, 314, 351
vitamin B complex 320, 351
vitamin B_{12} 315
vitamin C 352
vitamin K 320
vocal cords 277
voice 277
voluntary muscle 43, **115**

W

water, role in body 14-15, balance 145, 336, exretion **377-378**, loss 336, needs **336**
water-soluble vitamins *340-341*
Watson, J. D. and DNA *467*
white blood cells (leukocytes) *212*, **216**, types of **216**, 260-261
white matter 138, 151

X

X chromosome **480-482**
X-rays *92*

Y

Y chromosome **480-482**
yolk sac 449

Z

zygomatic arch 90
zygomatic bone 90
zygote 435, 445

Photo Credits

Introductory Photo. *Courtesy* Ontario Field and Track Association.

Unit I.

Opening photo. *Courtesy* Birgitte Nielsen.

Anatomical drawings from The Drawings of Leonardo da Vinci. From the Royal Library, Windsor Castle. Reproduced by gracious permission of Her Majesty Queen Elizabeth II.

Figures 1.2, 1.5, 1.7, 1.8, and 1.9. *Courtesy* The Ontario Science Centre, Toronto.

Figure 1.16. Photograph by Gordon S. Berry.

Interest Box. Early electron microscope. *Courtesy* The Ontario Science Centre, Toronto.

Figures 2.1 to 2.13. Photographs by Gordon S. Berry.

Unit II.

Opening photo. *Courtesy* Canadian Amateur Swimming Association - Ontario Section and Tony Duffy, photographer.

Figures 3.2 to 3.6. Photographs by Gordon S. Berry.

Unit III.

Opening photo. *Courtesy* Health and Welfare Canada.

Figure 4.3. *Courtesy* Department of Medical Genetics, Faculty of Medicine, The University of British Columbia.

Interest Box. Terry Fox. *Courtesy* Canadian Cancer Society.

Figure 4.4. Photograph by Gordon S. Berry.

Figure 4.6. *Courtesy* St. Joseph's Hospital, Peterborough, Ontario.
Figures 4.10(2), 4.13(2), 4.16, and 4.20. Photographs by Gordon S. Berry.

Figure 5.4. *Courtesy* Biology Department, Trent University.
Figures 5.8 and 5.10. Photographs by Harold S. Gopaul.

Unit IV.
Opening photo. *Courtesy* Ontario Volleyball Association.

Figure 6.1. Photograph by Gordon S. Berry.
Figure 6.6. Photograph by Harold S. Gopaul.

Unit V.
Opening photo. *Courtesy* Ontario Table Tennis Association.
Figure 9.3. Photograph by Gordon S. Berry.
Figure 9.4. Photograph by Harold S. Gopaul.

Figure 10.2. Photograph by Harold S. Gopaul.
Figure 10.5. *Courtesy* The Ontario Science Centre, Toronto.
Figures 10.12, 10.13, 10.14, and 10.16. Photographs by Gordon S. Berry.
Interest Box. Artificial heart. *Courtesy* Institute for Biomedical Engineering, University of Utah.
Interest Box. Cobalt bomb. *Courtesy* Atomic Energy of Canada Limited.

Figure 11.4. Photograph by Gordon S. Berry.

Unit VI.
Opening photo. *Courtesy* Health and Welfare Canada.
Case Study. Breathalyzer. *Courtesy* Peterborough Police Department.
Figure 12.3. *Courtesy* U.B.C. Pulmonary Research Laboratory, St. Paul's Hospital, Vancouver.
Figure 12.4. *Courtesy* Dr. James Hogg, Faculty of Medicine, The University of British Columbia.
Figure 12.7 (a). Photograph by Gordon S. Berry.
Figure 12.7 (b). Photograph by Harold S. Gopaul.
Figure 12.8. *Courtesy* Dr. James Hogg, Faculty of Medicine, The University of British Columbia.
Figure 12.11. *Courtesy* Canadian Cancer Society.
Case Study. Resuscitation. *Courtesy* Don Galbraith.

Unit VII.

Opening photo. *Courtesy* National Film Board of Canada.

Figures 15.2 to 15.6. Photographs by Gordon S. Berry.

Unit VIII.

Opening photo. *Courtesy* Ontario Sports Administrative Centre Inc. Joe Heim, Ontario Sportscene/Photographer.

Interest Box. *Courtesy* Sven Fletcher-Berg and Vancouver General Hospital.

Figure 16.1 (b). Photograph by Gordon S. Berry.

Unit IX.

Opening photo. *Courtesy* Health and Welfare Canada.

Figure 17.2 (a) and (b). *Courtesy* Dr. Wah Jun Tze, Department of Endocrinology, Faculty of Medicine, The University of British Columbia.

Figure 17.3 *Courtesy* Armed Forces Institute of Pathology, Washington, D.C. (Neg. No. 48221).

Figure 17.7. *Courtesy* Dr. Wah Jun Tze, Department of Endocrinology, Faculty of Medicine, The University of British Columbia.

Figure 17.8. *Courtesy* Armed Forces Institute of Pathology, Washington, D.C. (Neg. Nos. 57-10583 (a) and (b)).

Unit X.

Opening photo. *Courtesy* Health and Welfare Canada.

Figures 18.3 (b) and 18.7. Photographs by Gordon S. Berry.

Figure 18.12. Copyright © Lennart Nilsson, from *Behold Man*. Published in the United States by Little, Brown and Co., Boston, Mass., 1973.

Figures 19.5 (a), (b), and (c). *Courtesy* Department of Medical Genetics, Faculty of Medicine, The University of British Columbia.

Interest Box. Male and female karyotypes. *Courtesy* Department of Medical Genetics, Faculty of Medicine, The University of British Columbia.

Figure 20.14. *Courtesy* Department of Medical Genetics, Faculty of Medicine, The University of British Columbia.

Interest Box. Amniocentesis. *Courtesy* Department of Medical Genetics, Faculty of Medicine, The University of British Columbia.

Interest Box. Chromosomes (karyotype). *Courtesy* Department of Medical Genetics, Faculty of Medicine, The University of British Columbia.

Case Study. Genetic counselling. *Courtesy* Department of Medical Genetics, Vancouver General Hospital.

Text Credits

Interest Boxes on Dr. Gustave Gingras (page 126), Dr. Wilder Penfield (page 148), Dr. Wilfred Bigelow (page 240), Dr. Harold Johns (page 247), and Dr. Ghobind Khorana (page 486) are based on information in *The Mirrored Spectrum* (1973). Reproduced by permission of the Minister of Supply and Services Canada.

Chapter 6.
Table 6.1. Based on "Drinking and Drugs: Way to Cope" by Linda Laughlin, *Teen Generation*, March 1980.

Chapter 18
Information on sexually transmitted diseases is based on the pamphlet "Sexually Transmitted Diseases" published by Health and Welfare Canada (1978).